IET POWER AND E

The Control Tech
Drives and Co
Han

The Control Techniques Drives and Controls Handbook

2nd Edition

Bill Drury

The Institution of Engineering and Technology

Published by The Institution of Engineering and Technology, London, United Kingdom

© 2001 The Institution of Electrical Engineers
© 2009 The Institution of Engineering and Technology

First published 2001 (0 85296 793 4)
Second edition 2009
Reprinted 2009

This publication is copyright under the Berne Convention and the Universal Copyright Convention. All rights reserved. Apart from any fair dealing for the purposes of research or private study, or criticism or review, as permitted under the Copyright, Designs and Patents Act 1988, this publication may be reproduced, stored or transmitted, in any form or by any means, only with the prior permission in writing of the publishers, or in the case of reprographic reproduction in accordance with the terms of licences issued by the Copyright Licensing Agency. Enquiries concerning reproduction outside those terms should be sent to the publishers at the undermentioned address:

The Institution of Engineering and Technology
Michael Faraday House
Six Hills Way, Stevenage
Herts SG1 2AY, United Kingdom

www.theiet.org

While the author and publisher believe that the information and guidance given in this work are correct, all parties must rely upon their own skill and judgement when making use of them. Neither the author nor publisher assumes any liability to anyone for any loss or damage caused by any error or omission in the work, whether such an error or omission is the result of negligence or any other cause. Any and all such liability is disclaimed.

The moral rights of the author to be identified as author of this work have been asserted by him in accordance with the Copyright, Designs and Patents Act 1988.

British Library Cataloguing in Publication Data
A catalogue record for this product is available from the British Library

ISBN 978-1-84919-013-8 (hardback)
ISBN 978-1-84919-101-2 (PDF)

Typeset in India by Techset Composition Ltd, Chennai
Printed in the UK by Athenaeum Press Ltd, Gateshead, Tyne & Wear

Contents

Preface	xxv
Acknowledgements	xxxix

PART A	**DRIVE TYPES AND CORE TECHNOLOGY**			1
A1	**Industrial motors**			11
	A1.1	Introduction and basic electromagnetic principles		11
		A1.1.1	Magnetic circuits	11
		A1.1.2	Electromechanical energy conversion	16
			A1.1.2.1 The alignment of magnetic force/flux lines	16
			A1.1.2.2 The interaction between a magnetic field and a current-carrying conductor	18
	A1.2	D.C. motors		20
		A1.2.1	General	20
		A1.2.2	Operating principles	21
		A1.2.3	Fundamental equations of steady-state performance	25
			A1.2.3.1 The separately excited d.c. motor	25
			A1.2.3.2 The series d.c. motor	29
			A1.2.3.3 The shunt d.c. motor	30
			A1.2.3.4 The compound d.c. motor	30
		A1.2.4	Permanent magnet d.c. motor	31
		A1.2.5	Construction of the d.c. motor	32
			A1.2.5.1 D.C. motor frame	32
			A1.2.5.2 D.C. motor armature	33
			A1.2.5.3 Brush gear	34
			A1.2.5.4 Degree of protection and mounting	34
			A1.2.5.5 DCPM design	35
	A1.3	A.C. induction motors		36
		A1.3.1	General	36
		A1.3.2	Operating principles	36
			A1.3.2.1 Rotating magnetic field	37
			A1.3.2.2 Torque production	38

		A1.3.3	Fundamental equations of steady-state performance	43
			A1.3.3.1 Direct on line (DOL) starting current and torque	43
			A1.3.3.2 Starting current and torque when the motor is connected to a variable-frequency and/or variable-voltage supply	45
		A1.3.4	Voltage–frequency relationship	45
		A1.3.5	Slip-ring induction motor	48
		A1.3.6	Speed-changing motors	50
		A1.3.7	A.C. induction motor construction	50
	A1.4	A.C. synchronous motors		52
		A1.4.1	General	52
		A1.4.2	Operating principles	53
		A1.4.3	Fundamental equations of steady-state performance	54
			A1.4.3.1 General	54
			A1.4.3.2 Brushless PM servo motor	55
		A1.4.4	Limits of operation	57
		A1.4.5	Synchronous motor construction	58
			A1.4.5.1 Permanent-magnet servo motors	58
			A1.4.5.2 Permanent-magnet industrial motors	60
			A1.4.5.3 Wound-rotor synchronous motors	61
		A1.4.6	Starting of synchronous motors	61
	A1.5	Reluctance motors		62
	A1.6	A.C. commutator motors		63
	A1.7	Motors for special applications		64
		A1.7.1	Geared motors	64
		A1.7.2	Brake motors	64
		A1.7.3	Torque motors	64
	A1.8	Motors for hazardous locations		65
		A1.8.1	General	65
		A1.8.2	CENELEC	65
		A1.8.3	North American standards	69
		A1.8.4	Testing authorities	69
A2	**Drive converter circuit topologies**			**71**
	A2.1	Introduction		71
	A2.2	A.C. to d.c. power conversion		72
		A2.2.1	General	72
		A2.2.2	Converters for connection to a single-phase supply	73
			A2.2.2.1 Uncontrolled converters	73
			A2.2.2.2 Controlled converters	74
			A2.2.2.3 Sine-wave input converters	76
			A2.2.2.4 Summary of characteristics	76

		A2.2.3	Converters for connection to a three-phase supply	78

- A2.2.3 Converters for connection to a three-phase supply — 78
    - A2.2.3.1 Uncontrolled converters — 78
    - A2.2.3.2 Controlled converters — 78
    - A2.2.3.3 Summary of characteristics — 79
- A2.2.4 Converters for d.c. motor drive systems — 82
    - A2.2.4.1 Single-converter drives — 83
    - A2.2.4.2 Dual-converter drives — 84
    - A2.2.4.3 Field control — 85
- A2.3 D.C. to d.c. power conversion — 86
    - A2.3.1 General — 86
    - A2.3.2 Step down d.c. to d.c. converters — 87
        - A2.3.2.1 Single-quadrant d.c. to d.c. converter — 87
        - A2.3.2.2 Two-quadrant d.c. to d.c. converter — 88
        - A2.3.2.3 Four-quadrant d.c. to d.c. converter — 89
    - A2.3.3 Step-up d.c. to d.c. converters — 90
- A2.4 A.C. to a.c. power converters with intermediate d.c. link — 91
    - A2.4.1 General — 91
    - A2.4.2 Voltage source inverters — 91
        - A2.4.2.1 General characteristics — 91
        - A2.4.2.2 Six-step/quasi-square-wave inverter — 94
        - A2.4.2.3 Pulse-width modulated inverter — 96
        - A2.4.2.4 Multi-level inverter — 97
    - A2.4.3 Current source inverters — 99
        - A2.4.3.1 General characteristics — 99
        - A2.4.3.2 Converter-fed synchronous machine (LCI) — 100
        - A2.4.3.3 Converter-fed induction motor drive — 101
        - A2.4.3.4 Forced commutated induction motor drive — 101
        - A2.4.3.5 Static Kramer drive — 102
- A2.5 Direct a.c. to a.c. power converters — 103
    - A2.5.1 General — 103
    - A2.5.2 Soft starter/voltage regulator — 103
    - A2.5.3 Cycloconverter — 104
    - A2.5.4 Matrix converter — 106
    - A2.5.5 Static Scherbius drive — 107

A3 Power semiconductor devices — 109
- A3.1 General — 109
- A3.2 Diode — 114
    - A3.2.1 PN diode — 114
    - A3.2.2 PIN diode — 116
    - A3.2.3 Transient processes (reverse and forward recovery) — 118
        - A3.2.3.1 Reverse recovery — 118
        - A3.2.3.2 Forward recovery — 120
    - A3.2.4 Diode types — 121

A3.3	Thyristor (SCR)		122
	A3.3.1	Device description	122
	A3.3.2	Transient processes	124
		A3.3.2.1 Turn-on	125
		A3.3.2.2 Turn-off	126
	A3.3.3	Thyristor gating requirements	127
	A3.3.4	Thyristor types	127
A3.4	Triac		130
A3.5	Gate turn-off thyristor (GTO)		130
	A3.5.1	Device description	130
	A3.5.2	Switching characteristics and gate drive	132
		A3.5.2.1 Turn-on	133
		A3.5.2.2 Turn-off	134
	A3.5.3	Voltage and current ratings	135
A3.6	Integrated gate commutated thyristor (IGCT)		135
	A3.6.1	Device description	135
	A3.6.2	Switching behaviour and gate drive	136
	A3.6.3	Voltage and current ratings	137
A3.7	MOSFET		137
	A3.7.1	Device description	137
	A3.7.2	Principal features and applications	137
	A3.7.3	D.C. characteristics	139
	A3.7.4	Switching performance	140
	A3.7.5	Transient characteristics	141
		A3.7.5.1 Switching waveforms	141
		A3.7.5.2 Turn-on	142
		A3.7.5.3 Turn-off	143
	A3.7.6	Safe operating area (SOA)	143
		A3.7.6.1 Forward-bias safe operating area (FBSOA)	143
		A3.7.6.2 Reverse-bias safe operating area (RBSOA)	144
	A3.7.7	Parasitic diode	144
	A3.7.8	MOSFET gate drive requirements	145
		A3.7.8.1 Speed limitations	146
		A3.7.8.2 Driving paralleled MOSFETs	147
	A3.7.9	Voltage and current ratings	147
A3.8	Insulated gate bipolar transistor (IGBT)		147
	A3.8.1	Device description	147
	A3.8.2	Principal features and applications	148
	A3.8.3	D.C. characteristics	149
	A3.8.4	Punch-through versus non-punch-through structures (PT and NPT)	150
	A3.8.5	Switching performance	151
	A3.8.6	Transient characteristics	151
		A3.8.6.1 Switching waveforms	152
		A3.8.6.2 Turn-on	152
		A3.8.6.3 Turn-off	154

		A3.8.7	Safe operating area (SOA)		155
			A3.8.7.1	Forward-bias safe operating area (FBSOA)	155
			A3.8.7.2	Reverse-bias safe operating area (RBSOA)	156
		A3.8.8	Parasitic thyristor		157
		A3.8.9	IGBT gate drive requirements		157
			A3.8.9.1	IGBT switching speed limitations	157
			A3.8.9.2	Series and parallel operation	158
			A3.8.9.3	IGBT short-circuit performance	159
		A3.8.10	Voltage and current ratings		159
	A3.9	Bipolar junction transistor (BJT)			159
	A3.10	Other power devices and materials			160
		A3.10.1	MOS controlled thyristor (MCT)		160
		A3.10.2	MOS turn-off thyristor		161
		A3.10.3	Junction field-effect transistors (JFETs)		162
	A3.11	Materials			162
	A3.12	Power device packaging			163
		A3.12.1	General		163
		A3.12.2	Pressure contact packages		165
			A3.12.2.1	Construction	165
			A3.12.2.2	Features	166
		A3.12.3	Large wire-bonded packages for power modules		166
			A3.12.3.1	Construction	166
			A3.12.3.2	Package types	167
			A3.12.3.3	Features	168
		A3.12.4	Small wire-bonded packages for discrete devices		168
			A3.12.4.1	Construction	169
			A3.12.4.2	Package types	169
			A3.12.4.3	Features	169
A4	**Torque, speed and position control**				**171**
	A4.1	General principles			171
		A4.1.1	The ideal control system		171
		A4.1.2	Open-loop control		171
		A4.1.3	Closed-loop control		172
		A4.1.4	Criteria for assessing performance		173
	A4.2	Controllers in a drive			175
		A4.2.1	General		175
		A4.2.2	Torque control		176
		A4.2.3	Flux control		179
		A4.2.4	Speed control		179
			A4.2.4.1	Basic speed control	179
			A4.2.4.2	Setting speed controller gains	183
			A4.2.4.3	Speed control with torque feed-forward	185
		A4.2.5	Position control		186
			A4.2.5.1	Basic position control	186
			A4.2.5.2	Position control with speed feed-forward	190

	A4.3	D.C. motor drives			192
		A4.3.1	General		192
		A4.3.2	Torque control		192
		A4.3.3	Flux control		194
	A4.4	A.C. motor drives			195
		A4.4.1	Torque and flux control		195
			A4.4.1.1	Introduction	195
			A4.4.1.2	D.C. motor torque and flux control	196
			A4.4.1.3	Permanent magnet motor torque and flux control	197
			A4.4.1.4	Induction motor torque and flux control	203
			A4.4.1.5	Open-loop induction motor drive	205
		A4.4.2	Direct torque control		206
		A4.4.3	Performance summary		208
			A4.4.3.1	Permanent-magnet motor drives	209
			A4.4.3.2	Induction motor drives with closed-loop current control	210
			A4.4.3.3	Open-loop induction motor drives	210
A5	**Position and speed feedback**				**211**
	A5.1	General			211
		A5.1.1	Feedback quantity required		211
		A5.1.2	Absolute position feedback range		212
		A5.1.3	Position resolution		212
		A5.1.4	Position accuracy		214
		A5.1.5	Speed resolution		214
		A5.1.6	Speed accuracy		214
		A5.1.7	Environment		215
		A5.1.8	Maximum speed		215
		A5.1.9	Electrical noise immunity		215
		A5.1.10	Distance between the feedback device and the drive		216
		A5.1.11	Additional features		216
	A5.2	Speed feedback sensors			216
		A5.2.1	D.C. tacho-generator		216
	A5.3	Position feedback sensors			218
		A5.3.1	Resolver		218
		A5.3.2	Incremental encoder		221
		A5.3.3	Incremental encoder with commutation signals		223
		A5.3.4	Incremental encoder with commutation signals only		224
		A5.3.5	SINCOS encoder		224
		A5.3.6	Absolute SINCOS encoder		226
		A5.3.7	Absolute encoders		227
		A5.3.8	SINCOS encoders with serial communications		228
			A5.3.8.1	EnDat	228
			A5.3.8.2	Hiperface	229

			A5.3.8.3	SSI	229
			A5.3.8.4	Summary	230
		A5.3.9	Serial communications encoders		230
			A5.3.9.1	BiSS	230
			A5.3.9.2	EnDat	230
		A5.3.10	Wireless encoders		231
A6	**Motion control**				**233**
	A6.1	General			233
		A6.1.1	Position, speed, acceleration and jerk		234
			A6.1.1.1	Speed	234
			A6.1.1.2	Acceleration	234
			A6.1.1.3	Jerk	235
		A6.1.2	Possible configurations		236
	A6.2	Time-based profile			239
	A6.3	CAM profile			243
	A6.4	Electronic gearbox			248
	A6.5	Practical systems			249
		A6.5.1	Control Techniques' Advanced Position Controller		249
		A6.5.2	Control Techniques' Indexer		250
A7	**Voltage source inverter: four-quadrant operation**				**253**
	A7.1	General			253
	A7.2	Controlled deceleration			254
		A7.2.1	Performance and applications		255
			A7.2.1.1	Advantages	256
			A7.2.1.2	Disadvantages	256
	A7.3	Braking resistor			256
		A7.3.1	Performance and applications		257
			A7.3.1.1	Advantages	257
			A7.3.1.2	Disadvantages	257
	A7.4	Active rectifier			257
		A7.4.1	Performance and applications		259
			A7.4.1.1	Advantages	259
			A7.4.1.2	Disadvantages	260
A8	**Switched reluctance and stepper motor drives**				**261**
	A8.1	General			261
	A8.2	Switched reluctance motors and controllers			261
		A8.2.1	Basic principle of the switched reluctance motor		261
			A8.2.1.1	Operation as a motor	264
			A8.2.1.2	Operation as a brake or generator	265
			A8.2.1.3	Summary so far	265
			A8.2.1.4	Relationship between torque polarity and motoring and generating	267

xii *Contents*

	A8.2.2	Control of the machine in practice		267
		A8.2.2.1	Low-speed operation	267
		A8.2.2.2	What happens as speed is increased?	267
		A8.2.2.3	Medium-speed operation	268
		A8.2.2.4	How is performance maintained as speed increases?	269
		A8.2.2.5	High-speed operation	269
		A8.2.2.6	Summary of typical/practical control	270
		A8.2.2.7	Control of speed and position	271
	A8.2.3	Polyphase switched reluctance machines		272
	A8.2.4	Losses in the switched reluctance motor		273
	A8.2.5	Excitation frequency		274
	A8.2.6	Power electronics for the switched reluctance motor		275
		A8.2.6.1	Power supply and 'front end' bridge	275
		A8.2.6.2	Power switching stage	275
		A8.2.6.3	Single-switch-per-phase circuits	275
		A8.2.6.4	Multiple-phase operation	277
		A8.2.6.5	Single-switch circuit using bifilar winding	278
		A8.2.6.6	Two-switch asymmetrical bridge	278
	A8.2.7	Advantages of the switched reluctance system		279
		A8.2.7.1	Rotor construction	279
		A8.2.7.2	Stator construction	280
		A8.2.7.3	Electronics and system-level benefits	280
	A8.2.8	Disadvantages of the switched reluctance system		282
		A8.2.8.1	Torque ripple	282
		A8.2.8.2	Acoustic noise	283
A8.3	Stepper motor drives			284
	A8.3.1	Stepping motor principles		284
		A8.3.1.1	The permanent-magnet motor	284
		A8.3.1.2	The VR motor	285
		A8.3.1.3	The hybrid motor	286
	A8.3.2	Stepping motor drive circuits and logic modes		287
		A8.3.2.1	General	287
		A8.3.2.3	Unipolar switching	288
		A8.3.2.3	Bipolar switching	290
		A8.3.2.4	High-speed stepping: L/R drives	290
		A8.3.2.5	Chopper drives	292
		A8.3.2.6	Bilevel drives	292
	A8.3.3	Application notes		293
		A8.3.3.1	Effect of inertia	293
		A8.3.3.2	Resonance	293
		A8.3.3.3	Stepper/encoders	294

		Contents	xiii

PART B THE DRIVE IN ITS ENVIRONMENT 295

B1 The a.c. supply 299
    B1.1 General 299
    B1.2 Supply harmonics and other low-frequency disturbances 299
        B1.2.1 Overview 299
        B1.2.2 Regulations 300
            B1.2.2.1 Regulations for installations 301
            B1.2.2.2 Regulations and standards for equipment 301
        B1.2.3 Harmonic generation within variable-speed drives 302
            B1.2.3.1 A.C. drives 302
            B1.2.3.2 D.C. drives 304
        B1.2.4 The effects of harmonics 306
        B1.2.5 Calculation of harmonics 307
            B1.2.5.1 Individual drives: d.c. 307
            B1.2.5.2 Individual drives: a.c. 308
            B1.2.5.3 Systems 308
            B1.2.5.4 Isolated generators 310
        B1.2.6 Remedial techniques 310
            B1.2.6.1 Connect the equipment to a point with a high fault level (low impedance) 311
            B1.2.6.2 Use three-phase drives where possible 311
            B1.2.6.3 Use additional inductance 311
            B1.2.6.4 Use a lower value of d.c. smoothing capacitance 315
            B1.2.6.5 Use a higher pulse number (12 pulse or higher) 316
            B1.2.6.6 Use a drive with an active input converter 318
            B1.2.6.7 Use a harmonic filter 318
        B1.2.7 Typical harmonic current levels for a.c. drive arrangements 319
        B1.2.8 Additional notes on the application of harmonic standards 319
            B1.2.8.1 The effect of load 319
            B1.2.8.2 Choice of reference current: application of IEEE Std 519-1992 321
        B1.2.9 Interharmonics and emissions up to 9 kHz 321
        B1.2.10 Voltage notching 322
        B1.2.11 Voltage dips and flicker 323
    B1.3 Power factor 324
    B1.4 Supply imperfections 326
        B1.4.1 General 326

xiv Contents

	B1.4.2	Frequency variation	326
	B1.4.3	Voltage variation	326
	B1.4.4	Temporary and transient over-voltages between live conductors and earth	327
	B1.4.5	Voltage unbalance	327
	B1.4.6	Harmonic voltage	329
	B1.4.7	Supply voltage dips and short interruptions	329
	B1.4.8	Interharmonics and mains signalling	330
	B1.4.9	Voltage notching	331
	B1.4.10	EMC standards	333

B2 Interaction between drives and motors — 335
- B2.1 General — 335
- B2.2 Drive converter effects upon d.c. machines — 335
- B2.3 Drive converter effects upon a.c. machines — 336
  - B2.3.1 Introduction — 336
  - B2.3.2 Machine rating: thermal effects — 336
  - B2.3.3 Machine insulation — 337
    - B2.3.3.1 Current source inverters — 337
    - B2.3.3.2 Voltage source inverters — 337
  - B2.3.4 Bearing currents — 349
    - B2.3.4.1 Root causes of bearing currents — 349
    - B2.3.4.2 Good practices to reduce the risk of bearing currents — 351
  - B2.3.5 Overspeed — 352
- B2.4 Motors for hazardous (potentially flammable or explosive) locations — 353

B3 Physical environment — 355
- B3.1 Introduction — 355
- B3.2 Enclosure degree of protection — 355
  - B3.2.1 General — 355
  - B3.2.2 Motor — 356
    - B3.2.2.1 General — 356
    - B3.2.2.2 US practice — 356
  - B3.2.3 Drive — 356
- B3.3 Mounting arrangements — 360
  - B3.3.1 Motor — 360
    - B3.3.1.1 General — 360
    - B3.3.1.2 IEC 60034-7 standard enclosures — 360
    - B3.3.1.3 NEMA standard enclosures — 360
  - B3.3.2 Drive — 360
  - B3.3.3 Integrated motor drive — 363
- B3.4 Terminal markings and direction of rotation — 363
  - B3.4.1 Motor — 363
    - B3.4.1.1 General — 363

		B3.4.1.2	IEC 60034-8/EN 60034-8	364
		B3.4.1.3	NEMA	366
	B3.4.2	Drive		371
B3.5	Ambient temperature			371
	B3.5.1	Motor		371
	B3.5.2	Drive		372
		B3.5.2.1	Maximum operating temperature	372
		B3.5.2.2	Minimum operating temperature	372
B3.6	Humidity and condensation			373
	B3.6.1	Motor		373
	B3.6.2	Drive		373
B3.7	Noise			373
	B3.7.1	Motor		373
	B3.7.2	Drive		376
	B3.7.3	Motor noise when fed from a drive converter		376
B3.8	Vibration			378
	B3.8.1	Motor		378
	B3.8.2	Drive		380
B3.9	Altitude			380
B3.10	Corrosive gases			380
	B3.10.1	Motors		380
	B3.10.2	Drives		381

B4 Thermal management — 383

- B4.1 Introduction — 383
- B4.2 Motor cooling — 383
  - B4.2.1 General — 383
  - B4.2.2 D.C. motors — 385
    - B4.2.2.1 Air filters — 386
  - B4.2.3 A.C. industrial motors — 386
  - B4.2.4 High-performance/servo motors — 386
    - B4.2.4.1 Intermittent/peak torque limit — 388
    - B4.2.4.2 Forced-air (fan) cooling — 388
- B4.3 Drive cooling: the thermal design of enclosures — 389
  - B4.3.1 General — 389
  - B4.3.2 Calculating the size of a sealed enclosure — 389
  - B4.3.3 Calculating the air-flow in a ventilated enclosure — 391
  - B4.3.4 Through-panel mounting of drives — 392

B5 Drive system power management: common d.c. bus topologies — 393

- B5.1 Introduction — 393
- B5.2 Power circuit topology variations — 396
  - B5.2.1 General — 396
  - B5.2.2 Simple bulk uncontrolled external rectifier — 396
  - B5.2.3 A.C. input and d.c. bus paralleled — 397
  - B5.2.4 One host drive supplying d.c. bus to slave drives — 398

	B5.2.5	A bulk four-quadrant controlled rectifier feeding the d.c. bus	399
	B5.2.6	Active bulk rectifier	400
B5.3	Fusing policy		402
B5.4	Practical systems		402
	B5.4.1	Introduction	402
	B5.4.2	Variations in standard drive topology	403
	B5.4.3	Inrush/charging current	404
	B5.4.4	Continuous current	404
	B5.4.5	Implementation: essential knowledge	406
		B5.4.5.1 A.C. and d.c. terminals connected: drives of the same current rating only	406
		B5.4.5.2 A.C. and d.c. terminals connected: drives of different current ratings	407
		B5.4.5.3 One host drive supplying d.c. bus to slave drives	407
		B5.4.5.4 Simple bulk uncontrolled external rectifier	408
	B5.4.6	Practical examples	408
		B5.4.6.1 Winder/unwinder sharing energy via the d.c. bus	408
		B5.4.6.2 Four identical drives with a single dynamic braking circuit	409
	B5.4.7	Note on EMC filters for common d.c. bus systems	409

B6 Electromagnetic compatibility (EMC) — 411

B6.1	Introduction		411
	B6.1.1	General	411
	B6.1.2	Principles of EMC	411
	B6.1.3	EMC regulations	412
B6.2	Regulations and standards		412
	B6.2.1	Regulations and their application to drive modules	412
	B6.2.2	Standards	413
B6.3	EMC behaviour of variable-speed drives		414
	B6.3.1	Immunity	414
	B6.3.2	Low-frequency emission	414
	B6.3.3	High-frequency emission	415
B6.4	Installation rules		416
	B6.4.1	EMC risk assessment	416
	B6.4.2	Basic rules	417
		B6.4.2.1 Cable segregation	417
		B6.4.2.2 Control of return paths, minimising loop areas	417
		B6.4.2.3 Earthing	417
	B6.4.3	Simple precautions and 'fixes'	420
	B6.4.4	Full precautions	420

	B6.5	Theoretical background	422
		B6.5.1 Emission modes	422
		B6.5.2 Principles of input filters	424
		B6.5.3 Screened motor cables	425
		B6.5.4 Ferrite ring suppressors	425
		B6.5.5 Filter earth leakage current	426
		B6.5.6 Filter magnetic saturation	426
	B6.6	Additional guidance on cable screening for sensitive circuits	426
		B6.6.1 Cable screening action	426
		B6.6.2 Cable screen connections	428
		B6.6.3 Recommended cable arrangements	431
B7	**Protection**		**433**
	B7.1	Protection of the drive system and power supply infrastructure	433
		B7.1.1 General	433
		B7.1.2 Fuse types	433
		B7.1.3 Application of fuses to drive systems	434
		B7.1.4 Earth faults	435
		B7.1.5 IT supplies	435
		B7.1.6 Voltage transients	436
	B7.2	Motor thermal protection	438
		B7.2.1 General	438
		B7.2.2 Protection of line-connected motor	438
		B7.2.3 Protection of inverter-driven motor	439
		B7.2.4 Multiple motors	440
		B7.2.5 Servo motors	440
B8	**Mechanical vibration, critical speed and torsional dynamics**		**441**
	B8.1	General	441
	B8.2	Causes of shaft vibrations independent of variable-speed drives	443
		B8.2.1 Sub-synchronous vibrations	443
		B8.2.2 Synchronous vibrations	443
		B8.2.3 Super-synchronous vibrations	444
		B8.2.4 Critical speeds	444
	B8.3	Applications where torque ripple excites a resonance in the mechanical system	444
	B8.4	High-performance closed-loop applications	446
		B8.4.1 Limits to dynamic performance	446
		B8.4.2 System control loop instability	446
	B8.5	Measures for reducing vibration	446
B9	**Installation and maintenance of standard motors and drives**		**449**
	B9.1	Motors	449
		B9.1.1 General	449

xviii Contents

			B9.1.2	Storage	449
			B9.1.3	Installation	450
			B9.1.4	Maintenance guide	451
			B9.1.5	Brush gear maintenance	452
	B9.2	Electronic equipment			454
		B9.2.1	General		454
		B9.2.2	Location of equipment		454
		B9.2.3	Ventilation systems and filters		455
		B9.2.4	Condensation and humidity		455
		B9.2.5	Fuses		455

PART C   PRACTICAL APPLICATIONS   457

C1   Application and drive characteristics   461
C1.1   General   461
C1.2   Typical load characteristics and ratings   461
C1.3   Drive characteristics   472
    C1.3.1   General   472

C2   Duty cycles   477
C2.1   Introduction   477
C2.2   Continuous duty: S1   477
C2.3   Short-time duty: S2   478
C2.4   Intermittent duty: S3   479
C2.5   Intermittent duty with starting: S4   480
C2.6   Intermittent duty with starting and electric braking: S5   481
C2.7   Continuous operation periodic duty: S6   481
C2.8   Continuous operation periodic duty with electric braking: S7   482
C2.9   Continuous operation periodic duty with related load speed changes: S8   482
C2.10 Duty with non-periodic load and speed variations: S9   482
C2.11 Duty with discrete constant loads: S10   483

C3   Interfaces, communications and PC tools   485
C3.1   Introduction   485
C3.2   Overview of interface types   485
C3.3   Analogue signal circuits   486
    C3.3.1   General   486
    C3.3.2   Hardware implementations and wiring advice   487
        C3.3.2.1   General guidance on connecting analogue signal circuits   487
        C3.3.2.2   Single-ended circuits   490
        C3.3.2.3   Differential circuits   491
        C3.3.2.4   The case for 4–20 mA and other current loop circuits   496

		C3.3.2.5	The use of capacitors for connecting cable screens	496
	C3.3.3	Typical specifications for analogue inputs and outputs		497
C3.4	Digital signal circuits			499
	C3.4.1	Positive and negative logic		499
	C3.4.2	Digital input		500
	C3.4.3	Digital output		501
	C3.4.4	Relay contacts		501
C3.5	Digital serial communications			501
	C3.5.1	Introduction		501
	C3.5.2	Serial network basics		502
		C3.5.2.1	Physical layer	503
		C3.5.2.2	Data link layer	506
		C3.5.2.3	Application layer	508
		C3.5.2.4	Device profile	508
	C3.5.3	RS-232/RS-485 Modbus: A simple Fieldbus system		508
C3.6	Fieldbus systems			510
	C3.6.1	Introduction to Fieldbus		510
	C3.6.2	Centralised versus distributed control networks		512
		C3.6.2.1	Centralised network	512
		C3.6.2.2	Distributed network	513
		C3.6.2.3	Hybrid networks	514
	C3.6.3	Open and proprietary Fieldbus systems		516
		C3.6.3.1	Open networks	516
		C3.6.3.2	Proprietary networks	516
	C3.6.4	OPC technology		517
	C3.6.5	Industrial Fieldbus systems (non Ethernet)		517
		C3.6.5.1	Profibus DP	517
		C3.6.5.2	DeviceNet	518
		C3.6.5.3	CANopen	519
		C3.6.5.4	Interbus	520
		C3.6.5.5	LonWorks	520
		C3.6.5.6	BACnet	521
		C3.6.5.7	SERCOS II	522
	C3.6.6	Ethernet-based Fieldbuses		523
		C3.6.6.1	General	523
		C3.6.6.2	Modbus TCP/IP	523
		C3.6.6.3	EtherNet IP	524
		C3.6.6.4	PROFINET	525
		C3.6.6.5	EtherCAT	525
		C3.6.6.6	Powerlink	526
	C3.6.7	Company-specific Fieldbuses		526
		C3.6.7.1	CTNet	526
		C3.6.7.2	CTSync	527
	C3.6.8	Gateways		528

xx   *Contents*

	C3.7	PC tools	528
		C3.7.1 Engineering design tools	529
		C3.7.2 Drive commissioning and setup tools	529
		C3.7.3 Application configuration and setup tools	530
		C3.7.4 System configuration and setup tools	530
		C3.7.5 Monitoring tools	531

C4 Typical drive functions — 533

- C4.1 Introduction — 533
- C4.2 Speed or frequency reference/demand — 533
- C4.3 Ramps — 534
- C4.4 Frequency slaving — 535
- C4.5 Speed control — 535
- C4.6 Torque and current control — 535
  - C4.6.1 Open loop with scalar V/f control — 535
  - C4.6.2 Closed-loop and high-performance open loop — 536
- C4.7 Automatic tuning — 536
- C4.8 Second parameter sets — 537
- C4.9 Sequencer and clock — 537
- C4.10 Analogue and digital inputs and outputs — 537
- C4.11 Programmable logic — 537
- C4.12 Status and trips — 538
- C4.13 Intelligent drive programming: user-defined functionality — 539
- C4.14 Functional safety — 543
  - C4.14.1 Principles — 543
  - C4.14.2 Technical standards — 544
  - C4.14.3 Possible safety functions for drives — 546
    - C4.14.3.1 Safe torque off (STO) — 546
    - C4.14.3.2 Advanced drive-specific functions — 547
    - C4.14.3.3 Other machinery safety functions — 548
    - C4.14.3.4 Safety bus interfaces — 549
    - C4.14.3.5 Integration into a machine — 549
- C4.15 Summary — 549

C5 Common techniques — 551

- C5.1 General — 551
- C5.2 Speed control with particular reference to linear motion — 552
  - C5.2.1 Linear to rotary speed reference conversion — 555
- C5.3 Torque feed-forward — 555
- C5.4 Virtual master and sectional control — 556
- C5.5 Registration — 562
- C5.6 Load torque sharing — 567
  - C5.6.1 General — 567
  - C5.6.2 Open-loop systems — 568
  - C5.6.3 Paired d.c. motors — 570
  - C5.6.4 Paired a.c. motors — 572

		C5.6.4.1	Parallel motors	572
		C5.6.4.2	Frequency slaving	573
		C5.6.4.3	Current slaving	573
	C5.6.5	Torque slaving systems		574
	C5.6.6	Speed-controlled helper with fixed torque		575
	C5.6.7	Speed-controlled helper with shared torque		576
	C5.6.8	Full closed-loop systems		577
C5.7	Tension control			578
C5.8	Sectional control			579
C5.9	Winding			580
	C5.9.1	General		580
	C5.9.2	Drum winders		581
	C5.9.3	Centre-driven winders		582
C5.10	High-frequency inverters			589
	C5.10.1	General		589
	C5.10.2	Frequency control of a.c. induction motors		590
	C5.10.3	Purpose-designed high frequency motors		592
	C5.10.4	High-frequency inverters		593
	C5.10.5	High-frequency applications		594
C5.11	Special d.c. loads			594
	C5.11.1	Traction motor field control		595
	C5.11.2	Battery charging		595
	C5.11.3	Electrolytic processes		596
	C5.11.4	Electric heating and temperature control		596

C6 Industrial application examples — 599

C6.1	Introduction			599
C6.2	Centrifugal pumps			599
	C6.2.1	Single-pump systems		599
	C6.2.2	Multiple pump systems (duty-assist control)		605
		C6.2.2.1	Note on parallel operation of pumps	605
C6.3	Centrifugal fans and compressors			606
C6.4	Heating, ventilation, air conditioning and refrigeration (HVAC/R)			607
	C6.4.1	Introduction		607
	C6.4.2	Commercial buildings		608
		C6.4.2.1	Building automation systems	608
		C6.4.2.2	HVAC applications	609
	C6.4.3	Retail facilities		614
		C6.4.3.1	Refrigeration applications	615
	C6.4.4	Original equipment manufacturers		616
C6.5	Cranes and hoists			616
	C6.5.1	General		616
	C6.5.2	Overhead cranes		617
	C6.5.3	Port cranes		617
		C6.5.3.1	Ship-to-shore container cranes: grab ship unloaders	617

		C6.5.3.2	Rubber-tyred gantry cranes	618
		C6.5.3.3	Rail-mounted gantry cranes	618
	C6.5.4	Automated warehousing		620
	C6.5.5	Notes on crane control characteristics		620
		C6.5.5.1	Hoisting control	620
		C6.5.5.2	Slewing control	620
	C6.5.6	Retrofit applications		621
C6.6	Elevators and lifts			622
	C6.6.1	Lift system description		622
	C6.6.2	Speed profile generation		625
	C6.6.3	Load weighing devices		626
	C6.6.4	Block diagram of lift electrical system		627
C6.7	Metals and metal forming			627
	C6.7.1	Introduction		627
	C6.7.2	Steel		627
		C6.7.2.1	Main mill drives	628
		C6.7.2.2	Auxiliary drives	629
		C6.7.2.3	Strip rolling mills	630
		C6.7.2.4	Continuous casting	633
	C6.7.3	Wire and cable manufacture		635
		C6.7.3.1	Wire drawing machine	635
		C6.7.3.2	Twin carriage armourer	637
C6.8	Paper making			638
	C6.8.1	General		638
	C6.8.2	Sectional drives		639
	C6.8.3	Loads and load sharing		640
	C6.8.4	Control and instrumentation		642
	C6.8.5	Winder drives		644
	C6.8.6	Brake generator power and energy		645
	C6.8.7	Unwind brake generator control		647
	C6.8.8	Coating machines		648
C6.9	Plastics extrusion			649
	C6.9.1	General		649
	C6.9.2	Basic extruder components		652
	C6.9.3	Overall extruder performance		653
	C6.9.4	Energy considerations		654
	C6.9.5	Motors and controls		656
C6.10	Stage scenery: film and theatre			657
	C6.10.1	The Control Techniques orchestra		657

PART D APPENDICES 661

D1 Symbols and formulae 663

D1.1	SI units and symbols		663
	D1.1.1	SI base units	663
	D1.1.2	Derived units	664

				Contents xxiii

	D1.2	Electrical formulae			665
		D1.2.1	Electrical quantities		665
		D1.2.2	A.C. three-phase (assuming balanced symmetrical waveform)		666
		D1.2.3	A.C. single-phase		666
		D1.2.4	Three-phase induction motors		667
		D1.2.5	Loads (phase values)		667
		D1.2.6	Impedance		667
		D1.2.7	A.C. vector and impedance diagrams		667
		D1.2.8	Emf energy transfer		669
		D1.2.9	Mean and rms values, waveform		670
			D1.2.9.1	Principles	670
			D1.2.9.2	Mean d.c. value	671
			D1.2.9.3	rms value	672
			D1.2.9.4	Form factor	674
	D1.3	Mechanical formulae			674
		D1.3.1	Laws of motion		674
			D1.3.1.1	Linear motion	676
			D1.3.1.2	Rotational or angular motion	677
			D1.3.1.3	Relationship between linear and angular motion	678
			D1.3.1.4	The effect of gearing	679
			D1.3.1.5	Linear to rotary speed reference conversion	680
			D1.3.1.6	Friction and losses	681
			D1.3.1.7	Fluid flow	682
	D1.4	Worked examples of typical mechanical loads			684
		D1.4.1	Conveyor		684
		D1.4.2	Inclined conveyor		689
		D1.4.3	Hoist		689
		D1.4.4	Screw-feed loads		693
D2	**Conversion tables**				**695**
	D2.1	Mechanical conversion tables			695
	D2.2	General conversion tables			700
	D2.3	Power/torque/speed nomogram			706
D3	**World industrial electricity supplies (<1 kV)**				**707**
Bibliography					**715**
Index					**717**

Preface

With the rapid developments in the last 20 years in the area of industrial automation, it can be argued that the variable-speed drive has changed beyond all recognition. The functionality of a modern drive is now so diverse that its ability to rotate a motor is sometimes forgotten. Indeed, some customers buy drives not to control a motor but to utilise the powerfull auxiliary functionality that is built in. This is, however, unusual, and the drive remains a key component of the boom in all aspects of automation. Drives are also critical components in relation to energy saving. For over 30 years the case for energy saving through the use of variable-speed drives has been made by drive companies, and at last it seems that industry is moving quickly to adopting the technology. Consider the facts: 55–65 per cent of all electrical energy is used by electric motors. On average, fitting a variable-speed drive will save 30 per cent of the energy used by a fixed-speed motor, but today only 5 per cent of those motors are controlled by variable-speed drives. The opportunity is therefore enormous. Drives could save the world, or make a significant contribution to the cause. Before taking a brief look into the future it is helpful to look back at the relatively short history of drives and see how far and how quickly the technology has come.

1820   Oersted was the first to note that a compass needle is deflected when an electric current is applied to a wire close to the compass; this is the fundamental principle behind an electric motor.

1821   Faraday (Figure P.1), built two devices to produce what he called electromagnetic rotation: that is, a continuous circular motion from the circular magnetic force around a wire. This was the initial stage of his pioneering work.

1824   Arago discovered that if a copper disc is rotated rapidly beneath a suspended magnet, the magnet also rotates in the same direction as the disc.

1825   Babbage and Herschel demonstrated the inversion of Arago's experiment by rotating a magnet beneath a pivoted disc causing the disc to rotate. This was truly induced rotation and just a simple step away from the first induction motor, a step that was not then taken for half a century.

1831   Using an 'induction ring', Faraday made one of his greatest discoveries – electromagnetic induction. This was the induction of electricity in a wire by means of the electromagnetic effect of a current in another wire. The induction ring was the first electric transformer. In a second series of experiments in the

Figure P.1   Michael Faraday (1791–1867)

same year he discovered magneto-electric induction: the production of a steady electric current. To do this, Faraday attached two wires through a sliding contact to a copper disc, the first commutator; this was an approach suggested to him by Ampère. By rotating the disc between the poles of a horseshoe magnet he obtained a continuous direct current. This was the first generator. Faraday's scientific work laid the foundations for all subsequent electro-technology. From his experiments came devices that led directly to the modern electric motor, generator and transformer.

1832   Pixii produced the first magneto-electric machine.
1838   Lenz discovered that a d.c. generator could be used equally well as a motor. Jacobi used a battery-fed d.c. motor to propel a boat on the River Neva. Interestingly, Jacobi himself pointed out that batteries were inadequate for propulsion, a problem that is still being worked on today.
1845   Wheatstone and Cooke patented the use of electromagnets instead of permanent magnets for the field system of the dynamo. Over 20 years were to elapse before the principle of self-excitation was to be established by Wilde, Wheatstone, Varley and the Siemens brothers.
1870   Gramme introduced a ring armature that was somewhat more advanced than that proposed by Pacinotte in 1860, which led to the multi-bar commutator and the modern d.c. machine.
1873   Gramme demonstrated, at the Vienna Exhibition, the use of one machine as a generator supplying power over a distance of 1 km to drive a similar machine

as a motor. This simple experiment did a great deal to establish the credibility of the d.c. motor.

1879   Bailey developed a motor in which he replaced the rotating magnet of Babbage and Herschel by a rotating magnetic field, produced by switching of direct current at appropriately staggered intervals to four pole pieces. With its rotation induced by a rotating magnetic field it was thus the first commutatorless induction motor.

1885   Ferraris produced a motor in which a rotating magnetic field was established by passing single-phase alternating current through windings in space quadrature. This was the first alternating current commutatorless induction motor, a single-phase machine that Dobrowolsky later acknowledged as the inspiration for his polyphase machine.

1886   Tesla developed the first polyphase induction motor. He deliberately generated four-phase polyphase currents and supplied them to a machine with a four-phase stator. He used several types of rotor, including one with a soft-iron salient-pole construction (a reluctance motor) and one with two short-circuited windings in space quadrature (the polyphase induction motor).

1889   Dobrowlsky, working independently from Tesla, introduced the three-phase squirrel-cage induction motor.

1890   Dobrowlsky introduced a three-phase induction motor with a polyphase slip-ring rotor into which resistors could be connected for starting and control. The speed of these motors depends fundamentally upon its pole number and supply frequency. Rotor resistance control for the slip-ring motor was introduced immediately, but this is equivalent to armature resistance control of a d.c. machine and is inherently inefficient.

By 1890 there was a well established d.c.. motor, d.c. central generating stations, three-phase a.c. generation and a simple three-phase motor with enormous potential but which was inherently a single-speed machine. There was as yet no way of *efficiently* controlling the speed of a motor over the full range from zero to full speed.

1896   The words of Harry Ward Leonard first uttered on 18 November 1896 in his paper entitled 'Volts vs. ohms – speed regulation of electric motors' marked the birth of the efficient, wide-range, electrical variable-speed drive:

'The operation by means of electric motors of elevators, locomotives, printing presses, travelling cranes, turrets on men-of-war, pumps, ventilating fans, air compressors, horseless vehicles, and many other electric motor applications too numerous to mention in detail, all involve the desirability of operating an electric motor under perfect and economical control at any desired speed from rest to full speed.' (Figure P.2.)

The system he proposed was of course based upon the inherently variable-speed d.c. machine (which had hitherto been controlled by variable armature resistors). His work was not universally accepted at the time and attracted much criticism, understandably, as it required three machines of similar rating to do the job of one. Today, however, all d.c. drives are based upon

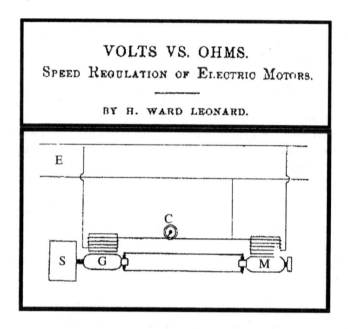

Figure P.2 110th Meeting of the American Institute of Electrical Engineers, New York, 18 November 1896

his control philosophy, only the implementation changing from multi-motor schemes through the era of grid controlled mercury arc rectifiers to thyristors and more recently, in demanding dynamic applications, to bipolar transistors, field-effect transistors (FETs), insulated gate bipolar transistors (IGBTs) and so on.

1904 Kramer made the first significant move with respect to frequency changing in 1904 by introducing a d.c. link between the slip rings and the a.c. supply. This involved the use of two a.c. ↔ d.c. motor sets. The d.c. link was later to become a familiar sight in many a.c. drive technologies. He published in 1908 (Figure P.3).

Subsequent advances in a.c. motor speed control was based upon purely electrical means of frequency and voltage conversion. Progress has followed the advances in the field of semiconductors (power and signal/control).

1911 Schrage introduced a system based upon an induction motor with a commutator on the rotor. This machine proved to be very popular, requiring no auxiliary machines and was very reliable. It found large markets, particularly in the textile industry and some other niche applications. It is still sold today but in rapidly reducing numbers.

The introduction of the ignitron made controlled rectification possible. The thyratron and grid controlled mercury rectifiers made life easier in 1928. This made possible the direct control of voltage applied to the armature of a

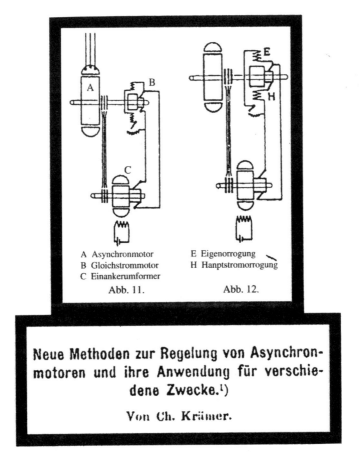

Figure P.3 Elektrotechnische Zeitschrift, *vol. 31, 30 July 1908*

        d.c. machine so as to apply the philosophy of Ward–Leonard control without additional machines.

1930   The ideas of inversion (d.c. to variable-frequency/voltage a.c., which is the basis for the present-day inverter) had been established, and the use of forced commutation by means of switched capacitors was introduced.

1931   Direct a.c. to a.c. conversion by means of cycloconverters was introduced for railway service.

1932   The Nyquist stability criterion was developed.

1938   The Bode stability criterion developed.

1950   The introduction of silicon into power switches replaced the bulky and relatively inefficient mercury arc rectifiers (MAR). By 1960, thyristors (SCRs) had become available and the key enabling technology for drives had arrived. D.C. drives and cycloconverters quickly embraced the new silicon technology at first using techniques with origins in the MAR forerunners.

The faster switching performance of the new silicon, however, opened many new doors, notably in the field of forced commutation. The way was clear for commercial variable-frequency drives (VFDs).

1957   The 'back to back' reversing d.c. drive introduced.

1960s   Power semiconductor voltage and current ratings grow and performance characteristics improve. Inverters became commercially viable, notably in industries such as textiles where a single (bulk) inverter was used to feed large numbers of induction motors (or reluctance motors, despite their low power factor, where synchronisation was required).

1963   Gain–bandwidth relationships of power converters were investigated.

1970   The 1970s saw a new and very significant revolution hit the variable-speed drives market – packaging. Up until this time the static variable-speed drive design process had essentially concentrated on performance/functionality. Both a.c. and d.c. drives of even low rating were broadly speaking custom built or hand crafted. This approach resulted in bulky, high-cost drives, the very uniqueness of which often compromised reliability and meant service support was difficult. The drives industry was not fulfilling its potential.

1970s   A.C. motor drives had made great advances in terms of performance but still lacked the dynamic performance to really challenge the d.c. drive in demanding process applications. Since the early 1970s considerable interest was being generated in field oriented control of a.c. machines. This technique, pioneered by Blaschke and further developed by Leonhard, opened up the opportunity for a.c. drives not only to match the performance of a d.c. drive but to improve upon it. The processing requirements were such that in its early days commercial exploitation was restricted to large drives such as mill motor drives and boiler feed pump drives. Siemens were very much in the forefront of commercialising field orientation. Siemens were also rationalising the numerous alternative drive topologies that had proliferated and, while stimulating to the academic, were confusing to drive users:

1. D.C. drives
   a. Single converter
   b. Double converter
      i. Circulating current free
      ii. Circulating current
2. A.C. drives
   a. Voltage (phase) control
   b. Voltage source inverters
      i. Quasi-square V/f
      ii. Quasi-square V/f with d.c. link chopper
      iii. Pulse width modulated (PWM)
   c. Current source inverters
      i. Induction motor
      ii. Synchronous machine
   d. Static Kramer drive
   e. Cycloconverter

Figure P.4   D.C. drive module [photograph courtesy of Control Techniques]

1972   Siemens launched the SIMOPAC integrated motor with ratings up to 70 kW. This was a d.c. motor with integrated converter including line reactors!

1973   A new approach to drives in terms of packaging. Utilising 19-in rack principles, a cubicle-mounting standard well used in the process industry, compact, high-specification ranges of d.c. drives in modular form (Figure P.4) became available off the shelf. Companies such as AEG, Thorn Automation, Mawdsley's and Control Techniques pioneered this work. A new era of drive design had started.

1979   Further advances in packaging design were made possible by the introduction of isolated thyristor packages.

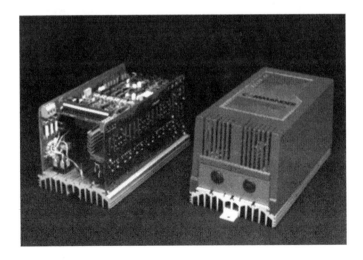

Figure P.5   Plastic mouldings introduced into drives [photograph courtesy of Control Techniques]

xxxii   *Preface*

1983   In 1983 plastic mouldings (Figure P.5), made their first significant impact in drives. Bipolar transistor technology also arrived, which eliminated bulky auxiliary commutation circuits.
1985   Takahashi and Noguchi published a paper on direct torque control (DTC) in the IEEE. This date is included not because of its technical significance but rather as a point of interest as DTC has received much commercial attention.
1986   Great advances were being made at this time in the field of microprocessors making possible cost-effective digital drives at low powers. Further drives were introduced containing application-specific integrated circuits (ASIC), which to that time had only been used in exceptionally large-volume/domestic applications. Further, new plastic materials were introduced that gave structural strength, weight, size, assembly and cost advantage (Figure P.6).
1988   IGBT technology was introduced to the drives market. IGBTs heralded the era of relatively quiet variable-speed drives (and introduced a few problems, some of which have led to substantial academic activity, and only a very few of which have required more pragmatic treatment).

Figure P.6   Digital d.c. drive with microprocessor and ASIC [photograph courtesy of Control Techniques]

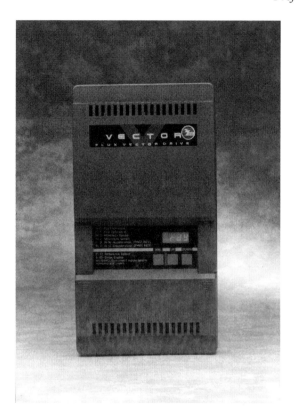

Figure P.7   Vector drive [photograph courtesy of Control Techniques]

1989   The first implementation of the field orientation or flux vector drive was introduced to the high-volume, lower-power market (Figure P.7). It found immediate application in machine tool spindle drives and has grown rapidly in application (and rating) since. It should be said that the name vector has been prostituted by some in the drives industry with 'voltage vector' and other such names/techniques, causing confusion and frustration to customers.

1990   The trend to smaller drive products, which were also simpler to design, was given a significant boost by Mitsubishi, who introduced intelligent power modules that integrated into the semiconductor package the necessary gate drive and protection functions.

1992   A new packaging trend emerged – the bookform shape (Figure P.8); this had previously been applied to servo drives and was now being applied to the broader industrial a.c. drives market. The trend continues today but there is not a consensus that this is the most suitable shape for all market segments.

1993   Another innovation in packaging arrived – at the low-power end of the spectrum when a DIN rail mounting 0.4 kW inverter package (Figure P.9), similar to that used widely in equipment such as contactors and control relays, was

Figure P.8   Bookform shape of drive [photograph courtesy of Control Techniques]

Figure P.9   DIN rail mounting drive with built in EMC filter [photograph courtesy of Control Techniques]

Figure P.10 Universal a.c drive modules [photograph courtesy of Control Techniques]

       launched. The first drive with a built-in supply-side filter fully compliant with the then impending EU regulations on conducted EMC was introduced.

1996   The first truly *universal drive* (Figure P.10) was launched that met the diverse requirements of a general-purpose open-loop vector drive, a closed-loop flux vector drive, a servodrive, and a sinusoidal supply converter with the selection purely by parameter selection. This was also the birth of what has become known as the *intelligent drive* with user-programmable functionality as well as a broad range of Fieldbus connectivity.

1998   The integrated d.c. motor launched in 1972 was not a great commercial success – much has been learnt since those days. In 1998 integrated a.c. motor drives were introduced onto the market (Figure P.11). These products are, for the most part, open-loop inverter-driven induction motors and were initially targeted on replacing mechanical variable-speed drives. Integrated servo motors followed.

1999   A radical servo drive was introduced with the position and speed loop embedded in the encoder housing on the motor itself (Figure P.12). This brought with it the advantage of processing the position information close to the source, thereby avoiding problems of noise etc, and allowed dramatic

Figure P.11   Integrated a.c. motor [photograph courtesy of Leroy–Somer]

improvements in control resolution, stiffness of the drive and reduced the number of wires between the drive and the motor.

2000   In the early years of the new millennium, rapid change continued. Those users who were looking to use drives as components in a larger control system, were looking for ever greater connectivity. The Fieldbus 'wars' were raging with passionate claims for many systems (I counted over 200 in a 12-month period), most of which have since disappeared. The war has now morphed into the Ethernet wars, with advocates of the different protocols all predicting dominance.

Figure P.12   Speed loop motor [photograph courtesy of Control Techniques]

Development is driven by component technology, design techniques and the vision of the industry. Power devices, notably Trench IGBTs, have driven improved efficiency, while improved microprocessor performance has yielded not only improved motor shaft performance, but facilitated further significant functionality. Ease of use has been, and remains, a key focus. Although there remain specific motor types and controls best suited to certain applications, users can buy a single product that can meet these different needs. The universal drive is truly a no compromise solution to a broad range of applications and grows in the market. Ease of use and optimising of setup is achieved by automatic tuning routines in drives, matching the drive to both motor and mechanical load.

Functionality has been greatly enhanced to a level where IEC 61131 compliant PLC functionality is available for users to programme very complex and demanding system applications. Drives can be synchronised together with control loop jitter of <2 μs without the need for a master controller.

Motor technology is changing. The brushless permanent magnet (BPM) motor, once only used in high-performance applications, is now being considered where efficiency or size is critical. Linear motors have made an appearance, but mainly at trade shows. For the most demanding applications, the quality of speed or position measurement is critical. The emergence of sine/cosine encoders has facilitated very-high-resolution position feedback, while all digital solutions such as EnDat point the way forward.

Much has indeed changed in the last ten years. Much will change in the next ten years and beyond. Component developments, particularly in the semiconductor industries, will continue to play a significant role in defining direction. Cost remains a driver of product development, and motor shaft performance improvements continue apace even if they do not always appear centre stage in the marketing brochures.

With the capabilities within a modern drive, users need to consider how to balance elegance and cost over what could be considered over-dependence on a single supplier. The improved interfacing technologies, including the emergence of Ethernet as an industrial backbone at the machine level, will certainly act to mitigate these concerns. Customer needs will of course be the key driver to future developments. Significant advances occur when users and drive designers get together and consider system solutions. Some drives are sold to customers who have no intention of connecting a motor to it! It has been purchased purely to use the comprehensive auxiliary functions.

The world of drives is therefore vibrant and dynamic, but the breadth of technology can be confusing. This book aims to de-mystify the technology and the way the products can be used to bring benefit in many applications. It covers the present state of development, or rather commercial exploitation of industrial a.c. and d.c. variable-speed drives and associated systems. It is intended primarily for the use of professional engineers who specify or design systems that incorporate drives. The theory of both the driven motor and the drive is explained in practical terms, with reference to fundamental theory being made only where appropriate for further illumination. Information on how to apply drive systems is included, as are examples of what can be found within commercially available drives and indications of what can be achieved using them.

Emphasis is placed on low-voltage (110 to 690 V) industrial drives in the range 0.37 kW to 1 MW.

The practical nature of the book has led to two unfortunate but I fear unavoidable consequences. First, some of the theory behind the technology contained in the book has had to be omitted or abridged in the interests of clarity and volume. Second, in such a practical book it has proved difficult to avoid some reference to proprietary equipment. In such circumstances a tendency towards referencing the products of Control Techniques is inevitable. It should be clear to readers that these products are described as examples to illustrate the technology. The IET, publisher of this book, does not endorse these products or their use in any way.

Professor Bill Drury

Acknowledgements

A book of this type relies upon the contribution and help of a great number of people. This edition of *The Control Techniques Drives and Controls Handbook* has been created with contributions from engineers both within Control Techniques itself as well as sister companies within the family that is Emerson. Accordingly, the contributions of all the following are acknowledged with great thanks.

The Control Techniques Drives and Controls Handbook (IET, 2009)

I would in particular like to thank Dr Mike Cade and Dr Colin Hargis, who have been a major source of help throughout the project, and have contributed to key areas of the book. Thanks are also due to Dr Volker Pickert, Dr Matthew Armstrong, Peter Worland, Kevin Manton, Jon Atkinson, Alex Harvey and Giannbattista Dubini for their help in specific areas of the book. Thanks are also due to our friends and colleagues at Leroy Somer, Heidenhain, Sick Stegmann and IC-Haus for the help and support they have given.

The Control Techniques Drives and Controls Handbook (IEE, 2001)

Dr Mike Cade, Dr Colin Hargis, Dr Peter Barrass, Ray Brister, Vikas Desai, Jim Lynch, John Orrells, Bleddyn Powell, Alex Rothwell, Michael Turner, and Peter Worland.

Drives and Servos Yearbook (Control Techniques, 1990)

J. Boden, K. Briggs, S. Buckley, A. Clark, R. Cottell, A. Davies, J. Day, C. French, B. Hardy-Bishop, C. Haspel, E. Introwicz, R.Lang, T. Miller, L. Mummery-Smith, J. Orrells, D. Reece, A. Richmond, A. Royle, N. Sewell, P. Sewell, R. Smith, G. Thomas, N. Vivian, G. West, and P. Worland.

Part A
Drive types and core technology

Introduction
A1 Industrial motors
A2 Drive converter circuit topologies
A3 Power semiconductor devices
A4 Torque, speed and position control
A5 Position and speed feedback
A6 Motion control
A7 Voltage source inverter: four-quadrant operation
A8 Switched reluctance and stepper motor drives

Introduction

The selection of a drive is determined by a great many factors. Section A of this book describes the core technology and performance of the different elements of a drive system. To put this into context, the main types of industrial drives are tabulated below with a brief summary of their features. Reference is made to areas of the book for further more detailed descriptions.

The number of topologies is so broad so for convenience the table has been split into the following:

- D.C. drives
- A.C. inverter drives
- Slip energy recovery and direct converter drives
- Soft starter, switched reluctance and stepper motor drives

Table A0.1 D.C. drives

Drive type	Single quadrant drive with single half controlled converter	Single quadrant[A] drive with single fully controlled converter	Four quadrant drive with single converter and armature or field reversal	Four quadrant drive with significant period without torque	D.C. chopper
Motor type	D.C. Separately Excited Motor (A1.2.3.1) or d.c. Permanent Magnet Motor (A1.2.4)				
Converter type	Single/three phase half controlled thyristor bridge. (A2.2.4/A2.2.2.2/A2.2.3.2)	Single/three phase fully controlled thyristor bridge. (A2.2.4/A2.2.2.2/A2.2.3.2)	Single/three phase fully controlled thyristor bridge. (A2.2.4/A2.2.2.2/A2.2.3.2)	Dual single/three phase fully controlled thyristor bridge. (A2.2.4/A2.2.2.2/A2.2.3.2)	Four quadrant d.c. chopper (A2.3.2.3) usually fed from uncontrolled converter (A2.2.2.1/A2.2.3.1)
Main switching power semiconductors	Thyristor/SCR (A3.3)				MOSFET (A3.7) [or IGBT (A3.8)]
Torque/speed quadrants of operation	Motoring in one direction only	Motoring in one direction only. Braking in the other direction	Motoring and braking in both directions		Two or Four quadrant versions
Method of speed control	Closed loop control of armature voltage with inner current control loop (A4.3)				
Method of torque control	Closed loop control of armature current (A4.3)				

Typical industrial power ratings	Up to 7.5 kW/10 HP	5 kW to 5 MW/7 to 7,000 HP	0.5 to 5 kW/0.7 to 7 HP. Traction >500 kW
Typical max. speed		Available to multi MW ratings but Power.Speed product limited to 3×10^6 kW·min^{-1}	
Typical min. speed		Good control possible down to standstill	
Notable features		Separately excited motors often used above base speed in a constant power mode (A2.2.4)	D.C./d.c. conversion
		Slow torque reversals (A2.2.4)	Fast torque reversals (A2.2.4)	Smooth torque possible
Note on industrial popularity		D.C. drives remain a significant part of the overall drives market. Popularity is diminishing and is focused mainly on applications where a d.c. motor already exists or very small simple drives or high power drives where the d.c. motor can still be competitive	
	Popular low cost solution for low power drives in simple applications	Popular single quadrant solution for retrofit to existing d.c. motors	Popular 4 quadrant solution for retrofit to existing d.c. motors. Motor + drive can be cost effective at higher powers	Once very popular as Servo drive with PM motor. Gradually loosing market to a.c. equivalents
		Used for applications with limited number of torque reversals and limited dynamic performance requirement		

Notes: [A] Whilst capable of two quadrants of operation this topology is frequently referred to as single quadrant

Table A0.2 A.C. inverter drives

Drive type	PWM inverter	Multi level PWM inverter	Square wave inverter with d.c. chopper	Current source inverter (induction motor)	Converter fed synchronous machine/LCI drive
Motor type	Induction Motor (A1.3) or Synchronous Motor (A1.4) Motor (A1.4.3.2) [or Reluctance Motor (A1.5)]		Brushless PM Servo (A1.4)	Induction motor (A1.3)	Synchronous machine (A1.4)
Converter type	PWM inverter (A2.4.2.3)	Multi level inverter (A2.4.2.4)	Six step inverter (A2.4.2.2)	Converter-fed induction motor (A2.4.3.3)	Three phase fully controlled bridges
Main switching power semiconductors	IGBT (A3.8)			IGBT (A3.8) or Thyristor/SCR	Thyristor/SCR
Torque/speed quadrants of operation	Motoring and Braking in both directions provided either a dual converter is provided in the supply side or a d.c. link chopper and brake resistor is provided			Motoring and Braking in both directions	
Method of speed control	Scalar and Flux Vector control variants available (A4.4)				
Method of torque control					
Typical industrial power ratings	up to 2 MW (400/690 V)/>10 MW (6600 V)		up to 2 MW (400/690 V)	up to 4 MW (400/690 V)/10 MW (6600 V)	2 to >20 MW (MV only)

Drive types and core technology 7

Typical max. speed	>80,000 min^{-1}	>10,000 min^{-1}	>80,000 min^{-1}	>10,000 min^{-1}	>10,000 min^{-1}
Typical min. speed	Control dependent = Generally open loop to 1 Hz; closed loop to standstill with full torque				
Notable features	Simple, robust, commonly available induction motor				High speed synchronous motors
	Low torque ripple. Excellent control dynamics	Low motor dV/dt (important at MV)	Capable of high output frequencies	Good efficiency	High efficiency. Dynamic control
Note on industrial popularity	Voltage source inverters are popular due to their ease of application on different motor types and loads			Once the topology of choice for single motor drives this topology has been largely superseded by the PWM inverter	The most popular solution for very high power (>5 MW) and/or very high speed applications
	The industrial workhorse. The most popular drive topology	Popular in medium voltage drives	Only used for high frequency applications. PWM inverters often default to this mode of operation at higher frequencies		

Table A0.3 Slip energy recovery and direct converter drives

Drive type	Static Kramer drive (A2.4.3.5)	Static Scherbious drive (A2.5.5)	Cycloconverter (A2.5.3)	Matrix converter (A2.5.4)
Motor type	Slip ring induction motor (A1.3.5)	Slip ring induction motor (A1.3.5)	Synchronous motor (A1.4) or induction motor (A1.3)	Synchronous motor (A1.4) or induction motor (A1.3)
Converter type	Diode bridge (A2.2.3.1) & 3 phase fully controlled bridge (A2.2.3.2)	Cycloconverter (A2.5.3)	Cycloconverter (A2.5.3)	Matrix converter (A2.5.4)
Main switching power semiconductors	Diode (A3.2) & Thyristors/SCR (A3.3)	Thyristor/SCR (A3.3) or IGBT (A3.8)	Thyristor/SCR (A3.3) or IGBT (A3.8)	IGBT (A3.8)
Torque/speed quadrants of operation	Motoring in one direction only	Motoring and Braking in both directions		
Method of speed control	Scalar and flux vector control variants available. (A4.4)			
Method of torque control				
Typical industrial power ratings	500 kW to 20 MW	500 kW to 20 MW	500 kW to >10 MW	50 kW to 1 MW
Typical max. speed	1470 min^{-1}	2000 min^{-1}	<30 Hz	<Supply Freq
Typical min. speed	900 min^{-1}			0 Hz
Notable Features	Economic high power solution		No d.c. link components. Should be very reliable	
Note on industrial popularity	Not prominent but still a very economic choice for limited speed range applications	Has much of the economic benefit of the static Kramer but with the ability to operate above synchronous speed	The topology of choice for very low speed high power applications	Has some niche areas of application notably integrated motors where its limited output voltage is not a problem

Table A0.4 Soft starter, switched reluctance and stepper motor drives

Drive type	Soft starter		Switched reluctance drive	Stepper motor
Motor type	Induction motor (A1.3)		Switched reluctance motor (A8.2)	Stepper motor (A8.3)
Converter type	Inverse parallel line switches		Specific topologies (See A8.2.6)	Specific topologies (See A8.3.2)
Main switching power semiconductors	Thyristor/SCR (A3.3) or IGBT (A3.8)		IGBT (A3.8)	MOSFET (A3.7) or IGBT (A3.8)
Torque/speed quadrants of operation	Motoring in one direction only		Motoring and Braking in both directions provided either a dual converter is provided in the supply side or a d.c. link chopper and brake resistor is provided	
Method of speed control			See A8.2	See A8.3
Method of torque control				
Typical industrial power ratings	1 kW to >1 MW		1 kW to >1 MW	10 W to >5 kW
Typical maximum speed	Not applicable		>10,000 min^{-1}	>10,000 min^{-1}
Typical minimum speed	Not applicable		Standstill	Standstill
Notable features			Simple motor construction	Simple motor and control strategy
Note on industrial popularity	The soft start has very specific application in reducing the starting current or controlling the starting torque of a DOL induction motor		Low speed/high torque applications	Open loop positioning systems

Chapter A1
Industrial motors

A1.1 Introduction and basic electromagnetic principles

All electric machines comprise coupled electric and magnetic circuits that convert electrical energy to and/or from mechanical energy. The term electric motor tends to be used to describe the machine used in an industrial drive, regardless of whether it is converting electrical energy to mechanical motion as a motor or if it is producing electrical energy from mechanical energy as a generator. Alternative designs of motor exist, but before considering these it is helpful to consider a few basic principles of magnetic circuits and electromechanical energy conversion.

A1.1.1 Magnetic circuits

Electric motors are all made up of coupled electric and magnetic circuits. An electric circuit is easily identified as a path in which electric current can flow. A magnetic circuit is similarly a path in which magnetic flux can flow.

A magnetic field is created either around any current-carrying conductor or by a permanent magnet.

Figure A1.1 presents the first scenario, in which the d.c. current I flowing through the wire produces a circular magnetic field strength H and a magnetic field B. The magnetic field strength is in a direction determined by the direction of the current in the wire, and its magnitude is inversely proportional to the distance away from the conductor:

$$H = I/2\pi r \qquad (A1.1)$$

The unit of magnetic field strength is A m^{-1}.

The magnetic field, frequently referred to as the flux density, is the flux per unit area at that point:

$$B = \mu H \qquad (A1.2)$$

where μ is the permeability of the magnetic circuit. The unit of the magnetic field is the Tesla (T).

12   The Control Techniques Drives and Controls Handbook

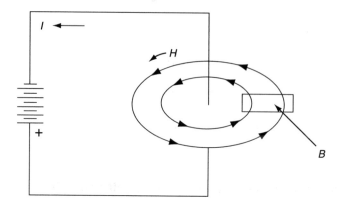

Figure A1.1   Magnetic field associated with a current in a conductor

The flux ω is given by

$$\omega = BA \qquad (A1.3)$$

where A is the area considered with a uniform flux density B. The unit of flux is the Weber (Wb).

The second scenario is where a magnetic field is caused by a permanent magnet, and this is determined by a number of factors including the B–H characteristic of the magnet and the position of the magnet in a magnetic circuit.

When more than one conductor carries current, and they are relatively closely positioned, the magnetic fluxes add together. Further, if the wire is wound in the form of a coil, its magnetic force is significantly intensified, and exhibits a characteristic very much like that of a permanent magnet (Figure A1.2).

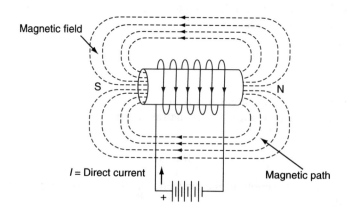

Figure A1.2   Magnetic field associated with current in a coil

In Figure A1.2 both the electrical and magnetic circuits are clearly identifiable, the magnetic circuit being the space in which the flux goes around the coil. The magnitude of the magnetic field is the product of the current I and the number of turns N:

$$\text{m.m.f.} = NI \tag{A1.4}$$

where mmf is the magneto-motive force.

The flux ω is given by

$$\omega = \text{m.m.f.}/R = NI/R \tag{A1.5}$$

where R is the reluctance of the magnetic circuit. It is useful to consider the equation of the magnetic circuit (A1.5) as being analogous to Ohms Law for an electrical circuit, where

- m.m.f. is the force to create flux in a coil,
- reluctance is the magnetic resistance, and
- flux is the flow of flux.

The reluctance of a magnetic circuit can be calculated from the geometry and physical properties of the material:

$$R = \text{length}/\mu \cdot \text{area} \tag{A1.6}$$

where μ is the permeability of the magnetic circuit ($\mu = 4\pi \times 10^{-7}$ N A^{-2} in a vacuum and is usually denoted μ_0). To improve the characteristics of the magnetic circuit, a magnetic core can be introduced into the circuit as shown in Figure A1.3.

Magnetic materials have a high relative permeability (relative to air), and give a magnetic circuit higher flux density for a given magnetic force (Table A1.1):

$$R = \text{length}/\mu_0 \mu_r \cdot \text{area} \tag{A1.7}$$

where μ_r is the relative permeability of the magnetic material. It is dimensionless.

The core material of an electric motor is usually ferromagnetic, and the relative permeability is not constant. The relationship between flux density B and the magnetic force H is known as the B–H or saturation curve.

When an a.c. current flows in the electric circuit, the direction of the magnetising force follows the current, and the flux (and flux density) reverses as the direction of the current changes. Ideally, the operating point of the magnetic circuit would move up and down the B–H curve shown in Figure A1.4 (and its mirror image). In practice, under a.c. excitation the magnetic circuit follows a path similar to that shown in Figure A1.5, which is commonly known as a hysteresis loop.

The enclosed area in the hysteresis loop is a measure of the energy lost in the magnetic core during that cycle. This loss occurs each time the cycle is traversed, so the loss can be seen to be dependent upon the frequency of the current. For most magnetic core materials the width of the B–H curve broadens with frequency.

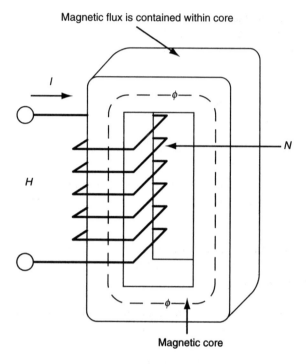

Figure A1.3   Magnetic circuit with a magnetic core

The inductance L of a coil is defined as the flux linkage per unit current:

$$L = N\omega/I \tag{A1.8}$$

where N is the number of turns in a coil and it can be assumed that all the flux passes through or links all the coils.

Table A1.1   Relative permeabilities of some materials

Material	Relative permeability	
Mu-metal	20 000	At 0.002 T
Permalloy	8 000	At 0.002 T
Transformer iron	4 000	$\rho = 0.01$ μΩ m at 0.002 T
Steel	700	At 0.002 T
Nickel	100	At 0.002 T
Soft ferrite	4 000	$\rho = 0.1$ Ω m at <0.1 T
	2 000	$\rho = 10$ Ω m at <0.1 T

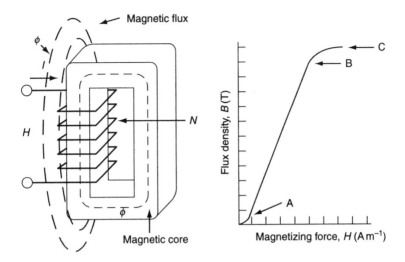

Figure A1.4   Effects of saturation of a magnetic core

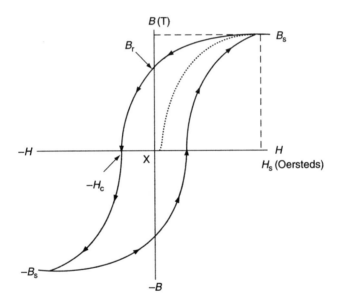

Figure A1.5   Typical hysteresis loop

We know from equation (A1.5) that $\omega = NI/R$ and so

$$L = N^2/R \tag{A1.9}$$

An electric motor contains an air gap between the rotor and the stator. An air gap appears as a high-reluctance element in an otherwise low-reluctance magnetic circuit.

16 The Control Techniques Drives and Controls Handbook

Because the reluctance of the air gap is usually so much higher than the rest of the magnetic circuit, it can be seen that the size of the air gap (invariably minimised) is critical to the performance of the motor.

A1.1.2 Electromechanical energy conversion

There are two basic mechanisms through which electric motors produce mechanical force. The first is the alignment of magnetic force/flux lines. The second arises from the interaction between a magnetic field and a current-carrying conductor.

A1.1.2.1 The alignment of magnetic force/flux lines

Most people are familiar with the way a simple permanent magnet attracts a metal part. What is happening is that the magnet is acting to decrease the reluctance path of the magnetic circuit. This mechanism forms the basis of reluctance motors, a very simple example of which is shown in Figure A1.6. The rotor is free to rotate between the pole pieces of the stationary 'stator'.

When the rotor is in the vertical position (Figure A1.6a), $v = 0$. In this condition equation (A1.7) shows the reluctance of the magnetic circuit to be

$$R_0 = 2g_0/\mu_0 A_0 \qquad (A1.10)$$

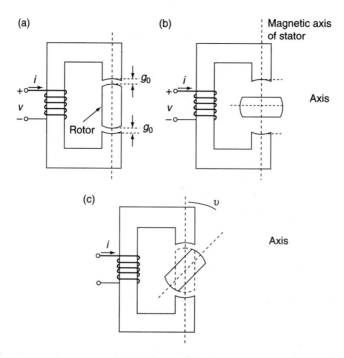

Figure A1.6 Reluctance-type motor

where g_0 is the air-gap length when $v = 0$ and A_0 is the effective area of the gap. Note that because the permeability of the stator core is much higher than the air gap, it can be ignored in this basic calculation.

When $v = \pi/2$, as shown in Figure A1.6b,

$$R_1 = 2g_1/\mu_0 A_1 \tag{A1.11}$$

where g_1 is the air-gap length and A_1 the effective area of the air gap. Because $g_1/A_1 > g_0/A_0$, $R_1 > R_0$. It can be seen that the reluctance is

- a minimum for values of $v = 0, \pi, 2\pi, 3\pi \ldots$
- a maximum for values of $v = \pi/2, 3\pi/2, 5\pi/2 \ldots$

The electrical circuit inductance is inversely proportional to the magnetic circuit reluctance (1.9), and it can therefore be seen that the inductance is

- a maximum L_{max} for values of $v = 0, \pi, 2\pi, 3\pi \ldots$
- a minimum L_{min} for values of $v = \pi/2, 3\pi/2, 5\pi/2 \ldots$

The variation of inductance with rotor position is as shown in Figure A1.7.

Motoring torque is produced when current i is flowing in the electric circuit and its inductance is increasing with respect to time. Braking torque is produced when the current is flowing in the electric circuit and the inductance is decreasing with respect to time.

It can be shown that

$$T_e = i^2 (dL/dv)/2 \tag{A1.12}$$

where T_e is the electromagnetic torque produced on the rotor.

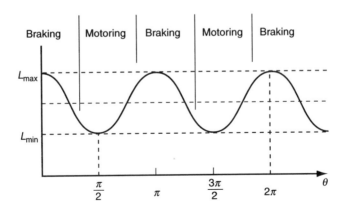

Figure A1.7   *Variation of inductance with angular position*

The circuit shown in Figure A1.6 could produce unidirectional torque if the current was pulsed. For example, if current only flowed for the period of rotation when the inductance was increasing, then only motoring torque would be produced. Alternatively, such a rotor construction when exposed to a rotating field in a uniform magnetic stator can produce relatively smooth torque.

This mechanism of electromechanical energy conversion within a motor is characteristic of reluctance motors. This group of motors includes reluctance motors, switched reluctance motors and stepper motors.

This mechanism can also be a useful secondary method of torque production in some synchronous motors.

A1.1.2.2 The interaction between a magnetic field and a current-carrying conductor

The Lorentz force equation states that a charged particle moving in a magnetic field experiences a force F in a direction perpendicular to the plane of movement of the charge and the field. Charge moving in a conductor is the current I flowing in the conductor of length l, and so the force acting upon that conductor can be given as

$$F = B \times I \times l \tag{A1.13}$$

The direction of the force can be found by the left-hand rule:

> Hold the **left** hand, as shown in Figure A1.8, in such a way that the forefinger points in the direction of the lines of flux, and the middle finger points in the direction of current flow in the conductor. The direction of motion is then in the direction in which the thumb is pointing.

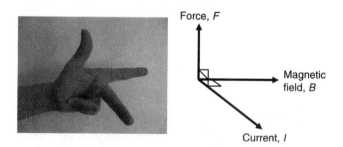

Figure A1.8 Fleming's left-hand rule (electrical to mechanical energy conversion)

If a conductor that is part of a circuit is moved through a magnetic field, a current is generated in the conductor in such a way as to oppose the motion. The direction of the current in this generator is given by Fleming's right-hand rule, which is illustrated in Figure A1.9.

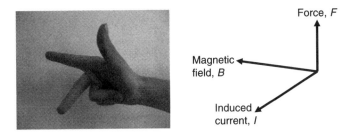

Figure A1.9   Fleming's right-hand rule (mechanical to electrical energy conversion)

Now consider the action of a coil when it is rotated in a magnetic field as shown in Figure A1.10. Faraday's law of electromagnetic induction shows that if a conductor of length l is moves at a velocity u through, and at right angles to, a magnetic field B, then a voltage v is induced:

$$v = Blu \tag{A1.14}$$

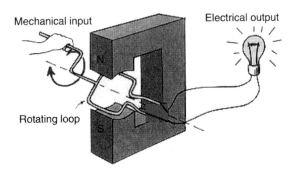

Figure A1.10   A simple generator (the wires to the light are fixed to the rotating loop)

Applying equation (A1.14) to a rectangular coil with N turns, of axial length l and radius of rotation r, rotating at constant velocity ω in a uniform magnetic field B, yields the following equation for the induced voltage:

$$v = B(2l)(r\omega \sin \omega t) \tag{A1.15}$$

The induced voltage, illustrated in Figure A1.11, is an alternating voltage with magnitude and frequency proportional to speed ω.

In the electric machine/generator as shown in Figure A1.10, where there is only a single pair of (magnetic) poles, ω is simply the mechanical rotational speed of the coil. In a typical industrial motor/generator there is more than a single pair of (magnetic) poles. Then it is clear that ω is a multiple (times the number of pole pairs) of

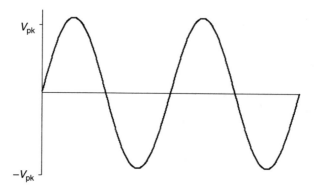

Figure A1.11 Induced voltage of a simple generator

the mechanical rotational speed. ω is commonly referred to as the electrical rotational speed.

This mechanism of electromechanical energy conversion within a motor is characteristic of most industrial motors, including d.c. motors, induction motors and synchronous motors (which embraces brushless servo motors). The operation of these motor types is described in more detail below.

A1.2   D.C. motors

A1.2.1   General

History will recognise the vital role played by d.c. motors in the development of industrial power transmission systems. The d.c. machine was the first practical device to convert electrical power into mechanical power (and vice versa in its generator form). Inherently straightforward operating characteristics, flexible performance and high efficiency encouraged the widespread use of d.c. motors in many types of industrial drive applications.

The later development of the lower cost a.c. cage motor and, more recently, of electronic variable frequency control have displaced the d.c. motor to some extent, particularly in the lower power ranges. Nevertheless, the advantages associated with the inherently stable and relatively simple to control d.c. machine are indisputable. In its most straightforward form, speed is approximately proportional to armature voltage, torque to armature current, and there is a one-to-one relationship between starting torque and starting current.

The modern d.c. motor under thyristor control and with sophisticated protection continues to provide very sound industrial variable-speed drive performance. In many demanding applications, such as high-performance test rigs and for the higher kilowatt ratings of drives for the printing and paper industries, for high-speed passenger lifts and drives subject to high transient loading in the metal and plastics industries, the d.c. motor with thyristor control is likely to be used for some considerable time, particularly in refurbishment programmes where a d.c. motor already exists. The task facing the a.c. drive in completely ousting its d.c. competitor is formidable.

The d.c. machine tool and servo drives based mainly on MOSFET chopper technology continue to offer high performance at low price at ratings up to approximately 5 kW (continuous), but here too a.c. technology is making significant inroads.

The introduction and development of electronic variable-speed drives continues to stimulate intensive development of motors, both d.c. and a.c. The performance capabilities of both are being extended as a result, and the d.c. motor is likely to find assured specialised applications for the foreseeable future.

The majority of standard d.c. motors, both wound-field and permanent magnet, are now designed specifically to take advantage of rectified a.c. power supplies. Square, fully laminated frame construction allows minimal shaft centre height for a given power rating, and affords reduced magnetic losses, which in turn greatly improves commutating ability.

Over the last few years the use of permanent magnet motors, usually in the fractional to 3 kW range, has become commonplace in general-purpose drive applications. In this design permanent magnets bonded into the motor frame replace the conventional wound field. The magnets have a curved face to offer a constant air gap to the conventional armature.

A1.2.2 Operating principles

The d.c. motor operates through the interaction of current-carrying conductors and electromagnetic fields. The fundamental method of torque production has been described earlier (Section A1.1.2.2). However, just as it has been shown that the terminal voltage of a simple coil rotating in a magnetic field is sinusoidal, it follows that the torque produced in such an arrangement would also be sinusoidal in nature.

A very important feature of industrial d.c. motors is the commutator (Figure A1.12). This is a mechanical device that switches the polarity of the rotating coil at a fixed mechanical position in its rotation.

The induced voltage is now unidirectional as shown in Figure A1.13. The voltage, although unidirectional, still has very high ripple. The industrial d.c. motor provides relatively smooth voltage and torque by having a number of armature coils, and instead of using the commutator to simply reverse the polarity of a single coil it

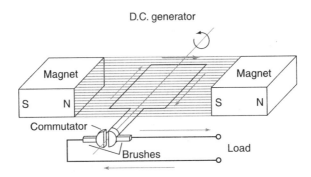

Figure A1.12   A simple commutator

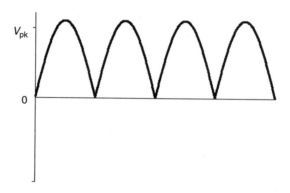

Figure A1.13   Induced voltage of a d.c. motor with a simple commutator

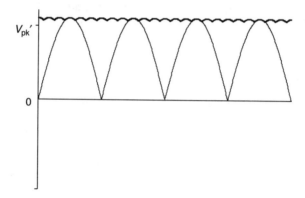

Figure A1.14   Induced voltage of a d.c. motor with multiple (six) armature coils

switches between coils. Figure A1.14 illustrates this effect for an armature with six coils. The switching between the coils, or windings, is known as commutation.

In practice, the windings in the armature of a d.c. motor are not independent windings that carry current only when the commutator connects to them. They are mesh connected, and carry current for most periods of operation. The d.c. voltage is consequently not simply the peak of the individual coil voltages as portrayed in Figure A1.14 but the vector product of the interconnected coils. The nature of the d.c. voltage is, however, the same.

An mmf results from current flowing in the armature windings. This mmf generates a flux, known as the armature reaction flux, which adds to the main flux in the machine air gap, skewing the total flux to the trailing edge of the poles. Although this armature reaction field is generally much lower than the main field, in practical industrial machines the effects can be important:

- The trailing edge of the pole becomes saturated and so the total field and therefore torque capability is reduced.

Industrial motors 23

- The armature windings under the trailing edge of the pole are in a higher field of higher flux density, which increases the winding voltage. At high speed that can result in insulation breakdown at the commutator.
- Flux is being pushed into the 'neutral' area of the air gap where commutation takes place. Any voltage in the winding at this time can result in arcing at the commutator.

There are three principle methods used to reduce armature reaction effects:

- pole shaping,
- interpoles, and
- compensating windings.

In *pole shaping*, the increased field at the trailing edge can be reduced if the air gap at the trailing edge is increased. This has a clear limitation if the machine has to operate in both directions of rotation, in which case both sides of the pole would have to have a longer air gap.

Pole shaping does not correct for armature reaction effects but serves to attenuate the harmful effects caused by field saturation.

Interpoles are additional field windings on the motor stator, usually 90° displaced from the main field winding, interposed between the main field poles (Figure A1.15).

The current flowing in the interpoles is the armature current and so the field produced by the interpole is proportional to the armature current and in a direction to oppose the distortion to the main field produced by the current flowing in the armature windings.

It follows that interpoles are most effective in their corrective action when the field flux is at its full level. The commonly used technique of field-weakening progressively

Figure A1.15   A d.c. machine stator with interpoles between the main field poles [photograph courtesy of Leroy-Somer]

reduces this self-corrective effect; hence the quality of commutation can be affected when a motor operates under the weak field conditions sometimes required in system applications.

Compensating windings (or pole face windings) are the third method of reducing the effect of the armature reaction, by neutralising the armature flux directly under the main pole faces. These windings are placed in slots cut into the main pole air gap faces and carry the full armature current, being connected in series with the armature circuit, and thus giving compensation with current loading. Compensating windings on the pole faces present problems of both mechanical security and electrical insulation, and a satisfactory solution often involves significant additional cost.

The use of a compensated machine tends to be restricted therefore to duties requiring a very wide range of field operation, such as in high-speed machine tool applications and others involving rapidly varying loads and speeds in which the commutating performance of the compensated motor can be extremely good when compared with that of the more normal non-compensated machine.

In some cases a combination of these three approaches is adopted by machine designers.

The idea of commutation was introduced earlier. Within the practical industrial d.c. machine, this commutation is undertaken by a mechanical commutator that is invariably made of copper, and the moving connection to this is made via 'brushes' made from carbon compounds.

As the armature winding passes through the 'neutral' region it has its current direction reversed. As adjacent commutator segments pass under a brush, as shown for winding A in Figure A1.16b, the winding is short circuited.

In the ideal case, a voltage should be induced in the shorted winding that is sufficient to reverse the current while it is still short circuited. In this case, when the brush moves fully to commutator segment c (Figure A1.16c) there will be no change in the current during the brush/commutator segment transition and so no sparking will occur.

As well as successfully compensating for the armature reaction effects described above, the brush position is critical. This is the mechanism for precisely positioning the winding to be commutated in the correct position in relation to the air gap flux. Industrial motors are usually designed to allow for the brush position to be changed in order to optimise the motor. This is usually factory set and should only ever be changed by a fully trained engineer.

The analysis and design of good commutation is difficult, because of the largely non-linear electrical characteristics of the carbon brush copper contact film. It should be noted that although the presence of sparking does give a very good indication of poor commutation, optimisation of commutation is a very skilled activity. Indifferent commutating performance places d.c. motors at some disadvantage when compared with a.c. cage induction motors, particularly regarding maintenance requirements and costs.

Commutator brushes do wear, and checking their condition should be part of a regular maintenance schedule. Motor manufacturers can advise on this.

Finally, there is a physical limit to the speed at which current can be commutated. This 'commutation limit' is often defined as the product of the motor shaft power

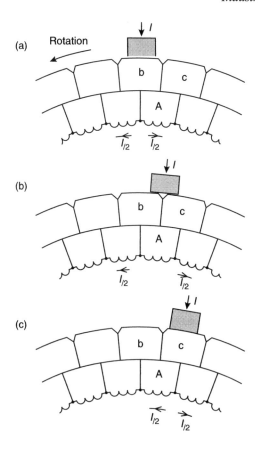

Figure A1.16 Reversal of current flow in one armature winding

and the rotational speed. The limit is widely accepted to be approximately 3×10^6 (kW min^{-1}). Where greater ratings are required, more than one armature can be fitted on a single shaft, or motors can be connected in series (provided the shafts are rated to transmit the total torque). It should be noted that the closer to this limit a motor is specified the harder it is for the motor designer to meet the requirements of performance, reliability and maintainability.

A1.2.3 Fundamental equations of steady-state performance

A1.2.3.1 The separately excited d.c. motor

The principles of operation of a d.c. motor described above reveal a system with two electrical circuit elements that can be considered as being independent. The first of these, the field circuit, determines the level of magnetic field in the motor air gap. The second, the armature circuit, determines the level of current flowing in the conductor(s) in the magnetic field. This can be represented schematically as in Figure A1.17. This circuit shows the armature, armature resistance R_a and field winding. The

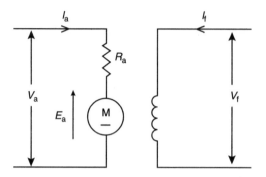

Figure A1.17 Schematic of the electrical circuits in a d.c. motor

armature voltage V_a is supplied typically from a controlled thyristor system and the field voltage V_f from a separate bridge rectifier.

As the armature rotates, an electro-motive force (EMF) E_a is induced in the armature circuit; this is termed the back-EMF because it opposes the applied voltage V_a and the flow of current produced by V_a. E_a is related to armature speed and main field flux by

$$E_a = k_1 n \phi \qquad (A1.16)$$

where n is the speed of rotation, ϕ is the field flux and k_1 is the motor constant. Also, the applied (or terminal) armature voltage V_a is given by

$$V_a = E_a + I_a R_a \qquad (A1.17)$$

where V_a is the applied armature voltage, I_a is the armature current and R_a is the armature resistance.

Multiplying each side of equation (A1.17) by I_a gives

$$V_a I_a = E_a I_a + I_a^2 R_a \qquad (A1.18)$$

In other words, total power supplied = power output + armature losses.

Interaction of the field flux and armature flux produces an armature torque:

$$\text{Torque } M = k_2 I_f I_a \qquad (A1.19)$$

where k_2 is a constant, I_f is the main field current and I_a is the armature current. This confirms the straightforward and linear characteristic of the d.c. motor, and consideration of these simple equations will show its controllability and inherent stability.

It is important to note from equation (A1.19) that the direction of torque can be reversed by reversing either the field current or the armature current. Because the field current is usually smaller than the armature current, it can be convenient to reverse the field current. However, the time constant of the field circuit (typically

1 s) is significantly higher than that of the armature circuit (typically 100 ms), which means that in many applications armature current reversal is preferred.

The speed characteristic of a motor is generally represented by curves of speed against input current or torque, and its shape can be derived from equations (A1.16) and (A1.17):

$$k_1 n\phi = V_a - (I_a R_a) \tag{A1.20}$$

If the flux is held constant, which is achieved by simply holding the field current constant, because interpoles/compensating windings compensate for armature reaction effects, then

$$n = k_3[V_a - (I_a R_a)] \tag{A1.21}$$

When the flux is held constant, the *no load* (no torque/no armature current) d.c. motor speed is proportional to the voltage applied to the armature V_a. As the load increases then the speed will fall. To consider the magnitude of this let us consider the practical industrial motors shown in Table A1.2.

For many applications, a droop in speed such as that in Figure A1.18, particularly for the smaller motor, is not acceptable, and the control system is required to

Table A1.2 Data for some practical industrial motors

Motor rating	1.45 kW 3 000 min^{-1}	59 kW 3 000 min^{-1}	155 kW 3 000 min^{-1}
$V_{a(rated)}$ (V)	460	460	460
$I_{a(rated)}$ (A)	3.7	140	355
R_a (Ω)	11.3	0.103	0.023

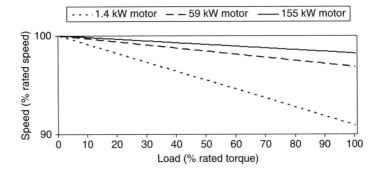

Figure A1.18 *Speed droop with load of a two-winding d.c. motor with constant rated armature voltage and field current*

compensate for this. It is most common to have a speed sensor, usually a tachogenerator, mounted on the motor shaft. The actual speed is then fed into the control loop of the associated drive, which then increases the armature voltage to compensate for the voltage drop over the armature resistance R_a. Using modern drive systems and precise speed-sensing encoders, very precise speed holding can be achieved. In applications where precise control is not needed, a speed sensor may not be needed and the controller can be designed to compensate the applied armature voltage for the armature resistance drop.

If the flux is NOT kept constant, rearranging equation (A1.20) yields

$$n = k_4[V_a - (I_a R_a)]/\phi \qquad (A1.22)$$

Clearly if the field is reduced the speed of the motor will increase. We know from equation (A1.19) that the torque is proportional to the product of the field (proportional to the field current as long as the magnetic circuit does not saturate) and the armature current. If the field is reduced the torque capability of the motor is therefore reduced. This region of operation is referred to as the field weakening range.

In many applications the field is held constant up to a certain speed, referred to as the base speed (the highest speed at which full torque can be produced for a given armature voltage), and then speed is increased further by reducing the field.

Figure A1.19 shows the operating regions of the separately excited d.c. motor. For clarity, the no load operating characteristic is shown. On load the speed droops as shown in Figure A1.18.

The d.c. motor described above with independent armature and field winding has a control characteristic that meets the requirements of a high proportion of industrial applications. There are, however, alternative configurations of d.c. motors that produce significantly different performance and control characteristics.

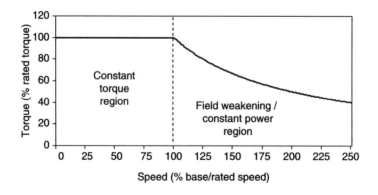

Figure A1.19   *Operating regions of a d.c. motor with constant field and reduced field above base speed*

A1.2.3.2 The series d.c. motor

A schematic of the series d.c. motor is shown in Figure A1.20. The series motor has, as the name suggests, the field winding in series with the armature winding. The field current I_f is therefore the same as the armature current I_a.

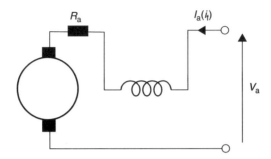

Figure A1.20   Schematic of series d.c. motor

The steady-state performance equations (A1.19) and (A1.20) can be rewritten for a series motor as

$$\text{Torque } M = k_5 I_a^2 \tag{A1.23}$$

$$\text{Speed } n = k_6 [V_a - (I_a R_a)]/I_a \tag{A1.24}$$

There is no longer a linear relationship between torque and armature current. Also there is not a defined no load speed.

Rearranging these equations, ignoring the armature resistance drop, shows the relationship between torque and speed to be

$$\text{Torque } M = k_7 (V_a/n)^2 \tag{A1.25}$$

This characteristic is shown in Figure A1.21 for a fixed value of V_a. Very high torque is therefore possible at low speed. In a practical motor, the magnetic circuit will saturate,

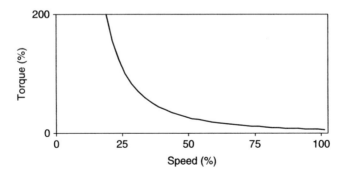

Figure A1.21   Performance characteristics of a typical series d.c. motor

and this, coupled by the limit of the commutator to satisfactorily commutate the current, will limit the available torque at low speed. However, the characteristic is still steep, which can be attractive in applications such as traction and some hoisting or some mixing duties where initial stiction is a dominant consideration.

If the load falls to a low value the speed increases dramatically, which may be hazardous, and the series motor should not normally be used.

A1.2.3.3 The shunt d.c. motor

The schematic of the shunt d.c. motor is shown in Figure A1.22. As the name suggests, the field winding is connected in parallel with the motor armature. It is usual that only the voltage applied to the field is controlled and so it is possible to design the shunt d.c. motor to operate with reasonably constant flux over a broad speed range.

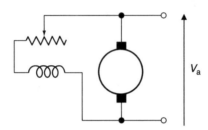

Figure A1.22   Schematic of a shunt d.c. motor

When operated as part of a system with modern power conversion equipment, the shunt d.c. motor offers little advantage over the separately excited d.c. motor and is rarely used.

A1.2.3.4 The compound d.c. motor

The compound d.c. machine combines the shunt and series characteristics. The circuit of the compound d.c. motor is shown in Figure A1.23. It is usual for the series field to be wound in such a way as to add to the field produced by the separate field winding. The resultant characteristic is a combination of the series and separately excited motors.

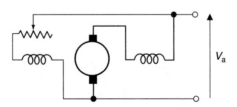

Figure A1.23   Compound d.c. motor

Under semiconductor converter control, with speed feedback from a tachogenerator, the shape of the speed/load curve is largely determined within the controller. The compound motor is therefore rarely used.

A1.2.4 Permanent magnet d.c. motor

The d.c. permanent magnet (DCPM) motors differ from a conventional separately excited motor in so far as the field is produced by permanent magnets, mounted on the stationary stator, rather than a current in a field winding.

The conventional DCPM motor tends to find application in high-performance servo applications and is consequently frequently designed to exhibit extremely good low-speed torque ripple characteristics. However, the high-speed characteristics of a DCPM motor are not ideal for all applications. Because of the mechanical and electrical constraints set by the commutator, increasing the motor speed with a constant load characteristic soon reveals the commutation limit of the motor. In practice, each motor is designed to work within a safe area of commutation where the available motor torque reduces as motor speed increases.

The torque/speed operating envelope for a conventional DCPM motor is shown in Figure A1.24. The area of continuous operation may be defined as the area where the motor can operate continuously at full load with an acceptable temperature rise.

The area known as the intermittent duty part of the motor characteristic defines the area of operation available on an intermittent basis for acceleration and deceleration periods. The commutation limit is most obviously exceeded if brush sparking becomes excessive. Severe overloading in this range will cause a complete ring of heavy sparking to run around the commutator circumference. This phenomenon is

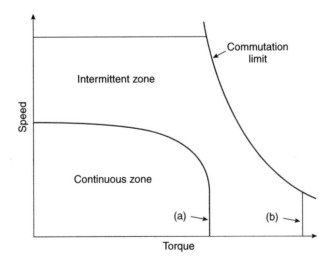

Figure A1.24   *Operating envelope of DCPM motor. (a) Peak torque. (b) Stall torque at 40 °C*

known as brush 'fire' or flash-over, and must be avoided because it damages both the commutator and brush gear, reducing the life expectancy of the motor considerably. Fortunately, the electronic controller supplying the motor may be easily specified to prevent overloading.

A1.2.5 Construction of the d.c. motor

A sectioned view of a typical square-frame d.c. machine stator is shown in Figure A1.25. The principal components of the d.c. machine are the rotating, shaft-mounted wound armature and connection made via the commutator, the field system surrounding the armature across the motor air gap, the supporting and enclosing end frames with shaft bearings, and brush gear to connect the commutator to external terminals; these are all clearly visible.

Many d.c. machines have more than one pair of field poles, and so more than one pair of commutator brushes. Figure A1.26 shows a four-pole d.c. motor stator.

A1.2.5.1 D.C. motor frame

The main frame of a typical modern four-pole d.c. motor is a square, laminated iron frame with a curved inner face. The main pole lamination and coil shunt field assembly are mounted on the inside face at 90° to each other, with interpole assemblies interposed at 45° between.

Square-frame motors offer reduced centre height compared with round-frame motors by virtue of the fact that the main field poles mounted on the four flat internal faces are of a 'pancake' design and that the four interpoles, mounted on diagonal axes, fit into the internal corners of the square.

Figure A1.25 Section of a typical d.c. motor [photograph courtesy of Leroy-Somer]

Industrial motors 33

Figure A1.26 A typical four-pole d.c. machine stator (with interpoles) [photograph courtesy of Leroy-Somer]

A1.2.5.2 D.C. motor armature

The armature consists of a cylindrical laminated iron core, shaft mounted, with a series of slots to accommodate the armature winding coils, insulated from each other and the core. The coil ends are connected to insulated segments of the commutator.

The commutator, a cylindrical copper segmented construction, is mounted at one end of the armature core, with wedge-section segment bars clamped between commutator end rings, insulated from each other and the metal of the armature. These segments are machined on the outer axial face to provide a smooth concentric surface on which the brushes can bear with a tensioned contact. An armature assembly is shown in Figure A1.27.

The armature winding slots are 'skewed' by about one slot pitch to reduce torque ripple (cogging effect), which results in significant speed ripple at low speeds and slot noise generation at high speed. Such a cogging effect superimposed on the otherwise steady rotation of the motor shaft would be unacceptable in many applications. For example, a machine tool axis drive with such a characteristic could leave undesirable surface patterning on the workpiece.

The precise method of skewing presents the motor manufacturer with two problems. The first is in determining the amount of skew for a particular design of motor. The second is that it is much more difficult to wind the armature on an automatic winding machine when the laminations are skewed in this fashion.

Figure A1.27 A d.c. motor armature assembly [photograph courtesy of Leroy-Somer]

Computer-aided design plays a large part in solving the first of these difficulties and most motor manufacturers have developed standard programmes to speed magnetic calculations. Modern wire insertion machines make relatively easy work of shooting windings into the skewed armature slots.

Armature winding coil ends are banded with glass-fibre and epoxy bonding for high-speed operation ($>6\,000$ min^{-1} for many machine tool applications) to prevent armature coil deflection under centrifugal stress. Two balancing discs, one behind each bearing, permit dynamic balancing to 'R' or the more stringent 'S' standard over the working speed range of the machine.

A1.2.5.3 Brush gear

The number of brushes is based upon the full-load current of the armature circuit. The working velocity of the commutator track and the working conditions generally govern the grade of carbon used. Brush pressure is generally fixed and unvarying over the permitted wear length, and is controlled by 'tensator' constant-tension springs. The importance of using the correct grade of brush cannot be over-stressed.

A1.2.5.4 Degree of protection and mounting

The typical d.c. motor shown in Figure A1.25 is a drip-proof (IP23) enclosed foot-mounted d.c. motor with side-mounted terminal box and with a feedback tachogenerator coupled to the non-drive end. Access to the brush gear is through the four inspection covers around the commutator end-shield, although in practice only three can be used. Any of these three inspection apertures can be used for mounting a force-ventilation fan to allow wide (typically 100:1) speed range operation at constant load torque.

The basic adaptability of the square-frame machine also allows the terminal box to be mounted on any of the three available faces of the machine, and for the machine itself to be mounted by end or foot flange. Because both end-shields have the mounting feet attached, they can readily be rotated through 90° to allow side foot mounting.

More details of alternative degree of protection and mounting arrangements and the associated international standards are given in Section B3.2.

A1.2.5.5 DCPM design

Industrial DCPM motors tend to offer a smaller frame size for a given kilowatt rating than their wound-field counterparts, although frame lengths can be greater.

A DCPM servo motor is conventional in most aspects of construction, but usually incorporates technical refinements to give high power density and high speed and position loop bandwidth. This is generally achieved by designing long low-inertia shafts and using premium magnetic materials.

Care should be taken not to apply very low-inertia servo motors to high-inertia loads as the inertia mismatch can lead to troublesome resonances (see Chapter B8).

Choice of magnetic material for the DCPM motor is a major deciding factor in the operational characteristic. Conventional DCPM motors use a low-cost ferrite magnetic material, which, although more than adequate for the majority of applications, does have some inherent disadvantages.

Mechanically, ferrite magnets are fragile, and so are usually bonded to the motor body, which is very secure. However, because ferrite magnets are also brittle, they are susceptible to mechanical shock and can be fractured unless the motor is handled with great care.

One of the more important disadvantages of low-cost magnets, however, is the ease with which they can become demagnetised. Ferrite magnets have a relatively low demagnetising level. Under normal circumstances, and with a correctly determined servo system, demagnetisation of the motor is not possible. However, if the system is not correctly determined it is possible to partially demagnetise the motor through over current in the armature or with temperatures in excess of the Curie point of the material, with the result of reduced torque and usually a higher than rated motor speed on light load, for a given input voltage.

Alternative types of magnetic material overcome these problems with very high demagnetisation levels and operating temperature ranges. Such a material is samarium cobalt, which is grouped with the rare earth category of magnetic materials. There is much development activity in this area and new, improved materials are appearing continually.

Currently, magnets made from rare earths are expensive and can double the cost of a motor, but for many applications the choice of such magnetic materials is unquestionably justified. The higher flux densities associated with rare earth magnets allow rotor diameter to be reduced and the rotor inertia can so be significantly reduced for a given rated output torque. This allows an improved acceleration/deceleration performance per unit armature current.

The search has been for a magnetic material with the cost advantages of the ordinary ferrite magnet and the performance advantages of the rare earth types. This is now realised with the advent of synthetic magnetic materials, principally neodymium iron boron, which is now fully acceptable to numerous industries, including the automotive industry, and provides extremely good performance characteristics without the large price premium of rare earth materials.

The magnet itself is formed from powders drawn from an ever-growing range of magnetic materials and is moulded into the required size, shape and profile using a high-temperature specialised sintering process.

The magnet is typically delivered to the motor manufacturer in an unmagnetised state, and is first bonded into the motor frame using a two-part adhesive with a suitably high melting point. The assembly is then magnetised by inserting a close-fitting metal conductor through the motor frame and introducing a high-current shock pulse through the assembly by discharging a capacitor bank through the conductor. The current pulse polarises the elements of the sintered magnets.

A1.3 A.C. induction motors

A1.3.1 General

The a.c. squirrel-cage induction motor is the basic, universal workhorse of industry, converting some 70–80 per cent of all electrical power into mechanical energy. This type of motor none-the-less exhibits some quite unattractive performance characteristics in spite of intensive development, notably instability and a non-linear load–current characteristic. It is almost without exception designed for fixed-speed operation, with larger ratings having such features as deep rotor bars to limit direct on line (DOL) starting currents. Electronic variable-speed converter technology is able to provide the necessary variable voltage/current, variable frequency supply that the three-phase a.c. machine requires for efficient, dynamic and stable variable speed control.

Modern electronic control technology is able not only to render the a.c. induction motor satisfactorily for many modern drive applications but also to greatly extend its application and enable advantage to be taken of its low capital and maintenance costs.

More striking still, microelectronic developments have made possible the highly dynamic operation of induction motors by the application of flux vector control. The practical effect of this is that it is now possible to drive an a.c. induction motor in such a way as to obtain a dynamic performance in all respects better than could be obtained with a phase-controlled d.c. drive combination. The various forms of a.c. induction motor control are fully described in Section A4.4.

It is also worth considering, as the demand for ever higher-efficiency motors increases, that designers are looking to reduce stator and rotor resistances. As we will show later this should result in such high starting currents if the motor were to be started DOL, that the use of an inverter becomes essential.

A1.3.2 Operating principles

Although it is not immediately obvious, the principle of operation and torque production in an induction motor is fundamentally the same as in the d.c. motor. There are, however, some striking differences.

In the d.c. motor the magnetic field is produced from d.c. current flowing in the field winding. The resulting magnetic field is stationary with respect to position within the motor. In an induction motor the field is produced from a.c. current and rotates around the air gap.

In a d.c. motor, torque is produced when current supplied to the armature cuts the magnetic field. In an induction motor, current is induced in the rotor by transformer action.

A1.3.2.1 Rotating magnetic field

The stator winding arrangements of three-phase a.c. machines are somewhat complex, but the net effect is to produce three windings evenly distributed (120° apart) over the stator periphery (see Figure A1.28).

Considering only the A-phase winding. If the winding has N turns and a sinusoidal current with an rms value I flows through that winding, then the mmf at any particular instant in time is rectangular in shape, and of value NI between the coil sides. Because the coil is split as shown in Figure A1.30, it is common to talk of mmf per pole, which would be $NI/2$. If a Fourier analysis is undertaken on the mmf distribution, the fundamental component can be shown to be

$$F_1(v,t) = 0.9NI \sin(\omega t)\cos(v) \tag{A1.26}$$

where ω is the frequency of the current waveform and v is the angular position in the air gap.

Figure A1.28   Effective winding arrangement of a three-phase induction motor

The three-phase induction motor has three pairs of poles (see Figure A1.28). The m.m.f. produced by each phase is sinusoidal in space, and displaced in time by 120° *electrically*.

The mmf produced in the three windings when balanced three-phase currents are applied is given by

$$F_A(v,t) = 0.9NI \sin(\omega t) \cos(v)$$
$$F_B(v,t) = 0.9NI \sin(\omega t - 120°) \cos(v - 120°) \quad (A1.27)$$
$$F_C(v,t) = 0.9NI \sin(\omega t + 120°) \cos(v + 120°)$$

Adding these three mmfs we obtain

$$F(v,t) = 1.35NI \sin(\omega t - v) \quad (A1.28)$$

This is a wave that travels around the air gap of the motor at speed ω rad^{-1}. A.C. machines are manufactured with different numbers of poles (always an even number 2, 4, 6, 8, and so on), which directly affects the relationship between the applied stator voltage/current frequency and the rotational frequency of the m.m.f. This relationship between revolutions per minute (min^{-1}) and applied frequency (Hz) is given by

$$n_s = 120f/p \quad (A1.29)$$

where p is the number of poles and f is the applied frequency ($= \omega/2\pi$). For example, for a four-pole machine with a 50 Hz supply, $n_s = 1\ 500$ min^{-1}.

It is important to note that increasing the number of poles means that the electrical/magnetic and mechanical angular displacement is no longer the same. This difference between electrical and mechanical degrees is important and can cause confusion.

The speed at which the field rotates is commonly referred to as the *synchronous speed*. This is the speed with which the field rotates in the motor; however, it is not possible for the induction motor to produce torque if the rotor was to rotate at this speed.

A1.3.2.2 Torque production

Torque in an induction motor is produced by the interaction of the field with currents induced in the rotor windings by transformer action. For current to be induced in the rotor the rotor windings have to cut the magnetic field (see Section A1.1.2.2). If the rotor were to rotate at the same speed as the field, i.e. at synchronous speed, there would be no relative movement, and so no current would be induced. It is therefore not possible for the induction motor to produce useful torque under this condition.

The operating speed n of an induction motor is usually close to the synchronous speed n_s, and the difference between the two is referred too as the *slip s*, where

$$s = (n_s - n)/n_s \tag{A1.30}$$

Slip s is dimensionless.

Be aware that some people like to talk of percentage slip, whereas others may refer to slip simply as $(n_s - n)$.

When the rotor is stationary, slip = 1 and the magnetic field couples the rotor windings at the frequency of the stator current. As described earlier, when the rotor rotates at synchronous speed, slip = 0 and there is no relative motion between the rotor windings and the field, so the frequency of the rotor current is zero. At intermediate rotor speed the rotor current frequency is proportional to the slip,

$$f_2 = sf_1 \tag{A1.31}$$

where f_2 is the frequency of the rotor current (Hz) (sometimes referred to as the slip frequency) and f_1 is the frequency of the stator current (Hz).

Considering the transformer action, the current induced in the rotor is given by

$$I_2 = (sE_2) \Big/ \sqrt{(R_2^2 + sX_2^2)} \tag{A1.32}$$

where E_2 is the Induced rotor voltage at standstill (V), R_2 is the rotor resistance per phase (Ω), and X_2 is the rotor leakage reactance per phase at standstill (Ω). Equation (A1.32) may be rewritten in the form

$$I_2 = E_2 \Big/ \sqrt{[(R_2/s)^2 + X_2^2]} \tag{A1.33}$$

which can be represented by the circuit diagram in Figure A1.29.

The stator circuit of the induction motor may be added to this equivalent circuit by considering the motor as a transformer, with the primary being the stator and the secondary the rotor. This is shown in Figure A1.30. As with a transformer, this circuit can be redrawn referring the rotor quantities to the stator as in Figure A1.31. The magnetising reactance of an induction motor, X_m, is significantly lower than would be the case in a normal transformer because of the air gap. By transforming the rotor variable to stator-based quantities, their direct physical relevance has been lost. R_2' and X_2' are the stator resistance and leakage reactance with respect to the stator.

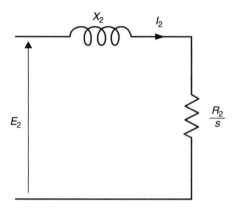

Figure A1.29 Equivalent circuit of an induction motor rotor

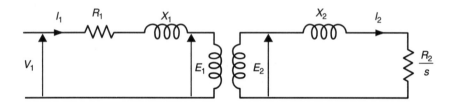

Figure A1.30 Transformer equivalent circuit of the induction motor [R_1 = stator resistance (Ω) and X_1 = stator leakage reactance (Ω)]

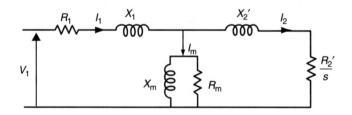

Figure A1.31 Equivalent circuit of the induction motor

It is usual, for reasons which will become apparent later, to split the rotor resistance in the above equivalent circuit as follows:

$$R_2'/s = R_2' + R_2'(1-s)/s \qquad (A1.34)$$

The equivalent circuit (Figure A1.31) can now be redrawn as in Figure A1.32.

The major benefit of an equivalent circuit is in the ability it gives to calculate performance. Remember, this equivalent circuit is made on a per-phase basis, and

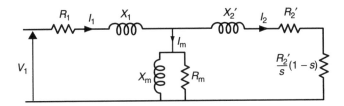

Figure A1.32 Revised equivalent circuit of the induction motor

assumes balanced operation of the motor and sinusoidal supplies – the effects of harmonics have not been taken into account. It is also a steady-state model and is not applicable to transient conditions.

In order to be able to use the equivalent circuit it is necessary to be able to determine the parameters for a particular motor. Modern drives undertake this evaluation automatically, but it is interesting to consider the traditional *approximate* method for this. The magnetising resistance R_m representing the iron losses is relatively large and is frequently ignored. Three tests yield the parameters: the d.c. voltage test, the synchronous test rotation and the locked rotor test.

- *The d.c. voltage test*: If a d.c. voltage is applied as V_1, the inductances of the equivalent circuit appear as a short circuit. The total impedance seen by the applied voltage is therefore R_1

$$R_1 = V_{1(d.c.)}/I_1 \quad (A1.35)$$

Remember this is a per-phase equivalent circuit so for a star-connected three-phase motor there will be two stator resistances in series.
- *Synchronous rotation test*: If the rotor of the motor is rotated at synchronous speed, then the slip is 0. We have seen in Figure A1.29 that the total rotor resistance of the rotor circuit is R_2/s, and so with a slip of 0 the rotor circuit appears open circuit. As we are ignoring R_m the equivalent circuit reduces to that shown in Figure A1.33. The total impedance seen by the applied voltage is therefore $R_1 + (X_1 + X_m)$.

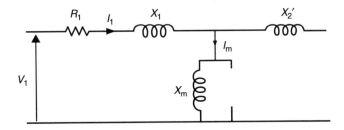

Figure A1.33 Induction motor equivalent circuit when slip = 0 (synchronous speed)

Applying rated a.c. voltage and frequency we measure the current. We already know the stator resistance, and so the total stator inductance $(X_1 + X_m)$ can be calculated:

$$(X_1 + X_m) = \sqrt{[(V_1/I_1)^2 - R_1^2]} \qquad (A1.36)$$

- *Locked rotor test*: If the rotor is locked in position, and an a.c. voltage at rated frequency is applied, the slip will be 1. The equivalent circuit of the rotor circuit therefore simplifies to $(R_2 + X_2)$. This is a much smaller impedance than the magnetising reactance, which can now be ignored. The equivalent circuit simplifies to that shown in Figure A1.34.

This is a low impedance circuit so a reduced voltage at rated frequency should be applied and the current and phase angle θ with respect to the voltage is measured.

$$R_2 = (V_1 \cos \theta / I_1) - R_1$$
$$X_1 + X_2 = V_1 \sin \theta / I_1 \qquad (A1.37)$$

It is usual to assume that the stator and rotor leakage inductances are equal; therefore

$$X_1 = X_2 = V_1 \sin \theta / 2I_1 \qquad (A1.38)$$

We have previously calculated $(X_1 + X_m)$ and so can now determine X_m.

This traditional approximate method of parameter calculation contains a significant number of simplifications. The parameter determination methods used within the auto-tuning routines of a modern drive are much more sophisticated, but the above routines serve to explain some of the requirements and the phenomenon of the modern tuning processes.

Some parameters such as rotor resistance can be significantly dependent upon temperature. Some of the reactances can be significantly dependent upon frequency. For this reason drive control algorithms need to be robust against such variations and/or perform continuous auto-tuning routines.

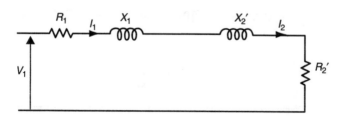

Figure A1.34 Induction motor equivalent circuit when slip = 1 (locked rotor)

A1.3.3 Fundamental equations of steady-state performance

When considering the performance of an induction motor, other than in efficiency calculations, it is usual to ignore R_m, the magnetising resistance, which represents the iron loss in the motor, which is mostly in the stator. If this is done, the power flow in the motor can be considered with clarity by observing the equivalent circuit in Figure A1.32, and is broken down as follows:

Input power to motor	$P_I = V_1 I_1 \cos\theta$
Power lost in stator resistance	$P_{SL} = I_1^2 R_1$
Power crossing the air gap	$P_G = I_2^2 R_2' + I_2^2 R_2'(1-s)/s$
Power dissipated in rotor resistance	$P_{RL} = I_2^2 R_2'$
Power converted to electromagnetic torque	$P_O = I_2^2 R_2'(1-s)/s$

The power converted to electromagnetic torque is therefore represented in the equivalent circuit of Figure A1.31 as the power dissipated in the resistor $R_2'(1-s)/s$.

The above represents the per-phase values only and so the total output power from three phases is given by

$$P_O = 3 I_2^2 R_2'(1-s)/s \tag{A1.39}$$

The torque produced is the power converted to electromagnetic torque divided by the rotational speed:

$$\text{Torque} = [3 I_2^2 R_2'(1-s)/s]/n = 3 I_2^2 R_2'/(s \cdot n_s) \tag{A1.40}$$

The above assumes a perfect lossless mechanical system. In practice there are mechanical losses within the machine that should be subtracted from the output power/torque. These losses in the motor itself are dominated by windage and friction. Combined these could comprise of the order of 1 per cent overall efficiency loss on a typical 30 kW motor.

As a general rule the lower the rated slip the higher the efficiency of an induction motor.

In order to produce the most efficient induction motors, some manufacturers use copper rotor bars rather than the usual aluminium, and efficiency improvements ranging from 1 to 7 per cent have been claimed. A word of caution, however. A good measure of the efficiency of an induction motor is its slip. The smaller the slip, the smaller the motor losses. The smaller the slip, the higher the speed. This means that on many loads including fans and pumps the load is increased. *Fitting a higher efficiency induction motor could result in higher energy consumption.* A further word of caution. Induction motors with very low slip may prove difficult to control with good stability.

A1.3.3.1 Direct on line (DOL) starting current and torque

High starting currents and torques occur when an induction motor is started by connecting it directly to the fixed-frequency supply. If an induction motor is at standstill and a fixed-frequency mains voltage supply is connected to its terminals a high inrush current results. The reason for this can be readily seen by consideration of the equivalent circuit in Figure A1.32.

At standstill slip $s = 1$. The resistance of the rotor circuit is therefore only R_2'. This low rotor resistance provides a very low impedance path, which results in high stator and rotor current. This initial starting current may be a factor of 6 or even more times the rated current of the motor, as shown in Figure A1.35.

The shaft torque during direct on line (DOL) starting does not follow a linear relationship, because the slip changes as the motor accelerates. It should noted that the starting current and torque speed characteristics for different machine designs and motor ratings vary widely. These characteristics are available from reputable motor manufacturers. These data may also show the guaranteed values and the manufacturing tolerance.

Induction motor designers take full account of the large stresses that occur during DOL starting. There are many applications/sites where such high starting currents are unacceptable.

It has been shown above that under normal operation, it is beneficial to have low rotor resistance (low slip and high efficiency), but to limit the starting current, without dramatically compromising the starting torque it would be helpful to have high rotor resistance. This seemingly impossible requirement can be met by careful design of the rotor bars:

- At standstill slip $= 1$ and the rotor current frequency is the same as the supply frequency.
- At rated speed the slip is the rated slip (at rated torque) and the rotor current frequency is the slip frequency, which is less than 1 Hz for a typical motor.

By selecting an appropriate shape or combination of rotor 'cages', motors can be designed that have an effective resistance at mains frequency several times their resistance at slip frequency. Such designs make use of the inductive effect of the slot

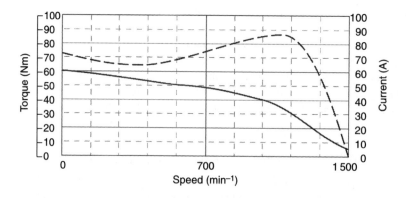

Figure A1.35 *Typical induction motor torque/speed and current/speed curves*

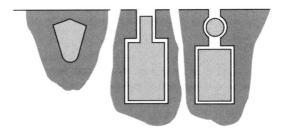

Figure A1.36   Examples of rotor bar profiles

leakage flux on the current distribution in the rotor bars. This is the same effect as the skin effect, which changes the resistance of a conductor in dependence on the current frequency.

Some examples of rotor bar geometries are shown in Figure A1.36.

A1.3.3.2   Starting current and torque when the motor is connected to a variable-frequency and/or variable-voltage supply

When operating with a variable-speed drive where the starting voltage and frequency are reduced, the troublesome starting current characteristics of an induction motor are no longer present. With appropriate control algorithms it is possible to start an induction motor with full rated shaft torque capability without exceeding the rated current of the motor.

Where the application does not demand full rated starting torque, power conversion equipment that varies the applied voltage while maintaining the fixed mains frequency can be used to substantially attenuate the starting currents. Such equipment goes under the generic term *soft starter* (see Section A2.5.2). The current can be limited by the controller to a level determined by the load on the motor shaft.

When used with either of these forms of power conversion equipment, motors do not need to be designed to withstand the stress of DOL starting. The competitive nature of the standard induction motors industry, however, dictates that it is rare for induction motors to be specifically designed without this capability.

A1.3.4   Voltage–frequency relationship

Nearly all commercially available industrial induction motors are wound for direct connection to and starting on the supply voltage and frequency prevailing in the country where they will be used. It is a relatively simple matter for the motor manufacturer to select the number of turns per coil and the wire size to match any voltage within a wide range. If it is desired to convert a constant-speed motor operating DOL to a variable-speed drive using an inverter it is necessary to consider the effect of frequency on flux and torque.

An induction motor on a normal supply operates with a rotating field set up by three-phase currents in the stator winding. The magnitude of the field is controlled broadly by the voltage impressed upon the winding by the supply. This is because the resistance of the winding results in only a small voltage drop, even at full load current, and therefore in the steady state the supply voltage must be balanced by the emf induced by the rotating field. This emf depends on the product of three factors:

- the total flux per pole (which is usually determined by the machine designer),
- the total number of turns per phase of the stator winding, and
- the rate of field rotation or frequency.

Exactly the same factors are valid for transformer design, except that the field is pulsating instead of rotating.

In a transformer it can be shown that the emf induced in a winding is given by

$$V = K\varphi F \qquad (A1.41)$$

where V is the emf induced in a coil, K is a constant, φ is the flux through the coil, and f is the frequency of the supply. This can be rearranged to give

$$\varphi = (V/f)/K \qquad (A1.42)$$

For constant flux, it can be seen that the ratio V/f must be held constant.

For inverter operation the speed of field rotation for which maximum voltage is appropriate is known as the 'base speed'.

If, again in the steady state, the voltage applied to the stator terminals is increased *without a corresponding increase in the frequency*, only the flux can vary to regain the balance between applied voltage and emf If the flux is forced to increase by applying excessive voltages, the iron core of the machine is driven progressively into saturation. This not only increases iron losses due to hysteresis and eddy currents, but can lead to a very marked increase in stator current, with corresponding resistive losses. Because most machines are designed to work with the minimum of material, their magnetic circuits are normally very close to saturation, and excessive stator voltage is a condition that must be carefully avoided.

The consequence of reducing the supply frequency can readily be deduced from the relationship described above. For the same flux the induced emf in the stator winding will be proportional to frequency, so the voltage supplied to the machine windings must be correspondingly reduced in order to avoid heavy saturation of the core. This is valid for changes in frequency over a wide range. *The voltage–frequency relationship should therefore be linear if a constant flux is to be maintained within the machine*, as the designer intended. If flux is constant so is the motor torque for a given stator current; hence, the drive has a constant torque characteristic.

Although constant V/f control is an important underlying principle, it is appropriate to point out departures from it that are essential if a wide speed range is to be covered. First, operation above base speed is easily achieved by increasing the output frequency of the inverter above the normal mains frequency; two or three times base speed is easily obtained, but the rating of the drive and the motor characteristic imposes an absolute torque limit. The output voltage of an inverter cannot usually be made higher than its input voltage, so the V/f characteristic is typically as shown in Figure A1.37a.

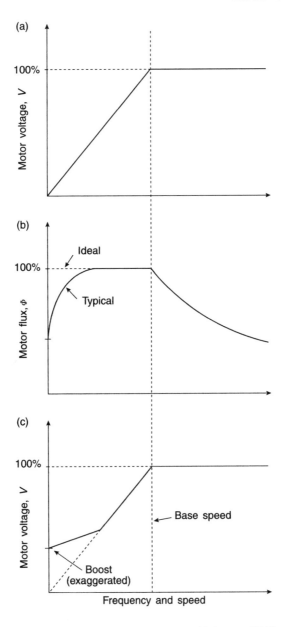

Figure A1.37 Voltage–frequency characteristics. (a) Linear V/f below base speed. (b) Typical motor flux with linear V/f (showing fall in flux at low frequency as well as above base speed). (c) Modified V/f characteristic with low frequency boost (to compensate for stator resistance effects in the steady state)

Because V is constant above base speed, the flux will fall as the frequency is increased after the output voltage limit is reached. The machine flux falls (Figure A1.37b) in direct proportion to the actual V/f ratio. Although this greatly reduces the core losses, the ability of the machine to produce torque is impaired and less mechanical load is needed to draw full load current from the inverter. The drive is said to have a constant power characteristic above base speed. Many applications not requiring full torque at high speeds can make use of this extended speed range.

The second operating condition where departure from a constant V/f is beneficial is at very low speeds, in which the voltage drop arising from the stator resistance becomes significantly large. This voltage drop is at the expense of flux, as shown in Figure A1.37b. To maintain a truly constant flux within the machine the terminal voltage must be increased above the constant V/f value to compensate for the stator resistance effect. Indeed, as output frequency approaches zero, the optimum voltage becomes the voltage equal to the stator IR drop. Compensation for stator resistance is normally referred to as 'voltage boost' and almost all inverters offer some form of adjustment so that the degree of voltage boost can be matched to the actual winding resistance. It is normal for the boost to be gradually tapered to zero as the frequency progresses towards base speed. Figure A1.37c shows a typical scheme for tapered boost. It is important to appreciate that the level of voltage boost should increase if a high starting torque is required, because in this case the IR drop will be greater by virtue of the increased stator current. In this case automatic load-dependent boost control is useful in obtaining the desired low speed characteristics. Such a strategy is referred to as constant V/f (or V/Hz) control and is a feature of most commercially available a.c. drives although more advanced open-loop strategies are increasingly being used (see Section A4.4).

So far the techniques described have been based on achieving constant flux within the air gap of the machine or, if that is not possible, then the maximum flux. Constant flux is the ideal condition if the largest capability of torque is required because the load cannot be predicted with certainty, or if the most rapid possible acceleration time is desired. A large number of inverters are used, however, for applications such as fans and centrifugal pumps where flow control is obtained by speed variation. The torque required by a fan follows a square law characteristic (Figure A1.38) with respect to speed, and reducing the speed of a fan/pump by 50 per cent will reduce its torque requirement to only 25 per cent of rated. As the load is entirely predictable there is no need for full torque capability and hence flux to be maintained, and higher motor efficiency can be obtained by operating at a reduced flux level. A further benefit is that acoustic noise, a major concern in air conditioning equipment, is significantly reduced. It is therefore common for inverters to have an alternative square law V/f characteristic or, ideally, a self-optimising economy feature so that rapid acceleration to meet a new speed demand is followed by settling to a highly efficient operating point.

A1.3.5 Slip-ring induction motor

The wound rotor (or slip-ring induction motor) is an induction motor with a rotor containing a three-phase winding of the same pole number as the stator. The rotor winding is brought out to usually three slip-rings through which connection can be made

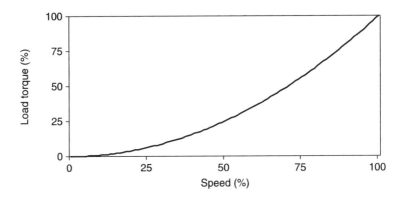

Figure A1.38   Typical square law characteristic of a fan or pump load

by stationary brushes usually connected to and within the stator housing. Although introducing the negative aspect of brushes, this arrangement does address some of the disadvantages of the cage induction motor.

With the correct value of (usually) resistance inserted in the rotor circuit, near-unity relationship between torque and supply current at starting can be achieved: 100 per cent full load torque (FLT), with 100 per cent full load current (FLC), 200 per cent FLT with 200 per cent FLC, and so on (i.e. comparable with the starting capability of the d.c. machine). The high starting efficiency and also the smooth controlled acceleration historically gave the slip-ring motor great popularity for lift, hoist and crane applications, although these are now largely being replaced by Flux Vector drives and will therefore not be considered further.

It has had similar popularity with fan engineers to provide a limited range of air volume control, either 2:1 or 3:1 reduction, at constant load, by the use of speed-regulating variable resistance in the rotor circuit. Although a fan possesses a square-law torque–speed characteristic, so that motor currents fall considerably with speed, losses in the rotor regulator at lower motor speeds are still relatively high, severely limiting the useful speed range.

Rotor slip-ring systems, used with this type of motor, offer a similar service life to that of the d.c. motor commutator brush system.

Efficient variable-speed control of slip-ring motors can be achieved by converters based upon the slip energy recovery principle first proposed by Kramer in 1904. Such schemes are based upon converting the slip frequency on the rotor to supply frequency. (These schemes are described in Sections A2.4.3.5 and A2.5.5.) Such applications are still economically attractive in some applications as the power converter in the rotor circuit to convert slip frequency to mains frequency need only be rated for a fraction (based on deviation from synchronous speed) of the motor rating. For this reason, slip-ring motors are finding significant application in wind energy schemes.

It is also possible to retrofit variable-frequency inverters to existing slip-ring motors. This can be done simply by shorting out the slip-ring terminations (ideally on the rotor, thereby eliminating the brushes) and treating the motor as a cage machine.

A1.3.6 Speed-changing motors

The cage rotor briefly described earlier is constructed with copper rotor bars brazed to shorting end-rings, or is aluminium-die-cast in a single operation. The construction is simple, cheap and robust. It has the further advantage that the same rotor can be used with different stator winding pole numbers, in which the sequence of current reversals in the rotor bars is altered, and the rotor end-rings provide a free path for the current to flow, adapting to the differing number of stator poles. Consequently, by designing stator windings of more than one pole combination to be wound on the same stator it is possible to realise a motor with more than one rated operating speed. Such motors are generically known as speed-changing motors and are commercially sold in two principle forms.

The simplest form is to combine two quite separate stator windings in the one machine, for example four and eight poles, providing a 2:1 speed selection with a constant torque relationship between the two speeds. This would suit many applications in materials handling and possibly in lift and crane drives, but would be wasteful of torque and therefore of cost for fan and pump drives, where the load torque requirement for low-speed operation falls as the square of the speed.

Alternatively, and often more commercially attractive, is the design of a single set of stator windings that can be arranged into different pole combinations, usually with contactors external to the motor. This is termed consequence pole switching and is restricted to 2:1 speed range combinations, i.e. 2/4 pole or 4/8 pole, few other 2:1 combinations being practicable. As this uses a single winding only, a constant torque relationship between the two speeds is available.

More than two speeds can be made available from a.c. squirrel cage motors by combining separately wound and consequence pole switching in the same machine. Pole amplitude modulated (PAM) speed change, as this method is known, is also available for machines of larger kilowatt ratings and is somewhat similar in principle to consequence pole switching.

A1.3.7 A.C. induction motor construction

The a.c. cage induction motor stator coils are wound on a laminated iron core formed into a pack, usually by seam welding. Semi-closed slots accommodate the windings, which are normally wound as a concentric three-phase arrangement with suitable insulation within each coil set, between phases and from phase to earth.

A laminated cylindrical rotor core is mounted centrally in the stator with the smallest practical radial air gap and with semi-closed or closed rotor slots around the periphery to accommodate the rotor winding. This winding comprises bare or lightly insulated aluminium or copper bars, occupying the slots and connected to shorting end-rings at each end-face of the rotor pack.

A typical die-cast aluminium rotor construction is shown (with stator) in Figure A1.39 for comparison with a copper bar cage rotor shown in Figure A1.40.

A die-cast aluminium rotor cage induction motor construction is generally used for applications up to about 50 kW and allows best economics in manufacture. Rotor

Figure A1.39   Typical squirrel cage induction motor with die cast rotor with a laminated stator [photograph courtesy of Leroy-Somer]

Figure A1.40   Typical squirrel cage induction motor rotor with inserted copper rotor bars [photograph courtesy of Leroy-Somer]

losses (and motor slip) with aluminium bars are higher than with copper, which tends to be used at ratings of approximately 50 kW and above.

The use of aluminium in the frame construction of smaller machines up to 37 kW rating is virtually the norm because of its better thermal conductivity when compared to cast iron. It offers the benefits of resistance to corrosion, ease of machining, reduced weight, and generally improved appearance. The better heat conduction of aluminium and the much improved force-ventilated cooling of the standard IP54 enclosed machine of current design are significant benefits, although such cooling still falls short of the effectiveness of drip-proof IP23 enclosures, which exchange the cooling air around the stator and rotor windings directly with the surrounding ambient air.

Above 37 kW, aluminium construction tends to be less of an advantage, particularly in terms of mechanical strength as required for the higher torque loading involved. Cast iron and steel construction becomes the standard, as does detailed attention to effective cooling of the IP54 enclosed machine. At these higher kilowatt ratings the inferior efficiency of the a.c. cage machine compared with the equivalent rating of d.c. motor and brushless synchronous a.c. motor, gives the larger a.c. cage machine significant power losses to dissipate.

However, a major advantage of the a.c. cage machine is that it is readily available in enclosures that meet the requirements of difficult and hazardous environments, and it can be specified to accept much higher levels of external vibration and shock loading than is possible for a comparable d.c. machine.

Standard cage induction machines are available to comply with a wide variety of international standards.

A1.4 A.C. synchronous motors

A1.4.1 General

In some respects a great mystery has developed around synchronous machines. This may be because of their very wide range of applications, from multi-megawatt power generation through to precision servo motors rated at a few watts. Although the applications and construction of these various forms vary considerably, the basic principles and characteristics of the motor itself are common. However, when combined with power electronics speed controllers, many of these characteristics are not immediately obvious to the user.

It should be noted that with the widespread use of synchronous motors in positioning and variable-speed applications, particularly permanent magnet synchronous motors in precise positioning applications (servos), the motor characteristics and performance equations are usually described in a specific manner not usually encountered in traditional synchronous motor theory. These are described and defined here.

One significant reason for some of the mystery surrounding permanent-magnet synchronous motors lies in the various names used for these motors and the variants with sinusoidal or trapezoidal winding distributions. Brushless servomotors are often called brushless d.c. servomotors because their structure is different from that of d.c. servomotors. Brushless servomotors switch/commutate current by means of transistor switching within the associated drive/amplifier, instead of a commutator as used in d.c. servomotors. To further confuse matters, brushless servomotors are also called a.c. servomotors because brushless servomotors of synchronous type with a permanent-magnet rotor detect the position of the rotational magnetic field to control the three-phase current of the armature. It is now widely recognised that *brushless a.c.* refers to a motor with a sinusoidal stator winding distribution that is designed for use on a sinusoidal or PWM inverter supply voltage. *Brushless d.c.* refers to a motor with a trapezoidal stator winding distribution, which is designed for use on a square-wave or block commutation inverter supply voltage. Sinusoidally distributed windings are the most common in industrial motors; we will only consider these here. The fundamental characteristics and relationships are very similar for trapezoidal motors.

A1.4.2 Operating principles

The stator of a typical synchronous machine has great similarity to that of the induction motor described in Section A1.3.2. There is usually a three-phase winding similar to that of an induction motor, but in this case it is not current in the stator winding that is responsible for creating the air-gap field. This field is instead established by the rotor.

The rotor of a synchronous machine is very different from that of an induction motor as it contains a means of producing a magnetic field, fixed in position relative to the rotor. This field is produced either by direct current flowing in a winding mounted on the rotor or by magnets attached to the rotor.

The method of torque production is also very different from that of an induction motor. The synchronous motor does not rely upon a difference in rotational speed of the rotor and the rotating magnetic field in order to produce torque. Instead the fundamental method of torque production is the cutting of the magnetic field by the current in the stator winding. If the stator current phase position can be caused to be always at right angles to the magnetic field, it becomes possible to make a highly efficient motor producing smooth torque without using brushes. This is why it is usual in synchronous-machine drive systems to have a rotor position sensor attached to the rotor. Advances in sensorless control schemes are being made in which the rotor position may be estimated.

The synchronous motor therefore stays in synchronism with the stator supply. There is, however, a limit to the maximum torque that can be developed, typically being between 1.5 and 4 times the continuously rated torque. The torque speed curve (Figure A1.41) is therefore simply a vertical line, which indicates that if we try to drive the machine above the synchronous speed it will become a generator.

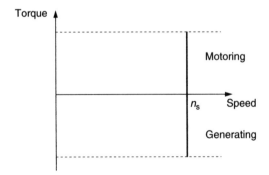

Figure A1.41 *Steady-state torque–speed curve for a synchronous motor supplied at constant frequency*

A1.4.3 Fundamental equations of steady-state performance

A1.4.3.1 General

If we consider a synchronous motor with a round symmetrical rotor, with a constant field current (or fixed magnets) rotating at an angular velocity ω, and a stator upon which there is a three-phase sinusoidally distributed winding, the voltages induced in the stator windings are given by

$$v_A = \sqrt{2} K_e \omega \sin(\omega t)$$
$$v_B = \sqrt{2} K_e \omega \sin(\omega t - 120°) \quad (A1.43)$$
$$v_C = \sqrt{2} K_e \omega \sin(\omega t + 120°)$$

where K_e is known as the back-emf constant. Note that the back-emf constant is usually expressed in units of V per krpm (where the voltage is the rms voltage and k min^{-1} is in thousands of revolutions per minute).

Unlike the induction motor, because the magnetic field is fixed in position to the rotor shaft, there is no difference between the speed of rotation of the rotor and the speed of rotation of the rotor shaft. There is no slip. The speed of rotation is in this case therefore the synchronous speed and is given by

$$n_s = 120f/p \quad (A1.44)$$

where p is the number of poles and f is the applied frequency ($\omega/2\pi$). For example, for a four-pole machine with a 50 Hz supply, $n_s = 1\,500$ min^{-1}.

We know from basic electrical theory and from considering sinusoidal waveforms that the power per phase developed by a synchronous motor (P_A) is given by

$$P_A = v_A i_A \cos v \quad (A1.45)$$

where v_A and i_A are the phase voltage and current, and v is the phase angle between the voltage and current. Note that $\cos v$ is also known as the power factor.

Under balanced conditions, the overall power (P_T) is given as

$$P_T = 3 v_A i_A \cos v \quad (A1.46)$$

In a star-connected machine, the terminal voltage $v_{AB} = \sqrt{3} v_A$ and so equation (A1.46) can be rewritten as

$$P_T = \sqrt{3} v_{AB} i_A \cos v \quad (A1.47)$$

The electromagnetic torque is the power divided by the rotational speed, so torque is given by

$$\text{Torque} = \sqrt{3} v_{AB} i_A \cos v p / 120f \quad (A1.48)$$

A1.4.3.2 Brushless PM servo motor

The analysis presented in Section A1.4.3.1 is generally applicable, but when specifying servo motors and matching drives to those motors a specific analysis and vocabulary is used to define the performance.

Torque constant K_t

In the armature of the motor of Figure A1.42, the current distribution is as illustrated. If the current flowing in the conductors to the right of the symmetrical axis OO' is in the direction (away from the reader), then current in the conductors to the left flows in the opposite direction (towards the reader). We now assume that all the conductors in the right half are under the north pole and all the conductors in the left half are under the south pole of the permanent magnet, and the magnetic flux density has an average value of B to simplify the discussion. Then the torque $RBIL$ should work on every conductor and the whole torque T around the axis will be

$$T = ZRBLI$$
$$T = (ZRBLI_a)/2 \tag{A1.49}$$

where Z is the total number of conductors, R is the radius of the rotor, B is the flux density linking the stator windings, L is the *I*nductance of the winding, and I_a is the current from the motor terminal, which is equal to $2I$.

In this model, the magnetic field is equal to

$$\varphi = \pi RLB \tag{A1.50}$$

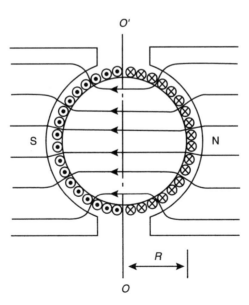

Figure A1.42 Field flux and current distribution in a rotor

Therefore, by substitution, we obtain

$$T = (Z/\pi)\varphi I_a/2 \tag{A1.51}$$

Now let us consider this equation. The number of conductors Z never changes in a finished motor. Because the magnetic flux φ is determined by the motor dimensions and state of magnetisation, $(Z/\pi)\varphi$ is a constant. Therefore, we can conclude that the torque T is proportional to the armature current I_a.

We can therefore define the *torque constant* K_t as

$$K_t = (Z/2\pi)\varphi \tag{A1.52}$$

Therefore, we obtain

$$T = K_t I_a \tag{A1.53}$$

The torque constant is usually expressed in units of N m A^{-1} (where the current is the rms current).

It should be noted that the above torque equation is identical to that of a d.c. motor with constant field [equation (A1.19) in Section A1.2.3.1].

Relationship between torque and back-emf constants

We have already defined the back-emf constant K_e in equation (A1.43). K_e can be expressed in terms of other parameters. If the rotor is revolving at a speed ω rad s^{-1}, the speed v of the conductor is

$$v = \omega R$$

Therefore, the back electromotive force e generated in a conductor is

$$e = \omega R B L$$

If the total number of conductors is Z, then the number of conductors in series connection is $Z/2$ and the total back-emf E at the motor terminals is

$$E = \omega R B L Z/2$$

By substitution we can express E in terms of the flux φ as

$$E = (\varphi Z/2\pi)\omega$$

We can therefore obtain the back-emf constant K_e

$$K_e = (Z/2\pi)\varphi \tag{A1.54}$$

Comparing equations (A1.54) and (A1.52) it can be seen that K_e and K_t are identical. However, this is only true when a self-consistent unit system is used. The international system of units (SI) is one such system. For example, if K_t is equal to 0.05 N m A^{-1}, then K_e is equal to 0.05 V s rad^{-1}. As stated earlier, although it is normal to express the torque constant in terms of N m A^{-1} it is more usual to express the back-emf constant in terms of V/kmin^{-1} (volts per thousand revolutions per minute).

A1.4.4 Limits of operation

The limits of operation for a brushless PM servo motor are summarised on the torque–speed characteristic in Figure A1.43. The individual limits shown in Figure A1.43 are as follows:

1. The torque/speed is limited by the temperature rise of the motor.
2. For the maximum theoretical field weakening possible with the selected drive, the intermittent drive peak current/power is considered as the constraint. Motor demagnetisation level must be carefully considered in this region.
3. The maximum motor speed must be selected considering the case of a drive trip (i.e. losing field-suppressing current), maximum voltage limit of the drive.
4. This intermittent limit due to available drive voltage (drive d.c. bus voltage) without field weakening operation.

The rms torque and speed of an operating profile of a PM servo motor gives an application 'operating point' that must be within the continuous region.

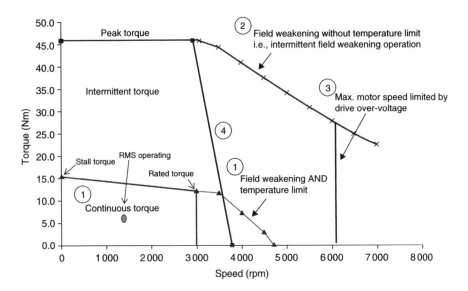

Figure A1.43 Limits of operation of a typical brushless PM servo motor

A1.4.5 Synchronous motor construction

Synchronous motors are available in a wide range of motor frames and types of construction. It is not possible to present all forms in this book and so those of greatest industrial importance will be considered. The forms taken are closely related to the areas of application.

Permanent magnet synchronous motors are readily available to powers in excess of 400 kW, but very high-power motors (tens of megawatts) have been designed for specific applications. The form of construction may be broadly split into high-performance servo motors and industrial motors.

Wound-rotor synchronous motors are available over a wide range of powers and are used frequently in very high-power applications.

A1.4.5.1 Permanent-magnet servo motors

Figure A1.44 shows the structure of a typical brushless servomotor. The windings are in the static armature, which is part of the stator. It is therefore possible to think of the rotor of a brushless servomotor to be equivalent to the stator of a d.c. motor. In other words, the magnetic field for generating torque is stationary in d.c. motors, but it rotates in brushless servomotors; conversely, the armature revolves in d.c. motors, but it is stationary as a stator in brushless servomotors.

As servomotors frequently require rapid acceleration and deceleration, or need to be able to maintain smooth motion in the presence of large load disturbances, the maximum torque has to be several times larger than the rated torque. As brushless servomotors, unlike d.c. motors, do not have a mechanical commutation limit, they can be operated up to the boundary of high-speed rotation without decreasing the maximum torque. Further, in a brushless servomotor, the primary area of heat dissipation occurs not on the rotating part but on the armature in the stator, because the permanent magnets are mounted on the axis of rotation. The heat dissipated in the stator diffuses into the air through the frame and importantly through the mounting flange (see Section B4.2.3). It is therefore relatively easy to cool brushless servomotors. Moreover, brushless servomotors provide more precise overload protection, because the temperature of the hottest part can be detected directly.

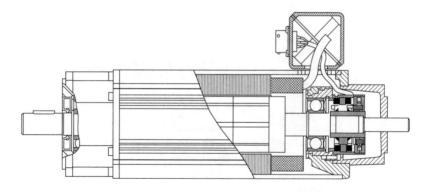

Figure A1.44   Structure of a brushless permanent-magnet servomotor

Stator structure

A typical stator consists of an armature core and armature windings. The armature core is made of laminated punched silicon steel sheet of 0.35–0.5 mm thickness (laminated core). In many cases, the armature core is slotted and skewed to reduce torque ripple, which results in speed ripple. The armature windings are similar to the one in the a.c. motor and are usually of the distributed three-phase type. The windings are usually designed according to the drive specification, which requires either a sinusoidal or trapezoidal back-emf waveform. The factors that govern the design are stator slot number, pole shape, windings coil pitch, rotor pole number and magnet shape.

Rotor structure

The above analysis assumes a perfect lossless mechanical system as well as sinusoidal balanced voltages and currents and smooth torque. It also assumes a round rotor with a symmetrical magnetic circuit. A symmetrical magnetic circuit is typical of synchronous motor designs that have surface-mounted magnets on the rotor, as shown in Figure A1.45. Such a construction is typical of permanent-magnet servo motors.

This symmetrical arrangement exhibits very little saliency. Salient-pole machines have a preferred direction of magnetisation determined by either protruding field poles or in the case of permanent-magnet motors an angle-dependent reluctance path when considering the stationary rotor construction. Typical examples of such forms of construction are shown in Figure A1.46.

A detailed analysis of salient-pole motors is beyond the scope of this book; however, it is worth noting that saliency can make a significant contribution to the torque output, perhaps as much as 30 per cent for some designs (for a discussion of reluctance torque see Section A1.1.2.1). However, magnetic leakage and saturation effects tend to limit the saliency ratio achieved in practice. In terms of specific output (power per unit volume) there is no real benefit in using saliency effects in permanent-magnet motors. Their use tends to be driven by considerations such as minimising magnet material, fault tolerance and in some applications demanding

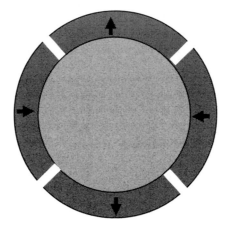

Figure A1.45   Surface-mounted magnet rotor structure

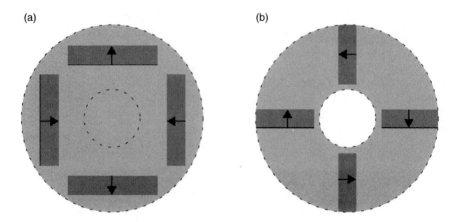

Figure A1.46   Permanent-magnet rotor structures exhibiting saliency. (a) Buried magnet with radial magnetisation. (b) Buried magnet with circumferential magnetisation

constant power operation. These characteristics explain why salient-pole permanent-magnet motors are frequently seen in applications such as electric vehicles.

Saliency in a machine is also used in some position-sensorless control schemes to determine the rotor position by means of on-line inductance measurement.

It should be noted that the fixing of permanent magnets to the rotor is critical in the design of robust brushless servomotors. Different techniques apply to the different types of rotor construction. Permanent-magnet servo motors tend to be designed with surface-mounted magnets. In this case, various techniques have been applied to the adhesion of magnets in order to prevent the destruction of motors resulting from centrifugal force in high-speed rotation or that caused by repetitive rapid acceleration and deceleration. Common methods used to prevent separation of the permanent magnet from the rotor surface include binding the outer surface of the permanent magnet with glass-fibre tape or yarn and using a thin stainless-steel cylinder as a sleeve to cover the outer surface of the permanent magnet. Adhesives are used in combination with either of these methods. The adhesives chosen have a linear expansion coefficient that is comparatively close to that of the permanent magnet and that of the axis of rotation. They also need to be stable against any thermal change.

For buried-magnet rotor designs, the rotor laminations can be used to retain the magnets as shown in Figure A1.47.

A1.4.5.2   Permanent-magnet industrial motors

Recent years have seen the emergence of permanent-magnet motors packaged in standard industrial motor (IEC or NEMA) housings (Figure A1.48). These motors are in general targeted at non-high-performance/servo applications.

This type of motor is today being marketed as a direct competitor of the induction motor in variable-speed applications. There is a higher cost associated with these

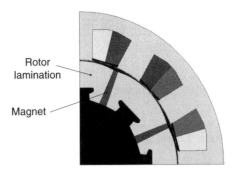

Figure A1.47   Magnet retention by laminations

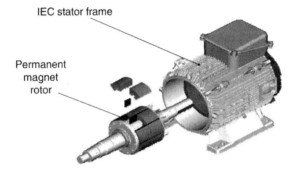

Figure A1.48   Permanent-magnet industrial motor in standard IEC frame

motors, but they do offer higher power density and improved efficiency. With having a permanent-magnet rotor, the rotor runs cooler that would be the case in a typical induction motor and so there may also be advantages in aspects such as bearing life.

A1.4.5.3   Wound-rotor synchronous motors

Wound-rotor synchronous machines are most commonly used at the highest powers and speeds. This is a specialist field and beyond the scope of this book. Some general points are of interest.

The synchronous motor stator is generally the same as for the induction motor. The rotor is constructed with rotor slots to accommodate a fully insulated wire or bar wound rotor winding of the same pole number as the associated stator with connections brought out to external terminals via slip rings and brushes or via a rotating rectifier arrangement on an auxiliary exciter.

A1.4.6   Starting of synchronous motors

Synchronous motors operate at synchronous speed and do not have an inherent starting capability when connected to a fixed-frequency supply. A typical way of starting synchronous motors is to mount an auxiliary motor on the rotor shaft to bring the motor to synchronous speed before making the stator connection to the mains supply.

A more elegant solution on wound-rotor motors is to design the machine with an amortisseur winding. This is similar in form to the rotor cage of an induction motor and behaves in just the same way to start the motor. Once the rotor has accelerated to near-synchronous speed, the field power is applied and the rotor is pulled into synchronism with the supply.

Of course by using a power electronic converter/drive, the motors are started by ramping the frequency from 0 Hz at standstill, maintaining synchronism at all times.

A1.5 Reluctance motors

The reluctance motor is arguably the simplest synchronous motor of all, the rotor consisting of a set of iron laminations shaped so that it tends to align itself with the field produced by the stator (Figure A1.49).

The stator winding is identical to that of a three-phase induction motor. The rotor is different in that it contains saliency (a preferred path for the flux). This is the feature that tends to align the rotor with the rotating magnetic field, making it a synchronous machine. The practical need to start the motor means that a form of 'starting cage' needs also to be incorporated into the rotor design, and the motor is started as an

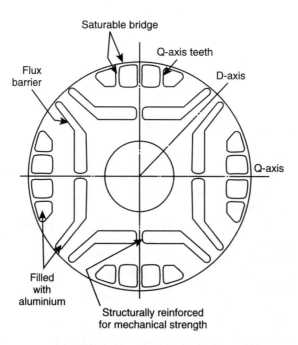

Figure A1.49 *Rotor lamination for a four-pole reluctance motor*

induction motor, so that the reluctance torque 'pulls in' the rotor to run synchronously in much the same way as a permanent-magnet rotor.

Reluctance motors may be used on both fixed-frequency (mains) supplies and inverter supplies. These motors tend to be one frame size larger than a similarly rated induction motor and have low power factor (perhaps as low as 0.4) and poor pull-in performance. As a result of these limitations their industrial use has not been widespread. Historically they were much favoured in applications such as textile machines, where large numbers of reluctance motors were connected to a single 'bulk' inverter to maintain synchronism. Today, as costs of power electronic equipment has fallen, bulk inverters are infrequently used and the reluctance motor is now rarely seen.

A1.6 A.C. commutator motors

Single-phase a.c. commutator motors of small kilowatt ratings are manufactured in large quantities, particularly for domestic applications and power tools. Development generally has been in the fields of improved commutation and reduced cost. The series a.c. machine is essentially similar to the series d.c. motor and remains the more important of the two. Although not capable of commutation to the standards required for industrial drives, the type can provide approximate speed control simply by crude voltage regulation, often facilitated by a single semiconductor switch.

An alternative design of some interest is the single-phase a.c. commutator repulsion motor, in which the armature brushes are shorted together, with the a.c. supply taken to the field windings only. Limited additional speed control can be achieved by angular brush shift – an inconvenient method, particularly in machines of low rating and small size. Starting torque is high for the repulsion motor, and its design survives nowadays in the form of the repulsion-start, induction-run single-phase machine, where a centrifugal device shorts out the commutator segments when the motor is up to speed. More important is the a.c. three-phase commutator motor, which over the years has provided outstanding service where variable speed is required.

Again, although many ingenious designs have been produced in the search for improved variable-speed performance, the most successful for ratings between 5 kW and 150 kW has been the Schrage or rotor-fed machine, followed by the stator-fed or induction regulator motor, which has been manufactured in ratings in excess of 2 000 kW.

Both designs combine commutator frequency changing and variable-speed motor in the same frame, generating the required slip frequency voltage in a primary winding, for injection into a secondary winding.

The kilowatt rating of the Schrage motor is restricted by the slip rings, carrying as they do the total power of the motor.

The stator-fed or induction regulator motor is a somewhat similar arrangement that converts the mains frequency injected voltage to slip frequency voltage for the rotor coils. A step-down variable-ratio transformer is generally connected between the line supply and the commutator connection to give voltage/phase control. The parallel transformer connection gives shunt characteristics and the series transformer connection, series characteristics. The stator power feed of this design allows, as has been mentioned, ratings over 2 000 kW to be designed successfully although commutators now appear to be an increasing disadvantage.

A1.7 Motors for special applications

A1.7.1 Geared motors

Standard industrial motors are often unsuitable as direct drives for low-speed applications. Moreover, the use of motors with low and medium ratings is uneconomical at low speeds. Geared motors are available for such applications. These units consist of a high-speed motor and a gear reducer assembled to form an integral unit. The hardened teeth of the gear wheel resist high stressing and ensure long life of the assembly.

Geared motors are widely used on single machines such as tower cranes, lifts/ elevators, construction machinery, in agriculture, and so on, as well as in industrial plants.

A1.7.2 Brake motors

Mechanical brakes are often used in conjunction with motors instead of, or as well as, electrical braking circuits. These units consist of a motor and a brake, assembled to form an integral unit. It is important to note that the brake may be rated to brake the motor and its load, or may be rated to provide a holding duty only. A holding only brake will be quickly destroyed if it is used to brake a load from speed. *It is common practice, particularly on brushless servo motors, that the brake be rated for holding duty only.*

In order to rate a brake motor correctly the following information about the application is necessary:

- the type of load and the type of duty of the unit,
- the frequency of braking cycles per hour,
- the total inertia (motor, brake, gearbox (if fitted) and load) referred to motor speed,
- the load torque as a function of speed, referred to the motor shaft,
- whether the load torque has a braking or accelerating effect, and
- the braking time and braking torque required.

Brake motors can be designed to be fail-safe. If the power is lost than the brake will automatically be applied.

A1.7.3 Torque motors

Torque motors have been developed from the basic designs of three-phase squirrel-cage induction motors. They are not designed for a definite output, but for a maximum torque, which they are capable of delivering at standstill and/or at low speed (when supplied from a fixed-frequency supply). They have a torque–speed curve of the form shown in Figure A1.50.

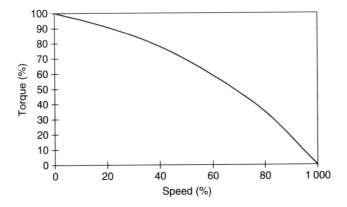

Figure A1.50   Torque–speed characteristic of a torque motor

A1.8   Motors for hazardous locations

A1.8.1   General

Manufacturing processes in many sectors of industry may be described as hazardous by the nature of the operating environment, where explosive gas–air mixtures may occur in dangerous concentrations. The decision as to whether an outdoor area or an enclosed location should be considered subject to an explosion hazard as defined by the relevant regulations and specifications rests entirely with the user, or in case of doubt, the competent inspecting authority. It should be realised that the degree of hazard is variable and this has led to the concept of area classification and the development of design techniques to ensure that electrical equipment would operate safely in the specified hazardous area zones.

Harmonised standards exist in Europe under the guidance of CENELEC. In North America, emphasis is also placed on test and accreditation, under the guidance of Underwriters' Laboratories (UL) and CSA; however, the nomenclature is different from European practice.

A1.8.2   CENELEC

The seven commonly recognised methods of protection, as published by CENELEC are as follows:

- EN 50 014   General
- EN 50 015   (Ex)o
- EN 50 016   (Ex)p
- EN 50 017   (Ex)q
- EN 50 018   (Ex)d
- EN 50 019   (Ex)e
- EN 50 014   (Ex)i

Although comprehensive guidance on the selection of explosion-protected equipment is contained within the standards, the information to be considered falls into four categories:

1. *The type of protection of the apparatus in relation to the zonal classification of the hazardous area.* The degree of protection required is dependent upon the presence of ignitable contaminations of inflatable gas or vapour in relation to the length of time that the explosive atmosphere may exist, and this is defined in Table A1.3.

Table A1.3   Protection zones in hazardous locations

Zone 0	A zone in which an explosive gas–air mixture is continuously present, or present for long periods
Zone 1	A zone in which an explosive gas–air mixture is likely to occur in normal operation
Zone 2	A zone in which an explosive gas–air mixture is not likely to occur in normal operation, and if it occurs will only exist for a short time

2. *The temperature classification of the apparatus in relation to the ignition temperature of the gasses and vapours involved.* Electrical apparatus must be selected to ensure that the maximum surface temperature is below the ignition temperature of the specified gas. EN50 014 gives temperature classifications for equipment as shown in Table A1.4.

Table A1.4   Temperature classifications

	Maximum surface temperature (°C)
T1	450
T2	300
T3	200
T4	135
T5	100
T6	85

These classifications can be related to the ignition temperatures as detailed in Table A1.5.

3. *The apparatus subgroup (if applicable) in relation to the relevant properties of the gases and vapours involved.* Explosion-protected electrical apparatus is divided into two main groups:
   - Group I: Mining applications
   - Group II: All non-mining applications

Table A1.5   Ignition temperatures

Example of compound	Ignition temp. (°C)	Suitable equipment regarding temperature classification					
Acetone	535	T1	T2	T3	T4	T5	T6
Butane	365		T2	T3	T4	T5	T6
Hydrogen sulphide	270			T3	T4	T5	T6
Diethyl ether	170				T4	T5	T6
Carbon disulphide	100					T5	T6

For some types of protection, notably flameproof enclosures, it is necessary to subdivide Group II according to the properties of the gases, vapours or liquids, because apparatus certified and tested for, say, a pentane–air mixture will not be safe in a more easily ignitable hydrogen–air mixture. This has led to apparatus subgrouping: IIA, IIB and IIC.

4. *The suitability of the apparatus for the proposed environment.* Explosion-protected apparatus appropriate to a particular zone is readily identified by reference to Table A1.6.

Table A1.6   Equipment suitable for different zones

Zone	Type of protection
0	(Ex)ia
	(Ex)s – specifically certified
1	Any of the above plus
	(Ex)d; (Ex)ib; (Ex)p; (Ex)e; (Ex)s
2	Any of the above plus
	(Ex)N or (Ex)n; (Ex)o; (Ex)q

These recognised types of protection are as follows:

- *(Ex)o – Oil immersed.* All or part of the *a*pparatus is immersed in oil to prevent ignition.
- *(Ex)p – Pressurised.* Because it is not practical to manufacture explosion-proof motors in large sizes, it is common practice to use pressurised motors for Zone 1 applications. These motors tend to be manufactured to normal industrial standards except that special attention is paid to the sealing of all removable covers and to shaft seals. The motors must be totally enclosed. Cooling must be by air-to-air or air-to-water heat exchangers.

Before the motor is energised it must be purged with at least five times its own volume of clean air to remove any flammable gasses with may be present.

- *(Ex)q – Powder filled.* This form of protection is very unusual in rotating machines.
- *(Ex)d – Explosion proof.* All parts of the motor where igniting arcs or sparks may be produced are housed in a flameproof enclosure. The sealing faces, cable entries, shaft glands and so on are made with comparatively large gap length and limited gap clearances to prevent ignition of the surrounding explosive atmosphere. During operation explosive mixtures penetrate only seldom into the enclosure. Should an internal explosion occur, it is prevented from spreading to the external atmosphere.
- *(Ex)e – Increased safety.* This type of construction is used for motors without a commutator or slip rings, which do not produce sparks during normal operation.

  This type of motor may be used in Zone 1, with some qualifications, and in Zone 2 areas. It is required that all surface temperatures are kept within the ignition temperature of the specified gas under all conditions of operation or fault.

  To avoid the danger of ignition in the event of a fault, suitable protective devices such as circuit breakers with matched thermal characteristics should be used to protect the motor against overheating.

  The worst abnormal condition that can occur without permanently damaging the motor is a stalled condition. Most motor designs are 'rotor critical', the rotor temperature increasing more rapidly than the stator under stalled conditions. The surface temperature of the rotor conductors is the critical and limiting factor in this type of motor.

  The t_e characteristic of (Ex)e motors is important and must be quoted on the nameplate. It is defined as the time taken for a winding, when carrying the worst-case current, to be heated up from the temperature under rated operating conditions to the limiting temperature. t_e must never be less than 5 s.

- *(Ex)i – Intrinsically safe.* The concept of intrinsic safety is based upon restricting the electrical energy within the apparatus and its associated wiring to prevent the occurrence in normal operation of incendive arcs, sparks or hot surfaces. It is necessary to ensure that high voltages cannot be induced into the intrinsically safe circuit. Shunt diodes are usually employed as barriers between the intrinsically safe and the hazardous areas.

  This method is used in signalling, measuring and control circuits but is not practical for motors.

- *(Ex)s – Special protection.* This protection concept allows for certification of equipment that does not comply with the specific requirements of the established forms of protection.
- *(Ex)N and (Ex)n.* In the designation of type-N apparatus, the upper case N is used in the UK, but the lower case n has been proposed for the European standard having a similar concept.

  Non-sparking motors, which are suitable for use in Zone 2 areas, are supplied in the UK to BS5000:Part 16. No part of the motor may exceed 200 °C during normal operation, but it may do so during starting.

A1.8.3 North American standards

The principles applied in North America are broadly similar to those in Europe. The key differences are as follows:

- *Temperature classification.* Although the basic temperature classifications of the European standards are retained, interpolation has occurred between some T classifications giving greater resolution. The resultant Table A1.7 also cross references to the European classification.

Table A1.7 United States and European temperature classification

European classification (EN50 014)	United States classification (NEC NFPA 70)	Maximum surface temperature (°C)
T1	T1	450
T2	T2	300
	T2A	280
	T2B	260
	T2C	230
	T2D	215
T3	T3	200
	T3A	180
	T3B	165
	T3C	160
T4	T4	135
	T4A	120
T5	T5	100
T6	T6	85

- *The apparatus subgroup.* In the United States, the system of area classification and gas grouping is again different to European practice. Here the hazardous area is divided into flammable gases or vapours and combustible dusts. The key classifications are given in Table A1.8.

Class I, Group D is approximately equivalent of the European Group IIA with temperature classification to suit the specific explosion hazard.

A1.8.4 Testing authorities

The main European and North American testing authorities are shown in Tables A1.9 and A1.10, respectively.

Table A1.8   North American hazardous area classification

Class I	Gas or vapour	Group C	Ethyl-ether, ethylene, cycle propane
		Group D	Gasoline, hexane, naphtha, benzene, butane, propane, alcohol, lacquer vapours and natural gas
Class II	Hazardous dusts	Group F	Carbon black, coal or coke dust
		Group G	Flour, starch or grain dust
Class III	Easily ignitable fibres		Fibres easily ignitable but not able to be suspended in air to produce ignitable mixtures, such as rayon, nylon, cotton, saw dust and wood chips

Table A1.9   Main EU approved testing authorities

Country	Name	Location
Belgium	INIEX	Paturages
Denmark	DEMKO	Herlev
France	Cerchar	Verneuil
	LCIE	Paris
Germany	BVS	Dortmund-Derne
	PTB	Braunschweig
Italy	CESI	Milan
UK	BASEEFA	Buxton

Table A1.10   Main North American testing authorities

Country	Name	Location
Canada	CSA	Toronto
USA	Factory Mutual	Norwood
	Underwriters' Laboratory	Northbrook

Chapter A2
Drive converter circuit topologies

A2.1 Introduction

The essence of any power electronics drive is the conversion of electrical energy from one voltage and frequency to another. Many circuit topologies exist to meet different requirements. Some of the topologies are specifically associated with the control of particular motor types, while others can be used to control various forms of electrical machine. Some topologies are associated with particular applications or speed ranges.

All of the topologies can be classified in terms of the type of conversion that they perform:

- a.c. to d.c.
- d.c. to d.c.
- a.c. to a.c. (with an intermediate d.c. link)
- a.c. to a.c. (direct conversion)

Some topologies comprise the series connection of some of these topologies. For example, some a.c. to a.c. topologies have an intermediate d.c. stage or link. A more complete classification of topology types would therefore be as follows:

- a.c. to d.c.
  - uncontrolled (producing a fixed d.c. output)
  - controlled (where d.c. voltage level can be controlled)
- d.c. to d.c.
  - step down (where output is at a lower voltage than the input)
  - step up (where output is at a higher voltage than the input)
- a.c. to a.c. (with an intermediate d.c. link)
  - where the magnitude of the output is controlled by the a.c. to d.c. conversion producing a variable d.c. link level, and the d.c. to a.c. conversion controls the output frequency only
  - where the a.c. to d.c. conversion is uncontrolled, producing a fixed-voltage d.c. link, and both the magnitude and frequency of the a.c. are controlled in the d.c. to a.c. conversion
- a.c. to a.c. (direct conversion)

Some converter topologies allow energy to flow in either direction, while others allow energy flow in one direction only. This can be an important consideration for some drive applications where braking torque is required.

A further form of classification in drive converter topologies is whether the converter is a voltage source or a current source. A *voltage source* imposes a voltage on the load/machine winding, the impedance of which determines the current. Voltage source inverters tend to have low impedance and the output current can change rapidly. A *current source* imposes a current on the load/machine winding, the impedence of which determines the voltage. Current source inverters tend to have inductive circuits which do not allow rapid change in current, even in the presence of rapidly changing load impedance.

Although alternative forms of control and type of switching power device do influence the dynamic and detailed characteristics of the circuits, the key steady-state performance characteristics are determined by the topology itself. In order to concentrate on these salient characteristics of the alternative power conversion circuits, it is convenient to make a number of simplifying assumptions. A somewhat idealised theory will therefore be presented. In the first place, practical aspects, such as switching delays, will only be discussed where they have practical significance. Second, the diversity of machines that can be used with alternative power converters will be limited to those seen as being of greatest practical importance.

A2.2 A.C. to d.c. power conversion

A2.2.1 General

Before considering the relative merits of alternative converters it is necessary to establish meaningful performance parameters. Useful to the user and system designer are the following parameters:

- the rms value of the a.c. current for a constant d.c. current I_d
- the overall a.c. power factor, defined as the ratio of mean power (W) to volt–amperes (VA). [Note that the power factor thus defined equals the product of the displacement factor (the fundamental power factor or $\cos v$, i.e. the phase shift of the fundamental current with respect to the a.c. supply) and the distortion factor (the ratio of the rms of the fundamental current to the rms of the total current).]
- the maximum attainable a.c. power factor that can be achieved using capacitors only to counter the fundamental VAR consumption of the converter
- the a.c. supply current harmonics for a constant d.c. current I_d
- the d.c. voltage as a function of a.c. voltage
- the d.c. power for a constant d.c. current I_d
- the voltage ripple – although form factor or peak–peak values are often used as a measure of voltage ripple. It is more useful in practice to consider a factor M, which is a measure of the volt-second integral of the voltage pulses. The

peak–peak *current* ripple Δi_d is then readily calculated from the following formula:

$$\Delta i_d = \frac{M V_{do}}{f \cdot L}$$

where V_{do} is the maximum attainable d.c. voltage from the converter, f is the a.c. frequency, and L is the d.c. circuit inductance.
- the pulse number p, the number of pulses of d.c. voltage during one complete a.c. cycle (1 cycle = 20 ms for a 50 Hz supply, 16.67 ms for a 60 Hz supply)

A2.2.2 Converters for connection to a single-phase supply

A2.2.2.1 Uncontrolled converters

Only two power circuits need be considered in this category: the half-wave configuration and the full-wave configuration.

Half-wave

This configuration (Figure A2.1) is not particularly useful for power applications, but is included as an introduction to semiconductor behaviour in bridge circuits.

The single device is available for conduction from 0° to 180° (the positive half cycle). A freewheeling diode may be added across the load to conduct the load current during the negative half cycle, and prevent it being reduced to zero. For long time-constant loads, the load current can be considered to be continuous, and is derived from the supply during the positive half cycle and carried by the diode during the negative half cycle.

When this circuit is used, a capacitor often replaces the freewheeling diode, maintaining the output during the idle half cycle. A notable application of this particular circuit is for high-frequency switched mode power supplies (SMPS) secondaries.

The detailed characteristics of this circuit will not be considered further.

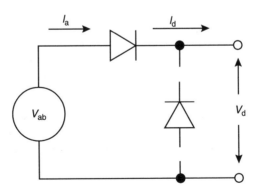

Figure A2.1   Single-phase half-wave uncontrolled bridge

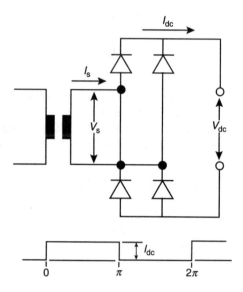

Figure A2.2   Single-phase full-wave uncontrolled bridge

Full-wave

Figure A2.2 shows the most popular method of full-wave rectification from an a.c. supply. The four diodes conduct in diagonal pairs during every alternate half cycle of a.c. line voltage.

A2.2.2.2 Controlled converters

Again only two power circuits are of practical importance: fully controlled and half controlled.

Fully controlled

Figure A2.3 shows the power circuit for the fully-controlled bridge together with associated a.c./d.c. relationships. Figure A2.4 shows how the d.c. voltage can be varied by adjusting the firing delay angle α. Because negative d.c. voltages are possible, energy flow from the d.c. side to the a.c. side is possible.

Half controlled

In this circuit, two of the thyristors shown in Figure A2.3 are replaced by diodes. A number of diode combinations are possible (Ap + Bp), (An + Bn), (Ap + An) or (Bp + Bn). Figure A2.5 shows typical waveforms for the case when Ap and Bp are diodes. Note the absence of any negative voltage, and the reduction in a.c. current periods. Negative voltages are not possible because of the conduction path provided by Ap when An is conducting and Bp when Bn is conducting. It should be noted that only energy flow from the a.c. to d.c. side is possible with this topology, only one polarity of both d.c. voltage and d.c. current being possible.

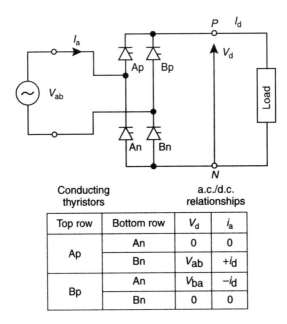

Figure A2.3   Single-phase fully controlled bridge

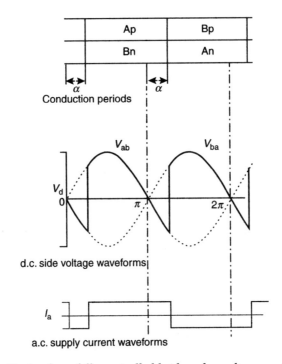

Figure A2.4   Single-phase fully controlled bridge: d.c. voltage control

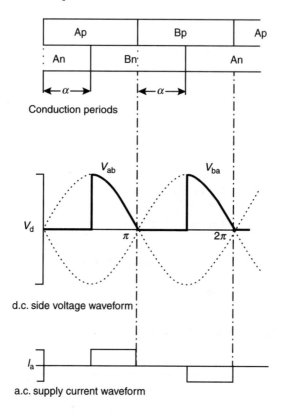

Figure A2.5   Single-phase half-controlled bridge: d.c. voltage control

A2.2.2.3 Sine-wave input converters

In some applications with relatively small power supplies or where sensitive equipment is connected to the same supply as the converter, stringent limits may be imposed on the harmonic currents drawn from the supply/the total harmonic distortion (THD) of the supply. Topologies are available together with custom control integrated circuits that draw near-sinusoidal currents from the a.c. supply at unity power factor. A typical circuit is shown in Figure A2.6.

It is expected that as greater regulation is introduced with respect to the allowable harmonic content of drive systems that this type of circuit will become more common. However, in such situations it is often more cost effective to have a single-supply converter with a common d.c. bus feeding multiple drives. This type of topology is therefore rarely used in industrial drives.

A2.2.2.4 Summary of characteristics

Table A2.1 shows the salient characteristics of the main single-phase a.c. to d.c. converters, assuming constant d.c. current.

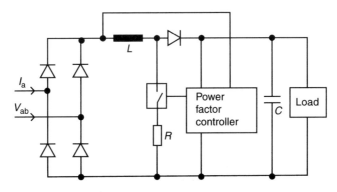

Figure A2.6 Single-phase sine-wave input converter

Table A2.1 Single-phase converter characteristics

Bridge	Fully controlled	Half controlled
Firing angle	α	α
V_{do}	$\dfrac{2\sqrt{2}}{\pi} V_s$	$\dfrac{2\sqrt{2}}{\pi} V_s$
P_{do}	$\dfrac{2\sqrt{2}}{\pi} V_s I_d$	$\dfrac{2\sqrt{2}}{\pi} V_s I_d$
V_d/V_{do}	$\cos \alpha$	$0.5(1+\cos\alpha)$
I_s/I_d	1	$\sqrt{[(\pi-\alpha)/\pi]}$
Overall power factor	$\dfrac{2\sqrt{2}}{\pi} \cos \alpha$	$\sqrt{(2/\pi)} \cdot \dfrac{(1+\cos\alpha)}{\sqrt{(\pi-\alpha)}}$
Maximum corrected power factor	$\dfrac{\cos\alpha}{\sqrt{[(\pi^2/8)-\sin^2\alpha]}}$	$\dfrac{1+\cos\alpha}{\sqrt{[\pi/2\cdot(\pi-\alpha)-\sin^2\alpha]}}$
Input power/P_{do}	$\cos \alpha$	$0.5(1+\cos\alpha)$
Input VARs/P_{do}	$\sin \alpha$	$0.5 \sin \alpha$
Supply current nth harmonic/I_d	0 for n even $0.9/n$ for n odd	0 for n even $0.64/n \cdot \sqrt{(1+\cos n\alpha)}$ for n odd
Phase of supply current harmonics	$n\alpha$	$n\alpha/2$

A2.2.3 Converters for connection to a three-phase supply

A2.2.3.1 Uncontrolled converters

A variety of uncontrolled converters are available; however, only one is of practical importance in drive systems: the full-wave bridge converter. This arrangement is shown in Figure A2.7.

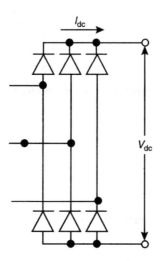

Figure A2.7 Three-phase full-wave uncontrolled bridge

In contrast to the single-phase bridge, where alternate pairs of devices conduct, switching in the three-phase bridge alternates between the upper and lower row of devices. This means that there are six conduction periods per a.c. cycle, each device conducting for a period of $2\pi/3$ (120° elect).

A2.2.3.2 Controlled converters

Two power circuits are of practical importance: full controlled and half controlled.

Fully controlled
This is by far the most important practical bridge arrangement. Figure A2.8 shows the power circuit together with associated a.c./d.c. relationships. Figure A2.9 shows how the d.c. voltage can be varied by adjusting the firing delay angle α. The pulse number p of this bridge equals 6 (there are six pulses of voltage in each complete cycle of the a.c. supply). Energy flow can be from a.c. to d.c. or d.c. to a.c. as the polarity of the d.c. voltage can change.

Half controlled
In this circuit either the top three devices of Figure A2.8 (Ap, Bp and Cp) or the bottom three devices (An, Bn and Cn) are replaced by diodes. The pulse number of this bridge equals 3 (there are three pulses of voltage in each complete cycle of the a.c. supply).

Drive converter circuit topologies 79

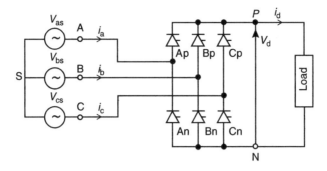

Conducting thyristors		a.c./d.c. relationships			
Top row	Bottom row	V_d	i_a	i_b	i_c
Ap	An	0	0	0	0
	Bn	V_{ab}	$+i_d$	$-i_d$	0
	Cn	V_{ac}	$+i_d$	0	$-i_d$
Bp	An	V_{ba}	$-i_d$	$+i_d$	0
	Bn	0	0	0	0
	Cn	V_{bc}	0	$+i_d$	$-i_d$
Cp	An	V_{ca}	$-i_d$	0	$+i_d$
	Bn	V_{cb}	0	$-i_d$	$+i_d$
	Cn	0	0	0	0

Figure A2.8   Three-phase fully controlled bridge

Only energy flow from a.c. to d.c. is possible as the d.c. voltage is unipolar. The voltage ripple is much greater than in the case of the fully controlled bridge, but the rms a.c. current drawn is lower at reduced d.c. voltage. Figures A2.10 and A2.11 show the half-controlled bridge.

A2.2.3.3 Summary of characteristics

Table A2.2 shows the salient characteristics of the three-phase a.c. to d.c. converters described above, assuming constant d.c. current.

Voltage ripple characteristics
The voltage ripple characteristics for the most significant bridge configurations are shown in Figure A2.12. It should be noted that these characteristics are for idealised conditions of zero supply impedance. Increases in supply impedance generally tend to result in somewhat lower d.c. voltage-ripple levels.

Practical effects
The characteristics presented above have, for the most part, been based upon idealised conditions of negligible a.c. inductance and constant d.c. current. Although these

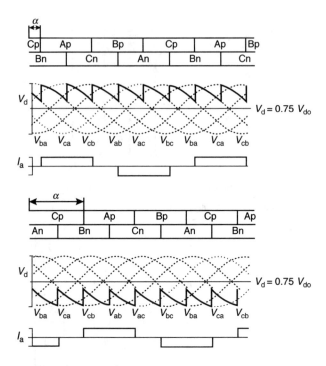

Figure A2.9 Three-phase fully controlled bridge: d.c. voltage control

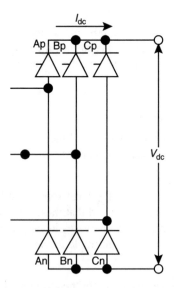

Figure A2.10 Three-phase half-controlled bridge

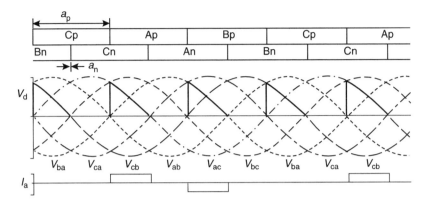

Figure A2.11 Three-phase half-controlled bridge: d.c. voltage control

Table A2.2 Three-phase converter characteristics

Bridge	Fully controlled	Half controlled	
Firing angle	α	α	
		$\alpha > \pi/3$	$\alpha < \pi/3$
V_{do}	$\dfrac{3\sqrt{2}}{\pi} V_s$	$\dfrac{3\sqrt{2}}{\pi} V_s$	
P_{do}	$\dfrac{3\sqrt{2}}{\pi} V_s I_d$	$\dfrac{3\sqrt{2}}{\pi} V_s I_d$	
V_d/V_{do}	$\cos\alpha$	$0.5(1+\cos\alpha)$	
I_s/I_d	$\sqrt{(3/2)}$	$\sqrt{[(\pi-\alpha)/\pi]}$	$\sqrt{(3/2)}$
Overall power factor	$(3/\pi)\cos\alpha$	$\dfrac{\sqrt{3(1+\cos\alpha)}}{\sqrt{(\pi-\alpha)}}$	$(3/2\pi)(1+\cos\alpha)$
Maximum corrected power factor	$\dfrac{\cos\alpha}{\sqrt{[(\pi^2/9)-\sin^2\alpha]}}$	$\dfrac{1+\cos\alpha}{\sqrt{[2\pi(\pi-\alpha)-\sin^2\alpha]}}$	$\dfrac{1+\cos\alpha}{\sqrt{(4\pi^2/9-\sin^2\alpha)}}$
Input power/P_{do}	$\cos\alpha$	$0.5(1+\cos\alpha)$	
Input VARs/P_{do}	$\sin\alpha$	$0.5\sin\alpha$	
Supply current nth harmonic/I_d	0 for $n = 3, 6, 9\ldots$ 0 for n even $\sqrt{6}/n\pi$ for n odd	0 for $n = 3, 6, 9\ldots$ $\sqrt{(3/n\pi)} \cdot \sqrt{(1-\cos n\alpha)}$ for n even $\sqrt{(3/n\pi)} \cdot \sqrt{(1+\cos n\alpha)}$ for n odd	
Phase of supply current harmonics	$n\alpha$	$n\alpha/2$	

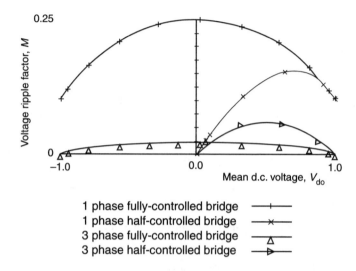

Figure A2.12 *'Ideal' voltage ripple characteristics*

assumptions provide a convenient means for comparision they are not often valid in practice. It is not practicable to consider all such effects here. The effect of d.c. link current ripple on a.c. supply harmonics is of great practical industrial importance mainly in relation to three-phase bridges (ignoring single-phase traction requirement). Practical experience has led to the adoption by many of the following values:

$$I_5 = 0.25I_1 \quad (\text{'Ideal'} = 0.2I_1)$$
$$I_7 = 0.13I_1 \quad (\text{'Ideal'} = 0.14I_1)$$
$$I_{11} = 0.09I_1 \quad (\text{'Ideal'} = 0.11I_1)$$
$$I_{13} = 0.07I_1 \quad (\text{'Ideal'} = 0.08I_1)$$

In general, the amplitudes of higher harmonics are rarely of significance regarding supply distortion. Under conditions of very high d.c. current ripple, the 5th harmonic can assume a considerably higher value than that quoted above. A practical example would be an application with a very capacitive d.c. load (e.g. a voltage source inverter); in such a case where no smoothing choke is used I_5 could be as high as $0.5I_1$.

A2.2.4 Converters for d.c. motor drive systems

In principle little has changed since 1896 when Harry Ward Leonard presented his historic paper 'Volts verses ohms – the speed regulation of electric motors'. In practice, however, many advances have been made from auxiliary machines through to mercury arc rectifiers and to thyristors.

The d.c. motor is still a versatile machine for variable-speed drive systems and is often the preferred choice when considerations such as freedom from maintenance or operation under adverse conditions are not paramount.

In Chapter A1 it has been shown that complete control of a d.c. machine can be achieved by controlling the armature voltage V_a and the field current I_f. Two power converters are employed for this purpose in most variable-speed drives that make use of the separately excited d.c. machine. (In refering to the number of converters in a drive, it is common to ignore the field converter, and this nomenclature will be adopted below.) It is relatively common in simple drives for the field converter to be a single-phase uncontrolled bridge, thereby applying fixed field voltage.

In applications where the variation in motor resistance with temperature results in unacceptable variations in field current, or indeed on sites with poorly regulated supplies, a controlled power converter is used with current control. Such field controllers are further discussed later as applied to field weakening control.

A2.2.4.1 Single-converter drives

Figure A2.13 shows a single-converter d.c. drive. In its most basic form the motor will drive the load in one direction only without braking or reverse running. It is said to be a 'single-quadrant drive', only operating in one quadrant of the torque–speed characteristic. Such drives have wide applications, from simple machine tools to fans, pumps, extruders, agitators, printing machines, and so on.

If the drive is required to operate in both the forward and reverse directions and/or provide regenerative braking a single fully-controlled converter can still be used; however, some means of reversing either the field or armature connections, as shown in Figure A2.14, must be added.

Reversal of armature current can involve bulky (high-current) reversing switches, but due to the low inductance of the armature circuit can be completed typically in 0.2 s. Field current reversal takes longer, typically in the order of 1 s; however, lower-cost reversing switches may be used. The field reversal time can be reduced by using higher-voltage field converters to force the current. Forcing voltages up to 4 per unit are used, but care must be taken not to over-stress the machine. Obviously, this

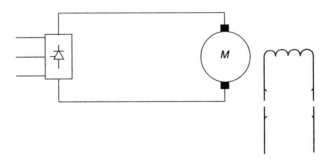

Figure A2.13   *Single-converter d.c. drive*

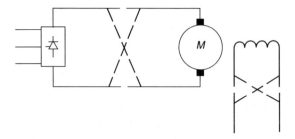

Figure A2.14   Single-converter reversing/regenerative d.c. drive

increased voltage cannot be applied continuously and requires either a switched a.c. supply or a controlled field converter.

Armature and field reversal techniques are used where torque reversals are infrequent, such as in hoists, presses, lathes and centrifuges.

A2.2.4.2   Dual-converter drives

When a four-quadrant drive is required to change the direction of torque rapidly, the delays associated with reversing switches described above may be unacceptable. A dual converter comprising two fully controlled power converters connected in inverse-parallel can be used as shown in Figure A2.15. Bridge 1 conducts when the armature current I_a is required to be positive, bridge 2 when it is required to be negative.

There are two common forms of dual converter. In the first, both bridges are controlled simultaneously to give the same mean output voltage. However, the instantaneous voltages from the rectifying and inverting bridges cannot be identical, and reactors L_p are included to limit the current circulating between them. The principal advantage of this system is that when the motor torque, and hence current, is required to change direction (polarity), there need be no delay between the conduction of one bridge and the other. This is the dual converter bridge *with circulating current*.

In the other, the circulating current-free dual converter only one bridge at a time is allowed to conduct. The cost and losses associated with the L_p reactors can then be

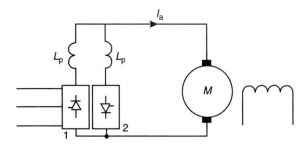

Figure A2.15   Single-phase dual-converter d.c. drive

eliminated, and economies can also be made in the drive control circuits. However, the penalty is a short time delay, as the current passes through zero, while it is ensured that the thyristors in one bridge have safely turned off before those in the second are fired. This delay introduces a 'torque-free' period of typically 10 ms. Speed reversal for a 3 kW drive of this type from $-1\,500$ to $+1\,500$ rev min^{-1} can still be achieved in approximately 200 ms.

This circulating current-free dual converter is by far the most common industrial four-quadrant drive and is used in many demanding applications, including paper, plastics and textile machines where rapid control of tension is required.

A2.2.4.3 Field control

The output power of a motor is the product of torque and speed. If torque reduces in proportion to an increase in speed, the motor is said to have a constant power characteristic.

In applications where material is coiled or uncoiled at constant tension, the torque required to produce that tension varies in proportion to coil diameter, whereas rotational speed required to maintain a constant peripheral speed (and therefore line speed) is inversely proportional to diameter. A motor having a constant power characteristic is well suited to this type of application, the advantage being that a smaller motor can be used than would otherwise be the case. Machine tool drives also make use of constant power operation, because loads are small at high speeds, and heavy work is done at low speed.

As explained in Chapter A1.2.3.1 the torque produced by a d.c. motor is proportional to the product of armature current and field flux. By weakening the field as speed increases, a constant power characteristic can be achieved.

In practice there are two major techniques for field weakening, both of which rely on a field controller, which itself is a simple thyristor converter operating in a current-control mode. In the first method, suitable for coiler and uncoiler applications, the field current reference is arranged to be inversely proportional to coil diameter (measured directly, or calculated from the ratio of line speed to motor speed). Because flux is not strictly proportional to field current, this method does not give a true constant-power characteristic unless compensation for the non-linear part of the motor field curve is applied. The second method is to use the field controller with an outer voltage loop having a fixed reference, and to use motor armature voltage as the feedback signal. At low speeds, the voltage loop saturates, providing maximum field current, because the armature voltage is below the set value. As speed increases, the armature voltage rises to the point where it matches the pre-set reference in the field controller. Above that speed, an error signal is produced by the voltage loop, which causes the field controller to weaken the motor field current and thereby restore armature voltage to the set point level. The resulting characteristics are shown in Figure A2.16.

As regenerative braking depends on the return of power from the motor to the mains, it cannot work if the mains supply fails due to a blown fuse or a power cut. Dynamic braking of 4Q drives is often encountered as a fail-safe means of stopping

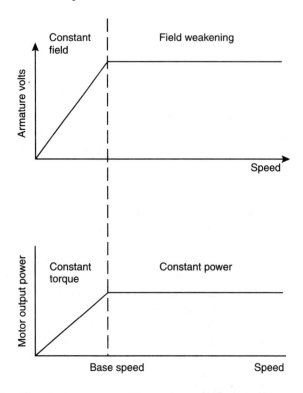

Figure A2.16   Constant power operation using a field controller

the motor and its load, and as the only means of (reverse) braking of single-ended drives. This involves switching in a resistor across the d.c. motor.

Because the kinetic energy of the motor and its load is converted into heat by the braking resistor, it is important to rate it correctly for the duty it is expected to perform, taking account of load inertia and the number of stops per hour.

A2.3   D.C. to d.c. power conversion

A2.3.1   General

D.C.–d.c. power converters, often referred to as 'choppers' in relation to motor drives and switched mode power supplies (SMPS) in relation to general power supplies, provide the means to convert energy from one d.c. voltage level to another. It is more usual for the conversion to be to a lower voltage, although step-up converter topologies are used.

D.C.–d.c. power converters are fed from a d.c. supply usually comprising an uncontrolled a.c. to d.c. converter or alternatively a battery supply; the controlled d.c. output can then be used to control a d.c. machine as in the case of the controlled a.c. to d.c. converters.

D.C. drives using controlled a.c. to d.c. converters have several important limitations, which are overcome by the d.c.–d.c. converter:

- The inability of a thyristor to interrupt current means that an alternating supply is necessary to commutate the converter; this precludes operation from a d.c. supply. This is a common requirement on battery vehicles and d.c. fed rail traction.
- The d.c. ripple frequency is determined by the a.c. supply and is, for a 50 Hz supply frequency, 100 Hz for single-phase and 300 Hz for three-phase fully-controlled bridges. This means that additional smoothing components are often required when using high-speed machines, permanent-magnet motors or other special motors with low armature inductance.
- As a result of the delay inherent in thyristor switching (3.3 ms in a 50 Hz three-phase converter) the current control loop band width of the converter is limited to approximately 100 Hz, which is too low for many servo drive applications.
- Thyristor-controlled a.c.–d.c. converters have an inherently poor input power factor at low output voltages. (Near unity power factor can be achieved using an uncontrolled rectifier feeding a d.c.–d.c. converter.)
- Electronic short-circuit protection is not economically possible with thyristor converters. Protection is normally accomplished by fuses.

D.C.–d.c. converters are, however, more complex and somewhat less efficient than a.c.–d.c. converters. They find application mainly in d.c. servo drives, rail traction drives and small fractional kilowatt drives using permanent magnet motors.

Because step-down converters are of greatest practical importance, emphasis will be placed on their consideration. For the purpose of illustration bipolar transistors will be considered; however, metal oxide semiconductor field effect transistors (MOSFETs) and insulated gate bipolar transistors (IGBTs) and at higher powers gate turn off thyristors (GTOs) and integrated gate commutated thyristors (IGCTs) are widely used.

A2.3.2 Step down d.c. to d.c. converters

A2.3.2.1 Single-quadrant d.c. to d.c. converter

The most basic d.c. to d.c. converter is shown in Figure A2.17. The output voltage is changed by pulse-width modulation (PWM), that is, by varying the time for which the transistor T is turned on. The voltage applied to the motor is therefore in the form of a square wave of varying periodicity. Because the motor is inductive the current waveform is smoothed, the flywheel diode D carrying the current while the transistor is turned off.

The basic formulae relating the variables in this circuit are as follows:

$$V_a = V_{dc} t f$$
$$\Delta I_a = V_{dc}/4L_a f$$

where f is the frequency of the transistor 'on pulse' (Hz), ΔI_a is the maximum deviation of armature current, L is motor inductance, t is on-pulse duration (s) and V_{dc} is source d.c. voltage.

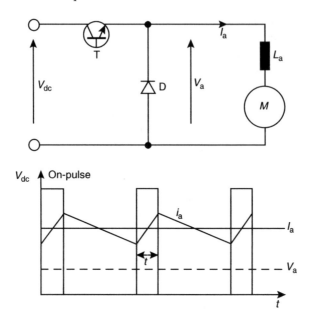

Figure A2.17   Single-quadrant d.c.–d.c. converter

The circuit is only capable of supplying unidirectional current and voltage to the motor and is therefore not capable of four-quadrant operation, that is reversing or regenerating.

Industrial applications for this circuit are normally limited to drives below 5 kW and simple variable-speed applications. In traction applications, however, drives of this fundamental type are designed at ratings of many hundreds of kilowatts.

A2.3.2.2   Two-quadrant d.c. to d.c. converter

In order to achieve full four-quadrant operation a converter must be capable of supplying reversible voltage and current to the motor. A circuit that is capable of two-quadrant operation – that is, motoring and braking in one direction only – is shown in Figure A2.18. This converter is able to reverse the current flow to the motor but unable to reverse the motor terminal voltage and hence the speed. During motoring, the converter operates as the basic 'chopper' with T1 and D2 carrying the current. During braking (or regeneration) T1 is inoperative and T2 controls the current. During its on periods, motor current builds up negatively, limited by motor inductance L. When T2 turns off, the only path for the current is via D1 back into the supply; hence the circuit is regenerative.

Because this circuit is not capable of motor speed reversal it is normally only used in unidirectional applications. However, because of its simplicity it is sometimes used in traction applications where reversing is carried out by means of a changeover switch to reverse the armature or field supply.

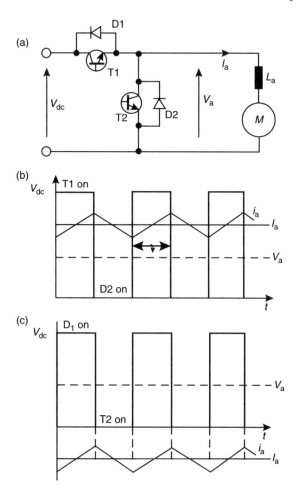

Figure A2.18   Two-quadrant d.c.–d.c. converter

A2.3.2.3   Four-quadrant d.c. to d.c. converter

Figure A2.19 shows a basic four-quadrant converter capable of supplying reversible voltage and current, in other words carrying out reversing and regeneration.

During motoring, positive output transistors T1 and T4 are switched on during the on period, while diodes D2 and D4 conduct during the off-period. When D2 and D4 conduct, the motor supply is reversed and consequently the voltage is reduced to zero at 50 per cent duty cycle. Any reduction of duty cycle below 50 per cent will cause the output voltage to reverse but with current in the same direction; hence the speed is reversed and the drive is regenerating. With transistors T2 and T3 conducting, the current is reversed and hence the full four-quadrant operation is obtained. More advanced switching sequences are adopted in many commercial systems in order to distribute the losses in the power devices more evenly.

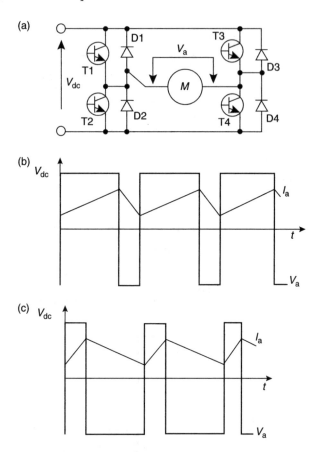

Figure A2.19 Four-quadrant d.c.–d.c. converter

One disadvantage of this converter is that the amplitude of the output ripple voltage is twice that of the simple converter, and the current ripple is therefore worse. This problem can be overcome by a technique known as double-edged modulation. With this technique the flywheel current is circulated via a transistor and a diode during the off period. For example, after T1 and T4 have been conducting, T4 is turned off and T3 on, so that the flywheel current circulates via T1 and D3. The net effect is a reduction in ripple voltage and a doubling of ripple frequency.

These four-quadrant converters are widely used in high-performance d.c. drives such as servos.

A2.3.3 Step-up d.c. to d.c. converters

As with step-down converters many alternative configurations exist for step-up converters. Although a full description is not necessary, the principle is of value.

Figure A2.20 shows a much-simplified arrangement of a step-up converter. When T is turned on, current builds up in inductor L. When T is turned off, the energy stored in L is transferred to capacitor C via D. When the capacitor voltage, which is the same as

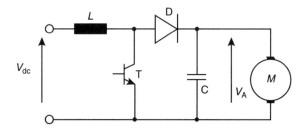

Figure A2.20   Step-up d.c.–d.c. converter

the motor armature voltage reaches the desired level, T is turned on once more. C cannot discharge via T as diode D is reverse biased.

In this way a stabilised voltage typically twice the input d.c. voltage can be obtained. This circuit is particularly useful when operating on low-voltage supplies and can lead to very cost-effective converter designs.

A2.4   A.C. to a.c. power converters with intermediate d.c. link

A2.4.1   General

This category of a.c. drive topology is by far the most widespread of industrial a.c. drives in use today. It is being considered here as a complete converter; however, the input or a.c. to d.c. stage has been considered earlier in isolation, and their individual characteristics so described are, of course, applicable. Alternative input stages to some of the drives are applicable.

Also, some converters may be used with a variety of machine types. Only practically important combinations are described.

The concept of these 'inverter' drives is straight forward: *rectification of fixed frequency, smoothing and then inverting to give variable frequency to feed an a.c. machine*. Within this broad concept two major categories of drive exist. First, the voltage source, in which the converter impresses a voltage on the machine, and the machine impedance determine the current. Second, the current source, in which the converter impresses a current on the machine, and the machine impedance determine the voltage. Each of these classes of converter have their own characteristics and will be considered separately.

A2.4.2   Voltage source inverters

A2.4.2.1   General characteristics

The fixed-frequency mains supply is a voltage source behind an impedance. Voltage source inverters can be similarly considered, and consequently are very flexible in their application. Major inherent features include the following:

- Multi-motor loads can be applied; this can be very economical in applications such as roller table drives, spinning machines, and so on.

- Inverter operation is not dependent upon the machine. Indeed, various machines (induction, synchronous or even reluctance) can be used provided the current drawn is within the current rating of the inverter. Care should be taken where a low-power-factor motor is used (e.g. reluctance) to ensure the inverter can provide the required VARs.
- Inherent open-circuit protection exists, which is very useful in applications where the cables between the inverter and motor are in some way insecure (e.g. fed via slip-rings, subject to damage).
- A facility to 'ride-through' mains dips can easily be provided by buffering the d.c. voltage link with capacitance or, where necessary, a battery.
- Motoring operation only, in both directions, is possible without the addition of resistive dumps for braking energy or expensive regenerative converters to feed energy back to the supply.

Design of the d.c. link

Conventionally, the input stage (a.c. to d.c.) and the output stage (d.c. to a.c.) are de-coupled by a d.c. link filter comprising an inductor and capacitor. The capacitor serves to smooth the six-pulse d.c. link voltage characteristic of the three-phase rectifier.

If a capacitor alone is used in the d.c. link, the current drawn from the supply will have very high current peaks where the supply voltage is at a higher potential than the d.c. link voltage (see Figure A2.21).

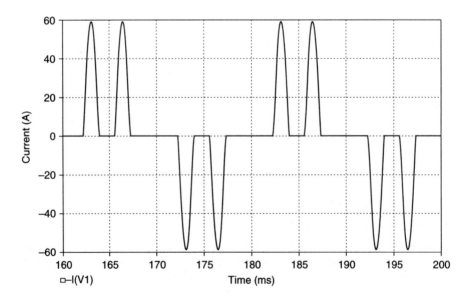

Figure A2.21   Supply current waveform with capacitance only in the d.c. link

The current flows in very short pulses at the peaks of the line-to-line voltage. The input current is very rich in harmonics. In the theoretical limit, all of the harmonics are 100 per cent of the fundamental and the THD is 400 per cent. In practice the supply impedance reduces these figures considerably. Because of the poor current waveform this arrangement is only used in small drives, e.g. below 4 kW.

A d.c. inductor can be added to the circuit to attenuate the current peaks. The inductor, which must carry the full rated d.c. current of the drive, is selected to improve the d.c. bus waveform and the a.c. input waveform. In this location the d.c. inductor will reduce the amount of a.c. ripple current on the d.c. bus, reduce the a.c. input line harmonics and offer protection against nuisance tripping due to voltage spikes such as those caused by capacitor switching. If a very high value of d.c. link inductance is chosen, the resultant supply current would be as shown in Figure A2.22.

This approaches the standard textbook idealised case, where the current has a quasi-square waveform. The harmonic magnitudes follow the simple $1/h$ rule; for example, the 5th harmonic is 20 per cent, and the THD is 31 per cent. In practice a lower value of inductance is used in industrial drives.

D.C. link inductors can be used individually, typically on the positive d.c. bus, or in pairs with one each on both the positive and negative bus. When two d.c. inductors are used on the bus, the inductance is additive.

An a.c. inductor can be used to perform the same function, but the voltage drop across these inductors reduces the d.c. link voltage and thereby the maximum available motor voltage. For best performance, combine the use of both an a.c. input inductor and a d.c. link inductor. This is discussed further in relation to harmonic currents in Chapter B1.1.

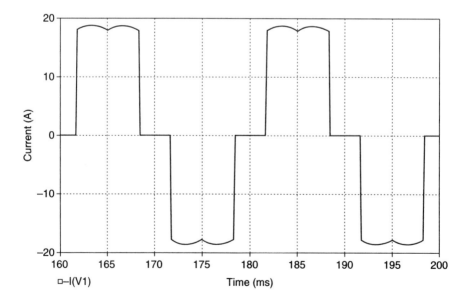

Figure A2.22 Supply current waveform with a high d.c. link inductance

A2.4.2.2 Six-step/quasi-square-wave inverter

A typical d.c. link square-wave voltage-fed inverter drive is shown in Figure A2.23. The three-phase a.c. supply is converted to d.c. in the phase-controlled rectifier stage. The rectified d.c. power is then filtered and fed to the inverter.

Note that the d.c. link reactor is small compared to those used in current source designs. Indeed in drives up to about 4 kW it is not practically necessary. Some manufacturers omit the reactor in designs to 400 kW and even above; however, this has a significant effect upon supply harmonics and unduly stresses the rectifier and filter capacitor.

The inverter switching elements shown as transistors TR1 to TR6 are gated at 60° intervals in the sequence in which they are numbered in the diagram, and each transistor conducts for 180°. The feedback diodes D1 to D6 are connected in inverse-parallel with the transistors, and permit the return of energy from reactive or regenerative loads through the inverter to the d.c. link.

For a star-connected motor, the synthesis of inverter output voltage waveforms is shown in Figure A2.24. The phase-to-neutral voltage of the inverter has a six-step waveshape and the corresponding phase-to-phase voltage has a 120° conduction angle. The output frequency is controlled by the rate at which the inverter transistors are triggered into conduction by the inverter control circuitry. Reversing the firing sequence of transistors in the inverter changes the direction of rotation of the output three-phase waveform and thereby the motor, and no switching of power leads, to the motor itself, is necessary.

The phase-controlled rectifier regulates the d.c. link voltage and this, in turn, determines the magnitude of the output voltage from the inverter. Thus, the output voltage/frequency relationship may be controlled to regulate the steady-state motor flux in the desired manner.

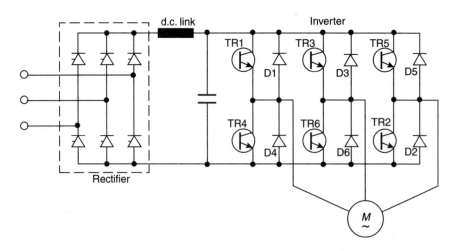

Figure A2.23   Square-wave voltage-fed inverter

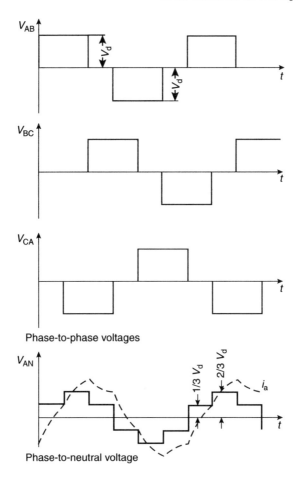

Figure A2.24 Square-wave voltage-fed inverter: output voltage and current

The advantages of the square-wave inverter are high efficiency (as switching losses in the d.c. to a.c. converter are minimised), suitability for standard motors, potential good reliability and high-speed capability. However, it suffers from low-speed torque pulsations and possible low-speed instability.

In a square-wave inverter, each harmonic voltage amplitude is inversely proportional to the harmonic order and hence there are no pronounced high-order harmonics. These are filtered by the motor leakage inductances.

Very high-speed motor operation is possible by increasing the output frequency. Faster switching devices such as MOS transistors and insulated gate bipolar transistors (IGBT) can be used to achieve this performance.

It is known that the square-wave inverter gives objectionable torque pulsations at low-frequency operation, below approximately 5 Hz. This pulsating torque is due to the low-order harmonics causing a stepping or cogging motion to the rotor at low speed. Hence, the pulsating torque limits the low-frequency operation of the square-wave inverter. Appropriate feedback control techniques or flux weakening can attenuate the low-speed pulsating torque problems.

The existence of a phase-controlled rectifier to control the voltage of the inverter as illustrated in Figure A2.23 is an inherent weakness of this circuit. The phase-controlled rectifier will present a low power factor to the a.c. supply, at low speeds, and the d.c. link filter capacitor needs to be large. This reduces the response time of the system to voltage and hence speed changes. If the drive system is one for which regenerative braking operation is a requirement, the rectifier has to be of inverse-parallel type (see A2.2.4.2). The input power factor and response time of the drive can be improved by replacing the phase-controlled rectifier with a diode rectifier feeding a d.c. chopper, which regulates the input voltage to the inverter. For recovering regenerative energy of the load, a two-quadrant chopper will be necessary. The alternative supply converter arrangement of a diode bridge plus chopper also provides a fixed voltage link, which is more economically buffered, if mains dip ride-through is required.

The voltage-fed square-wave drive is usually used in low-power industrial applications where the speed range is limited to 10:1 and dynamic performance is not important. Recently, this type of drive has largely been superseded by PWM-type voltage-fed inverters. Nevertheless, the voltage-fed square-wave inverter can be easily adapted to multi-motor drives where the speed of a number of induction motors can be closely tracked. It is also used in some high-frequency (>1 kHz) applications.

A2.4.2.3 Pulse-width modulated inverter

In the PWM inverter drive, the d.c. link voltage is uncontrolled and derived from a simple diode bridge. The output voltage can be controlled electronically within the inverter by using PWM techniques. In this method, the transistors are switched on and off many times within a half cycle to generate a variable-voltage output, which is normally low in harmonic content. A PWM waveform is illustrated in Figure A2.25.

A large number of PWM techniques exist, each having different performance, notably in respect to the stability and audible noise of the driven motor.

Using the PWM technique, low-speed torque pulsations are virtually eliminated because negligible low-order harmonics are present. Hence, this is an ideal solution where a drive system is to be used across a wide speed range.

Because voltage and frequency are both controlled with the PWM, quick response to changes in demand voltage and frequency can be achieved. Furthermore, with a diode rectifier as the input circuit, a high power factor (approaching unity) is offered to the incoming a.c. supply over the entire speed and load range.

MOSFETs are used in PWM inverters of small ratings, but the bulk of industrial drives are today IGBT-based. MOSFET inverters tend to operate with very high switching frequencies of 16–20 kHz, which is outside the audible range. IGBT-based inverters operate at switching frequencies typically at 2–16 kHz.

PWM inverter drive efficiency is typically around 98 per cent, but this figure is heavily affected by the choice of switching frequency – the higher the switching frequency, the higher the losses in the drive. In practice the maximum fundamental output

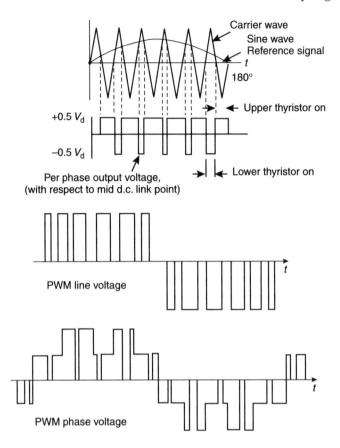

Figure A2.25   Sinusoidal PWM line and phase voltages

frequency is usually restricted to 100 Hz in the case of GTOs or about 1 kHz for a transistor-based system. The upper frequency limit may be improved by making a transition to a less sophisticated PWM waveform with a lower switching frequency and ultimately to a square wave if the application demands it. However, with the introduction of faster-switching power semiconductors, these restrictions to switching frequency and minimum pulse-width have been eased.

In general, a motor with a large leakage reactance is desirable to limit the flow of harmonic currents and thereby minimise losses.

A2.4.2.4   Multi-level inverter

The PWM inverter topology described above is widely used in drive systems rated to 2 MW and above at voltages up to 690 V. At higher powers switching the current in the devices proves more problematic in terms of the losses, and so switching frequencies have to be reduced. At higher voltages the impact of the rate of change of voltage on the motor insulation causes switching times to be extended and hence losses

98 *The Control Techniques Drives and Controls Handbook*

increased. Furthermore, although higher-voltage power semiconductors have been developed, they tend to be relatively expensive and so consideration is given to the series connection of devices. Voltage sharing between series-connected devices is a problem due to the huge impact of any small difference in the switching characteristic of the devices.

Multi-level converters have been developed that address these issues. A typical topology is shown in Figure A2.26. In this example of the multi-level topology each arm of the bridge has four series-connected IGBTs with anti-parallel diodes. In addition, diodes are placed from intermediate d.c. link voltage levels to the connections between series IGBTs. By careful consideration of the switching/modulation of the IGBTs it is possible to produce an output waveform of the type shown in Figure A2.27.

Multi-level inverters have advantages over the conventional two-level PWM inverter described in Section A2.4.2.3:

- They have higher effective output switching frequency for a given PWM frequency. Smaller filter components are required, with reduced size and cost.
- Electromagnetic compatibility (EMC) is improved due to a lower dV/dt at output terminals.
- Higher d.c. link voltages are achievable for medium voltage applications due to voltage sharing of power devices within each inverter leg.
- Lower voltage rated power devices can be used, which significantly reduces the cost per device, and in MV applications the cost/kVA.

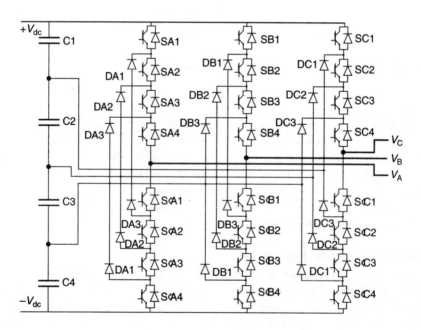

Figure A2.26   Multi-level inverter (d.c. to a.c. stage only)

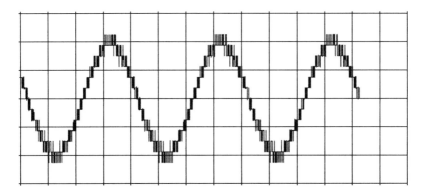

Figure A2.27   Multi-level inverter output voltage waveform

However, there are drawbacks:

- The number of power devices is increased by at least a factor of two, with each IGBT requiring a gate drive and power supply to the gate drive.
- Additional voltage clamping diodes are required.
- The number of d.c. bus capacitors may increase. This is unlikely to be a practical problem as lower-voltage capacitors in series are likely to be the most cost-effective solution.
- At five levels and above, minimisation of inductance between d.c. capacitors and the power switches is difficult and expensive to achieve. Complicated laminated bus bar arrangements are necessary. More devices require snubbers.
- The production of balanced and controlled d.c. voltages is difficult and/or expensive.
- Modulation schemes are more complicated.

For low-voltage drives (690 V and below) the disadvantages of using a multi-level inverter far outweigh the advantages, with even the simplest three-level scheme being prohibitively expensive to implement.

A2.4.3   Current source inverters

A2.4.3.1   General characteristics

Whereas each voltage-fed inverter can be used with most forms of a.c. machine, a different design of current fed inverter is usually adopted for synchronous and induction motors. Current source drives are usually, but not always, single-motor systems, and because current is controlled, there is simple short-circuit protection.

In contrast to voltage source inverters, full four-quadrant operation is inherently possible.

A2.4.3.2 Converter-fed synchronous machine (LCI)

Once rotating, a synchronous machine generates a.c. voltages, which can be used for the natural commutation of a converter connected to its terminals. Indeed the connected synchronous machine behaves as the mains in respect to the a.c. to d.c. converters described earlier. This is why it is frequently refered to as a load commutated inverter (LCI).

Figure A2.28 shows the basic components of the drive system. A low-impedance or 'stiff' d.c. current source is required and is obtained from a controlled rectifier and a series reactor. With a stiff current source, the output current wave is not greatly affected by the magnitude of the load.

The synchronous machine can be approximately represented by a counter-emf in series with an equivalent leakage inductance. The d.c. current is switched through the inverter thyristors so as to establish three-phase, six-stepped symmetrical line current waves. Each thyristor conducts for 120° and at any instant one upper thyristor and one lower thyristor remain in conduction.

It is necessary to maintain an approximately constant angular relationship between the rotor and stator mmfs and hence automatically maintain the correct inverter frequency. This is an important point. The inverter does not impose a frequency upon the machine; rather the machine itself determines the frequency. The motor cannot therefore pole-slip. The drive is accelerated by increasing the current fed to the motor, which then accelerates and thereby increases the frequency.

As in the d.c. drives, the a.c. supply power factor is poor at low speeds. Full four-quadrant operation is possible without additional components.

Special procedures are necessary for starting these drives because at standstill the machine voltage is not available to commutate the current. In essence this is usually achieved by momentarily switching off the d.c. link current every sixth of a cycle. This allows the thyristors in the inverter to turn off so that the next pair can be fired. Above approximately 5 per cent of rated speed the machine generates sufficient voltage for natural commutation and control is undertaken in a similar manner to that of a d.c. drive.

Relatively conventional synchronous motors/alternators can be used, with a requirement for low sub-transient reactance (typically <0.1 pu) meaning that damper bars are not fit, to facilitate commutation in a reasonable time, which impacts overall

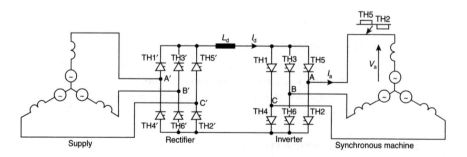

Figure A2.28 Converter-fed synchronous machine

efficiency. A simple position sensor is needed to ensure that current is injected into the right motor phase windings during the starting process.

Applications for this type of drive fall into two main categories: (1) starting converters for large synchronous machines, the converter being rated only for a fraction of the machine rating; and (2) large high-power (and sometimes high-speed) variable-speed drives for a variety of applications. Power ratings, typically from 2 to 100 MW at speeds up to 8 000 rpm are available, although at powers up to 5 MW the voltage source inverters are now proving to be more popular. Also of importance is the fact that high-voltage drives are offered with supply voltages up to 12 kV typical, but systems over 25 kV are in service where the high-voltage converter technology is similar to that used for HVDC power converters.

A2.4.3.3 Converter-fed induction motor drive

Unlike the synchronous machine, the induction motor is unable to provide the VARs, or terminal voltage to commutate a converter connected to its terminals. Commercial schemes are available, however, that are closely based upon the converter-fed synchronous machine drive with additional components to provide VAR compensation.

Figure A2.29 shows a basic power circuit. The diagram somewhat belies the potential complexity of the VAR compensator. In its simplest form this could comprise capacitors plus appropriate switches. Control of such a system is somewhat involved. It is often better to use a cycloconverter or even an auxiliary synchronous machine to provide the commutation, and motor VARs.

This system is rarely used and only appropriate for high-power, high-voltage drives, generally above 4 MW, where an induction motor is preferred.

A2.4.3.4 Forced commutated induction motor drive

This topology was strongly favoured for single-motor applications for a long period, and was available at power levels in the range 50–3 500 kW at voltages up to normally

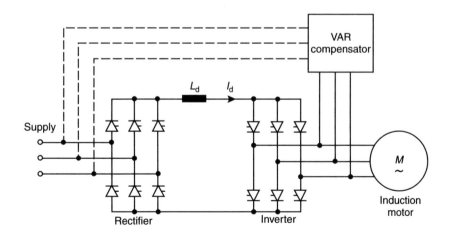

Figure A2.29   Converter-fed induction motor

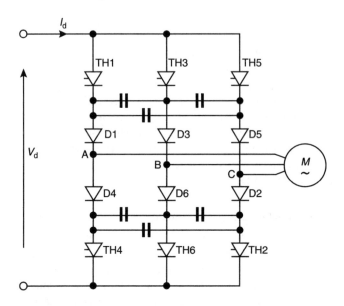

Figure A2.30 Forced commutated induction motor

690 V. High-voltage versions of 3.3/6.6 kV have also been developed, but they have not proved to be economically attractive. Today this design is not seen as having merit and has disappeared from most companies' product portfolios. A brief description is included here for interest only.

Figure A2.30 shows the inverter and motor of the drive. The d.c. link current I_d, taken from a stiff current source, is sequentially switched at the required frequency into the stator windings of the induction motor. The motor voltage waveform is approximately sinusoidal, apart from the superposition of voltage spikes caused by the rise and fall of machine current at each commutation. Further distortion is caused by the effects of slot ripple and d.c. current ripple.

The operating frequency range is typically 5–60 Hz, the upper limit being set by the relatively slow commutation process. Special motors with low leakage inductance do offer advantages with this converter and allow reduced capacitance in the inverter and/or higher operating frequency. Below 5 Hz, torque pulsations can be problematic but PWM of the *current* can be used at low frequencies to ease the problem.

This system is most commonly used for single-motor applications such as fans, pumps, extruders compressors, where very good dynamic performance is not necessary and a supply power factor, which decreases with speed, is acceptable.

A2.4.3.5 Static Kramer drive

The static Kramer drive is shown in Figure A2.31. The drive comprises a slip ring (wound rotor) induction motor together with an uncontrolled converter, d.c. smoothing reactor and a fully controlled converter in the rotor circuit.

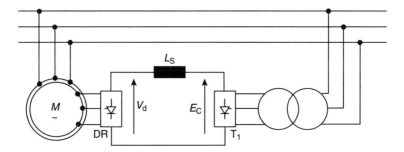

Figure A2.31   Static Kramer drive

The diode bridge gives an output voltage V_d that is proportional to the slip of the motor. V_d is opposed by the d.c. voltage of the fully controlled bridge, a small potential difference being sufficient to circulate current corresponding to the required load torque. Ideally, neglecting losses, the fully controlled bridge d.c. voltage sets the speed to which the motor will accelerate. Control is therefore very similar to a d.c. drive.

Power can flow in only one direction via the diode bridge, which means that motoring torque can be developed only at sub-synchronous speeds. For reverse running it is necessary to reverse the phase sequence of the stator supply.

This drive can be very economical when designed for operation over a limited speed range below synchronous speed (this is the useful operating region for fans, pumps, and so on). The converter bridges required for operation at such a limited speed range need only be rated at a fraction of that of the machine it is controlling. It is necessary in such designs to provide a starter, usually a resistance to run the motor up to the lowest controllable speed. This means that should there be a fault with the converter equipment, the system can be easily designed to run at full speed without the controller.

Note that the supply harmonic currents and VARs associated with the converter part of the drive may be substantially reduced by adopting a limited speed range solution.

The static Kramer drive finds application mainly at ratings between 1 and 20 MW, with slip-ring induction motors with four or more poles. (Stability problems exist with two-pole motors, which can only be resolved with care.) Speed ranges of 30 per cent are typical (i.e. 70–100 per cent rated speed). The induction motor stator can be wound for any conventional voltage, e.g. 6.6 kV, 11 kV.

A2.5   Direct a.c. to a.c. power converters

A2.5.1   General

This final category of power converters converts the fixed-frequency, fixed-voltage a.c. supply to variable frequency and/or variable voltage without an intermediate d.c. link.

A2.5.2   Soft starter/voltage regulator

Soft starters are of importance in the motor control arena as there are many instances where a motor is not required to operate at variable speed or in precise control, but

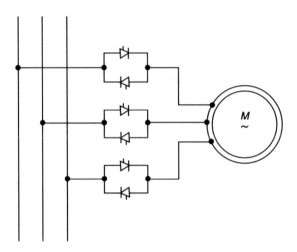

Figure A2.32   Typical soft start

direct on line (DOL) starting is not possible because of either the high DOL currents drawn from the supply or the mechanical stress on the load resulting from the starting torque.

The fundamental mode of control for a basic soft starter is control of the magnitude only of the voltage applied to the motor. The applied frequency is not controlled. The voltage is ramped up either on a simple timed ramp or with a current control loop, ensuring that a prescribed current level, pattern or envelope is followed.

Figure A2.32 shows a typical soft start comprising inverse-parallel-connected thyristors in each supply line to an induction motor. Alternative connections are available but the principles are similar.

Because the stator frequency is unchanged, a reduced running voltage, and hence flux, equates to a large slip, which results in additional rotor losses. Care must therefore be taken in its application.

In a number of specialised cases, purpose-designed, high-resistance rotors (or slip-ring motors with external rotor resistors) are used to form a variable-speed drive. The rational for such a system is based more on history than technology.

More recently such converters have been used as combined soft starters/power-factor controllers/energy-saving devices. The case for significant energy saving using this form of converter is, however, often hard to prove.

A crude form of frequency control is possible by modulating (varying cyclically) the thyristor firing angles at the required output frequency. Although commercial systems are available, they are of limited value because supply current and motor torque are of poor quality.

A2.5.3   Cycloconverter

A typical scheme for a cycloconverter drive is shown in Figure A2.33. Each motor phase is supplied, in effect, from a dual a.c. to d.c. converter as described earlier. It

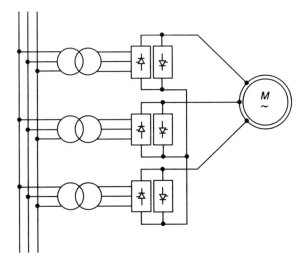

Figure A2.33   Cycloconverter

is usual to use circulating current-free converters. To avoid line-to-line short circuits, isolating transformers are used on the supply side. By modulating the firing angles of the dual bridge converters, a controllable three-phase set of voltages can be produced comprising sections of the supply voltage. One phase voltage and the resulting phase current is shown in Figure A2.34.

The drive is inherently four quadrant. The maximum output frequency is limited to approximately half the supply frequency by considerations related to harmonics in the motor currents and torque, stability and dimensions of the drive components. The cycloconverter therefore finds application in low-speed drives. The complexity of the drive also means that only high-power systems (>1 MW) or specialised applications (e.g. conveyor drives for use in hazardous environments) are of economic value.

Such drives are used on large ball mills, minewinders, and so on. As a voltage source converter, they are also capable of multi-motor loads.

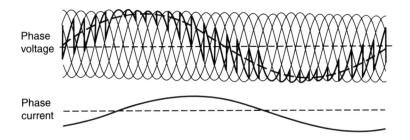

Figure A2.34   Cycloconverter phase voltage and current

Owing to the modulation of the converter firing angles, the harmonic content of the a.c. supply is complex, and designs for appropriate harmonic filters somewhat involved.

The cycloconverter is suitable for feeding both induction and synchronous machines. In specialised applications such as wind generators, cycloconverters have been placed in the rotor circuit of a slip-ring induction motor. Such a system, known as a static Scherbius drive is described in Section A2.5.5.

A2.5.4 Matrix converter

Recently, attention has been refocused on the matrix converter shown in Figure A2.35. Although the basic circuit is not new, recent advances in power devices offer the potential to overcome many of the drawbacks inherent in the circuit when the switches comprise inverse parallel thyristors. Limitations in the maximum output voltage (86 per cent input) means that its application in the commercial industrial market could be problematic. There are prospects in relation to integrated motors and some servo systems where machine voltage is not seen as critical.

Commercial systems are available only for very specialised applications at present. It has yet to be proven to be a practical and cost-effective industrial drive.

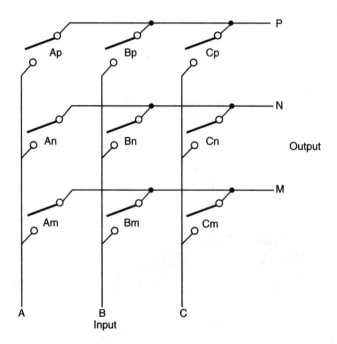

Figure A2.35   Matrix converter

A2.5.5 Static Scherbius drive

The static Scherbius drive is closely related to the static Kramer drive, with the single quadrant diode bridge in the rotor circuit replaced by a cycloconverter.

The cycloconverter is used as the voltage and frequency changer between the rotor and the supply. The cycloconverter is inherently regenerative, and the output can be controlled up to half the supply frequency in both phase sequences. It is thus possible for the system to operate as a full four-quadrant drive. For a given converter rating the range of speed control is therefore twice that of a static Kramer drive.

The relative complexity of the drive limits its application to somewhat specialised high-power applications where a very limited speed range only is required and perhaps stringent harmonic current limits have been imposed by the supply authority.

Chapter A3
Power semiconductor devices

A3.1 General

All a.c. and d.c. drives use power semiconductor devices to convert and control electrical power. This section reviews important characteristics of the most conventional power devices in drives applications.

It is common to operate semiconductor devices in switched mode operation. This mode of operation implies that the device is either fully on or fully off, and power dissipation (product of I and V in Figure A3.1) is therefore low compared to that encountered in the linear mode of operation. It is this feature that makes switched mode operation the key to achieving high efficiency.

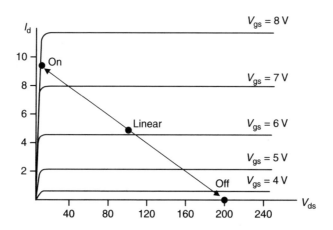

Figure A3.1   Linear versus switched mode operation

The practically important power semiconductor devices in relation to motor drives are as follows:

- the diode,
- the thyristor (also called the silicon controlled rectifier, SCR),
- the triode thyristor (Triac),
- the gate turn-off thyristor (GTO),

Table A3.1 Typical power device characteristics

Property	Thyristor/SCR	Triac	GTO	IGCT	MOSFET	IGBT
Self-commutation ability	No	No	Yes	Yes	Yes	Yes
Maximum rms current rating (A)	5 000	400	2 000	1 700	300	2 400
Maximum voltage rating (V)	12 000	1 200	6 000	5 500	1 500	6 500
Maximum switched VA rating	30 MVA	240 kVA	30 MVA	12 MVA	30 kVA	4 MVA
Surge current ability	Excellent ($15 \times I_{rms}$)	Good ($10 \times I_{rms}$)	Excellent ($15 \times I_{rms}$)	Excellent ($15 \times I_{rms}$)	Limited ($4 \times I_{rms}$)	Limited ($2 \times I_{rms}$)
Operating current density at rated device voltage	140 A cm^{-2} at 2 kV	85 A cm^{-2} at 1.2 kV	30 A cm^{-2} at 4.5 kV	30 A cm^{-2} at 4.5 kV	75 A cm^{-2} at 200 V 15 A cm^{-2} at 800 V	140 A cm^{-2} at 1 200 V 35 A cm^{-2} at 3.3 kV
Maximum junction temperature (°C)	125 °C	125 °C	125 °C	115 °C	175 °C	175 °C
On-state losses	Low	Low	Medium	Low	High	Medium
Switching losses	Very high	High	Very high	Medium	Very low	Low
Turn-on ability	Medium (di/dt limit)	Medium (di/dt limit)	Medium (di/dt limit)	Medium (di/dt limit)	Very good	Very good

Power semiconductor devices 111

Turn-off ability	None via gate	Medium	Poor – slow and lossy	Good	Very good	Very good
Turn-off safe operating area (percentage of rated voltage at rated rms current)	NA	Medium (60%)	Poor (50%)	Medium (70%)	Excellent (100%)	Excellent (100%)
Load short-circuit turn-off ability	None	None	Poor ($2 \times I_{rms}$)	Poor ($2 \times I_{rms}$)	Medium ($4 \times I_{rms}$)	Excellent ($10 \times I_{rms}$)
Snubbers usually required	Yes	Yes	Yes	No	No	No
Minimum on or off time	10–100 μs	10–50 μs	10–50 μs	10 μs	<100 ns	<1 μs
Maximum switching frequency	250	250	500	500	100 000	10 000
Switching time controllable from drive circuit	No	No	No	No	Yes	Yes
Drive circuit power	Low	Medium	High	High	Low	Low
Drive circuit complexity	Low	Medium	High	High	Low	Low
Series and parallel operation	Device selection and passive components required	Device selection and passive components required	Very difficult series or parallel	Fairly simple in series, more difficult in parallel	Fairly simple series and parallel	Fairly simple series and parallel, selection may be needed for parallel

Table A3.2 Application of power semiconductor devices

Application	Supply voltage and equipment VA rating		
	Up to 240 V a.c., 400 V d.c. Up to 1 kVA	From 240 V a.c., 400 V d.c. up to 690 V a.c., 1 200 V d.c. From 1 kVA up to 1 MVA	Above 690 V a.c., 1 200 V d.c. Above 1 MVA
A.C. motor drives			
Voltage source inverter	MOSFET, IGBT	IGBT, BJT	IGBT, IGCT, GTO
Current source inverter		Thyristor/SCR, BJT, GTO	Thyristor/SCR, GTO, IGCT
Cycloconverter			Thyristor/SCR
Soft-starters	Thyristor/SCR, Triac	Thyristor/SCR	Thyristor/SCR
D.C. motor drives			
Line commutated	Thyristor/SCR, Triac	Thyristor/SCR	Thyristor/SCR
Force commutated	MOSFET, IGBT, BJT	IGBT, BJT	IGBT, IGCT, GTO

Power semiconductor devices 113

- the integrated gate commutated thyristor (IGCT),
- the metal-oxide semiconductor field-effect transistor (MOSFET),
- the insulated gate bipolar transistor (IGBT), and
- the bipolar junction transistor (BJT) (although this device has largely been superseded by the other devices shown in the list).

Table A3.1 shows typical power device characteristics. Clearly there is no outstanding device that has high voltage or current ratings, great VA ratings, easy controllability and simple driver circuit complexity. For example the MOSFET is the fastest switching device but is limited in achieving high blocking voltage. The unique characteristic of each power device makes them suitable for some drive applications but not all. Table A3.2 provides an overview of typical applications of power semiconductor devices.

All power devices come in different packages allowing different mounting methods. The type of packaging has an impact on the performance of the device. Table A3.3 summarises power device performances for different packages.

Table A3.3   Availability of power device packages

Property	Discrete	Power module	Pressure pack
Available devices	Thyristor/SCR, GTO, BJT, MOSFET, IGBT	Thyristor/SCR, BJT, MOSFET, IGBT	Thyristor/SCR, GTO, BJT, IGBT, IGCT
Maximum voltage rating (V)	2 000	6 500	12 000
Maximum current rating (A)	100	2 500	6 000
Usual electrical failure mode	Open circuit	Open circuit	Short circuit
Power circuit connections	Solder	Solder, screw, or pressure contact	Pressure contact
Control circuit connections	Solder	Solder, screw, or pressure contact	Flying leads
Mounting method	Solder, screw or clip	Screw	Pressure plate
Cooling method	Convection to air, conduction to PCB or heatsink	Single-sided conduction to heatsink	Double-sided conduction to heatsink
Isolation from heatsink	Only with selected packages	Yes	No
Package rupture current	Low	Medium	High

Power switching devices require electronic 'gate drive' circuits for turning the device on and off. These, called 'driver circuits', are in general complex and also include protection features such as over-current protection. For this reason, details of these circuits have for the most been part limited to a description of the requirements to gate the devices.

A3.2 Diode

A3.2.1 PN diode

The diode is the simplest power device. A diode may be considered as an electronic switch with a conduction state depending on the polarity of an externally applied voltage. When a sufficiently high positive voltage is applied to the anode with respect to the cathode, current will flow in a forward direction, the device acting as a closed switch. During conduction a voltage drop of the order of 0.7–1 V applies across the device. Conversely, when a negative voltage is applied, current flow is prevented and the diode is able to block voltages up to a certain level.

The principles of operation of a diode can be obtained by studying the PN diode. A junction is formed at the interface between two dissimilar semiconductor types: one side is p-type (doped with acceptors – usually group III elements) and other side is n-type (doped with donors – usually group V elements). In silicon, at room temperature, the acceptors and donors are ionised, giving rise to positively and negatively charged mobile carriers (holes and electrons). With no external bias applied across the junction, a redistribution of charge occurs at the interface. The affected region around the junction is known as the depletion region (see Figure A3.2).

In power devices, it is typical to find one side of the junction much more heavily doped than the other (a factor of more than 1 000 is not uncommon). Such structures are referred to as one-sided junctions and the depletion region will extend much further into the lightly doped region.

The voltage–current characteristic of the PN junction diode is shown in Figure A3.3, illustrating the two modes of operation (forward conducting and reverse breakdown).

The ideal diode characteristic obeys an exponential law:

$$I_d = I_{sat}\left[\exp\left(\frac{qV_d}{kT}\right) - 1\right] \quad (A3.1)$$

where I_{sat} is the saturation current, q is the electron charge (1.602×10^{-19} C), V_d the voltage across the diode, k the Boltzmann's constant (1.380×10^{-23} J K^{-1}) and T is the temperature given in kelvin.

At negative voltages I_d becomes $-I_{sat}$, which is known as the saturation current or leakage current. As the applied reverse bias is increased the diode will breakdown. The breakdown is caused either by the breakdown of the silicon because of the high electric field, or the edge of the depletion reaches the edge of the lightly doped region. This latter process is known as punch-through.

Power semiconductor devices 115

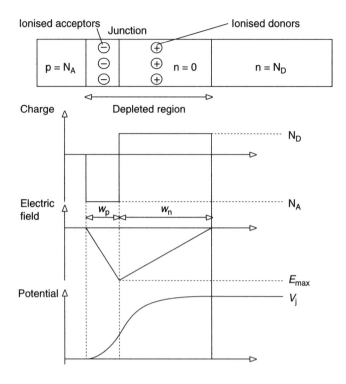

Figure A3.2 The PN junction under reverse bias

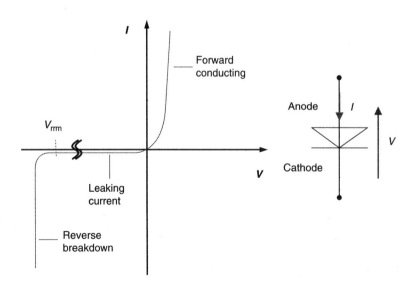

Figure A3.3 Symbol, PN junction and I/V curve of a diode

A PN diode, however, is not able to operate at high reverse blocking voltage. To increase the reverse bias an additional layer is integrated between the p and n types; this is called the intrinsic layer, and will be discussed in the next section.

A3.2.2 PIN diode

This device makes use of a wide, lightly doped (I stands for intrinsic) region sandwiched between heavily doped n-type and p-type regions (hence PIN). Under reverse-bias conditions the lightly doped region punches through at relatively low voltages. Breakdown does not, however, occur, because of the heavily doped buffer layer. Instead a high value of electric field is able to build up across the entire width of the lightly doped region (the slope of the electric field is proportional to the doping density). In the limiting case the theoretical maximum voltage supported by the structure is almost twice that of a standard diode of the same width:

$$V_{br} = E_{crit} w_m \tag{A3.2}$$

where V_{br} is the breakdown voltage, E_{crit} the critical electric field and w_m the length of the intrinsic layer. Figure A3.4 shows the structure of a PIN diode, including the distribution of the electric field.

In a PIN diode the idealised exponential characteristic described earlier is only a good approximation at relatively small levels of forward bias. For high levels of forward bias, additional physical effects in the device serve to increase the on-state voltage for a particular current level. At very high current densities the diode takes on an almost resistive characteristic (Figure A3.5).

The mechanism of conduction in a PIN diode involves the injection of carriers into the lightly doped region from the heavily doped regions; this process is known as conductivity modulation. Figure A3.6 shows representative electron and hole density distributions for the on and off states of a PIN diode. Note that the resulting

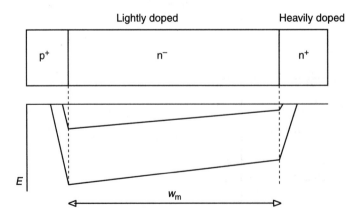

Figure A3.4 Electric field distribution in a PIN diode

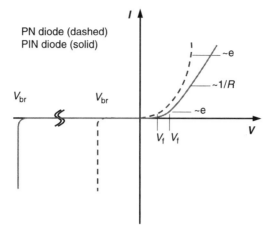

Figure A3.5   Practical diode characteristics

large quantity of carriers that have been injected into the lightly doped region of the device now appear as a stored charge. The electron and hole concentrations are almost equal, and both are several orders of magnitude greater than the background doping level. As a result, the resistivity of the conductivity-modulated region is much lower than that afforded by the background doping level. In operational terms this translates to a much reduced on-state voltage. An approximate relationship between the quantity of stored charge Q_{TOT} and the diode current density J may be derived:

$$Q_{TOT} = J\tau_{HL} \qquad (A3.3)$$

where τ_{HL} is the lifetime of the electrons and holes in the lightly doped region. The quantity of charge clearly increases in proportion to the current.

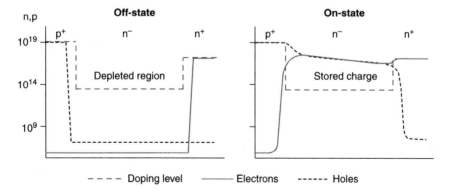

Figure A3.6   Carrier distributions in a PIN diode under reverse and forward bias

From equation (A3.3) one might assume that the quantity of charge would be the same for all diodes at a given level of current density. This is, however, only true if the carrier lifetime is the same for each device. In practice, higher-voltage diodes must use longer carrier lifetimes to limit the on-state voltage drop to a low level. It may be shown that

$$\tau_{HL} > w_m^2/4D_a \tag{A3.4}$$

Here D_a is a material parameter known as the ambipolar diffusion coefficient. The quantity of stored charge is thus higher in high-voltage diodes, which have a wider voltage-blocking layer.

A3.2.3 Transient processes (reverse and forward recovery)

The transient process of a diode can be split into reverse recovery (when the diode turns off) and forward recovery (when the diode turns on). Sections A3.2.3.1 and A3.2.3.2 describe the two processes and show that very rapid changes in diode current during switching transitions can cause significant transient over-voltages and, in the case of reverse recovery, a considerable reverse current. This results in increased switching stresses in devices throughout the circuit, a possible increase in switching losses, and in extreme cases failure of the diode and associated switch. The faster the imposed transitions the worse the situation becomes. All transitions should therefore be controlled, either by judicious choice of switch drive conditions or, where this is impracticable, by application of snubber circuits.

A3.2.3.1 Reverse recovery

From Figure A3.6 it is clear that transition between the on and off states requires the removal of a considerable quantity of stored charge from the lightly doped region of the device. If forward current is suddenly removed from a diode (e.g. by disconnecting it), the level of charge will not fall immediately but will decay with a well-defined time constant. During this period the diode will be unable to support any significant reverse bias. Any attempt to apply reverse bias before the charge has been reduced to a low level will result in substantial reverse current flow. Note that the amount of stored charge increases with the level of current and voltage rating of the diode.

The circuit shown in Figure A3.7 is used by device manufacturers in the characterisation of diodes under reverse recovery conditions. Switch S controls the path taken by a constant current i_0, and diode D (the device under test, DUT), diode D_{cl} and V_{cl} form a clamp circuit for the switch voltage.

The analysis presented here will consider a simple case in which L_s alone determines the rate of change of current (i.e. the switch can be considered ideal). In practice, switch S, L_s and diode D have a complex interaction.

At $t = 0$ switch S is closed and the voltage across inductance L_s rises rapidly to V_{dc}. The current in D thus begins to fall linearly with time (Figure A3.8):

$$L_s(di_d/dt) = -v_{dc} \tag{A3.5}$$

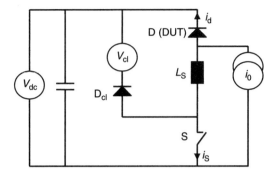

Figure A3.7   Typical diode recovery test circuit

At time t_1 the current in the diode reverses and conduction continues with little reverse voltage across the diode. The reason for this apparently anomalous behaviour lies in the presence of the on-state stored charge (Figure A3.6), which provides the carriers needed to carry the reverse current. As the reverse current flows, the level of stored charge is gradually reduced (by natural recombination and the action of the reverse current) until a high electric field can once again be supported within the diode. At this time (close to t_2 in Figure A3.8), a significant reverse voltage begins to be established. Note that the level of reverse current must continue to increase until the voltage across inductance L_s reverses (time t_2). The peak current attained is known as the reverse recovery current I_{rr}.

The period from t_2 to t_3 is known as the tail period. During this time the decaying inductor current produces a voltage spike across the diode. The height of this spike is determined by the detailed physical structure of the diode and the inductor value. The period from t_1 to t_3 is the reverse recovery time t_{rr} and is terminated when the diode reverse current falls to 25 per cent of its peak value. A final parameter used to describe the reverse recovery process is the reverse recovered charge Q_{rr}. This is the total charge recovered during the reverse recovery time (t_1 to t_3).

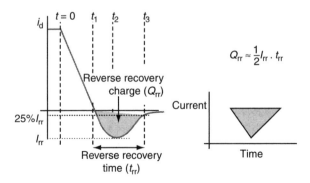

Figure A3.8   Typical reverse recovery waveforms

An approximate expression linking the reverse recovery time, peak current and charge may be obtained by assuming a triangular form for the reverse current between t_1 and t_3:

$$L_s(di_d/dt) = -v_{dc} \qquad (A3.6)$$

The energy dissipated in the diode during this event can be estimated from the product of the tail charge Q_{rr} and the supply voltage V_{dc}:

$$E_{off} \approx \frac{1}{2} Q_{rr} V_{dc} \quad \text{with} \quad Q_{rr} \approx \frac{1}{2} I_{rr} t_{rr} \qquad (A3.7)$$

In general, faster rates of fall of current lead to higher peak reverse currents and larger amounts of recovered charge. Diode designers typically aim for a low quantity of stored charge (and hence small reverse recovery current) coupled with a controlled tail period to limit the amplitude and sharpness of the voltage spike. A very fast reverse recovery is not desirable in many applications. Note that the desire to reduce the level of stored charge is in direct conflict with the requirement for a low conduction voltage drop.

The reverse recovery time may be reduced by careful design of the doping profile of the PIN junction and measures such as doping with particular elements or irradiating the junction with an electron beam. These features are designed to reduce the number of charge carriers in the diode and also reduce their lifetime so that I_{rr} and t_{rr} are both reduced. A side effect of this is that the forward voltage drop increases so there is a trade-off between speed and forward voltage drop.

A3.2.3.2 Forward recovery

Transition from conditions of reverse bias to those of strong forward bias requires the injection of large quantities of charge into the lightly doped region of the diode. Exactly how much charge depends on the current rating (device area), voltage rating (width and carrier lifetime) and the detailed design of the device. It will usually be of the same order of magnitude as the reverse recovered charge Q_{rr}.

A typical forward recovery event is illustrated in Figure A3.9.

Utilising the circuit of Figure A3.7, switch S is opened and the voltage across the diode collapses. This requires the supply of a relatively small amount of depletion charge (the background doping level is several orders of magnitude less than the final on-state carrier concentrations). Next, a flow of forward current is established. This does not happen instantaneously because of the limiting effect of inductance L_s and the turn-on speed of the switch. If, however, the rate of rise of current is fast enough, the level of charge in the diode, and hence the degree of conductivity modulation, will still be quite low. Thus the forward voltage of the diode will rise to a higher than usual level. A rough guide to the possible magnitude of the voltage spike can be estimated by comparing the steady on-state charge level to the quantity of injected charge at any time in the transient. Transient voltages over 10 V are not uncommon.

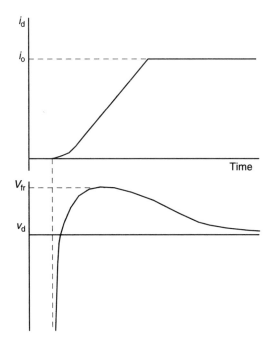

Figure A3.9   Typical forward recovery waveforms

Finally, the continued injection of charge results in a gradual fall in voltage to its steady-state level.

A3.2.4   Diode types

Commonly there are three diode types available. Converter rectifier diodes are used in low-frequency rectification of a.c. to d.c. power. They have a long t_{rr} and high Q_{rr} optimised for minimum forward voltage drop. Diodes of this type are available in ratings as high as 9 000 V and 6 000 A. Diodes that have fast characteristics, i.e. short t_{rr} and low Q_{rr}, are referred to as fast-recovery diodes. As these have been optimised for speed they tend to have higher forward volt drops, which restricts their current rating for a given chip size. Fast-recovery diodes find their main use in freewheel functions (in which they must quickly commutate current from and to primary switching devices) and high-frequency rectification. Schottky diodes are mainly used in low-voltage switched-mode power supplies (Table A3.4).

Converter rectifier diodes and fast/ultrafast recovery diodes are mostly made from PIN diodes. The Schottky diode has a different semiconductor structure. The junction in Schottky diodes is formed by metal–semiconductor contact and not by semiconductor–semiconductor contact. A metal–semiconductor combination is based on the field effect, which is different to the semiconductor process involved in the PN junction. Schottky diodes are classified as majority carrier devices. PIN

Table A3.4 Applications for the most common diode types

	Voltage/current range	Principal features	Relative cost	Typical application
Schottky	$V < 100$ V $I < 40$ A	Low forward voltage at moderate current, very fast switching performance	High	Output rectifier in low-voltage SMPS
Converter rectifier	$V < 9$ kV $I < 6\,000$ A	Low forward voltage, high surge current capability, poor switching performance	Low	Line frequency rectification/conversion
Fast/ultrafast recovery	$V < 4.5$ kV $I < 4$ kA	Moderate on-state voltage, high surge current capability, good switching performance	Moderate	High-frequency power electronic switching

and PN diodes are classified as minority carrier devices. In general, majority carrier devices have very fast turn-on and turn-off capabilities, resulting in fast switching frequencies. Minority devices have slow switching frequencies.

Most commercially available diodes are silicon based. Some of the new IGBT standard three-phase bridge inverters replace the anti-parallel-connected silicon diodes with diodes made from silicon carbide. Silicon carbide diodes have a very low reverse recovery effect, which places them in the category fast/ultrafast recovery diodes. The cost of silicon carbide diodes are currently high, making them cost-effective in a few niche drives applications only. It is, however, expected that the cost will fall progressively in the future, making them an attractive component.

A3.3 Thyristor (SCR)

A3.3.1 Device description

Thyristors are used at very high powers (ratings to 5 kV and 4 kA) and in applications where their latching characteristics or low cost are particularly advantageous. They fall into two categories: thyristors (or silicon-controlled rectifier, SCR) having no gate turn-off capability, and gate turn-off thyristors (GTOs). Thyristor applications include HVDC transmission systems, static VAR compensators and large industrial converters (typically >1 MW). In drives, thyristors are used in current-fed inverters and small d.c. drives. Applications of GTOs include high-power traction (inverter

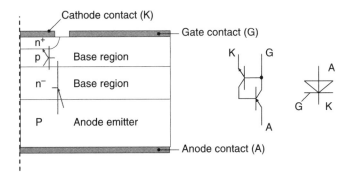

Figure A3.10 Thyristor structure and symbol

motor drives and choppers) and high-power industrial converters for welding and induction heating. GTO ratings can go up to 12 kV and 5 kA.

The thyristor is a four-layer PNPN device, as shown in Figure A3.10. Thyristor action can be best explained in terms of the two-transistor model. The p-anode emitter, n-base and p-base regions form the emitter, base and collector of a p–n–p transistor, while the n^+, p-base and n-base form the emitter, base and collector of an n–p–n transistor. Note that the collector region of one transistor forms the base of the other. Application of a gate signal to the p-base region initiates transistor action in the n–p–n device. Its collector then drives the n-base of the p–n–p transistor. This in turn provides base current for the n–p–n transistor. Effectively a regenerative, positive feedback loop is established.

One consequence of this regenerative behaviour is that thyristors can take massive pulse currents without significant increases in forward voltage. They are typically rated to carry surges many times their steady-state rating and can usually be protected by simple fusing.

Another result of this latching behaviour is the absence of any steady-state, active portion in the thyristor characteristic (Figure A3.11). Transition between the forward blocking and forward conduction occurs suddenly and in an essentially uncontrolled manner.

Figure A3.11 shows that the reverse characteristic (cathode positive with respect to anode) is similar to that of the diode; however, the forward characteristic exhibits no current flow other than leakage current until a gate current is injected. The anode current I_A is then able to flow, limited solely by the external load and supply capacity.

The forward breakover voltage is equal in magnitude to the reverse voltage, because in the blocking state the emitter–n-base junction supports almost all the voltage. Once breakover in the forward direction occurs, the thyristor behaves rather like a diode. The overall forward volt drop is between 1.5 and 2 V.

With the thyristor forward biased it is normally turned on by injecting a positive pulse of current into the gate, I_G. Once the anode current has exceeded the latching

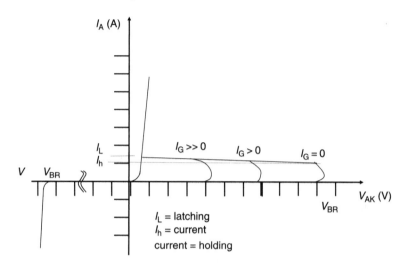

Figure A3.11 Thyristor characteristics

current I_L, the gate pulse can be removed. For the thyristor to remain in the conducting state, the anode current must reach the latching current level I_L and not fall below the holding current I_H.

The thyristor is normally turned off by forcing the anode current to zero by applying a reverse voltage for a minimum period of time before it can regain its forward blocking state. The total turn-off time t_q is an important parameter for thyristors in fast switching applications. The transients of a thyristor are discussed in the next section.

A3.3.2 Transient processes

Unlike the MOSFET and IGBT, the thyristor is a current-controlled device. Switching transitions are initiated by the injection of base charge from the gated region of the device. The latching on-state characteristic gives rise to two important thyristor features:

- Drive signals are required only to establish an anode current, not to maintain it.
- Turn-off can only be accomplished by forcing the anode current to zero in the external circuit.

Figure A3.12 shows a typical line commutated thyristor converter circuit. The load is assumed to be highly inductive. Figure A3.13 shows typical turn-on waveforms for the thyristor. The turn-off transition is almost identical to that encountered in the power PIN diode.

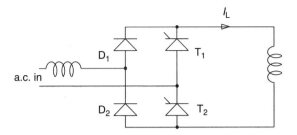

Figure A3.12 Half-controlled thyristor bridge

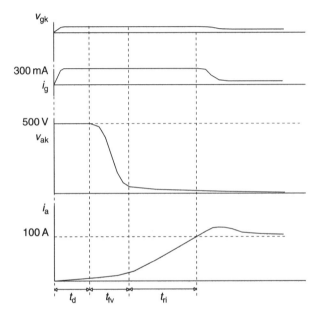

Figure A3.13 Typical turn-on waveforms for a thyristor embedded in a line-commutated converter

A3.3.2.1 Turn-on

This requires a pulse of gate current to initiate the transition from the forward blocking state to the forward conduction state. Once started, the transition will maintain itself through thyristor action.

Turn-on delay period t_d

Gate current causes injection of electrons into the n-base from the cathode and, through transistor action, injection of holes into the n-base from the anode.

Build-up of base region stored charge continues until it becomes sufficient to allow collapse of the depletion layer. Typical gate currents are in the order of 20 mA for a 10 A device and 300 mA for a 3 000 A device.

Voltage fall period t_{fv}

Both n–p–n and p–n–p transistor sections of the thyristor traverse their active regions as the depletion layer collapses. The anode voltage falls and the thyristor latches as the level of stored charge gradually increases (voltage fall occurs before current rise in this circuit). Note that the latched region initially extends only over a relatively small area close to the gate contact. Regions of the device remote from the gate contact carry little current and have relatively small levels of stored charge. This applies particularly to larger devices.

Current rise period t_{ri}

Inductance in the supply lines limits the rate of rise of forward thyristor current. The on-region of the thyristor extends outward from the gate contact as stored charge builds up throughout the device. This is often termed plasma spreading. If the rate of rise of anode current is not limited by the external circuit, then the small initial on-region may suffer excessive levels of current density, leading to device failure (limits for particular devices range from 100 to 2 000 A μs^{-1}). A turn-on snubber circuit may be required in some applications, particularly those with a capacitively decoupled voltage source such as a d.c. chopper.

A3.3.2.2 Turn-off

Turn-off in the line-commutated converter occurs when the thyristor current naturally falls below the holding current level. Once this occurs the thyristor can revert to its forward or reverse blocking state. For the circuit of Figure A3.14 the thyristor blocks in the reverse direction after commutation. The usual considerations for diode reverse recovery apply to this transition.

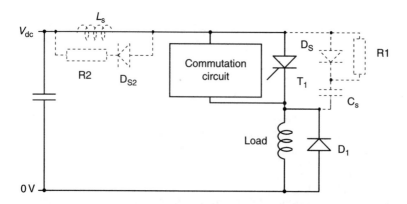

Figure A3.14   Thyristor chopper employing an external commutation circuit (snubber components shown dotted)

Turn-off can also be effected by using an external commutation circuit. Applications using this technique include choppers, voltage source inverters and other d.c.-supplied circuits. The basic principle involves the application of a reverse current to the thyristor, which effectively cancels the load current for a period of time sufficiently long for the thyristor to recover its forward blocking characteristics. During and after recovery, the rate of rise of off-state voltage must be limited to prevent re-triggering of the thyristor (often referred to as the re-applied dv/dt effect). In simple terms the changing voltage produces a flow of current through the thyristor as a result of changes in the depletion layer charge distribution. If the dv/dt is unlimited this current can become sufficient to drive the thyristor on again (typical limits between 100 and 2 000 V μs^{-1}). The situation can be avoided by using a turn-off snubber circuit.

A3.3.3 Thyristor gating requirements

The gate cathode characteristic of a thyristor resembles that of a poor PN junction and will vary between production batches for a given type. To be certain of turning on the thyristor, the gate current and voltage must attain minimum levels, dependent on junction temperature, that lie between the upper and lower resistance limits shown in Figure A3.15a. It is also necessary to ensure that the peak gate power ($V_G \times I_G$) is not exceeded. Figure A3.15b shows a typical gating characteristic, illustrating the boundary conditions.

Typically, only a small pulse of current is required to turn a thyristor on. Once in the on-state, no further drive is required. For small thyristors, therefore, it is usual to employ small pulse transformers with pulse current capabilities in the mA/A range. For larger devices, a larger pulse current may be required and the use of a directly connected, high-current gate circuit is recommended. A simplified example of a pulse transformer-based firing circuit is shown in Figure A3.16. Resistor R1 limits the gate current while R2 provides a low impedance across the gate to attenuate any gate voltage when the thyristor is in the off-state. To achieve short turn-on times, the gate current is required to rise at a minimum of 1 A μs^{-1}. A succession of gate pulses (Figure A3.16b), supplied by the gate drive circuit, causes firing to occur when external conditions are suitable for conduction. The turn-on di/dt capability of thyristors can be improved by utilising a higher pulse current during the turn-on period as this helps to promote plasma spreading.

A3.3.4 Thyristor types

Three types of thyristor can be distinguished:

- *Thyristor for a.c. line commutation (phase control thyristor)*. Thyristors used in a.c. applications are turned off, or commutated, naturally by the existence of the a.c. supply, which changes polarity in alternate half cycles. The thyristor is designed to have a low on-state voltage, thereby maximising current rating at the expense of relatively long turn-off times (typically 100 to 200 μs). This

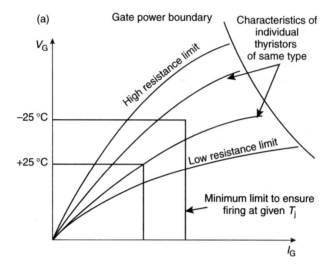
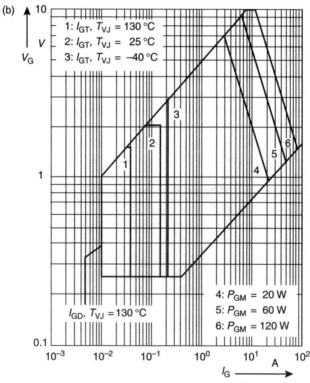

Figure A3.15 Thyristor gate curves: (a) range of characteristics and limits; (b) example characteristics

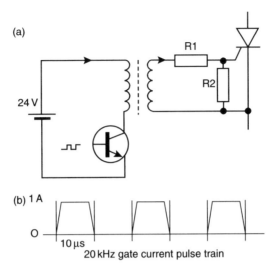

Figure A3.16   Thyristor gate circuit

does not matter because the thyristor switching frequency is low. Equal forward and reverse voltages up to 12 000 V are possible for large phase-control thyristors. For applications on a.c. supplies up to 500 V a.c., it is usual to specify 1 400 V types, to allow for an overload factor of two. It is common practice to use RC networks and varistors across the thyristor to give additional protection.

- *Fast thyristors.* These devices are generally used in d.c. circuits such as choppers or inverters, although their use is now less frequent as more modern devices such as IGBTs have replaced them in many applications. Within d.c. circuits there is no natural reversal of the supply for thyristor commutation, so it must be derived by external circuits. The process of turn off under these conditions is called forced commutation, in contrast to line commutation. Typical commutating circuits are expensive because they consist of inductors, capacitors and auxiliary thyristors; however, their size can be reduced if t_q, the total turn-off time, is kept to a minimum. The design of the thyristor is therefore optimised for low t_q (typical values, 15 to 30 μs) but unfortunately this has the undesirable effect of increasing the on-state volt drop, which consequently lowers the current rating.
- *Asymmetric thyristor.* In many fast-switching applications the reverse blocking capability of the thyristor is not required because an anti-parallel diode must be connected across the device for reactive current conduction. Manufacturers have exploited this relaxation by offering the asymmetric thyristor or ASCR, which has even lower t_q times than the fast thyristor but at the expense of very limited reverse blocking. Turn-off times as low as 8 μs are possible while still retaining an acceptable 15 V reverse blocking. Another technique to enhance the performance of both ASCRs and fast thyristors is to use an interdigitated gate structure, which considerably increases the device di/dt rating at turn-on. This technique

effectively enlarges the turn-on area of silicon available at the start of gate firing, thus preventing excessive current density near the gate, which could lead to device failure.

A3.4 Triac

The triode thyristor (Triac) is a five-layer device that behaves like a bi-directional thyristor and is commonly used as an a.c. switch (controlled turn-on but not turn-off) or for low-cost, low-power, low-performance motor controllers. Triac ratings are up to 1 200 V and 40 A (rms). The Triac has three terminals: the control terminal and two power terminals (T1, T2). Figure A3.17 shows the symbol and characteristic of the device.

The polarity of the firing current and voltage across the Triac can be positive and negative. However, different polarities have different requirements for gate current and gate voltage that must be satisfied. Figure A3.17 shows the four quadrants in which a Triac operates.

A3.5 Gate turn-off thyristor (GTO)

A3.5.1 Device description

Attempts at controlled gate turn-off using normal thyristors typically meet with failure. If a negative gate current is employed it will tend to turn off only those areas in the immediate vicinity of the gate contact, leaving other areas of the device in the latched on-state. Use of larger negative gate currents will force the conduction area to become smaller but will provoke reverse breakdown of the device gate–cathode junction in the area closest to the gate. Turn-off can only be guaranteed if the access resistance from the gate to all parts of the device cathode is very low. In the GTO this is achieved by dividing the cathode into a large number of small islands

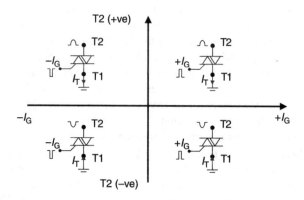

Figure A3.17   Triac firing quadrants

surrounded by a sea of gate metallisation. This patterned structure prevents conduction continuing in the cathode islands between the gate contacts. The pattern also ensures an even current density across the die during turn-off. A 3 000 A device may have over 1 200 cathode islands typically 0.2–0.4 mm wide and around 3 mm in length. In all GTOs the use of a turn-off snubber is essential to limit the rate at which the forward voltage is re-applied.

The GTO can replace the fast thyristor and its associated commutation circuits in d.c. switching applications. The circuit symbol and the more complex GTO structure are shown in Figure A3.18. Although the physical operation of the GTO is very complex, it is helpful to refer to the two-transistor model that was introduced for

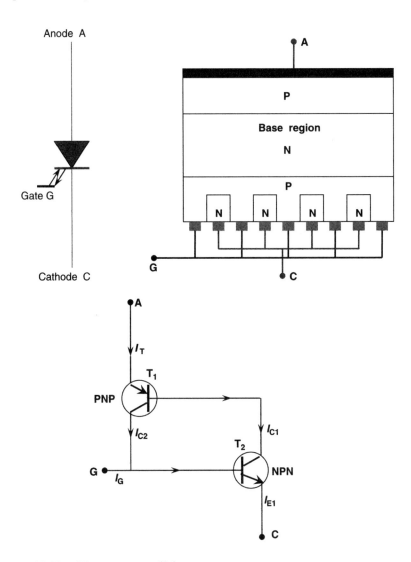

Figure A3.18   The gate turn-off thyristor

the thyristor to understand how turn-off is achieved. The devices may be considered as two interconnected transistors that have regenerative action: the collector current of one feeds the base current for the other transistor. It can be shown that the anode current I_A is given by

$$I_A = \frac{I_G \alpha_{NPN}}{1 - (\alpha_{NPN} + \alpha_{PNP})} \tag{A3.8}$$

where I_G is the gate current, α_{NPN} and α_{PNP} are the common base gains, where $\alpha = \beta/(1+\beta)$ and β is transistor current gain, I_C/I_B. The current gains are dependant on the collector current and increase as the current increases from zero. Conduction of the thyristor is initiated by a gate current pulse that raises the loop gain ($\alpha_{NPN} + \alpha_{PNP}$) to unity, so from equation (A3.8) I_A is infinite.

In practice the anode current is limited by the load. Turn-off action is produced by extracting sufficient current from the gate to cause the loop gain to fall to a point where regenerative action ceases. The turn-off gain β_{Off} is the ratio of anode current being controlled to negative gate current required to produce turn-off, and is an important parameter. Typical values of β_{Off} lie between 3 and 5. To reduce the loop gain, and hence increase the turn-off gain, the gate is often connected to the cathode with a low resistance within the package, 'emitter shorts'; this has the side effect of making the gate less sensitive for turn-on. Unfortunately, there is a limit to the maximum anode current that can be switched off, generally this being about four times the average current.

A3.5.2 Switching characteristics and gate drive

Figure A3.19 explains the turn-on and turn-off characteristics of a GTO. GTOs need snubber circuits. Two types of snubber circuits are shown here: a turn-on snubber (L_s, R_2, D_{S2}) and a turn-off snubber (C_s, R_1, D_S).

Typical switching waveforms for the chopper circuit of Figure A3.19 are shown in Figure A3.20. The voltage and current waveforms are heavily influenced by the

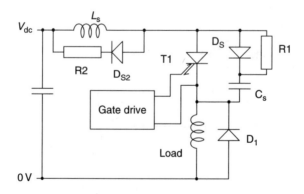

Figure A3.19   GTO chopper circuit with turn-on and turn-off snubbers

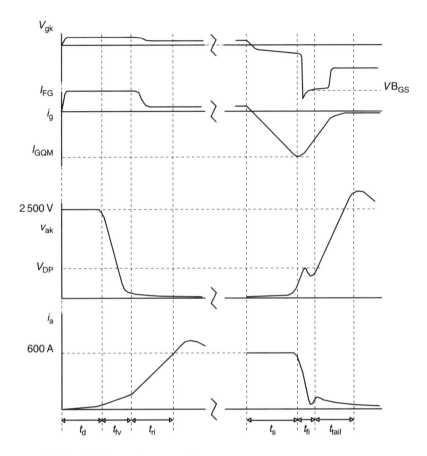

Figure A3.20   GTO switching waveforms

snubber circuits, as will be described in the next section. The switching frequency of GTOs are normally limited to 2 kHz, due to considerable power dissipation during switch-on and switch-off.

A3.5.2.1   Turn-on

During the turn-on process the GTO undergoes a similar sequence as described for the thyristor.

- *Turn-on delay period* t_d. It is assumed that the GTO is off and a positive gate current is applied. As in a thyristor, the gate current causes injection of electrons into the n-base from the cathode and, through transistor action, injection of holes into the n-base from the anode. The depletion layer will collapse once enough stored charge has been built up in the base region. In order to provide a good and equal conductivity of all cathode islands the maximum gate current pulse should have a high di_g/dt (3 to 5 µs) and a relatively large amplitude I_{FG}.

- *Voltage fall period* t_{fv}. With a continuous supplied maximum gate peak current both transistor sections traverse their active regions. The result is the fall of the anode voltage until the GTO latches. The voltage drop is caused by the increase of stored charge.
- *Current rise period* t_{ri}. The rate of rise of the anode current is limited by the inductance in the supply lines. The on-region of the GTO extends outward from the gate contact as stored charge builds up throughout the device; this is often called plasma spreading. There is a practical limit to the rate of rise of anode current that can be manipulated by the use of an additional external inductor (also called turn-on snubber). A turn-on snubber is needed in particular for applications with a capacitively decoupled voltage source such as the d.c. drive chopper. If in this case an inductor is not included then the small initial on-region may suffer excessive levels of current density, leading to device failure. After turn-on, the gate current may fall to a lower level to minimise the on-state voltage drop. If a continuous gate drive were not provided, the voltage drop would tend to increase at low anode currents due to the fall in loop gain.

A3.5.2.2 Turn-off

During the turn-off process the GTO undergoes the following sequence:

- *Storage time period* t_s. To turn the GTO off, the forward gate current is removed and a negative voltage source (-12 to -15 V) is applied. The applied gate current must be in the order of 0.2 of the anode current, which, depending on the anode current, can be very high. This driver requirement is often seen as a drawback when using GTOs. The negative current removes the stored charges within the semiconductor layers between the cathode islands. The anode current starts to fall once sufficient amount of stored charges has been removed. It is important that the negative di_G/dt is within the manufacturer specifications. On the one hand the di_g/dt should be high in order to reduce switching losses; on the other hand a small rate of change in the gate current will result in a long anode tail current, which also generates losses and increases the recovery time.
- *Current fall period* t_{fi}. With the removal of a relatively large amount of stored charge within the semiconductor layers between the cathode islands, the gate–cathode junction starts to recover its reverse-blocking capability. The growth in the reverse-blocking, however, generates two effects on the gate side of the GTO. The first effect is a rapid fall in the gate current and the second effect is a fall of the gate–cathode voltage from a positive value to a negative level. At negative levels the gate–cathode junction goes into an avalanche breakdown. With the avalanche in place the stored charge in the semiconductor layer, which is connected to the gate, is swept out.
- *Tail current period* t_{tail}. At the end of the avalanche effect some of the excess stored charge will still remain. This excess charge will recombine within the gate–cathode junction. This recombination process is reflected in a continuous anode current flow, which is often termed tail current. The current is driven by

the voltage difference between gate and anode. During the period of the tail current, significant losses apply in the switch. It is therefore desirable to keep the tail period short. The stray inductances in the power circuit lead to a large anode–cathode over-voltage, as shown in Figure A3.20. The GTO has to be protected against over-voltage and a turn-off snubber must be applied across the GTO device.

A3.5.3 Voltage and current ratings

Like the thyristor, the GTO can be designed to have equal forward and reverse blocking capabilities, or limited reverse blocking with the advantage of improved turn-off times. The latter type is sometimes referred to as an anode-short GTO and is designed to be used with an anti-parallel diode. The forward on-state voltage drop is of the order of 2 V, giving current ratings similar to the fast thyristor, namely up to 6 kV and 2 kA.

A3.6 Integrated gate commutated thyristor (IGCT)

A3.6.1 Device description

The integrated gate commutated thyristor (IGCT) is a development of the GTO and has essentially the same device structure. By operating with a turn-off gain of 1, rather than 3 to 5 for a GTO, faster turn-off is achieved and the requirements for snubbers much reduced. In order to have the correct conditions for turn-off the gate circuit inductance must be very low. To achieve this the gate circuit is integrated into the device – hence the name integrated gate commutated thyristor. IGCTs are available with or without an anti-parallel diode. The symbol for the IGCT is shown in Figure A3.21.

As with a GTO the IGCT is a four-layer device with a highly inter-digitated gate structure. The device can be triggered into the on-state by applying a positive gate current allowing current to flow between the anode and cathode. Once switched on the current is determined only by the external circuit. The on-state voltage is typically 3 V for a 4 500 V rated device. Turn-off may be achieved by reverse biasing the main power circuit, or more normally by extracting current from the gate. Sufficient current is extracted from the gate so that no current flows across the gate to cathode junction,

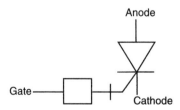

Figure A3.21   IGCT symbol

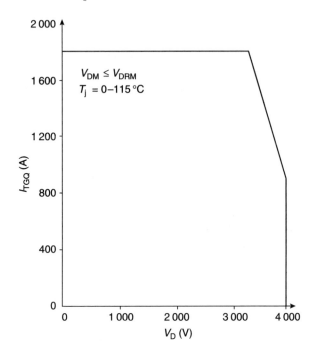

Figure A3.22 IGCT safe operating area

and the device behaves as a PNP transistor. This gives fast turn-off with a safe operating area (SOA) similar to a bipolar transistor (Figure A3.22).

A3.6.2 Switching behaviour and gate drive

Turn-on is initiated by application of a positive gate current, typically up to 100 A. Lower gate currents may be used but this increases the turn-on time and limits di/dt capability. Once conduction is initiated the gate current may be reduced to the 'back porch' current of a few amps. As with a GTO this is necessary due to the low device loop gain at low anode currents.

At turn-on anode current di/dt must be limited to prevent hot spots as the conduction spreads out from the gate, and also in applications with a freewheel diode to limit the reverse recovery di/dt. However, once turn-on is initiated through the gate, the anode current di/dt cannot be controlled via the gate, and so an inductor and associated snubber must be added.

In order to turn the device off sufficient gate current must be extracted from the device for the regenerative action to stop. To allow operation with little or no snubber and minimise switch-off losses, a turn-off gain of 1 is used (i.e. all anode current is diverted out of the gate), and this must be accomplished fast enough to prevent current redistribution in the device, which would lead to hot-spots and a reduction in device turn-off rating. In a standard GTO package the gate inductance is of the order of 50 nH. To divert an anode current of 2 000 A out of the gate in 1 μs would require 100 V. This would lead to very high gate drive power losses, and also exceeds the gate reverse breakdown voltage of approximately 20 V. By integrating the gate drive unit into the IGCT assembly a gate circuit inductance of 5 nH is achieved. Now only 10 V is required to achieve the required gate current di/dt, gate drive power is reduced, and the breakdown voltage is not exceeded.

To switch the large gate current requires an array of low-voltage MOSFETs coupled to a bank of low-impedance electrolytic capacitors. Gate drive power consumption can be significant, approximately 80 W for a 700 A device switching at 500 Hz.

As with a GTO an IGCT has a minimum on and off time. The minimum times of 10 μs are determined by a combination of time required for current density across the chip to stabilise and for the gate drive circuit to prepare for the next switching event. A typical maximum average switching frequency is 500 Hz, although this may be increased to 2 kHz with a suitable gate unit and some reduction in current rating. In the event of a short circuit fault the device must be switched off before the anode current rises above the maximum controllable current or control will be lost and the device destroyed.

A3.6.3 Voltage and current ratings

The maximum repetitive voltage that can be blocked is represented by V_{DRM}, with devices available up to 5 500 V. This voltage is greater than the maximum voltage that can be sustained during switch off. The on-state average and rms current ratings, I_{TAVE} and I_{TRMS}, determine the maximum load current. In the on-state the device can withstand high surge currents, but the device cannot switch off these currents. The maximum current that can be turned off under defined conditions is I_{TGQM}. Devices are available with ratings up to 4 000 A. The maximum controllable current is approximately twice the on-state current.

A3.7 MOSFET

A3.7.1 Device description

The metal oxide semiconductor field-effect transistor (MOSFET) was developed into a useful fast switching power device in the 1980s. Both N-channel and P-channel devices are available, but for power applications N-channel devices predominate due to their lower losses. The device symbol for an N-channel MOSFET is shown in Figure A3.23.

A3.7.2 Principal features and applications

Unlike the junction p–n diode, the MOSFET is a majority carrier device (for its structure see Figure A3.24). There is no injection of minority carriers into the lightly doped region of the device and as a result the density of mobile carriers able to take part in

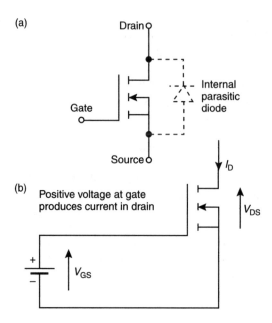

Figure A3.23   MOSFET: (a) circuit symbol; (b) electrical circuit

the conduction process is limited to the doping density. This lack of conductivity modulation means that the on-state voltages are, in general, relatively high. On the other hand there is no significant stored charge to slow switching transitions. The MOSFET, therefore, finds much application in high-speed circuitry and in particular at voltages below 400 V.

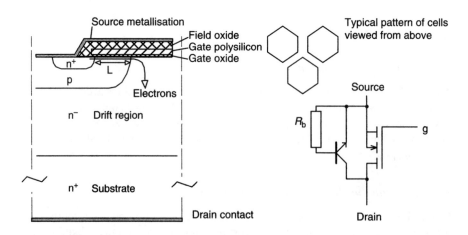

Figure A3.24   MOSFET structure

A3.7.3 D.C. characteristics

Figure A3.25 shows the d.c. characteristics of a typical power MOSFET. The MOSFET is a voltage-controlled device and has negligible gate current requirements when operated under steady-state conditions (very small oxide leakage current).

Application of a positive voltage V_{GS} between the gate and source terminals will set up an electric field in the oxide under the gate. If the field is high enough (approximately 1×10^6 V cm^{-1}), an inversion layer is formed just under the oxide surface. The field then modulates the resistance between the drain and source, permitting a current to flow from source to drain. The gate voltage at which this occurs is known as the threshold voltage and is typically between 1.5 and 3 V.

When zero V_{GS} is applied, any reverse voltage applied will be clamped by the presence of a parasitic diode in the device structure. The MOSFET then behaves like a p$^+$–n–n$^+$ diode, which is reverse biased for $V_{DS} > 0$ (usual operating condition). A positive voltage will be blocked at the drain until the breakdown limit is reached. Breakdown is determined by the p$^+$–n–n$^+$ diode, the behaviour of which resembles that of a p–i–n diode (not exactly the same because the drift region doping density is relatively high). Higher-voltage devices thus require a wider drift region with lower drift region doping.

If V_{GS} is raised above zero, no drain current will flow until the gate threshold voltage is reached. Once V_{GS} is raised above the threshold voltage, the characteristic has two distinct regions: a constant-resistance region with channel resistance $R_{DS(on)}$, and a constant-current region where the trans-conductance of the device is

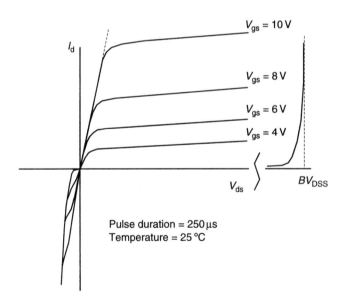

Figure A3.25 D.C. MOSFET characteristics showing operation in first and third quadrants

almost constant. $R_{DS(on)}$ is a key parameter and will determine the forward volt drop ($R_{DS(on)} \times I_D$) and ultimately the current rating of the device. Operation within the current-region is normally avoided (to minimise conduction losses) by setting V_{GS} high enough for the load current; a value of 10 V is usually sufficient.

With V_{GS} above threshold voltage the channel is able to conduct current in the reverse direction ($V_{DS} < 0$) as well as in the normal forward direction. This feature can be exploited by using MOSFETs as very-low-voltage drop diodes. In this situation, the parasitic diode forward-biased current flows in the reverse direction (from the source to the drain), so the device may operate in the third quadrant (see Figure A3.25). Operation of MOSFETs in the third quadrant permits the construction of controlled rectifiers with a low on-state voltage drop. Such devices have limited application in low-voltage systems where the on-state voltage drops of standard rectifiers can be unacceptable.

A3.7.4 Switching performance

In contrast to the diode, the MOSFET conduction mechanism does not rely on stored charge. Also, the MOSFET has two very important differences to the BJT. First, it is a voltage-controlled device rather than a current-controlled device. Second, it is a majority carrier device rather than a minority carrier device.

Being voltage controlled with an oxide insulated gate, very little power is needed to control the device, current only being required to charge and discharge the gate capacitance (Figure A3.26). A majority carrier device only conducts due to intrinsic charge carriers. This enables the device to switch off very quickly as there is no time required for removal and recombination of minority charge carriers, as there is in a bipolar transistor or thyristor. Switching behaviour is, instead, determined by the oxide and junction depletion capacitances (Figure A3.26). Fast switching allows the MOSFET to be used in applications with switching frequencies of 100 kHz or

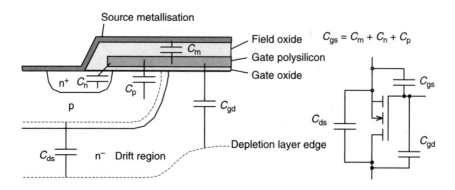

Figure A3.26   *Oxide and depletion capacitances in a power MOSFET and their representation in the equivalent circuit*

more. The down side of only having intrinsic charge carriers available for conduction is that the conductivity of the silicon is lower and so a larger chip is required for a given current rating.

A3.7.5 Transient characteristics

During switching transitions, charge must be transferred between the MOSFET capacitances and the external circuit; for example, to take a MOSFET from the off-state to the on-state will require charging of C_{gs} (increase in gate–source voltage V_{gs}) and discharging of C_{gd} and C_{ds} (decrease in drain–source voltage). Rapid switching of the device requires correspondingly high levels of gate current to ensure rapid charge/discharge of C_{gs} and C_{gd}. All three device capacitances display changes in value with applied bias. It is therefore impossible to characterise the switching performance of the device by considering a single value of capacitance. Instead, manufacturers typically specify the total quantity of gate charge Q_g associated with a particular switching event. The gate current requirements for a particular transition time can then be estimated:

$$Q_g = \int_0^{t_{switch}} i_g dt \approx i_g t_{switch} \tag{A3.9}$$

where i_g is the gate current and t_{switch} the transition time.

A3.7.5.1 Switching waveforms

Figure A3.27 shows a typical clamped inductive load switching circuit employing a power MOSFET and a clamp diode (assumed ideal) as its active components. The corresponding turn-on and turn-off transitions are detailed in Figure A3.28.

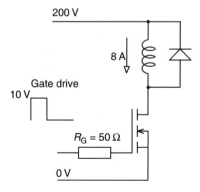

Figure A3.27 Inductive load switching circuit

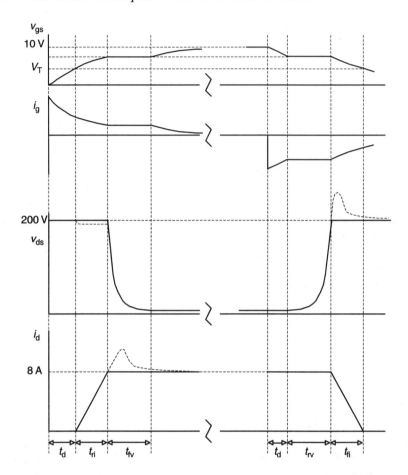

Figure A3.28   Switching waveforms for a MOSFET with clamped inductive load

A3.7.5.2   Turn-on

At time $t = 0$ a 10 V step is applied to the gate of the MOSFET through a 50 Ω resistor. The initial conditions are such that the diode is conducting the full inductive load current of 8 A and the voltage across the MOSFET may, therefore, be assumed to be around 200 V.

- *Turn-on delay period* t_d. Charge is supplied to the gate–source capacitance and gate–drain capacitance. Voltage and current follow an exponential characteristic as a consequence of the R–C network. During this period there is no change in either drain voltage or drain current.
- *Current rise period* t_{ri}. Once the gate voltage reaches the threshold level V_T, a channel begins to form between the source and the drift region. Although current begins to be diverted from the diode, the drain voltage remains high

until the (ideal) diode current falls to zero. During this time the MOSFET is operating in the active region of its characteristic. At the end of the rise period the gate voltage is at a level to support load current.

- *Voltage fall period* t_{fi}. At the end of t_{ri} the current reaches the load level allowing the (ideal) diode to support a reverse bias (in practice a reverse recovery current would flow – dotted line in Figure A3.16). The drain voltage falls as the MOSFET moves through the active region under conditions of virtually constant drain current (see Figure A3.11). As a result, the required change in gate–source voltage is relatively small. Thus the gate voltage remains almost constant while the drain voltage collapses at a rate determined by the current supplied at the gate. Once all the mobile charge in the drift region has been restored, the gate voltage will once again begin to increase (effective input capacitance is now greater as a result of increased contribution from C_{dg}). A further, small decrease in drain voltage will occur as the remainder of the channel is driven into strong inversion.

A3.7.5.3 Turn-off

- *Turn-off delay period* t_d. Charge is removed from the gate–source and gate–drain capacitances, leading to a reduction in the channel inversion charge and a fall in gate–source voltage. At the end of the turn-off delay period the MOSFET is just on the edge of the active region.
- *Voltage rise period* t_{rv}. The gate–source voltage changes only slightly and the MOSFET moves through the active region of its characteristic. The variation of the drain voltage with time mirrors that of the voltage fall period during turn-on.
- *Current fall period* t_{fi}. Once the drain voltage reaches the clamped level (200 V), the diode may conduct, allowing the MOSFET drain current to fall. The gate–source voltage falls at a rate determined principally by the R_g–C_{gs} combination. At the end of the current fall period the gate–source voltage crosses the threshold voltage and the channel is fully cut off.

A3.7.6 Safe operating area (SOA)

A3.7.6.1 Forward-bias safe operating area (FBSOA)

This typically consists of a set of curves, plotted using logarithmic scales on the i–v plane for a range of pulse durations. Each curve indicates the limits of reliable on-state operation for each pulse duration. Figure A3.29 shows the safe operating area (SOA) for a typical power MOSFET.

The MOSFET does not exhibit the phenomenon of second breakdown, which means that the SOA curve for all operating modes is square. It extends to V_{DSS} along the voltage axis and up to four times I_D in the current axis. Under steady-state (d.c.) conditions the limit of the device SOA is determined by the package power dissipation constraints (typically 50 to 125 W for MOSFETs). For short pulse durations (<10 ms) the MOSFETs SAO is limited by the transient thermal impedance of the

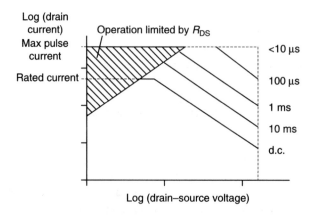

Figure A3.29 Typical safe operating area for a power MOSFET

MOSFET die and package, while for pulses of less than 10 μs the SOA extends to the breakdown voltage limit for all levels of drain current.

A3.7.6.2 Reverse-bias safe operating area (RBSOA)

Also known as the turn-off SOA, this gives the range of off-state voltage and on-state current for which safe switching can be achieved. In many cases this is influenced by the type of load (inductive/resistive) and the gate drive conditions employed. Care should be taken to identify the specified conditions.

The MOSFET benefits from virtually unrestricted safe turn-on and safe turn-off transition areas; i.e. it is possible for the device $i-v$ trajectory to pass through any region of the SOA rectangle during switching, no matter how fast the applied gate drive. This is certainly not true of bipolar (minority carrier) devices, which are prone to latch up and failure. Most MOSFETs can also endure short periods of forward avalanche during turn-off of unclamped inductive loads (this can also apply where stray or leakage inductances are significant). In this case the MOSFET is rated in terms of the avalanche energy associated with a single turn-off event.

A3.7.7 Parasitic diode

The parasitic or body-drain diode that exists within the MOSFET structure has a slow switching characteristic compared with the MOSFET channel itself. The switching frequency of circuits that make use of the diode (e.g. ultrasonic inverters) may be limited solely by the diode and not the MOSFET. This is mainly due to very significant diode switching losses, which are a function of reverse recovery charge, and operating frequency. In some applications the body diode is actually bypassed by a fast diode as shown in Figure A3.30, and the series-connected diode prevents current from flowing in the body diode.

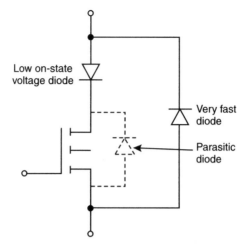

Figure A3.30   MOSFET with anti-parallel diode

A3.7.8   MOSFET gate drive requirements

MOSFET gate drive requirements under steady-state conditions are negligible due to the isolation barrier provided by the gate oxide. A low-impedance drive must, however, be provided to achieve fast switching speeds and to guard against false turn-on resulting from rapidly applied off-state drain voltages. A simple, effective and relatively low-cost solution is to use parallel connected CMOS logic gates (e.g. the hex Schmitt inverter). To achieve faster switching speeds, more gates are simply used (see Figure A3.31).

For high performance it is necessary to use a buffered push–pull circuit (Figure A3.32). Note that the loop comprising gate drive output devices, power MOSFET gate and source connections and gate drive decoupling capacitances should be as short as possible.

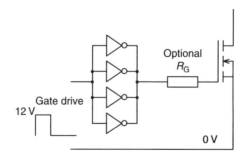

Figure A3.31   Simple MOSFET/IGBT gate drive based on parallel connected CMOS inverters

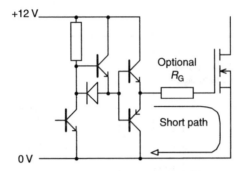

Figure A3.32 High-performance MOSFET/IGBT gate drive

A3.7.8.1 Speed limitations

There are two main limitations to speed for a typical power MOSFET. One is the intrinsic resistance of the polysilicon used to form the gate. In combination with the gate–source and gate–drain capacitances, this distributed resistance forms a low-pass filter that limits the voltage available at each MOSFET cell gate. The distributed nature of both resistance and capacitance means that cells furthest away from the gate will respond more slowly to rapid changes in applied gate voltage. This can lead to excessive current in those cells closest to the gate bond pad. Power dissipation due to current flow in the polysilicon can also become a problem. Devices intended for radio-frequency use typically have a metal gate construction to avoid these problems.

The other limitation is the source bond wire inductance, which appears simultaneously in the gate and drain circuits (Figure A3.33). Voltage appears across L_S due to the rate of change of drain current. This reduces the effective gate voltage available from the drive and is particularly important for larger die MOSFETs. One solution is to use a separate gate return path from the source (so-called kelvin source connection). The source inductance for a typical TO220 package is around 5–10 nH.

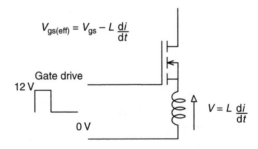

Figure A3.33 Degradation of MOSFET switching speed by package source inductance ($L = L_s$)

A3.7.8.2 Driving paralleled MOSFETs

It is easy to drive parallel MOSFET devices, provided a separate series gate resistance (typically 10 Ω) is included between the gate drive circuit and each MOSFET gate. This will prevent oscillations occurring between the paralleled devices during switching. Alternatively, and preferably, it is possible to use a separate gate drive for each device. However, this is clearly a more expensive solution.

A3.7.9 Voltage and current ratings

Current ratings are usually given for case temperatures of 25 °C; at the more practical temperature of 100 °C the current rating is reduced because of an increase in $R_{DS(on)}$. The variation in $R_{DS(on)}$ with temperature depends on the voltage rating of the device. Lower-voltage devices have a lower temperature dependence, typically a factor of 1.2 from 25 °C to 125 °C, and for high-voltage devices the ratio may be 2 or more.

The positive temperature coefficient of the on resistance has the benefit that devices may be connected in parallel (having the same applied bias at gate and drain) and will share current equally. Any increase in the current flowing in one device of a parallel combination will produce a temperature rise in that device that will tend to oppose the increase in current. The system is, therefore, inherently stable. A chip is composed of thousands of MOSFET 'cells' connected in parallel.

In conventional high-voltage MOSFETs (>200 V) the lower doping density and thicker die required for higher voltages has resulted in devices that have an on resistance proportional to $V_{DSS}^{2.5}$, where V_{DSS} is the voltage rating. This is why there are relatively few devices rated above 600 V, and devices rated above 1 000 V are very rare. A process technology, marketed as 'CoolMOS' by Infineon, uses a new '3D' doping profile that allows the doping density to be increased and chip thickness to be reduced compared to conventional MOSFETs so that the on resistance only increases linearly with voltage rating. This allows a smaller die to be used for a given rating, although the more complicated chip fabrication increases costs.

For low-voltage MOSFETs (<50 V) the channel resistance is significant, as opposed to high-voltage devices where the body or 'drift region' dominates. To reduce the channel resistance another process technology has been used: the trench gate structure. A trench gate device, as its name suggests, has the gate etched down into the chip, rather than being a planar feature near the surface of the chip. This reduces the channel length and therefore resistance too, and also allows the cell size to be reduced to give more cells per unit area and hence reduced resistance per unit area.

A3.8 Insulated gate bipolar transistor (IGBT)

A3.8.1 Device description

The insulated gate bipolar transistor (IGBT) combines the best features of a MOSFET and a BJT to give a voltage-controlled device with low on-state losses. The circuit symbol and terminal designations are shown in Figure A3.34.

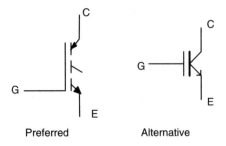

Figure A3.34   *IGBT circuit symbol*

A3.8.2 *Principal features and applications*

The structure of an IGBT is essentially that of the MOSFET with an additional p^+ layer at the drain (usually termed the collector in IGBTs) (Figure A3.35). Although the on resistance of a high-voltage MOSFET is mostly due to the drain drift region, as explained in the previous section, in an IGBT the resistivity of this region is substantially reduced by the injection of charge carriers in the additional semiconductor layer. This acts in the on-state to inject minority holes into the n^- drift region. The resulting conductivity modulation reduces the on-state voltage (reduced drift region voltage drop), and thus results in on-state losses comparable to the BJT. However, this does mean the presence of significant stored charge (slow transients). The IGBT is a minority carrier device. It may also be modelled as a p–n–p transistor driven by a complementary n-channel MOSFET. The additional semiconductor layer is the emitter of the PNP transistor, which is also the collector of the IGBT.

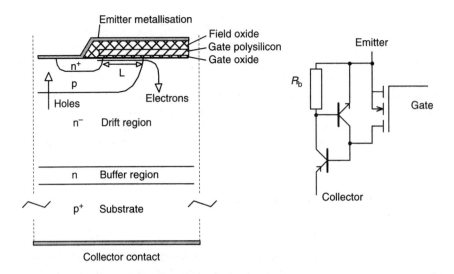

Figure A3.35   *Physical structure and equivalent circuit of a punch-through IGBT*

A3.8.3 D.C. characteristics

The d.c. characteristics of the IGBT are shown in Figure A3.36. Note the offset of V_{CE} due to the forward voltage drop of the p^+–n junction at the collector. The on-state voltage $V_{CE(sat)}$ of the IGBT has three components:

$$V_{CE(sat)} = V_{BE} + I_D \times R_{DRIFT} + I_D \times R_{CH} \qquad (A3.10)$$

where V_{BE} is the forward voltage drop of the PNP transistor, R_{DRIFT} is the drift region resistance (which is much smaller than in a equivalent MOSFET due to the conductivity modulation), and R_{CH} is the MOSFET channel resistance. Typical values for $V_{CE(sat)}$ are 2–3 V at rated current and 25 °C for a 1 200 V rated device.

Trench gate IGBTs have been developed recently, which, with an optimised geometry, reduce both R_{DRIFT} and R_{CH}, giving a lower $V_{CE(sat)}$ for a given current density.

Breakdown in the forward direction occurs in the same manner as for the MOSFET (punch-through structure). With a reverse bias applied between collector and emitter, the p^+–n junction at the emitter is driven into reverse bias. As a result the reverse blocking capability of the IGBT is very limited. The IGBT does not, therefore, display the reverse conduction characteristic of the MOSFET.

Like the pin diode, the IGBT conduction mechanism relies upon stored charge in the on-state (Figure A3.37). The quantity of stored charge, and hence the degree of conductivity modulation, depends on the carrier lifetime τ_{HL}. A device with a high carrier lifetime will thus exhibit high levels of stored charge, low on-state voltages,

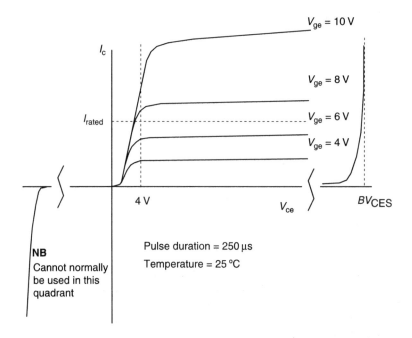

Figure A3.36   Static IGBT characteristics

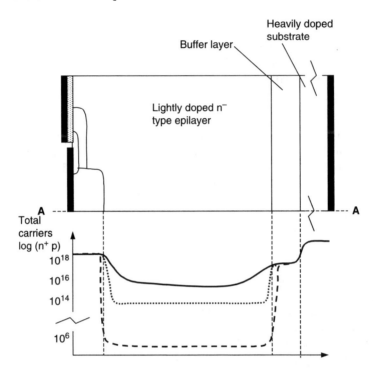

Figure A3.37   Cross-section through an IGBT half-cell (symmetrical about line AA) and charge distribution (measured as total free carriers) along line AA. The solid line shows on-state distribution, the dashed line off-state distribution. The dotted line shows the charge level in an equivalent MOSFET structure.

but rather slow transient performance. Lower carrier lifetimes yield faster devices at the expense of increased forward voltage drop. Some manufacturers produce several grades of IGBT ranging from 'low-sat' types (low on-state voltage but slow) through to 'ultrafast' (fast but high on-state voltage). A further consequence of the requirement for adequate conductivity modulation is that the carrier lifetime must increase with device voltage rating. Higher voltage devices will, therefore, be inherently slower.

A3.8.4   Punch-through versus non-punch-through structures (PT and NPT)

There are two basic types of IGBT, punch-through (PT) and non-punch-through (NPT). All early IGBTs were of the PT type. More recently, device processing techniques have allowed the production of NPT devices (Figure A3.38). The difference relates to the doping and thickness of the drift region and the collector layer, PT types having a thinner drift region and higher collector doping density than NPT types. Devices of this type are cheaper to produce and can be fabricated with voltage-blocking

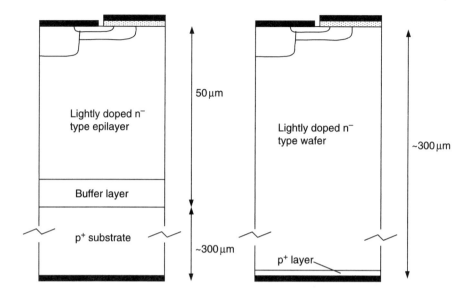

Figure A3.38 Punch-through (PT) and non-punch-through (NPT) IGBT structures

capabilities exceeding 3 kV. Note, however, the presence of the very wide lightly doped region. This requires a relatively long carrier lifetime, even for relatively low-voltage devices, which has important consequences for the switching characteristics. NPT devices generally have faster temperature-independent switching, and a positive temperature coefficient of $V_{CE(sat)}$, whereas PT devices have a strong temperature dependence of switching loss and a negative temperature coefficient of $V_{CE(sat)}$.

A3.8.5 Switching performance

Figure A3.39 shows the parasitic capacitances present within the IGBT structure. Of these the gate–emitter capacitance C_{ge}, gate–collector capacitance C_{gc} and collector–emitter depletion capacitance C_{ce} are analogous to those present in the MOSFET. The capacitance C_{cer} represents the quantity of stored charge in the lightly doped region of the device. Its value varies markedly with the applied bias and with the level of collector current. Unlike the depletion capacitance, C_{cer} has no component associated with the gate contact. In the off-state there is no stored charge and C_{cer} is effectively zero. During turn-on the level of stored charge increases and C_{cer} rises to its full on-state value. At turn-off C_{cer} is discharged by the action of the collector current.

A3.8.6 Transient characteristics

Like the MOSFET, the IGBT is voltage controlled and the gate presents a capacitive load to the drive circuit. Turn-on takes place when the capacitance has been charged to above the gate–emitter threshold voltage, which is usually 4–5 V. Typically a

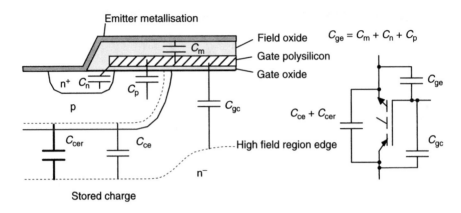

Figure A3.39   IGBT capacitances

1 200 V, 100 A IGBT will turn on in less than 200 ns. During the conduction time, the gate–emitter voltage is held between 13 and 17 V so that $V_{CE(sat)}$ is kept as low as possible to minimise conduction losses.

Turn-off is initiated by discharging the gate emitter capacitance. The MOSFET structure turns off first, allowing I_C to fall rapidly to an intermediate level. A slower fall of current then follows as the PNP structure turns off. Unfortunately, the gate drive circuit can only control the MOSFET turn-off and has no influence on PNP behaviour. Turn-off delay time and current fall time are much shorter than for equivalent bipolar transistors due to the low gain of the PNP structure and processing steps taken to reduce the carrier life-time. Both these measures have the effect of increasing $V_{CE(sat)}$ so there is a trade-off between low switching loss and low conduction loss. To give optimum performance in low- and high-frequency applications, IGBTs are available in different 'families', optimised for either low conduction or low switching loss.

A3.8.6.1   Switching waveforms

Figure A3.40 shows a typical clamped inductive load switching circuit employing a power IGBT and a clamp diode (assumed ideal) as its active components. The corresponding turn-on and turn-off transitions, under conditions of a rapid gate drive, are detailed in Figure A3.41.

A3.8.6.2   Turn-on

At time $t = 0$ a 15 V step is applied to the gate of the IGBT through a 10 Ω resistor. The initial conditions are such that the diode is conducting the full inductive load current of 40 A and the voltage across the IGBT may, therefore, be assumed to be around 1 000 V. The steps are then as follows.

- *Turn-on delay period* t_d. Charging of gate–emitter capacitance. No change in collector voltage or current.

Power semiconductor devices 153

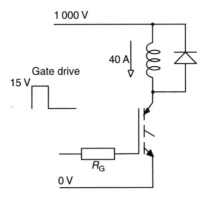

Figure A3.40 *IGBT clamped inductive load switching circuit*

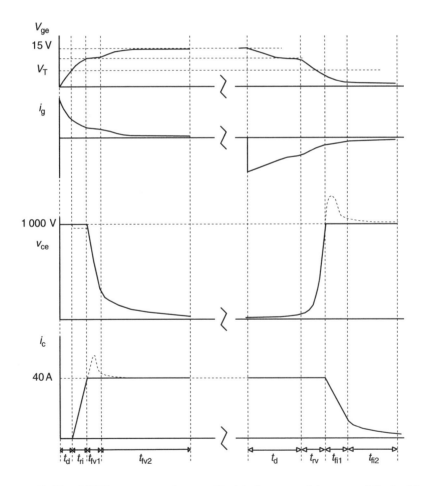

Figure A3.41 *IGBT turn-on and turn-off with fast gate drive; speed limited by bipolar transistor stored charge*

- *Current rise period* t_{ri}. Once the gate–emitter voltage exceeds the threshold voltage the MOS channel begins to conduct. Stored charge builds up during this period to a level limited by the presence of a high electric field.
- *First voltage fall period* t_{fv1}. Once the full load current has been established the collector voltage can begin to fall. As the high electric field region collapses the IGBT traverses its active region under conditions of constant collector current.
- *Second voltage fall period* t_{fv2}. The MOSFET channel is now operating in the ohmic region. As the gate voltage continues to rise, more charge is supplied, thus allowing a greater current to reach the bipolar transistor. The build up of stored charge is, however, a relatively slow process and is reflected in the rather slow rate of fall of collector voltage during this period.

A3.8.6.3 Turn-off

Turn-off is achieved by cutting off the MOS channel, thus stopping the supply of current to the base (n⁻) region of the p–n–p transistor. Provided a fast gate drive is applied the MOS channel will be cut off before any significant rise in collector voltage occurs (Figure A3.42).

- *Turn-off delay phase* t_d. If the rate of extraction of charge from the gate is large the MOS channel may become completely cut off before any significant increase in collector voltage has occurred. Although the p–n–p transistor base current has been removed it will continue to conduct current at a low collector voltage because of the high level of residual base charge. Only when sufficient base charge has recombined will the collector voltage begin to rise.
- *Voltage rise phase* t_{rv}. Extraction of charge from the wide n⁻ region (capacitance C_{cer}) occurs through the removal of electrons at the collector end and through the removal of holes at the emitter. A region of high electric field with a relatively low level of carrier concentration extends into the base region as the voltage increases (Figure A3.42). Because, to a first approximation, the level of stored charge and its rate of extraction increase in proportion to the collector current, the rate of rise of voltage is almost independent of current level (in practice an increase in dv/dt is observed). The resultant lateral flow of holes through the p-body region may become large enough to stimulate the injection of electrons from the emitter. This acts as a base current for the parasitic n–p–n transistor and may lead to triggering of the p⁺–n–n⁻–p–n⁺ thyristor structure and subsequent device destruction (avoid fast gate drives at high collector current). Note the absence of a distinct Miller plateau in Figure A3.41.
- *First current fall phase*. Once the clamp voltage level has been reached the collector current level can begin to fall. The initial fall is very rapid, being determined by the relatively small amounts of charge that must be extracted from the high field region and the p⁺–n junction.
- *Second current fall (current tail) period* t_{fi2}. Most of the charge remaining in the IGBT is now trapped at the collector end of the base region. The amount of charge depends principally on the applied voltage level – a low voltage means more

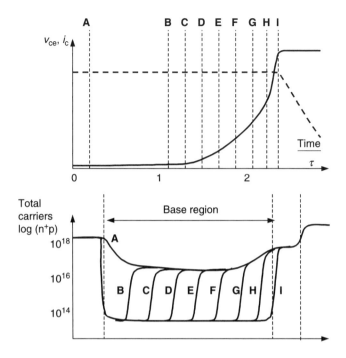

Figure A3.42 Charge distribution in base region of IGBT and corresponding collector waveforms during hard-switched turn-off. Successive curves denote the extent of the high field region at each of the labelled points on the collector waveforms. The charge remaining in the high field region ($\sim 5 \times 10^{13} cm^{-3}$) sustains the flow of collector current.

charge. Recombination of the stored charge produces a tail current, which may typically be of the order of 10 to 25 per cent of the initial collector current (PT structure) and which decays exponentially at a rate determined by the carrier lifetime. For NPT structures, the long lifetimes required might indicate unacceptable switching losses. However, careful engineering of the p^+ emitter region can produce a device with very low initial tail current levels (typically less than 5 per cent). The net effect is that NPT devices can exhibit turn-off losses that are similar or even lower than PT devices.

A3.8.7 Safe operating area (SOA)

A3.8.7.1 Forward-bias safe operating area (FBSOA)

Figure A3.43 shows a typical FBSOA for an IGBT. Note that the use of a higher gate voltage permits higher pulse currents but for a shorter duration. Operation on this part of the characteristic is usually only encountered under fault conditions and is often referred to as the 'short-circuit safe operating area'. Some IGBTs can be operated

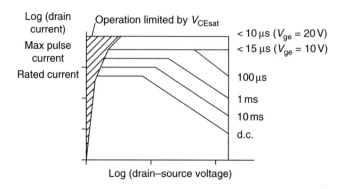

Figure A3.43 Typical forward-bias safe operating area for an IGBT

under pulsed conditions over the full possible range of collector currents at all levels of gate voltage (e.g. Figure A3.36), but others cannot.

The FBSOA curve for an IGBT chip is square shaped, and bounded by the rated breakdown voltage, and the pulsed collector current I_{DM} is usually twice the rated d.c. current I_D. The square SOA makes the IGBT a very robust device and allows operation without snubber circuits, thus reducing system losses and size.

A3.8.7.2 Reverse-bias safe operating area (RBSOA)

In general, the maximum current that can be safely turned off will be reduced at collector–emitter clamp/d.c. supply voltages approaching the rated voltage of the device. It should be noted that stray inductance can often lead to significant spikes of voltage at turn-off, which may take the switching trajectory outside the SSOA. This can become particularly important when considering turn-off from overload conditions. An example switching safe operating area is shown in Figure A3.44.

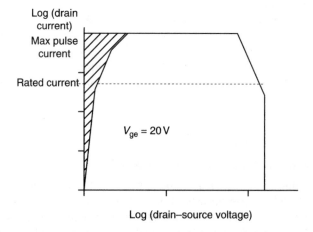

Figure A3.44 Typical reverse-bias safe operating area for an IGBT

Although the chip RBSOA is also square in shape, the device RBSOA may have some reduction in V_{CE} at high currents. This is due to wiring inductance within the package, which during turn-off increases the voltage at the chip above that measured at the terminals. Modern IGBTs do not suffer from secondary breakdown, unlike BJTs.

A3.8.8 Parasitic thyristor

Unlike the MOSFET, the IGBT does not have a parasitic diode inherent within its structure, so a suitable anti-parallel fast diode may be selected to match the speed of the IGBT. However, the addition of an injecting junction at the collector side of the IGBT serves to form a parasitic thyristor (p^+–n–n^-–p–n^+ structure). In normal operation the shunt resistance R_b across the p–n^+ body–emitter junction prevents the triggering of the n–p–n transistor. If the collector current density is high enough, however, the flow of lateral hole current around the n^+ emitter may become sufficient to stimulate the injection of electrons from the emitter into the p-body (lateral flow causes voltage drop and forward bias of the p–n^+ junction at the edge of the channel). This causes a limit to the maximum collector current that may flow without latch-up. The normal way of avoiding this problem is to limit the gate voltage such that the device enters the active region.

A3.8.9 IGBT gate drive requirements

Both turn-on and turn-off times can be adjusted by selection of gate resistor value; this controls the rate at which the input capacitance is charged or discharged. This is a very important feature as it allows turn-on to be set at a rate that suits the reverse recovery characteristic of the freewheeling diode, and allows the rate of turn-off to be reduced if required to limit inductive voltage over-shoot. During the off-state the IGBT gate emitter is normally held at a minimum of -5 V to ensure the device cannot be spuriously turned on.

A3.8.9.1 IGBT switching speed limitations

From the above it is clear that the turn-on speed of the IGBT is essentially limited only by the speed of drive supplied to the MOS gate (as for the MOSFET). Only the second, slow fall in collector voltage, which occurs as a result of the build-up of stored charge in the n^- region, remains unaffected. The final level of stored charge in the device, and hence the on-state voltage, is determined by the final level of applied gate voltage. Typical turn-on times are in the region of 100–200 ns and are largely independent of device type (e.g. 'low-sat'). For small IGBTs the turn-on rise time can be as low as 10 ns.

At turn-off the behaviour of the device depends largely on the characteristic lifetime of carriers in the n^- region. In a 'low-sat' device the lifetime is large and turn-off, even with moderate gate drive levels, will be controlled by the decay of stored charge rather than the MOS channel. For 'fast' or 'ultrafast' devices, turn-off speed can be limited by the level of applied gate drive. This is characterised by the appearance of a 'Miller' plateau in the gate voltage during the collector voltage rise phase (as for the MOSFET).

A3.8.9.2 Series and parallel operation

Although devices are available up to 6 500 V, devices above 1 700 V are somewhat specialised. For some applications this is not sufficient and so devices must be connected in series. This is easier to achieve with IGBTs than many other power devices due to the following combination of characteristics: voltage control, fast switching and square SOA.

Figure A3.45 shows how two IGBTs may be connected in series. The resistors maintain steady-state voltage balance by compensating for differences in device leakage currents. Very fast transients are balanced by the capacitors, which can be much smaller than for a similar circuit using BJTs or thyristors. During switching, differences in delay from one device to another will tend to lead to unbalanced voltages. The active clamp circuit, formed by the zener string and IGBT, limits V_{CE} to a value a little above the zener breakdown voltage. The voltage-controlled gate means that little current is needed and fast switching ensures a fast response. The IGBT is able to survive this operation because of the square SOA.

Unlike the power MOSFET, the on-state voltage of the IGBT does not have a guaranteed positive temperature coefficient. Thus it is possible that parallel connected devices will not share current evenly unless they are specifically selected (e.g. from the same silicon wafer). Most IGBTs are now manufactured with a positive $V_{CE(sat)}$ temperature coefficient to ease parallel connection and yield improved short-circuit characteristics. Devices may be connected in parallel to make up power switches with ratings of many kiloamperes.

When using devices with a positive temperature coefficient, little or no de-rating is required. However, it is important to note that when using devices with a negative

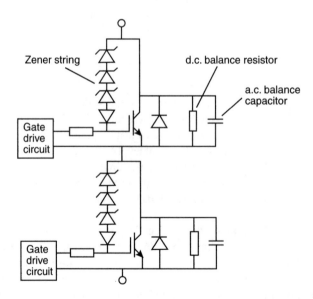

Figure A3.45   Series-connection IGBT with active clamp

temperature coefficient it is usual to select devices with the same $V_{CE(sat)}$ and apply a de-rating factor.

A3.8.9.3 IGBT short-circuit performance

Most IGBTs produced for high-power applications are short-circuit rated. This means that the device is able to withstand a short period of very high dissipation without damage. A typical device will be rated for a short-circuit duration of 10 μs with test conditions of a current of ten times I_D, with a collector–emitter voltage of half rated voltage, and 125 °C. During this fault condition the junction temperature may exceed 300 °C. The control and gate drive circuit must detect the over-current condition and switch off the IGBT within the rated time to avoid damage. The short-circuit current may be controlled by adjusting the gate voltage. For the Infineon 'DN2' series devices a V_{GE} between 10 and 17 V will produce short-circuit currents of between 2 and 10 times I_D. To reduce the inductive voltage over-shoot when turning off the large short-circuit current, the gate drive circuit may slow down the turn-off under these conditions by limiting the gate discharge current.

A3.8.10 Voltage and current ratings

The principal voltage rating of the IGBT is the collector–emitter breakdown voltage V_{CES}, specified with zero gate–emitter voltage. Devices are available with ratings from 250 up to 6 500 V. Although devices are available with terminal current ratings up to 2 400 A these devices use many IGBT chips connected in parallel. Single chips typically have a maximum current rating of 300 A.

A3.9 Bipolar junction transistor (BJT)

The bipolar junction transistor (BJT) was one of the earliest common switching device, but today has largely been superseded by the IGBT, although BJTs are still used in existing equipment. A three-layer NPN device is shown in Figure A3.46. PNP types are available but they tend to have inferior voltage and current ratings.

To conduct a collector current I_C, the transistor must be supplied with a continuous base current I_B, depending upon the voltage level between collector and emitter. The ratio I_C/I_B is called the current gain h_{FE}, and may be less than 10 for a 1 000 V transistor. The gain can be greatly improved if the base current is obtained from another transistor using the Darlington connection shown in Figure A3.47. The two transistor stages are integrated on the same silicon chip, giving an overall gain of several hundred.

The BJT ratings are in the order of 1 800 V and 1 000 A (not simultaneously in a single device). The main disadvantages of a BJT are the need to supply a significant proportion of the total collector current as a base drive current, slow turn-off, limited SOAs and high on-state voltages in Darlington arrangements.

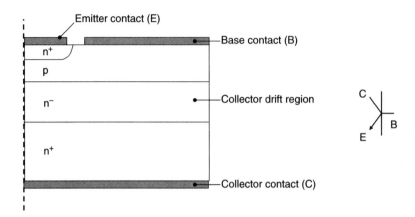

Figure A3.46   BJT structure and symbol

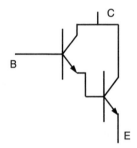

Figure A3.47   Two-stage Darlington arrangement

A3.10   Other power devices and materials

This section briefly describes emerging power devices that have the potential to be used in future drives. It also includes a brief introduction to the semiconductor materials that may replace silicon in the future.

A3.10.1   MOS controlled thyristor (MCT)

The MOS controlled thyristor (MCT) was developed to exploit the low conduction loss of a thyristor and the low gate drive power and fully controlled behaviour of a MOSFET. The symbol and simplified equivalent circuit of an MCT are shown in Figure A3.48.

Operation is most easily understood from the equivalent circuit. In the off-state the turn-off MOSFET is held on, keeping the PNP device off, and the MCT can block positive anode cathode voltage. To turn on, the 'turn-off' FET is switched off and the 'turn-on' FET switched on. This provides base current to the NPN transistor and regenerative action then takes place as with a standard thyristor. Once in the

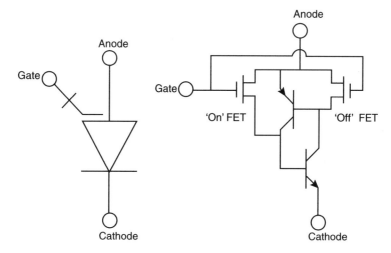

Figure A3.48   MOS controlled thyristor symbol and equivalent circuit

on-state, the MCT has a similar surge capability to a thyristor, with a low voltage drop of 1 to 2 V. To turn off the MCT the 'turn-off' FET is again turned on. This short-circuits the PNP transistor base and so regenerative action stops and the device turns off in a manner very similar to a GTO.

Although the MCT offers low on-state loss and low gate drive power, there are important limitations that have prevented its widespread use: the gate bias must be maintained at all times to ensure the device remains off, and the switching SOA is limited to half rated voltage at rated current. Devices have been made with maximum ratings of approximately 1 500 V, 100 A.

A3.10.2   MOS turn-off thyristor

The MOS turn-off thyristor (MTO) is very similar in many respects to the MCT and IGCT. The circuit symbol and equivalent circuit are shown in Figure A3.49. In contrast to other power devices the MTO is a four-terminal device; in addition to the main power terminals there are two gate terminals, one for turn-on and another for turn-off.

As can be seen from the equivalent circuit, the MTO comprises a GTO, which is responsible for turn-on and conduction, and a MOSFET, which is only used during turn-off. To turn on a gate, current of several tens of amps is injected in the turn-on gate. Once conduction is initiated, regenerative action starts and the anode current is limited only by the external circuit. As with GTOs and IGCTs, a current of several amps is required to ensure the device stays on with minimum voltage drop. To turn off, a positive voltage is applied to the turn-off gate terminal. This switches on the MOSFET, which diverts current out of the gate, forcing regenerative action to stop and the GTO to turn off. As with an IGCT, sufficient gate current is removed to give a turn-off gain of near unity, which gives much faster turn-off than for a GTO operating at a turn-off gain of 3 to 5. As the turn-off MOSFET is integrated into

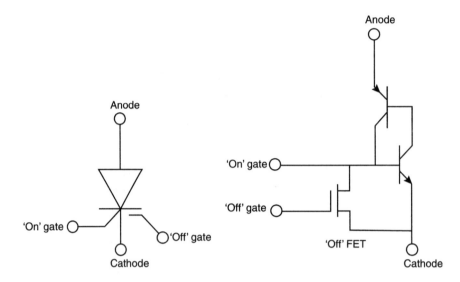

Figure A3.49   MOS turn-off thyristor symbol and equivalent circuit

the MTO the inductance is low enough to ensure that the gate current can be removed fast enough to prevent current redistribution during turn-off.

Devices have been made with ratings up to 4 500 V and 500 A, with plans for ratings up to those of conventional GTOs.

A3.10.3   Junction field-effect transistors (JFETs)

Junction field-effect transistors (JFETs) are available in silicon or silicon carbide (SiC). Silicon-based JFETs are commercially available and used in low-power applications (below 2 W). SiC JFETs are starting to become available. SiC JFETs are normally-on devices, where the channel is 'open', conducting significant current when no gate voltage is present. This is a problem in conventional drives applications as it is easy to generate a short-circuit in one phase due to the top-bottom arrangement of the switches in the inverter. Therefore, SiC JFET devices can come with a series connection with a silicon MOSFET, packaged in one housing. This arrangement is called a 'cascode' (Figure A3.50). The advantages in using a cascode are that the switch is normally-off, and a standard MOSFET gate drive circuit can be applied. The disadvantages with this circuit are that the possibilities of the SiC JFET are not fully utilized and the silicon MOSFET contributes to the on-resistance.

SiC JFETs have been made with 1 500 V, 35 A ratings.

A3.11   Materials

Although nearly all commercially available power devices are based on silicon technology, there are several semiconductor materials that offer potential improvements in

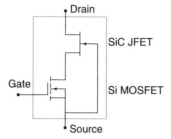

Figure A3.50 JFET cascode

performance, including higher blocking voltages and extended operating temperature range. These include gallium arsenide (GaAs), SiC and diamond. Silicon has a maximum junction temperature of 125–175 °C. GaAs can operate at temperatures of 200 °C and SiC operates with junction temperatures of 250–350 °C. The electric field strength of SiC is in the order of magnitudes larger than GaAs or silicon. Diamond is the material with the highest semiconductor properties.

In all cases, the main problem is immaturity of materials technology and the high cost of both the starting material and processing. So, far GaAs and SiC have been exploited commercially to produce very fast diodes with low reverse recovery effect. SiC diodes become more and more popular in replacing silicon-based anti-parallel diodes in standard three-phase drives. The cost of these SiC diodes is, however, still high. SiC switching devices are not available on the market, but samples are provided by a few semiconductor manufacturers. The only SiC switching device available at the time of writing is the JFET.

A3.12 Power device packaging

A3.12.1 General

Every power semiconductor device must eventually be packaged. Depending upon the application, the packaging may be responsible for the following important functions:

- physical containment for one or more basic component building blocks e.g. semiconductor dies, capacitors, inductors, resistors,
- protection from the environment, e.g. ingress of liquids, dust,
- thermal management (dissipation/transfer of heat generated),
- circuit interconnections either directly as part of the package structure or through internal bus-bars, printed circuit boards (PCBs), direct bonded copper (DBC) ceramic substrates, or metal insulated substrates (MIS),
- electromagnetic management (EMC issues), and
- packages for power semiconductor dies.

For this reason, power devices are available in a very wide range of packages, a selection of which is shown in Figure A3.51. The very smallest packages are the 'wafer-scale' packages, which only have a layer of passivation on the surface of the chip. These must be soldered to a PCB or other substrate and switch ratings are limited to a few hundred VA. Packaged devices, normally housing a single chip, for through hole or surface mounting are available in industry standard sizes such as D2-Pak, TO-220, TO-247 in ratings up to 10 kVA. For larger devices with ratings up to 1 MVA isolated base modules with moulded plastic cases are used. There is a

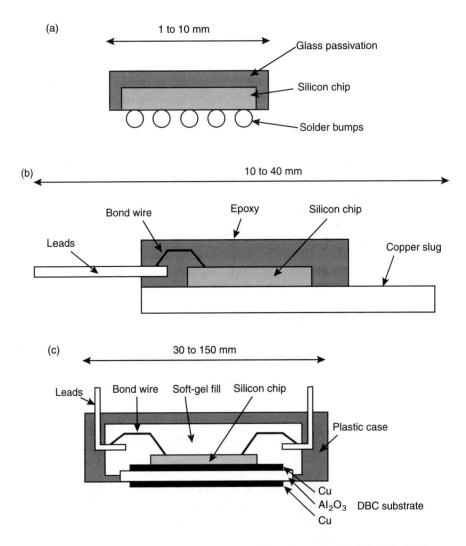

Figure A3.51 Package type cross-sections: (a) wafer scale, (b) TO-220, (c) direct-bonded copper (DBC), (d) copper base, (e) pressure

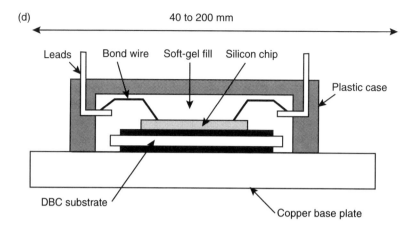

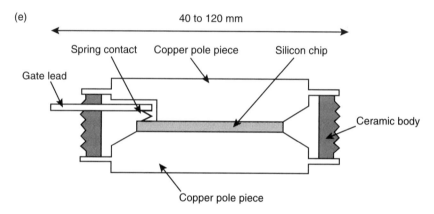

Figure A3.51   Continued

wide variety of packages containing one or more devices but with little standardisation in sizes. For the highest power devices, up to 10 MVA, pressure contact packages are used. These only house a single device.

A3.12.2   Pressure contact packages

Pressure contact packages are in general only used for very-high-power applications. Devices rated above 4.5 kV or 2.5 kA are in general only available in pressure contact packages. They are well suited to devices with a large single chip such as thyristors, although they are also now being applied to multi-chip devices such as IGBTs.

A3.12.2.1   Construction

As shown in Figure A3.51e a pressure contact package consists of two large copper pole pieces between which the chip is sandwiched. Externally applied pressure ensures contact between the chip and the contact plates. For the gate connection an internal pressure contact is provided. A ceramic body provides the rest of the enclosure and may be ribbed to increase the creepage distance over the surface. After assembly the joints are welded, and the package evacuated and sealed. In use, the device is clamped between two plates, usually the heatsinks, and a known pressure applied. As the pole pieces are in direct contact with the chip the heatsinks and mounting plates are all live. Great care must be taken to ensure correct even contact pressure; this makes the mechanics of pressure contact devices quite complex.

A3.12.2.2 Features

Pressure contact devices have several features that make them well suited to very-high-power, high-reliability applications.

- Double-sided cooling gives significantly reduced thermal impedance compared to single-sided cooling. This allows a device to operate at a high loss per unit area.
- The two main causes of power device wear-out are the failure of wire bonds and soldered contacts due to thermally induced mechanical stress. As a pressure contact device has no wire bonds or soldered joints it has a very good thermal cycling capability, which is especially important in applications such as railway traction.
- In the event of a failure the device will go short circuit. This allows redundancy to be built into high-voltage applications, which have several devices connected in series.
- The package has a high rupturing I^2t so that with correct fusing it is possible to prevent rupture of the device in the event of failure. This is very important in high-voltage applications (>3 kV) to limit damage to other equipment.

A3.12.3 Large wire-bonded packages for power modules

Large wire-bonded packages are used for power modules housing single devices rated over 50 A or multiple devices rated over 10 A. Maximum ratings are typically 4.5 kV and 2.5 kA. They are very widely used and are much cheaper than the equivalent device in a pressure pack.

A3.12.3.1 Construction

There are two basic types of package, either with or without a copper base plate, as shown in Figure A3.51c and d. The copper base plate gives better transient thermal impedance, aids heat spreading across the heatsink, and makes the device less prone to damage due to incorrect mounting.

The chip or chips are first soldered to the DBC substrate. This consists of two layers of copper, between which there is an aluminium oxide (Al_2O_3) or aluminium nitride (AlN) insulator. The DBC isolates the chips and the power circuit from the base plate of the device and so the heatsink may be earthed. The top layer of copper is etched to form an interconnect pattern similar to a PCB. Wire bonding is then used to connect the substrate, chips and package terminals. A plastic package supports the power terminals and provides mechanical protection. To provide electrical insulation and environmental protection the chips and bond wires are covered in an insulating gel.

In use the package is fixed to a heatsink using screws with electrical contacts using either soldered or screw terminals.

A3.12.3.2 Package types

Single power devices with screw power terminals are available in ratings from approximately 50 to 2 500 A. For currents above 800 to 1 000 A, several parallel contacts are used. There are a wide variety of package sizes, from 20×92 mm^2 to 140×190 mm^2, some sizes being adopted as a de facto standard.

Screw terminal packages are also available with multiple power devices. These may range from a half-bridge (two power switches) to a three-phase inverter. Ratings are available up to 450 A, 1 200 V as a three-phase inverter.

For packages that contain multiple devices at current ratings up to 150 A, all-solder terminals tend to be the most cost-effective (Figure A3.52). Although module packaging is frequently supplier proprietary, some standardisation has occurred, notably in the Econo2 and Econo3 packages. Care must be taken, however, as the fact that the package is standard does not mean that the components are electrically interchangeable. These modules may house various topologies, such as a complete three-phase rectifier or a three-phase rectifier and brake chopper or a three-phase d.c.–a.c. inverter or a rectifier, inverter and brake chopper. The latter type is often known as a power integration module (PIM) or converter inverter brake (CIB) module.

Packages containing multiple devices are also available with screw terminals for the power connections or pressure contacts (Figure A3.53).

In addition to the main power devices, the control, monitoring and protection circuits may also be mounted within the module (Figure A3.54). These parts are commonly known as intelligent or integrated power modules (IPMs). The additional circuits are normally assembled onto a small PCB, which is wire-bonded to the main power devices.

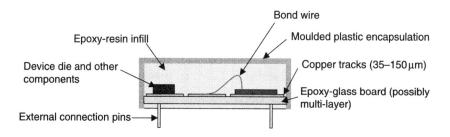

Figure A3.52   Cross-section of a typical d.c.–d.c. converter module

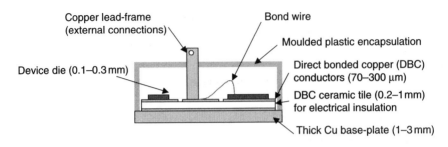

Figure A3.53   Typical power device module schematic

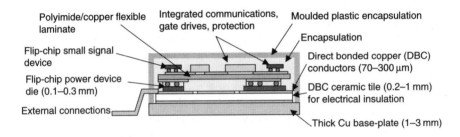

Figure A3.54   Flip-chip on flex integrated power module

A3.12.3.3   Features

Large wire-bonded packages with moulded plastic housings are the most flexible of power device packages. The moulded case allows many terminals to be placed in a variety of locations, and features may be incorporated to increase the creepage and clearance distances between terminals.

The DBC substrate allows several chips to be placed within the package and connected in a variety of ways to form either high-current single switches or complete power circuits, as in the case of a PIM or CIB module. Where additional control or monitoring circuits are needed, these may be easily incorporated in the DBC or a separate PCB within the housing.

Devices with solder terminals are only designed to be soldered to a PCB. Screw terminal packages are more flexible as they may be screwed to a PCB for low- or medium-current applications or to bus-bars for higher currents.

Compared to pressure packs, wire-bonded packages have some disadvantages for very-high-power applications. Surge current rating may be wire-bond limited; in the event of a failure the device will go open circuit as the wire bonds blow off, and in the event of a major device failure the package may rupture.

A3.12.4   Small wire-bonded packages for discrete devices

Small wire-bonded packages for discrete devices are made in very large volumes and have the lowest production costs. In general they only contain a single chip, although some packages are available with two chips, such as an IGBT with anti-parallel diode. The ratings of these devices are limited by the available chip area and the current rating of the leads. Surface-mount (SM) packages are commonly available up to 30 A, with through-hole devices up to 70 A.

A3.12.4.1 Construction

The chip is first soldered to a copper 'slug' before wire-bonding to the lead-frame. An epoxy resin is then moulded over the chip to provide mechanical support for the leads, electrical isolation and environmental protection.

A3.12.4.2 Package types

There are a wide variety of packages, many of which are industry standard. Packages that are deigned for screw mounting, such as TO-220 or TO-247, can also be clip mounted. Using a clip saves parts and labour, and gives more even contact pressure to the heatsink. By having a package without a mounting hole, the popularity of clip mounting is being exploited to fit larger chips within a given footprint.

In many power applications it is desirable to have isolation for the heatsink. Traditionally this has been achieved by putting an insulating washer between the device and heatsink. By moulding the epoxy resin around the whole copper slug and chip, a fully insulated package can be produced. As the epoxy is a poor thermal conductor, device ratings are reduced. To overcome this limitation, devices are also available that have a small DCB substrate rather than a copper slug. This provides electrical isolation and much better thermal conductivity than epoxy resin.

A3.12.4.3 Features

Packages are soldered to a PCB, which in the case of SM packages provides the cooling path in addition to electrical connection.

Chapter A4
Torque, speed and position control

A4.1 General principles

A4.1.1 The ideal control system

Many applications exist where something has to be controlled to follow a reference quantity. For example, the speed of a large motor may be set from a low-power control signal. This can be achieved using a variable-speed drive as described in the following.

Ideally, the relationship between the reference and the motor speed should be linear, and the speed should change instantly with changes in the reference. Any control system can be represented as in Figure A4.1b, with an input reference signal, a transfer function F and an output. For the system to be ideal, the transfer function F would be a simple constant, so that the output would be proportional to the reference with no delay.

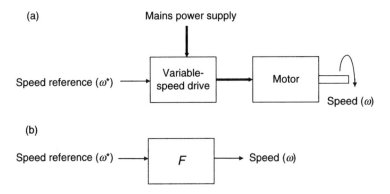

Figure A4.1   Variable-speed drive and motor

A4.1.2 Open-loop control

Unfortunately, the transfer function of many practical systems is not a constant, and so without any form of feedback from the output to correct for the non-ideal nature of the transfer function, the output does not follow the demand as required. Using an induction motor supplied by a simple open-loop variable-speed drive as an example, the following illustrates some unwanted effects that can occur in practical systems:

- *Speed regulation.* The output of a simple open-loop drive is a fixed frequency that is proportional to the speed reference, and so the frequency applied to the motor remains constant for a constant speed reference. The speed of the motor drops as load is applied because of the slip characteristic of the motor, and so the speed does not remain at the required level.
- *Instability.* It is possible under certain load conditions and at certain frequencies for the motor speed to oscillate around the required speed, even though the applied frequency is constant. Another major source of instability in rotating mechanical systems is low-loss elastic couplings and shafts.
- *Non-linearity.* There are many possible sources of non-linearity. If, for example, the motor is connected to a gearbox, the speed at the output of the gearbox could be affected by backlash between the gears.
- *Variations with temperature.* Some aspects of the system transfer function may vary with temperature. For example, the slip of an induction motor increases as the motor heats up, and so for a given load the motor speed may reduce from the starting speed when the motor was cold.
- *Delay.* With a simple open-loop inverter and induction motor there can be a delay before the motor speed reaches the demanded level after a change in the speed reference. In very simple applications such as controlling the speed of a conveyor belt, this type of delay may not be a problem. In more complex systems, such as on a machine tool axis, delays have a significant effect on the quality of the system.

These are just some of the unwanted effects that can be produced if an open-loop control system is used. One method that improves the quality of the controller is to use a measure of the output quantity to apply some feedback to give closed-loop control.

A4.1.3 Closed-loop control

The simple open-loop drive of Section A4.1.2 can be replaced with a control system as in Figure A4.2. This control system not only provides a means to correct for any error in the output variable, but also enable a stable response characteristic.

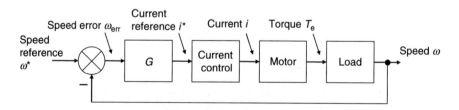

Figure A4.2 *Closed-loop control system*

The speed of the motor shaft is measured and compared with the speed reference to give a speed error. The error is modified by a transfer function G to give a current reference i^* at the input to the current control block. Various methods of current control for motors are discussed in this chapter; however, for now it should be assumed that the motor current can be controlled to give a torque that is proportional to the current reference. If the speed of the motor varies from the reference level a speed error is produced and the torque applied to the load is modified to bring the speed back to the required level.

It is necessary to choose a suitable transfer function G to obtain the required performance from the closed-loop control system. The function could be a simple gain, therefore the current reference $i^* = K_p \times \omega_{err}$. This would give some degree of control over the output speed, but the speed error must have a non-zero value if any torque is required to hold the motor speed. If the speed error is not zero, then the speed would not be at the required reference level, and so the speed would vary with load. By adding an integral term, so that the current reference $i^* = K_p \times \omega_{err} + K_i \int \omega_{err} \, dt$, it is no longer necessary to have any speed error even when torque is required to drive the load at the reference speed. The integral term accumulates any speed error over time and builds up a current reference to provide the necessary torque. A closed-loop control system with proportional and integral terms is called a PI controller. Although there are many types of closed-loop controller, the PI controller is the most commonly used because it is simple to implement, relatively easy to set up and is well understood by most engineers.

A4.1.4 Criteria for assessing performance

The step response, where a step change is made on the demand and the output response monitored, is one method of assessing the performance of a closed-loop control system. Some example step responses are shown in Figure A4.3 for a simple second-order system.

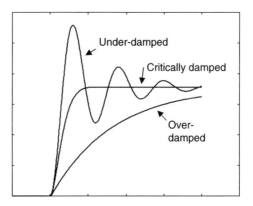

Figure A4.3 *Step responses*

- If the output reaches the reference in the shortest possible time without any overshoot, the response is described as critically damped. If overshoot is not acceptable, then this represents the best possible response giving the minimum delay between the input and output of the system.
- If the system damping is increased, the response becomes slower and is described as over-damped.
- If the system is under-damped the response includes some overshoot and may oscillate about the required reference before settling.

These results, which are for a simple-second order system, show that increasing the damping reduces overshoot and slows down the system response. As will be demonstrated later, real systems can be more complex, and increasing the damping does not always give this result.

The step response may be the closed-loop response of the system, where a change to the reference level in a minimum time is required. Alternatively, the step response may show the change of output to some other stimulus, such as a load torque transient. In this case the response should usually be as small as possible.

The closed-loop step response can be used to assess the performance when the control system is used in isolation. However, if the controller itself is to be included within the closed control loop of another system, the gain and delay are important as they affect the performance of the outer system. The gain and delay can be measured by producing a Bode plot of the gain and phase response against frequency. An ideal controller would have unity gain and zero phase shift at all frequencies; however, in a real system the gain reduces and the phase delay increases at higher frequencies.

A measure of these effects is the bandwidth of the system, which is often defined as the -3 dB point of the gain characteristic. In the example given in Figure A4.4 this occurs at 670 rad s^{-1}. The corresponding phase delay varies depending on the order of the system. A first-order system has a delay of $45°$ at the -3 dB point, whereas a second-order system like the one in the example has a delay of $60°$ at the -3 dB point. The transport delays associated with digital systems can further increase the delay at the -3 dB point.

In many cases the bandwidth is quoted as an indication of the dynamic performance of a control system; in other words, the higher the bandwidth, the better the performance. *This bandwidth is usually quoted as the frequency at the -3 dB point of the gain characteristic, and it is very important to note that, particularly with a digital implementation, there may be no indication of the quality of the control system at all.* If the controller is to be included within another closed-loop control system the phase delay is important. If the delay is too large it may be necessary to de-tune the outer loop to maintain stability. The amount of overshoot in the step response is also important in many applications. Increasing the frequency of the -3 dB point of the gain characteristic may result in unacceptable overshoot.

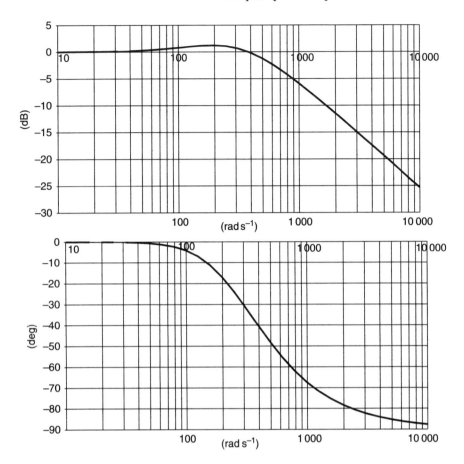

Figure A4.4   Bode plot of system gain and phase

A4.2   Controllers in a drive

A4.2.1   General

Although a modern variable-speed drive includes many features, the basic function of the drive is to control torque (or force), speed or position. Before proceeding to the specific details of how different types of variable-speed drive function, the theory of control for each of these quantities is discussed. A position control system is shown in Figure A4.5. This includes an inner speed controller, and within the speed controller there is an inner torque controller. It is possible to create a system where the position controller determines the mechanical torque that is applied to the load directly without the inner speed and torque loops. However, the position controller would need to be able to control the complex combined transfer function of the motor windings, the mechanical load and the conversion from speed to position.

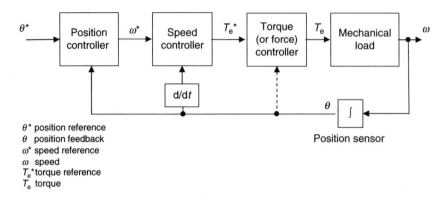

θ^* position reference
θ position feedback
ω^* speed reference
ω speed
T_e^* torque reference
T_e torque

Figure A4.5   Position control system

Therefore it is more usual to use the format shown in Figure A4.5. The other advantage of this approach is that limits can be applied to the range or rate of change of speed and torque between each of the controllers. When a system is required to control speed only, the position controller is omitted, and when a system is required to control torque only, the position and speed controllers are omitted.

A position sensor is shown providing feedback for the system, but this may be replaced by a speed sensor or it may be omitted altogether as follows.

- Position information is required by the torque controller to function in an a.c. motor drive (see the dotted line). If position feedback is provided the speed feedback is derived as the change of speed over a fixed sample period. Sensorless schemes are possible for speed and torque control of a.c. motors, in which case the sensor is not required.
- Position feedback is not necessary for the torque controller in a d.c. motor drive, so a speed feedback device such as a tacho-generator can be used to provide the feedback for the speed controller. Again, sensorless schemes are possible where a speed feedback device is not required.

A4.2.2   Torque control

A torque controller for a rotary motor, or a force controller for a linear motor, is the basic inner loop of most variable-speed drives. Only torque control is discussed here, but the principles also apply to force control for a linear application. In order to explain the principles of torque control, the simple d.c. motor system in Figure A4.6 is used as an example. The analysis of torque control in an a.c. motor can be done in exactly the same way, provided suitable transformations are carried out in the drive. These transformations will be discussed later.

The torque demand or reference (T_e^*) is converted by the torque controller into a current in the motor armature, and the motor itself converts the current into torque

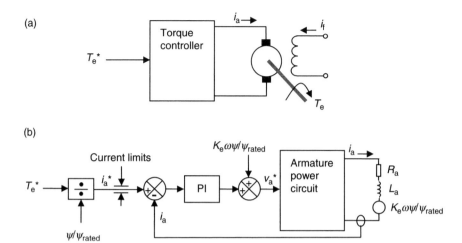

Figure A4.6 Torque and current controllers in a d.c. motor drive: (a) torque control; (b) current control

to drive the mechanical load. Figure A4.6b shows the system required to convert the torque reference into motor current. The torque reference (T_e^*) is first transformed into a current reference (i_a^*) by including the scaling effect of the motor flux. The motor flux, controlled by the motor field current (i_f), is normally reduced from its rated level at higher speeds when the terminal voltage would exceed the maximum possible output voltage of the power circuit without this adjustment. Current limits are then applied to the current reference so that the required current does not exceed the capabilities of the drive. The current reference (limited to a maximum level) becomes the input for the PI controller. The electrical equivalent circuit of the motor consists of a resistance (R_a), an inductance (L_a) and a back emf that is proportional to flux and speed ($K_e \omega \psi / \psi_{rated}$).

The PI controller alone could successfully control the current in this circuit because as the speed increases, the voltage required to overcome the back emf would be provided by the integral term. The integral control is likely to be relatively slow, so to improve the performance during transient speed changes a voltage feed-forward term equivalent to $K_e \omega \psi / \psi_{rated}$ is included. The combined output of the PI controller and the voltage feed-forward term form the voltage reference (v_a^*), and in response to this the power circuit applies a voltage (v_a) to the motor's electrical circuit to give a current (i_a). The current is measured by a sensor and used as feedback for the current controller.

As well as the linear components shown in Figure A4.6, the current control loop in a digital drive includes sample delays as well as delays caused by the power circuit. In practice, the response of the controller is dominated by the proportional gain. In particular, if a voltage feed-forward term is used, the integral term has very little effect on the transient response.

Setting of the control loop gains is clearly very important in optimising the performance of the control loop. One of the simplest methods to determine a suitable proportional gain is to use the following equation:

$$K_p = K \times L_a / T_s \tag{A4.1}$$

where L_a is the motor inductance and T_s the current controller sample time. K is a constant that is related to the current and voltage scaling, and the delays present in the control system and power circuit. Most modern variable-speed drives include auto-tuning algorithms based on measurement of the electrical parameters of the motor taken by the drive itself, and so the user does not normally need to adjust the current controller gains.

It is useful to know the closed-loop transfer function of the torque controller (i.e. T_e/T_e^*) so that the response of a stand-alone torque controller, or the effect of an inner torque controller on outer loops such as a speed controller, can be predicted. As the response is dominated by the system delays it is appropriate to represent the closed-loop response as simple gains and a unity gain transport delay as shown in Figure A4.7.

The torque reference could be in N m, but it is more conventional to use a value that is a percentage of the rated motor torque. Figure A4.7a gives the transfer function when the torque controller is used alone. K_t is the torque constant of the motor in N m A^{-1}. If the torque controller is used with an outer speed controller a slightly different representation must be used, as in Figure A4.7b. The speed controller produces a torque reference where a value of unity corresponds to a current level that is specified for the size or rating of the drive. From a control perspective it is unimportant whether this is the maximum current capability of the drive, the rated current or some other level. The actual level used is defined as K_c (in amperes), and should be included in the transfer function as shown. These simple models allow the drive user to predict the performance of a stand-alone torque controller or a torque controller with an outer speed loop.

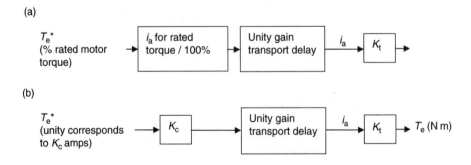

Figure A4.7 *Torque controller closed-loop response: (a) stand-alone torque controller; (b) torque controller with outer control loop*

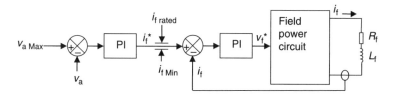

Figure A4.8   Flux controller

A4.2.3   Flux control

The motor flux and hence the motor terminal voltage for a given speed are defined by the flux producing current. In the example of a simple d.c. motor drive used previously, the motor flux level is set by the field current, i_f. The flux controller (Figure A4.8) includes an inner current loop and an outer loop that maintains rated flux in the motor until the armature terminal voltage reaches its maximum limit. When the motor speed increases above rated speed it then controls the field current and hence the flux, so that the armature voltage remains at the maximum required level.

The torque and flux controller are both shown in a simplified form in Figure A4.9. Both have an inner current controller that generates a voltage reference and a power circuit that converts this into a current. The system shown is for a d.c. motor drive, but vector-based transformations can be used to simulate this for an a.c. motor and form the basis of a vector control system for an a.c. motor drive.

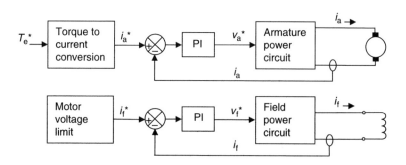

Figure A4.9   Torque and flux controllers

A4.2.4   Speed control

A4.2.4.1   Basic speed control

Closed-loop speed control can be achieved by applying a simple PI controller around the torque controller described previously. For the purposes of this analysis it is assumed that the load is an inertia J, with a torque T_d that is not related to speed (friction is neglected). The resulting system is shown in Figure A4.10.

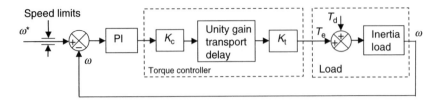

Figure A4.10   Speed controller

If the PI controller is represented as $K_p + K_i/s$, the torque controller is assumed to be ideal with no delays so that the unity transport delay can be neglected, and the inertia load is represented as $1/Js$ then the forward loop gain in the s domain is given by

$$G(s) = (K_p + K_i/s) \times K_c K_t \times 1/Js \tag{A4.2}$$

The closed-loop transfer function in the s domain $\omega(s)/\omega^*(s)$ is given by $G(s)/[1 + G(s)]$. Substituting for $G(s)$ and rearranging gives

$$\omega(s)/\omega^*(s) = (sK_p/K_i + 1)/[s^2 J/(K_c K_t K_i) + sK_p/K_i + 1] \tag{A4.3}$$

If the natural frequency of the system is defined as $\omega_n = \sqrt{(K_c K_t K_i/J)}$ and the damping factor is defined as $\xi = \omega_n K_p/(2K_i)$ then

$$\omega(s)/\omega^*(s) = (s2\xi/\omega_n + 1)/(s^2/\omega_n^2 + s2\xi/\omega_n + 1) \tag{A4.4}$$

As with the torque controller, it is useful to know the closed-loop response so that the response of a stand-alone speed controller, or the effect of an inner speed controller on an outer position loop, can be predicted. If a moderate response is required from the speed controller it is not significantly affected by system delays, and a linear transfer function such as equation (A4.4) can be used. All the constants in these equations and the delays associated with the current controllers are normally provided to users so that calculations and/or simulations can be carried out to predict the performance of the speed controller.

In addition to providing the required closed-loop step response, it is important for the system to be able to prevent unwanted movement as the result of an applied torque transient. This could be because a load is suddenly applied or because of an uneven load. The ability to prevent unwanted movement is referred to as stiffness. The compliance angle of the system is a measure of stiffness and will be discussed later (A4.2.4.2).

The following equations can be derived from Figure A4.10:

$$\omega(s) = [T_e(s) + T_d(s)]/sJ \tag{A4.5}$$

and

$$T_e(s) = -\omega(s) \times K_c K_t (K_p + K_i/s) \quad (A4.6)$$

Combining these equations gives

$$\omega(s)/T_d(s) = [1/K_c K_t K_i] \times [s/(s^2 J/K_c K_t K_i + sK_p/K_i + 1)] \quad (A4.7)$$

This is the speed response to a load torque transient. Dividing both sides by s is equivalent to integrating in the time domain, giving the position response to a torque transient:

$$\theta(s)/T_d(s) = [1/K_c K_t K_i] \times [1/(s^2 J/K_c K_t K_i + sK_p/K_i + 1)] \quad (A4.8)$$

Again substituting for the natural frequency and damping factor gives

$$\theta(s)/T_d(s) = [1/K_c K_t K_i] \times [1/(s^2/w_n^2 + s2\xi/w_n + 1)] \quad (A4.9)$$

The step response of the closed-loop transfer function (equation A4.4) and the response to a torque transient (equation A4.9) are shown in Figure A4.11. These examples are for a Control Techniques Unidrive SP1404 ($K_c = 5.8$ A) and a Control Techniques

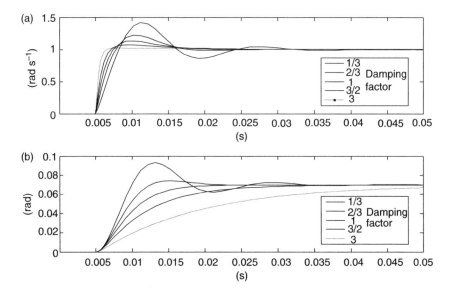

Figure A4.11   Responses of an ideal speed controller: (a) closed-loop step response; (b) response to a torque transient

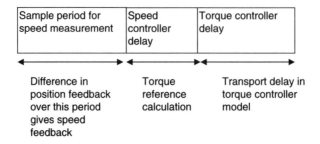

Figure A4.12   Unwanted delays in a practical digital drive

Dynamics 115UMC 3 000 rpm servo motor ($K_t = 1.6$ N m A^{-1}, $J = 0.00078$ kg m^2) with the speed controller gains set to $K_p = 0.0693\xi$ and $K_i = 14.32$.

As the damping factor is increased, the closed-loop response overshoot is reduced and the speed of response improves. The closed-loop response includes 10 per cent overshoot with a damping factor of unity because of the s term in its numerator.

As the damping factor is increased, the overshoot of the response to a torque transient is reduced and the response becomes slower. In this case there is no s term in the numerator and the response includes no overshoot with a damping factor of unity.

It would appear from these results that the higher the proportional gain, and hence the higher the damping factor the better the responses; however, the results so far assume an ideal torque controller and no additional unwanted delays. In a real digital drive system the delays given in Figure A4.12 are likely to be present. A delay is included to represent the sample period for speed measurement, but this is only relevant if the speed feedback is derived from a position feedback device such as an encoder and is measured as a change of position over a fixed sample period.

The effect of the unwanted delays can be seen in the closed-loop step response for a real system as shown in Figure A4.13. In each case the response of the real system has more overshoot than the ideal system. If the damping factor is set to unity then the overshoot may be acceptable, but with a damping factor of 1.25 the response is quite oscillatory and is likely to be unacceptable. The effect of the unwanted delays is more pronounced the longer the delay and also as the set bandwidth of the speed controller is increased.

The effect of the additional delays can be seen in the Bode plot of the closed-loop response of the speed controller set up to give unity damping factor (Figure A4.14). The frequency at the -3 dB point of the gain characteristic has increased significantly from the ideal speed controller, whereas the frequency at the 60° point of the phase characteristic is almost unchanged. If this is to be used as a stand-alone controller the gain characteristic could be used to predict the bandwidth, although it should be noted that the gain is greater than unity at some frequencies. Often the bandwidth based on the gain characteristic is the only bandwidth that is quoted, because this makes the performance appear to be better, in this case 2 000 rad s^{-1}. However, if the speed controller is to be included within an outer position controller, the bandwidth

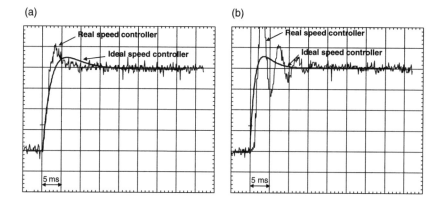

Figure A4.13 Effect of delays on a closed-loop step response: (a) damping factor = 1; (b) damping factor = 1.25

based on the phase delay (672 rad s^{-1} for this example) must be used, as this affects the performance of the outer loop.

Unwanted delays limit the performance of the speed controller. The quantised nature of speed feedback when it is derived from a position sensor as the change of position over a fixed sample period can also limit this. A high proportional gain in the speed controller, and hence high bandwidth, generate high-frequency torque ripple and acoustic noise from the quantised speed feedback.

A4.2.4.2 Setting speed controller gains

Provided the load inertia is known, it is possible to choose proportional and integral gains for the speed controller based on the required bandwidth or the required compliance angle using the equations for the ideal speed controller. If the resulting bandwidth of the control loop is fairly conservative, the controller will behave approximately as predicted by the ideal model equations.

Selection based on bandwidth

If the bandwidth is defined as the frequency where the closed-loop gain is -3 dB (i.e. $1/\sqrt{2}$), then substituting into equation (A4.4) gives

$$1/\sqrt{2} = (s2\xi/\omega_n + 1)/(s^2/\omega_n^2 + s2\xi/\omega_n + 1) \qquad (A4.10)$$

If the bandwidth is defined by the frequency ω_{bw} and $s = j\omega_{bw}$ then substituting for s in equation (A4.10) and rearranging gives

$$\omega_{bw}/\omega_n = \sqrt{\{(2\xi^2 + 1) \pm \sqrt{[(2\xi^2 + 1)^2 + 1]}\}} \qquad (A4.11)$$

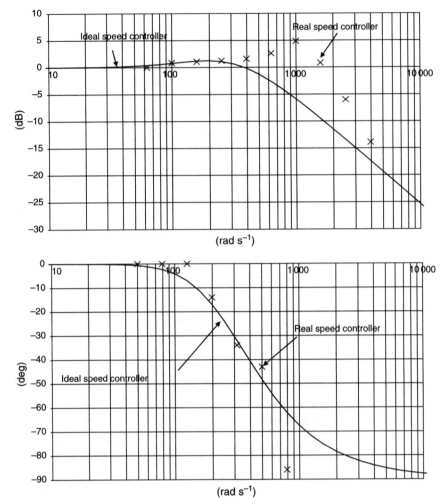

Figure A4.14 Bode plot of closed-loop response of a speed controller

The characteristic defined by equation (4.11) is shown in Figure A4.15. The required damping factor must first be selected, and from this the ratio ω_{bw}/ω_n is taken from the graph. For example, if a damping factor of unity is required, the value of ω_{bw}/ω_n is 2.5.

From the definition of natural frequency as $\omega_n = \sqrt{(K_c K_t K_i / J)}$

$$\omega_{bw} = \omega_n \times (\omega_{bw}/\omega_n) = \sqrt{(K_c K_t K_i / J)} \times (\omega_{bw}/\omega_n) \qquad (A4.12)$$

Rearranging this gives the required value for the integral gain

$$K_i = J/(K_c K_t) \times [\omega_{bw}/(\omega_{bw}/\omega_n)]^2 \qquad (A4.13)$$

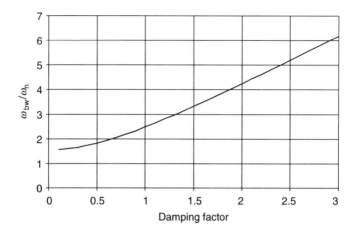

Figure A4.15 Effect of damping factor on bandwidth

The definition of damping factor is $\xi = \omega_n K_p/(2K_i)$. By rearranging this and substituting for natural frequency, a suitable value for the proportional gain can be derived:

$$K_p = 2\xi \times \sqrt{[(K_i J)/(K_c K_t)]} \qquad (A4.14)$$

Selection based on compliance angle

From equation (A4.9) the steady-state response to a torque transient can be derived by setting $s = 0$. The resulting change of output angle for a given steady-state torque T_d is

$$\theta = T_d/K_c K_t K_i \qquad (A4.15)$$

If the compliance angle is defined as the change of output angle with a steady-state torque equal to $K_c K_t$ (i.e. the torque produced by the motor with a torque producing current equal to K_c) then a suitable value of integral gain can be selected for a given compliance angle θ_c as

$$K_i = 1/\theta_c \qquad (A4.16)$$

The proportional gain can then be determined in the same way as for the selection based on bandwidth using equation (A4.14).

If the load (including motor) inertia is known, or it is possible for the drive to measure the load inertia using an auto-tuning algorithm, then the user need only enter the required damping factor and the speed loop bandwidth or compliance angle into the drive parameters. The drive can then automatically set suitable values for the speed controller gains.

A4.2.4.3 Speed control with torque feed-forward

There are situations where it is desirable to have low-speed controller gains, for example in a winder application where the inertia of the load may change in time,

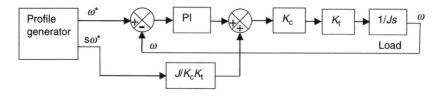

Figure A4.16 Speed controller with torque feed-forward

or an elevator where high gains would excite resonances in the system, making it unstable. The disadvantage of using low gains is that the response of the speed controller is degraded. If the load torque characteristic is known, it is possible to use a torque feed-forward term to significantly improve the response and still use low-speed controller gains. Figure A4.16 shows a system with a known inertia load using torque feed-forward.

The profile generator produces the required speed reference ω^*, which may be a simple linear ramp that accelerates and decelerates the load. It also produces the rate of change of speed or acceleration $s\omega^*$ ($d\omega^*/dt$ in the time domain). When this is scaled to include the inertia and the system constants, it gives a current reference that should accelerate and decelerate the load as required. Now the speed controller is only required as a trim to compensate for inaccuracies in the torque feed-forward. The response of the system is only limited by the response of the current controllers and the sample rate of the profile generator, and not by the response of the speed controller.

A4.2.5 Position control

A4.2.5.1 Basic position control

If position control is required, an additional outer loop is applied to the speed controller as shown in Figure A4.17. The position control loop only includes a proportional term K_v. An integral term is not generally required as any static error gives a speed reference ω^*, and the integral term in the speed controller forces the load to move until the position error is removed. K_v normally has units of mm s^{-1}/mm, ms^{-1}/m, rad s^{-1}/s etc. Provided the position units in the speed and position parts are the same (i.e. rad and rad s^{-1}) then the following analysis applies. From

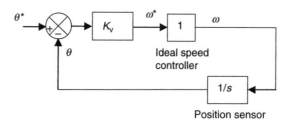

Figure A4.17 Position controller

Figure A4.17 the closed-loop response can be calculated in the same way as for the speed controller:

$$\theta(s)/\theta^*(s) = K_v/s/(1 + K_v/s) = K_v/(s + K_v) \quad (A4.17)$$

Applying a unity step change of position reference gives the step response in the s domain:

$$\theta(s) = 1/s \times K_v/(s + K_v) = 1/s + [1/(s + K_v)] \quad (A4.18)$$

The response in the time domain is given by

$$\theta = 1 - e^{-K_v t} \quad (A4.19)$$

This shows that the basic position controller has a first-order step response with a time constant that is defined by K_v, as shown in Figure A4.18. If K_v is set to 10 the response is relatively poor with a time constant of 0.1 s and a bandwidth of 10 rad s^{-1} (1.6 HZ). If K_v is set to 50 the response is considered quite good with a time constant of 20 ms and a bandwidth of 50 rad s^{-1} (8 HZ). The maximum value of K_v is constrained by the bandwidth of the inner speed controller and the system sample times.

A position controller is often used to ensure that the load follows a position profile that includes periods of acceleration and deceleration. A measure of the quality of the system is the error between the position reference and the actual position throughout the profile, often referred to as the following error. Figure A4.19 gives the position error for a basic position controller (Figure A4.17) that is trying to follow a position profile that causes the motor to accelerate to 100 rad s^{-1} and then decelerate back to standstill. The position controller gain has been set to $K_v = 50$.

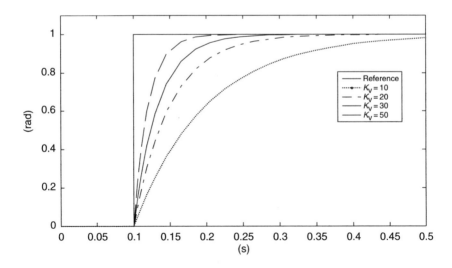

Figure A4.18 Position controller step response

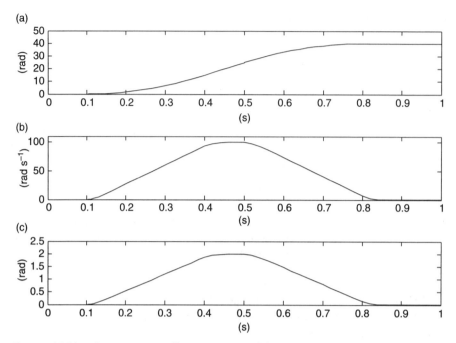

Figure A4.19   Position controller response with linear ramped profile: (a) position; (b) speed; (c) position error

The position controller only includes a proportional gain, and so the response includes regulation because there must be a position error to generate the speed reference. The position error is proportional to speed and is given by $\theta_{\text{error}} = \omega/K_v$. In this case the maximum speed is 100 rad s^{-1}, and so the maximum position error is 2 rad.

To look at the response of a position controller to a load transient the inner speed controller must also be included in the model. Figure A4.20a shows the s domain representation of the outer position controller and an inner ideal speed controller.

If the feedback loop is considered alone, it can be simplified as shown in Figure A4.20b. This is very similar to the speed controller feedback loop, but with the additional $(1 + K_v/s)$ term. Therefore

$$\omega(s) = [T_e(s) + T_d(s)]/sJ \qquad (A4.20)$$

and

$$T_e(s) = K_c K_t (1 + K_v/s)(K_p + K_i/s) \times -\omega(s) \qquad (A4.21)$$

Combining these equations gives

$$\omega(s)/T_d(s) = (1/K_c K_t K_i) \times \{s^2/[s^3 J/K_c K_t K_i + s^2 K_p/K_i \\ + s(1 + K_v K_p/K_i) + K_v]\}7 \qquad (A4.22)$$

Torque, speed and position control 189

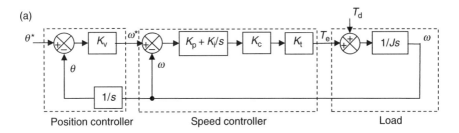

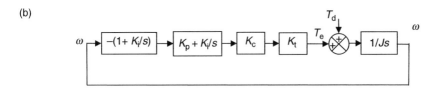

Figure A4.20 s-Domain representation of position controller with load transient: (a) full position controller; (b) simplified feedback loop

This is the speed response to a load torque transient. Dividing both sides by s is equivalent to integrating in the time domain, giving the position response to a torque transient.

$$\theta(s)/T_d(s) = (1/K_c K_t K_i) \times \{s/[s^3 J/K_c K_t K_i + s^2 K_p/K_i \\ + s(1 + K_v K_p/K_i) + K_v]\} \quad (A4.23)$$

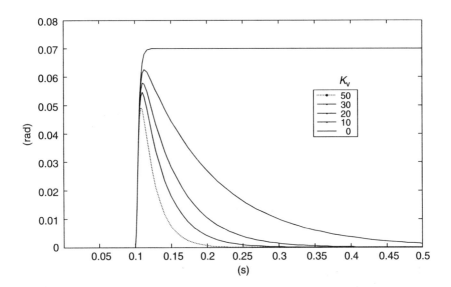

Figure A4.21 Position controller with speed feed-forward

The responses in Figure A4.21 are based on the same speed controller used to give the response to a torque transient (see Figure A4.11), with the gains set up to give a damping factor of unity. When K_v is set to zero the response is the same as the speed controller. When K_v has a non-zero value the position controller compensates for the position change caused by the torque transient and the final position error is zero. As would be expected, the higher the gain the faster the response. It is important to note that with a speed controller alone there will always be a position error when a torque transient is applied because the controller has limited stiffness. When an outer position loop is used the system becomes infinitely stiff in the steady state.

A4.2.5.2 Position control with speed feed-forward

It has been shown that a position error proportional to speed occurs during position changes with a basic position controller (Figure A4.19). One method of overcoming this problem is to use speed feed-forward in addition to the position controller as shown in Figure A4.22.

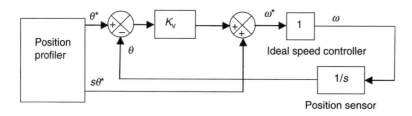

Figure A4.22 Position controller with speed feed-forward

The profiler produces the position reference θ^* and the rate of change of position reference $s\theta^*$ ($d\theta^*/dt$ in the time domain). If the speed feed-forward term is used alone without the position controller and the load is a pure inertia, the position follows the reference, but with a position error related to the stiffness of the speed controller because torque is required to accelerate and decelerate the load. The stiffness of the speed controller is given by equation (A4.15). The torque required to accelerate or decelerate the load is $J \times d\omega/dt$, so the position error during the speed changes is given by

$$\theta_{error} = (J \times d\omega/dt)/(K_c K_t K_i) \tag{A4.24}$$

Figure A4.23 shows the results of using the speed feed-forward term alone with a position error of 0.002 rad as defined by equation (A4.24) during acceleration and deceleration ($J = 0.00078$ kg m^2, $K_c = 5.8$ A, $K_t = 1.6$ A N^{-1} m^{-1}, $K_i = 14.32$). If a position controller is added then the position error during acceleration and deceleration is reduced to zero after a delay defined by the response time of the position controller. This can be seen in Figure A4.24, where the position controller has now

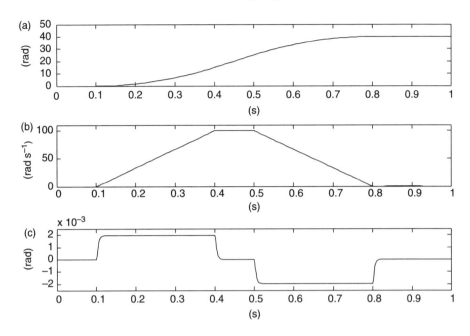

Figure A4.23 Speed feed-forward only: (a) position; (b) speed; (c) position error

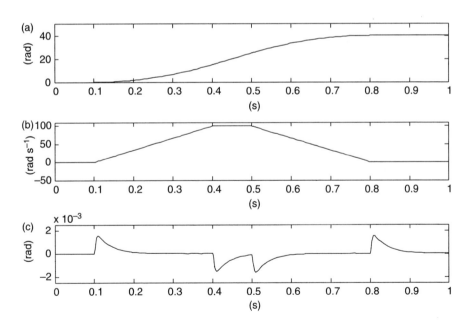

Figure A4.24 Position control and speed feed forward: (a) position; (b) speed; (c) position error

been added with K_v set to 30. The 'torque transients' caused by the acceleration and deceleration have been modified in the same way as shown in the torque transient response (Figure A4.21). If K_v is reduced, the peak of the position error transients will not be reduced so much and the time to recover to zero error will be longer. Conversely if K_v is increased, the peak of the position error and recovery time is reduced.

Throughout this section it has been assumed that the system sample times and current controller delays do not affect the response of the system. This is true provided the required response from the position controller is not too fast and the controllers are operated with reasonable sample times.

A4.3 D.C. motor drives

A4.3.1 General

Most medium and large industrial d.c. motor drives are based on a separately excited motor. The flux is generated by a field winding and the torque by a higher current armature winding fed via the commutator. These two windings are completely independent, and so the flux and torque can be controlled independently. Although the field winding usually has a long time constant, the armature winding time constant is normally fairly short, allowing fast changes of armature current and hence torque. The general principles of torque, speed and position control have already been covered using an example of a simple d.c. motor drive. The position and speed control loops need no further explanation as the descriptions given previously cover most types of variable-speed drive. In this section the specific details of the torque and flux controllers for a medium/large industrial d.c. motor drive are given. Figure A4.9 gives a generic torque and flux control scheme, Figure A4.25 shows more detail specifically related to an industrial d.c. motor drive.

A4.3.2 Torque control

For four-quadrant operation, two thyristor bridges are used for the armature converter. Both bridges can apply positive or negative voltage to the motor, but the positive bridge can only supply positive current and the negative bridge negative current. Therefore, the positive bridge conducts when positive torque is required and the negative bridge when negative torque is required. The bridges are controlled to apply the voltage demanded by the reference v_a^* to the motor. Owing to high voltage ripple in the converter output and the unidirectional nature of thyristors, the current in the armature can be continuous or discontinuous. While the current is continuous, the relationship between the voltage reference and the actual applied voltage is a cosine function where the voltage reference is used directly to control the firing angle of the converter. When the current is discontinuous the relationship is highly non-linear and varies with the voltage level applied to the motor. A typical industrial drive stores the relationship between the firing angle and motor current for different output voltage levels, and

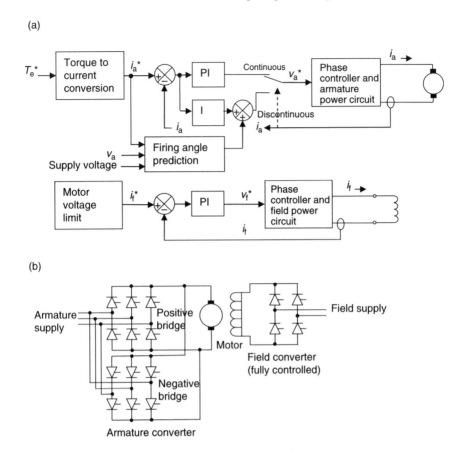

Figure A4.25 Separately excited d.c. motor drive: (a) torque and flux control; (b) power circuit

during discontinuous current operation, the correct firing angle is selected by the firing angle prediction block for a given current reference. Any errors are trimmed out by an integrator operating on the current error.

When a change in direction of torque is required, one bridge must stop conducting and the other bridge must become active. It is clear from the power circuit diagram that only one bridge may conduct at a time during this changeover to avoid a short circuit across the supply. It is important that this changeover occurs as quickly as possible to give good dynamic torque control. Modern microprocessor controlled drives enable intelligent methods to be used to keep the bridge changeover delay as short as possible.

Commutation of current from one thyristor to another during continuous-current operation is achieved by applying firing pulses to a non-conducting thyristor that has a more positive phase supply voltage than the thyristor presently conducting

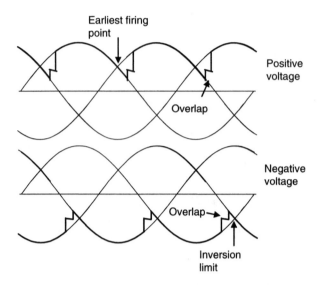

Figure A4.26   Voltage produced by the top row of thyristors

in the top row (or more negative in the bottom row). The supply has some inductance, and so the current does not transfer instantaneously from one thyristor to the other, but takes a finite time known as the overlap time. During this period the voltage applied to the motor is derived from the average of the two conducting phases. This presents no problem during motoring, when the bridge operates as a rectifier, except for distortion of the supply known as notching. However, during braking the bridge operates as an inverter and care must be taken to ensure that commutation is complete before the inversion limit shown in Figure A4.26. If it is not complete by this point the current builds up again in the thyristor that should turn off, and a large pulse of current flows in the motor. This is known as inversion failure. The traditional way to prevent inversion failure is to limit the firing angle to ensure commutation is completed before the inversion limit. However, this limits the motor voltage during braking so that it is lower than the limit when motoring. Many modern industrial d.c. drives include algorithms that monitor the situations that lead to inversion failure, allowing the firing angle to move right back to the inversion limit under some conditions.

A4.3.3   Flux control

The field converter is either a fully or half-controlled thyristor bridge. The half-controlled bridge can only apply positive voltage to the field winding, and so the current in the winding can be increased quickly, but decays relatively slowly. The fully controlled converter can apply positive or negative voltages, and so the performance is the same whether the current is increasing or decreasing.

A4.4 A.C. motor drives

The general principles of torque, speed and position control have already been covered using a simple d.c. drive as an example. The position and speed loops need no further explanation as the descriptions given previously cover most types of a.c. motor drive. In this section the torque and flux controllers in an a.c. drive for a permanent magnet or induction motor are described. An alternative, so called open-loop scheme, is also described that is often used with induction motors. The principles described may be applied to current or voltage source inverters, but only voltage source inverters are covered here.

A4.4.1 Torque and flux control

A4.4.1.1 Introduction

The simple d.c. motor drive system previously shown in Figure A4.9 is used here along with the equivalent system for an a.c. motor drive (Figure A4.27). A d.c. motor has separate windings to control torque and flux. Although separate windings do not exist in an a.c. motor, separate control of torque and flux are still required. There are several different conventions used to identify the currents and voltages associated with these functions in a.c. motor drives. Here, the subscript T is used

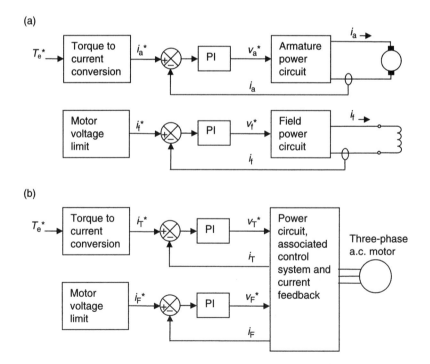

Figure A4.27 Torque and flux controllers: (a) d.c. motor drive; (b) a.c. motor drive

for quantities associated with controlling torque and F for quantities associated with controlling flux.

The armature and field windings are no longer separate and are now combined into one set of three phase a.c. windings. A power circuit and associated control system must be provided that can convert the outputs from the torque and flux producing current controllers (v_T^* and v_F^*) into the appropriate motor terminal voltages. A current feedback system is also required that can resolve the three-phase motor currents into torque and flux producing current feedback (i_T and i_F). Before considering how this could be implemented for an a.c. motor drive it is useful to look in more detail at torque and flux control in a d.c. motor.

A4.4.1.2 D.C. motor torque and flux control

Figure A4.28 gives a simplified representation of a d.c. motor where the windings are represented as one conductor. The field (stator) winding is producing flux ψ_f, and so there is a force on the armature (rotor) conductors because of the interaction of this flux and the armature current i_a. The flux is a physical effect in the motor and it is clear that it can be represented by a space vector as shown. The armature current i_a can also be represented by a space vector, but in reality it is the vector representing the flux ψ_a produced by the current. The force on the rotor conductors applies a torque to the rotor in an anti-clockwise direction. If the rotor is free to move it will turn in this direction. As it turns the torque reduces until the rotor has turned through 90°, at which point there will be no more force and it will stop. The function of the commutator is to ensure that current flows in the armature winding with the physical relationship shown in Figure A4.28, even when the rotor turns so that maximum torque is always produced.

It is easier to understand the force produced on the armature conductors by considering the flux produced by the field (ψ_f) and the current in the armature conductors ($\mathbf{i_a}$). In reality both windings are producing flux as shown in the vector diagram. The vector sum of the field and armature fluxes ψ is all the flux that passes through the rotor and the stator generated by the field and the armature windings neglecting the effects of flux leakage.

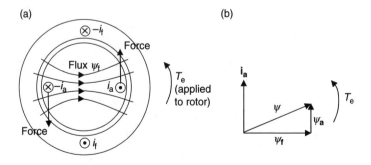

Figure A4.28 Current and flux in a d.c. motor: (a) physical representation; (b) vector representation

A4.4.1.3 Permanent magnet motor torque and flux control

The d.c. motor described above has a field (flux controlling) winding on its stator and an armature (torque controlling) winding on its rotor. In most a.c. permanent magnet motors the flux is provided by magnets on the rotor and the torque is controlled by current in the three-phase windings on the stator. (It is also possible to control the level of the flux with current in the stator windings.) A physical representation of a permanent magnet motor is given in Figure A4.29a, which shows the interaction of the rotor magnet flux with the stator U-phase winding only. (The interaction of the flux with the other windings will be considered later.) If the rotor is positioned as shown, and the U-phase current is at its maximum (Figure A4.29b), then the force on the rotor, and hence the torque applied to the rotor, will be the maximum possible for a given level of current in the windings. If the U-phase current remains fixed at this level and the rotor is free to move it will rotate anti-clockwise. The torque will be reduced as it rotates until it has rotated by 90°. At this point the torque will have reduced to zero and the rotor will stop.

Before proceeding to describe how it is possible to change the current in the stator windings in order to prevent the torque from reducing as the rotor turns, the effect of the V and W windings must be considered. To do this the current in all three phases is represented as a vector. In Figure A4.30 the instantaneous level of current in each phase winding is added together, including the effect of the phase shift between the windings to form a single vector i_T. The current levels shown are those where the U-phase current is at its positive peak and the other two phase currents are at half their negative peak (as in Figure A4.29b). The V- and W-phase currents also contribute to the torque applied to the rotor. It should be noted that current cannot really be expressed as a space vector and the vector actually represents the flux produced by the torque-producing current shown as ψ_T in the diagram.

In Figure A4.30 the torque-producing current vector (i_T) and rotor flux vector (ψ_F) are at right angles to each other and the motor will produce the maximum torque for the level of current in the stator windings. To maintain maximum torque, the torque-producing current vector must move as the rotor, and hence the rotor flux moves. In

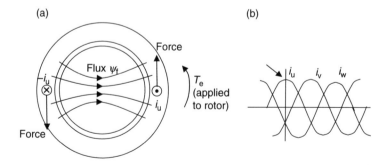

Figure A4.29 Current and flux in a permanent magnet a.c. motor: (a) physical representation; (b) motor phase currents

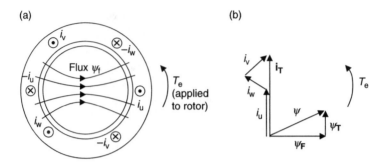

Figure A4.30   *Vector representation including all three-phase currents: (a) physical representation; (b) vector representation*

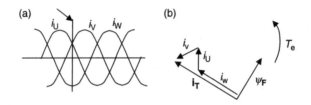

Figure A4.31   *Current vector movement: (a) motor phase currents; (b) vector representation*

Figure A4.31 the point in the current waveforms is displaced by 60° from the example used previously and now i_W is at its negative peak. The vector representing the current has moved anti-clockwise (forwards) by 60°. This shows that as the phase angle of the currents changes, the angle of the vector changes. If the current vector moves at the same rate as the rotor then the required angle between the current and the flux is maintained to give maximum torque for a given level of current.

The torque produced by the motor is proportional to the product of the magnitude of the torque-producing current vector i_T and the magnitude of the flux vector ψ_F. If the correct angle is maintained between these two vectors the torque produced by the motor can be changed quickly to the required level by changing the magnitude of the current vector. Therefore the torque in a permanent magnet motor can be controlled in the same way as in a d.c. motor. As well as being able to control the torque, it is also possible to control the flux level in the motor. If the angle of the current is displaced as shown in Figure A4.32 the vector now has a component that produces torque (i_T) and a component that modifies the level of the flux (i_F) that is at right angles to the torque-producing component. This is the basis for vector control of an a.c. motor giving fast independent control of torque and flux. To be able to implement such a system in practice, a set of axes referred to as a reference frame is defined with the x-axis aligned to the rotor flux space vector. The current component in the x-axis controls the motor flux and the component in the y-axis controls the torque.

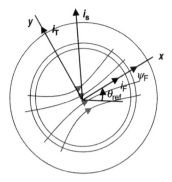

Figure A4.32 Reference frame

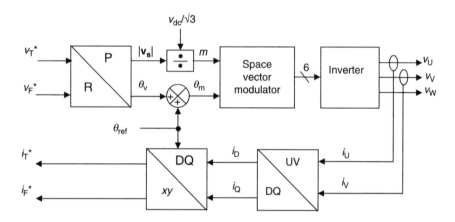

Figure A4.33 Reference frame translations

The block diagram in Figure A4.33 shows a practical implementation of the system required to transform or translate the voltage references in the reference frame (v_T^* and v_F^*) into the physical motor terminal voltages, and to convert the physical motor phase currents into the current feedback components in the reference frame (i_T and i_F). This provides the functionality of the block shown in Figure A4.27 as 'Power circuit, associated control system and current feedback'.

In this system, the voltage references are first converted from rectangular components into a polar representation of magnitude and phase:

$$|v_s| = \sqrt{(v_T^{*2} + v_F^{*2})}$$
$$\theta_v = \tan^{-1}(v_T^*/v_F^*)$$
(A4.25)

An inverter is used to produce the motor terminal voltages. The inverter is fed with a d.c. supply (v_{dc}) and the output of the inverter is dependent on this voltage. To remove

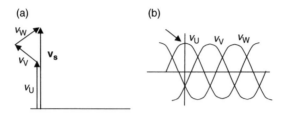

Figure A4.34 Voltages represented by a vector: (a) voltage vector; (b) motor terminal voltages

this dependency the voltage magnitude is scaled and divided by v_{dc} to give m, the 'modulation index'. θ_v is the angle of a voltage vector in the reference frame. This angle is converted to an angle with respect to the stator of the motor θ_m:

$$\theta_m = \theta_{ref} + \theta_v \tag{A4.26}$$

m and θ_m can be used as the inputs to a space vector modulator, which produces the control signals for the inverter. It should be clear that although a voltage vector has been defined, this is only conceptual, as an actual voltage space vector does not exist in the motor. The voltage vector is a way of representing the required magnitude and phase of the motor terminal voltages as in Figure A4.34. At the point shown, the U-phase voltage is at its peak and the V- and W-phase voltages are negative at half their peak levels. As the vector rotates, balanced sinusoidal voltages should be produced by the inverter.

The voltage vector concept is used by the space vector modulator to derive the inverter switching signals. There are eight possible states for the devices in the inverter, excluding those that result in shoot-through, as shown in Figure A4.35. These states give six possible active voltage vectors (u_1 to u_6), and the remaining two states (u_0 and u_7) give zero voltage vectors where all three output phases are at the same voltage level. By switching rapidly between the various states, the average output from the inverter can be a voltage vector at any angle with the required magnitude. In the example given below, a voltage vector at angle α could be synthesised by producing u_1 and u_2, each for 50 per cent of the time. Changing the ratio of the time during which u_1 and u_2 are active allows the vector to be moved from u_1 to u_2 with a locus that forms one side of a hexagon connecting the tips of u_1 and u_2. The magnitude of the vector can be controlled to be anywhere inside the hexagon connecting the tips of the active vectors by introducing periods of zero voltage with vectors u_0 and u_7. The active vector periods and inverter output waveforms for the required vector u_s are shown in Figure A4.35b.

In addition to translating the voltage references from the reference frame components to the inverter output voltages, the inverter output phase currents must be translated back into reference frame components. In the first stage, the three 120° displaced phase currents (i_U, i_V, i_W) are converted to two quadrature 90° displaced

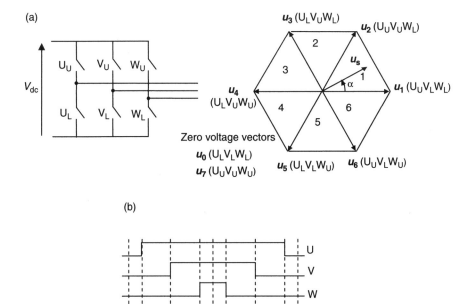

Figure A4.35   Space vector modulator: (a) space vectors; (b) inverter output voltages

currents (i_D, i_Q). As there is no neutral connection $i_U + i_V + i_W = 0$, and only two of the phase currents are required in order to describe all the currents in the motor. If i_U and i_V are used, then substituting for $i_W = -(i_U + i_V)$ gives

$$i_D = (2/3)(i_U - i_V/2 - i_W/2) = i_U$$
$$i_Q = (2/3)[(\sqrt{3/2})i_V - (\sqrt{3/2})i_W] = (i_U + 2i_V)/\sqrt{3} \qquad (A4.27)$$

The quadrature components are then translated into components in the reference frame:

$$i_F = i_D \cos \theta_{ref} + i_Q \sin \theta_{ref}$$
$$i_T = i_Q \cos \theta_{ref} - i_D \sin \theta_{ref} \qquad (A4.28)$$

Figure A4.36 gives an example of the transformation of the currents in a drive. These results are for an induction motor and include a component of flux-producing current to demonstrate the control of current in both axes (a permanent-magnet motor would have no flux-producing current at standstill because the flux is produced by the magnets). The flux-producing current is set to 30 A at the start to increase the flux in the motor and then it is reduced down to the rated level of 10 A at approximately 0.1 s. At 0.2 s a torque-producing current of 20 A is applied until 0.4 s to accelerate

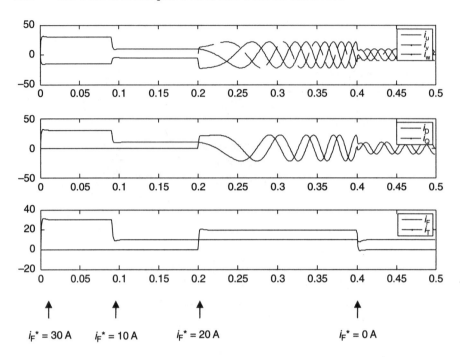

Figure A4.36 Transformation of current waveforms

the motor. After this point the torque-producing current is reduced and the motor stops accelerating. This example shows how the reference frame transformation converts the three-phase a.c. currents into d.c. components in the reference frame.

Note that a scaling value of 2/3 has been used in equation (A4.27) so that the current components in the reference frame are in peak-phase current units. This can be seen in Figure A4.36 when the motor is rotating with $i_F = 10$ A and $i_T = 0$ A. During this period the phase currents have a peak of 10 A. Using a value of 2/3 is convenient in making the transformed values equivalent to the peak-phase values, but this is a non-power invariant transformation and care must be taken to include additional constants when calculating power or torque from transformed voltages and currents.

If a position sensor is present, the reference frame angle can be derived for the two-pole permanent magnet motor that has been used as an example by simply taking the rotor position from the sensor, because the flux position is directly related to the rotor position. If the motor has more than two poles then the reference frame position is given by

$$\theta_{\text{ref}} = P \times \theta \tag{A4.29}$$

where θ is the position feedback from the sensor and P is the number of pole pairs. Various types of position sensor are described in Section A5.3, but it is clear from equation (A4.29) that if a sensor is to be suitable for permanent-magnet motor control it must give a unique absolute position within each electrical revolution. For example with a six-pole motor the position must be uniquely defined within each 120° of mechanical rotation of the motor.

If a position sensor is not used then the reference frame angle can be derived from the motor voltages and currents. This is referred to as a model-based sensorless control scheme. The quadrature components of current and voltage in a fixed reference frame locked to the stator, referred to as the stationary reference frame, are required for the model. The currents have been derived already from the current feedback as defined by equation (A4.27). It would not be cost-effective to measure the actual motor voltages, so these are derived from the voltage references. Provided the current controllers are controlling the currents successfully, the voltage references will match the motor voltages. Referring to Figure A4.33, $|v_s|$ and θ_m represent the magnitude and phase of the motor voltages. These can be converted to quadrature components in the stationary reference frame as follows:

$$v_{sD} = |v_s|\cos\theta_m$$
$$v_{sQ} = |v_s|\sin\theta_m \qquad (A4.30)$$

The quadrature components of flux produced by the rotor magnets in the stationary reference frame are given by the following equations, the derivation of which is beyond the scope of this book:

$$\psi_{FD} = \int (v_{sD} - R_s i_{sD})dt - L_s i_{sD}$$
$$\psi_{FQ} = \int (v_{sQ} - R_s i_{sQ})dt - L_s i_{sQ} \qquad (A4.31)$$

where R_s and L_s are the stator winding resistance and inductance, respectively. The reference frame angle is derived from the flux components as

$$\theta_{ref} = \tan^{-1}(\psi_{FQ}/\psi_{FD}) \qquad (A4.32)$$

A model-based scheme breaks down at very low frequencies because the voltage components become very small and must be supplemented for low-speed operation by an alternative such as a measurement scheme using injected currents.

A4.4.1.4 Induction motor torque and flux control

The torque and flux can be controlled in an induction motor in the same way as described for a permanent-magnet motor in the previous section. A squirrel-cage induction motor has a rotor cage that appears as a set of three-phase rotor windings. Currents in these windings are induced by the stator currents and produce the rotor flux that would be produced by the magnets in a permanent-magnet motor. The stator currents always include a reactive component necessary to create the flux in the motor. Figure A4.37 shows the current and the flux vectors in an induction motor. The stator flux ψ_s is the flux linking the stator windings. This is composed of the rotor flux ψ_r and the flux produced by the stator currents flowing in a combination of the motor inductances referred to as the transient inductance ρL_s.

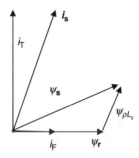

Figure A4.37   Vector representation of an induction motor

The vector representing the rotor flux is not locked to the rotor position and the angle of the vector in the stationary reference frame can be derived from the feedback angle θ from a position sensor:

$$\theta_{\text{ref}} = \theta + (L_m/T_r) \times \int (i_T^*/\psi_F)dt \tag{A4.33}$$

where L_m and T_r are the magnetising inductance and rotor time constant, respectively, of the motor.

It should be noted that the rotor time constant includes the rotor winding resistance, which varies with rotor temperature. Any difference between the real rotor time constant and the value used by the drive causes the reference frame to be misaligned. If this happens, the flux and torque control are no longer completely decoupled, which results in sub-optimum performance and possible instability. It is possible, however, to include additional systems in the drive to estimate the rotor time constant and correct for this problem. As the rotor of an induction motor is symmetrical, only the relative position of the rotor is required in equation (A4.33), so only the incremental position is required from the position sensor.

Rotor flux control (RFC) is a form of model-based sensorless control for induction motors. This is based on a slightly modified form of equation (A4.31):

$$\psi_{\text{FD}} = \int (v_{\text{sD}} - R_s i_{\text{sD}})dt - \rho L_s i_{\text{sD}}$$
$$\psi_{\text{FQ}} = \int (v_{\text{sQ}} - R_s i_{\text{sQ}})dt - \rho L_s i_{\text{sQ}} \tag{A4.34}$$

The stator winding inductance has been replaced by the transient inductance ρL_s. The control system operates in the same way as the model-based control for a permanent-magnet motor. Again the model performance breaks down at very low frequencies and an alternative model must be used for low-speed operation.

Earlier in this chapter equations were given to calculate suitable values for the speed controller. These equations used K_t, the torque constant of the motor. This value is normally provided by the manufacturers of permanent-magnet motors, but not for induction motors. However, the drive can measure the parameters of an induction motor using various auto-tuning algorithms and it can then automatically calculate a suitable value for K_t.

A4.4.1.5 Open-loop induction motor drive

The earliest voltage source inverters used an open-loop control method where the inverter output frequency is derived directly from the frequency reference via a ramp that limits the rate of change of frequency to give smooth acceleration and deceleration. The voltage applied to the motor is derived from the frequency with a linear change of voltage from zero to rated frequency and a constant voltage above rated frequency. An offset, usually referred to as boost, is applied at low frequencies to overcome the stator resistance voltage drop. This type of control is simple and robust and can be used to supply different types of motors including induction motors and reluctance motors. Alternatively, it can be used as a variable-voltage, variable-frequency power supply for feeding multiple motors as well as applications other than motors.

The low-frequency boost characteristic will only set the motor flux to the correct level under specific load conditions. This can lead to over- or under-fluxing of the motor as the load changes. Figure A4.38 shows a more modern adaptation of this scheme that is based on the vector controller described previously and gives better control of an induction motor, particularly at low speeds.

The frequency reference after the ramps is integrated to define the angle of a reference frame that rotates at the required frequency. The voltages in the torque- and flux-producing axes of the reference frame are compensated to take account of the stator resistance voltage drop, which has the effect of aligning the reference frame with the stator flux in the motor as shown in Figure A4.39. This does not give the same transient performance as the rotor flux control described previously, because the reference frame alignment only applies under steady-state and not transient conditions, but this type of control requires fewer motor parameters to operate correctly. It should be noted that with open-loop control it is possible for instability to occur with some motors, particularly under lightly loaded conditions.

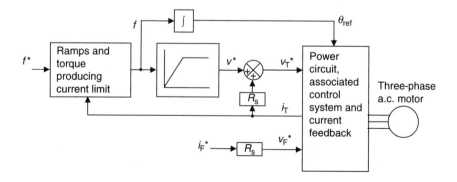

Figure A4.38   Open-loop sensorless control

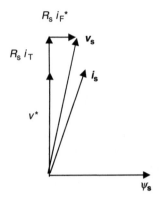

Figure A4.39   Vector representation of open-loop control

A4.4.2 Direct torque control

The torque and flux controllers described so far are based on control of the torque- and flux-producing currents and use a space vector modulation to derive the inverter control signals. As its name suggests, direct torque control (DTC) is an alternative method that uses a hysteresis method for direct control of torque and flux in a motor. As was shown in Figure A4.35, the inverter can only produce one of eight possible voltage vectors. At each sample point DTC selects one of the eight voltage vectors to change the stator flux and torque to the reference values; it does this as quickly as possible. The stator flux is given by the following equation:

$$\psi_s = \int (v_s - R_s i_s) dt \qquad (A4.35)$$

where v_s and i_s are the vectors representing the motor terminal voltages and phase currents, respectively.

Figure A4.40 gives an example of direct torque control. The corresponding voltage vectors that can be selected and their effect on the stator flux and the torque are given in Table A4.1.

The aim is to hold the magnitude of the stator flux, ψ_s, within the hysteresis band indicated by the two dotted lines, so at the instance shown, one of the voltage vectors u_3, u_4 or u_5 must be selected to reduce the magnitude of ψ_s. The torque produced by the motor is proportional to the rotor flux multiplied by the component of stator current at right angles to the rotor flux. If the angle between the rotor flux and the stator flux is increased in the direction of rotation, the component of stator current leading the rotor flux by 90° must increase, because the difference between the rotor and stator flux vectors is $\rho L_s i_s$. Therefore, advancing the stator flux in front of the rotor flux increases the accelerating torque, and retarding the stator flux behind the rotor flux increases decelerating torque. The selection of the voltage vector is further constrained by the change in torque required. If the torque is to be increased at the instance shown in

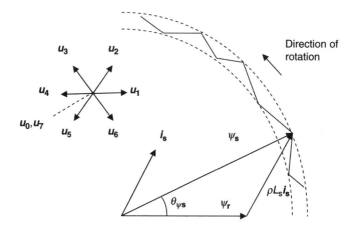

Figure A4.40 Direct torque control

the example, then voltage vector u_3 or u_4 must be selected or if the torque is to be reduced u_5 must be selected. Torque is also controlled by a hysteresis method. A switching table is constructed that contains the voltage vectors to be selected to control the stator flux and the torque to stay within their hysteresis limits.

An example of a torque and flux control system based on DTC is shown in Figure A4.41. An estimate of the stator flux and the torque must be provided as feedback for the hysteresis comparators. Also, if speed control is required, an estimate of rotor speed is necessary for the speed controller. These are derived from a model using the motor currents and an estimate of the motor voltages from the switching table state and the inverter d.c. input voltage. It is important to note that, although the principle of direct torque control appears simple and does not depend on estimates of motor parameters, the motor model used to derive the estimates of torque, flux and speed

Table A4.1 Voltage vectors and their effect on stator flux and torque for the arrangement of Figure A4.40

Voltage vector	Effect on magnitude of ψ_s ($|\psi_s|$)	Effect on the angle of ψ_s ($\theta_{\varphi s}$)
u_1	Increase	Retard
u_2	Increase	Advance
u_3	Decrease	Advance
u_4	Decrease	Advance
u_5	Decrease	Retard
u_6	Increase	Retard
u_0 or u_7	Almost no change	Almost no change

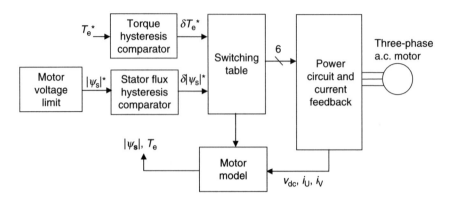

Figure A4.41   Direct torque control system

is similar to that required with other sensorless schemes and is dependent on the motor parameters. The following list compares the features of direct torque control with a control system using current feedback and a space vector modulator:

- A direct torque control drive inherently delivers a change in torque in the shortest possible time within the limits of the sample rate. Because of the sampling and calculation delays usually associated with space vector modulator-based systems, a change in torque can take several samples. However, deadbeat-type algorithms can be used with space vector modulator systems, giving a performance that compares well with that of DTC.
- The calculations for the current controllers, reference frame translation and space vector modulator are more complex than the DTC hysteresis comparators and switching table. However, the sample rate required for DTC (typically 40 kHz) is much higher than that for a space vector modulator (6–12 kHz), because DTC uses a hysteresis method.
- Because DTC is based on hysteresis controllers, the inverter has a continuously variable switching frequency. This is considered to be an advantage in spreading the spectrum of the audible noise from the motor. However, the range must be controlled so that it does not exceed the maximum allowed by the power electronics of the inverter. Furthermore, care must also be taken with DTC to ensure that changes from one voltage vector to another more than 60° away do not occur repetitively, as this can substantially increase the stress on motor insulation.
- As well as supplying a single induction motor, most induction motor drives can also supply more than one motor in parallel where the motors are different sizes, or the drive can be used as a general-purpose variable-frequency/variable-voltage power supply. DTC cannot be used in these applications, and so a DTC drive must also be able operate with space vector modulation for these applications.

A4.4.3   Performance summary

The different types of a.c. motor drive have different characteristics (Table A4.2) making them suitable for different applications. These will be described in the following sections.

Table A4.2 Comparison of different induction motor control schemes

	No position feedback				Position feedback
	Basic open-loop mode	**Open-loop compensated for R_s drop**	**Direct torque control**	**Rotor flux control**	**Closed-loop control**
Multi-motor connection	Yes	No	No	No	No
Torque response (ms)	10	<10	1–5	<0.5	<0.5
Speed recovery time (ms)	100	100		<20	<10
Minimum speed with 100% torque (Hz)	1–3	1	0.5	0.8	Standstill
Maximum torque at 1 Hz (%)	>150	>150	>175	>175	>175
Light load stability	Moderate	Moderate	Excellent	Excellent	Excellent
Torque control capability			Good	Good	Excellent
User-selectable switching frequency	Yes	Yes	No	Yes	Yes
Speed loop response (Hz)	10	10		50–100	150

Note: The performance of a closed-loop permanent-magnet drive is comparable to that of the closed-loop induction motor drive.

A4.4.3.1 Permanent-magnet motor drives

Permanent-magnet motors can be constructed with low-inertia rotors for use in high-performance servo applications such as machine tools or pick-and-place applications where fast precise movements are required. Alternatively, permanent-magnet motors with high pole numbers can be used in low-speed applications such as gearless lift systems. The performance characteristics of these drives are summarised as follows:

- There is good dynamic performance at speeds down to standstill when position feedback is used.

- Position feedback is required that defines the absolute position uniquely within an electrical revolution of the motor. This can be provided with a position sensor, or alternatively a sensorless scheme can be used. The performance of a sensorless scheme will be lower than when a position sensor is used.
- Field weakening of permanent-magnet motors is possible to extend their speed range, but reducing the flux requires additional motor current, and so the motor becomes less efficient in the field-weakening range. Also, care must be taken not to de-magnetise the magnets.
- Permanent-magnet motors exhibit an effect called cogging related to the geometry of the motor design, which results in torque ripple. This effect can be minimised by good motor design, but can still be a problem.
- Permanent-magnet motors can be very efficient as the rotor losses are very small.

A4.4.3.2 Induction motor drives with closed-loop current control

Induction motor drives with closed-loop current control are used in similar applications to d.c. motor drives (i.e. cranes and hoists, winders and unwinders, paper and pulp processing, metal rolling, etc.). These drives are also particularly suited to applications that must operate at very high speeds with a high level of field weakening, for example spindle motors. The performance characteristics of these drives are summarised as follows:

- There is good dynamic performance at speeds down to standstill when position feedback is used.
- Only incremental position feedback is required. This can be provided with a position sensor, or alternatively a sensorless scheme can be used. The transient performance of a sensorless scheme will be lower than when a position sensor is used and lower torque is produced at very low speeds.
- Induction motors are particularly suited to applications that require field weakening. The motor current reduces as the speed is increased and the flux is reduced.
- Induction motors are generally less efficient than permanent-magnet motors because of their additional rotor losses.

A4.4.3.3 Open-loop induction motor drives

Open-loop induction motor drives are used in applications that require moderate performance (i.e. fans and pumps, conveyors, centrifuges, etc.). The performance characteristics of these drives are summarised as follows:

- There is moderate transient performance with full torque production down to approximately 2 per cent of rated speed.
- Although a good estimate of stator resistance improves torque production at low speeds, the control system will work with an inaccurate estimate, albeit with reduced torque.
- Although a good estimate of motor slip improves the ability of the drive to hold the reference speed, the control system will work with an inaccurate estimate, albeit with poorer speed holding.

Chapter A5
Position and speed feedback

A5.1 General

In systems where precise control of position or speed is important, a position or speed sensor is required. Although there are many different types of position and speed sensors available, the devices described in this chapter are limited to those that are likely to be used with a modern variable-speed drive. To control an a.c. motor it is necessary to determine the speed of the motor and the position of the flux in the motor. Various sensorless control schemes are available, which can estimate these quantities by measuring the motor currents; however, for higher-quality performance a position or speed feedback device is normally used. To control a d.c. motor it is only necessary to determine the speed of the motor. Again, sensorless control is possible by measuring the armature current and voltage, but for accurate speed control a speed feedback device is still required.

It should be remembered that although sensorless control has many attractions, in a large number of practical systems it is necessary to have a direct measurement of the motor shaft motion as part of an outer control loop or safety case. Sensors are therefore an important element of many drive systems and play a critical part in determining the performance of high-performance systems. It should also be noted that a significant number of site problems with drive systems are associated with either the selection or installation of the position or speed sensor!

Different feedback devices have different features, and these should be considered when designing a system. There is often poor understanding of the characteristics of speed and position sensors and so these features are described in some detail in the following paragraphs.

A5.1.1 Feedback quantity required

If both position and speed feedback are required, as is the case in a.c. motor control, it is possible to use a sensor that gives position feedback and then derive speed feedback as a change of position over a fixed sample time. If speed feedback alone is required, as is the case in d.c. motor control, it is possible to use a device that gives speed feedback only. Most types of position feedback sensor are available to measure either rotary or linear movement.

For rotary motor control a rotary feedback device is normally used; however, for linear motor control a linear feedback device is more suitable. Throughout the rest of this chapter, the descriptions given relate mostly to the rotary version of each type of position sensor.

A5.1.2 Absolute position feedback range

The absolute position feedback range defines the movement over which it must be possible to uniquely determine the position. Table A5.1 gives some examples of the absolute position feedback range required for different applications.

Some position feedback sensors can provide absolute position information as soon as they are powered up. Others may need to be moved to a home position and then track the change of position from the homing point to give absolute position. In this case the absolute position is not available immediately when the device is powered up.

A5.1.3 Position resolution

The position resolution and position accuracy of a feedback device should not be confused. The resolution defines the nominal position movement required for the device to detect a change of position. The accuracy, on the other hand, is a measure of the maximum deviation of the feedback position from the actual position. An

Table A5.1 Encoder requirements for different motor control schemes

Application	Absolute position feedback range
Rotary induction motor control	Incremental position only. Absolute position not required.
Rotary synchronous motor control including permanent-magnet servo	Absolute position range equivalent to one electrical revolution (i.e. 120° of mechanical rotation for a six-pole motor).
Rotary position control	Absolute position range equivalent to the movement between all the required positions. If this involves more than one turn then a multi-turn position sensor is required.
Linear induction motor	Incremental position only. Absolute position not required.
Linear synchronous motor control	Absolute linear position range equivalent to one motor pole pitch.
Linear position control	Absolute linear position range equivalent to the movement between all the required positions.

analogue feedback signal from a potentiometer used to measure position has almost infinite resolution, but the resolution of an encoder is limited by the number of pulses produced for a given movement. It is important to note that the position accuracy is almost always lower than the resolution.

Probably the most notable limit imposed by position resolution is the maximum possible gain in a speed control system where the speed is derived from a position feedback device by calculating the position change over a fixed sample period. An encoder rotating at a constant speed may not necessarily produce an exact integer number of pulses over each sample period. The result is that the pulses counted each time will vary between the integer value above and below the mean number of counts per sample. In the example given in Figure A5.1 the mean number of counts per period is 4.3, but either 4 or 5 counts are seen during each period. This gives rise to a ripple in the speed feedback that is equivalent to a speed that gives one encoder count per sample period. The speed controller tries to correct for the ripple seen in the feedback and generates a high-frequency torque component that produces acoustic noise. The ripple and noise increase as the gain of the speed controller is increased. The speed that gives one count per sample period is equivalent to a movement of $R/360$ revolutions per period, where R is the resolution of the feedback device in degrees. Therefore if the sample time (T_s) is in seconds this speed is given by

$$\text{Speed in rps} = (R/360)/T_s \qquad (A5.1)$$

and so

$$\text{Speed in rpm} = [(R/360)/T_s] \times 60 = R/(6 \times T_s) \qquad (A5.2)$$

The speed given in equation (A5.2) is the speed feedback ripple in $\min^{-1}$ from a position feedback device with a resolution of R degrees per revolution when the speed measurement is made with a sample time of T_s seconds. It is a popular misconception

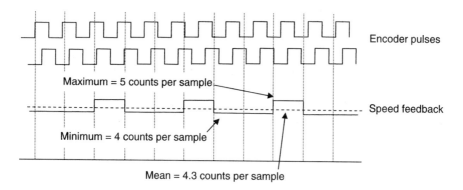

Figure A5.1   Speed feedback derived from an encoder

that this speed ripple changes with actual speed, and that it is worse as the speed approaches zero. This is not the case because the speed ripple is always defined by (A5.2) except in the unlikely case when the speed remains absolutely constant at a level where an exact integer number of counts occur during each sample period.

As can be seen from (A5.2), speed feedback ripple can be reduced by increasing the sample time, which has a detrimental effect on the speed controller response, or increasing the resolution of the position sensor, which tends to add cost. Pulse width measurement is sometimes used in an attempt to reduce speed feedback ripple; however, this is not recommended as the system becomes non-deterministic at low speeds and the electrical noise that is usually present on the pulse edges due to power electronic switching in the drive can give rise to large fluctuations in the speed feedback.

A5.1.4 Position accuracy

Position accuracy is a measure of the maximum deviation of the measured position from the actual position. As would be expected, this limits the accuracy of a position control system using the feedback device. However, deviations in the position, in addition to those generated by limited resolution, can contribute to speed feedback ripple when position change over a fixed sample period is used to derive speed feedback. This becomes more noticeable with a high precision feedback device such as a SINCOS encoder.

A5.1.5 Speed resolution

The only speed feedback device that will be considered is a d.c. tacho-generator. This is an analogue feedback device, and so speed feedback resolution of the device itself is not a problem. However, most modern variable-speed drives use digital control, and so the speed feedback signal is fed into the speed controller via an analogue-to-digital (A to D) converter, which imposes a limit on the speed feedback resolution. Therefore the speed resolution of the system is limited by the A to D converter resolution in the drive.

It may seem counter-intuitive, but the resolution of the mean speed feedback derived by measuring position change over a fixed sample time from a position feedback device such as an encoder can give extremely high mean speed feedback resolution. If a speed controller that includes an integral gain is used, the integral term accumulates the speed error between the speed reference and speed feedback. This has the effect of extending the sample period over which the speed is measured to the time for which the system has been enabled and extending the sample time increases the resolution of the mean speed feedback. The resolution of such a system is only limited by the resolution of the speed reference and not the speed feedback.

A5.1.6 Speed accuracy

Speed feedback accuracy is usually worse than speed feedback resolution. The speed accuracy of a speed feedback device such as a tacho-generator is defined by its absolute accuracy and non-linearity. Where speed feedback is derived from a position sensor, such as an encoder, the main effect on the speed accuracy is the accuracy of the sample period, which in a digital system is defined by the system clock. Where this clock is produced by a quartz crystal, accuracies of 100 ppm (0.01 per cent) are easily achievable. It should be noted that this is a percentage of the actual speed and not full scale speed, so the accuracy in rpm or rad s^{-1} improves as the actual speed is reduced.

A5.1.7 Environment

Consideration must be given to the environment in which a position or speed sensor is to operate. It is obvious that if the device is to be mounted on a motor, it is likely that the environment will be hot and subject to mechanical vibration. Other effects such as axial movement of the motor shaft and radial eccentricity of the mounting should also be taken into account. Although most industrial feedback devices are sealed, it is possible for contamination to occur. This can be a problem for optical encoders with fine lines on the glass disc, where gases and dust can degrade the performance of the device. In the presence of moisture or corrosive gases, corrosion of the sensor and any associated signal conditioning electronics can be an issue. Special designs of sensor are available that are protected against some of the milder corrosive contaminants.

Providing a seal against contamination is more difficult with a linear encoder, and so care has to be taken in the design and mounting of the device. Unlike an encoder, a resolver can be used without problems in a hot environment with high levels of vibration and contamination.

A5.1.8 Maximum speed

There is always a maximum mechanical operating speed for any position or speed sensor above which the device would be damaged.

The accuracy of the feedback from some sensors such as tacho-generators or SINCOS encoders can degrade above a speed that is lower than the maximum mechanical operating speed.

Note also that the processing electronics in a drive may give reduced accuracy or cease to work at all as the frequency of the signals from a digital encoder, SINCOS encoder or resolver increases with the mechanical speed of the sensor. The maximum frequency for an encoder input will be defined by drive manufacturers.

A5.1.9 Electrical noise immunity

Considerable electrical noise can be generated by the switching action of the power electronic devices in a variable-speed drive. Careful system design can prevent this noise from affecting other equipment including position or speed sensors. Even so, the noise immunity of the different types of feedback device should be considered. Analogue devices are likely to be more prone to disturbance than digital devices, and so a SINCOS encoder with 1 V p–p sine-wave output signals is less immune to electrical noise than a digital encoder producing 5 V square waves.

Although a tacho-generator produces analogue signals, the type of device traditionally used with a d.c. motor drive generates a relatively high voltage, and so it is reasonably noise immune.

A5.1.10 Distance between the feedback device and the drive

The following list details some problems that may occur as the distance between the feedback device and the drive is increased.

- Active sensors such as encoders require a significant amount of power supply current to drive their internal circuits and the terminations on their output signals at the receiving end. The voltage drop in the power supply conductors may reduce the voltage at the encoder to a level where the device does not function correctly.
  - The resolution of a SINCOS encoder is reduced as the sine-wave signal magnitude is reduced. Again, voltage drops in the conductors can cause these signals to be reduced.
- Sensors that produce sine-wave outputs such as resolvers will suffer from phase shifting if very long cabling is used.
- Some encoders use synchronous digital communications. As the clock frequency and distance are increased, skew between the clock and data can become a problem. The clock is generated by the drive and the skew occurs on the data transmitted back from the encoder. It is possible to electronically measure the line length between the drive and the encoder and adjust the sampling point for the data at the drive end to counteract this problem. Otherwise, the distance between the drive and the encoder must be limited as the clock and data rate are increased.

A5.1.11 Additional features

The following additional features are available in some encoders:

- automatic recognition of the encoder,
- non-volatile storage within the encoder, which allows the user to store data such as motor or machine parameters.
- advanced error detection providing information about the state of the encoder, and
- facility to offset the encoder zero position electronically.

A5.2 Speed feedback sensors

A5.2.1 D.C. tacho-generator

A d.c. tacho-generator, often referred to as a tacho, is a speed feedback device that produces an analogue voltage in proportion to its rotational speed with a polarity that depends on the direction of rotation (Figure A5.2). The principle of operation is

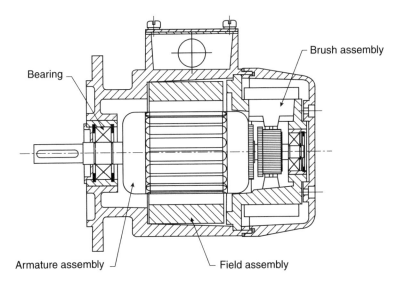

Figure A5.2   Typical construction of a d.c. tacho-generator

the same as a d.c. motor, but the field is normally provided by permanent magnets (see Section A1.2.4).

The following characteristics should be considered when selecting a d.c. tacho-generator.

- *Output voltage ripple.* The main component of voltage ripple is defined by the number of armature coils and is usually specified as a peak-to-peak voltage. Additional electrical noise can be present on the tacho-generator output due to varying contact resistance between the brushes and the commutator. To overcome this distortion, high-quality brushes are used, having low contact resistance and sufficient brush force to maintain positive contact stability.
- *Temperature effects.* The temperature coefficient for a tacho-generator specifies the percentage change of the output voltage for a given change in temperature. The lower this value the more stable the speed feedback is with variations in temperature. Generally the cost of a tacho-generator increases with improved temperature coefficient. Care should be taken not to exceed the maximum allowed loading on the device as this will elevate the internal temperature and cause higher than expected deviation from the nominal output voltage. Table A5.2 gives the characteristics of different designs.
- *Linearity and accuracy.* The speed linearity, typically 0.1 to 0.2 per cent, is specified up to the maximum operating speed. If this range is exceeded the linearity degrades because of aerodynamic lift of the brushes, hysteresis losses, armature reaction and saturation. The speed accuracy is normally lower than the linearity, and is typically in the range from 1 to 2 per cent.

Table A5.2   D.C. tachogenerator characteristics

Type	Typical temperature coefficient per degree K (%)	Description
Uncompensated	0.2	Basic design using low-cost magnets
Compensated	0.05	Thermistor-based compensation is used to improve a low-cost uncompensated design. The output impedance is higher than more expensive types and the temperature compensation range is normally limited.
Stable	0.02	More stable magnets are used
Ultra-stable	0.01	More stable magnets are used with a compensating alloy in the magnetic field circuit to improve the temperature stability

- *Maximum operating speed.* As already mentioned, if the maximum specified operating speed is exceeded, the output voltage linearity is degraded. Operating above this speed can also result in damage to the commutator and brushes.

In summary, d.c. tacho-generators have the following characteristics. They are robust against vibration and shock loads. They have a wide operating temperature range, and are the most common speed feedback device used with industrial d.c. motor drives. With the exception of some low-voltage high-linearity types of design used with some servo drives, they are not normally used with a.c. drives, where position feedback devices such as resolvers and encoders are much more common. d.c. tacho-generators are relatively expensive and the brushes require maintenance. It is possible to use a.c. tacho-generators that do not have a commutator, but which produce an a.c. output that can be rectified to give a d.c. signal proportional to speed. These can be used for unidirectional applications where only moderate performance is required, as the rectifier diodes give a voltage error and the speed signal contains significant ripple.

A5.3   Position feedback sensors

A5.3.1   Resolver

A resolver has one rotor winding and two 90° displaced stator windings. The rotor winding is fed with a high-frequency signal, typically in the range 6–8 kHz, and the feedback signals are taken from the stator windings. The connection to the rotor can be made via slip rings and brushes, but a brushless design is more common where a transformer arrangement is used to excite the rotor winding across the air gap between the stator and the rotor (Figure A5.3).

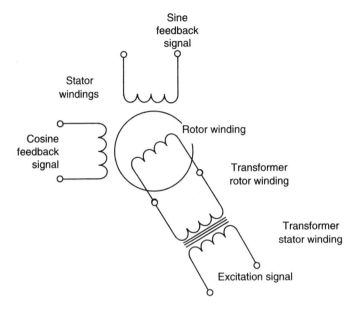

Figure A5.3   Brushless resolver

The sine and cosine feedback voltages contain the excitation frequency modulated by the sine and the cosine of the rotor position, respectively, as shown in Figure A5.4. For a two-pole resolver, which is the most common type, the signals are modulated once per revolution and so any position within a revolution can be determined uniquely. For basic control of a permanent-magnet motor it is only necessary to determine the position uniquely within one electrical revolution (i.e. 120° for a six-pole motor), and so resolvers with more than two poles can be used, giving better position accuracy. Although different combinations of resolver and motor

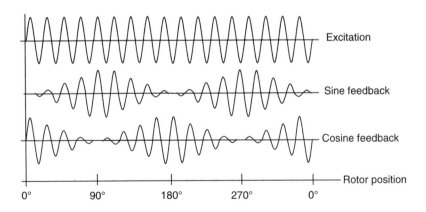

Figure A5.4   Resolver signals

poles are possible, where a resolver with more than two poles is used it normally has the same number of poles as the motor.

Various methods can be used to obtain the rotor position from the sine and cosine feedback signals. Monolithic semiconductor resolver to digital converters (R to D converters) are available that can track the rotor position and give a digital feedback value. An example of a tracking-type R to D converter is shown in Figure A5.5.

The estimated position is accumulated in the counter and this is used to produce the sine and cosine of the estimated position. These values are combined with the sine and cosine feedback signals as shown to give the excitation frequency with a magnitude defined by the sine of the error between the actual and estimated position [$\sin(\omega_e t)$ $\sin(\theta - \phi)$]. The sign of the excitation signal is used in the phase-sensitive detector to remove the excitation signal, giving an error value that drives a voltage-controlled oscillator so that the estimated position ϕ tracks the actual position θ. This method can give the position as a binary value with a resolution of 14 bits for the movement equivalent to one resolver pole (i.e. 14 bits per revolution for a two-pole resolver). It should be noted, however, that the combined accuracy of a typical resolver and R to D converter is likely to be much poorer than the resolution, and may be in the order of 1° (360 bits per revolution).

In summary, resolvers are robust against vibration and shock loads as well as having a wide operating temperature range. They can provide moderately high-resolution absolute position feedback within one revolution, which is useful for motor control, including that of permanent-magnet motors; however, the absolute accuracy is very limited. The analogue sine and cosine feedback signals usually have a magnitude of about 2 V_{rms}, giving moderate noise immunity. Although the resolver itself is fairly low cost, the R to D converter can be expensive. Also, because the R to D converter includes a control loop with limited bandwidth, it introduces an additional delay in the control system that can limit the performance when compared to an encoder.

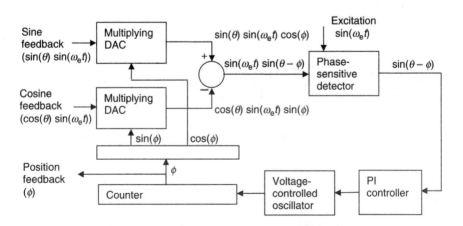

Figure A5.5 Tracking resolver to digital converter (note: θ is the actual position, ϕ is the estimated position, and ω_e is the excitation frequency)

A5.3.2 Incremental encoder

An incremental encoder is a position feedback device that produces pulses as the device rotates. The pulses are then accumulated by counting (usually within the drive) to give the position. At present, optical technology is still used in many encoders, although other techniques that are discussed later are starting to replace this. Optical encoders are based on the Moiré principle where light from a source shines through a fixed element and a glass disc on the rotor before being detected by a photosensor (Figure A5.6). There are gratings with equally spaced lines on the disc and the stationary element. As the gratings move with respect to each other, periodic fluctuations in brightness are seen by the photosensor. These fluctuations are approximately sinusoidal in shape.

This technology can be used with grating periods as small as 10 μm, which gives a practical limit on the number of lines per revolution for a rotary encoder. Although it is possible to have 50 000 lines per revolution, more cost-effective devices have a maximum of 4 096 lines per revolution. The outputs from an incremental encoder are normally differential EIA-485 standard signals, and so the sinusoidal brightness variations from the photosensor need to be squared and then converted to differential signals with suitable line drivers. To allow relative movement in either direction to be detected, two photosensors are used with separate fixed gratings that are displaced by a quarter of a grating period. This gives two signals displaced by 90°. These are often referred to as quadrature signals, and the phase relationship between these can be used to sense the direction of rotation.

By using a counter in the drive that either increments or decrements on each of the A and B channel signal edges, a relative position can be obtained with a resolution

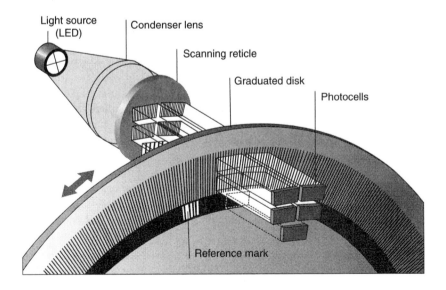

Figure A5.6   Optical encoder based on Moiré principle

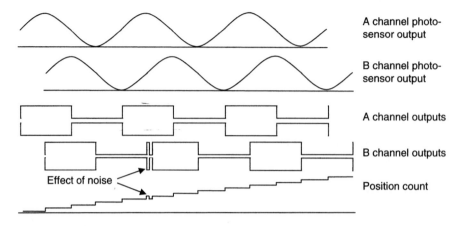

Figure A5.7   Incremental encoder signal processing

equal to four times the number of lines per revolution (i.e. a 4 096-line encoder gives a binary position value with 14-bit resolution). The counting principle shown in Figure A5.7 demonstrates the additional noise immunity provided by using quadrature signals. Noise is most likely to cause multiple edges as the encoder signals change state. The effect of this noise is to simply cause the position count to increase by one extra count and then decrease again with no accumulated error. Other schemes such as frequency and direction signals do not offer the same level of noise immunity, because multiple edges cause accumulated position errors. Also the use of differential signals is important as this gives good noise immunity against external influences such as switching transients generated by the power electronics within a drive. It is still important, however, to follow good wiring practices to avoid problems. It should be noted that it is possible for coupling between the A and B channels to generate transients on the other channel in the centre of each pulse. If balanced differential signals are used this is not a problem because the transient appears as a common mode disturbance and is rejected. However, some encoders use single-ended and not differential drivers, and in this case the transients coupled from the other channel can cause false counting and position feedback drift. In summary, quadrature signals with differential drivers should always be used where possible as these provide the best possible noise immunity.

The accuracy within an optical encoder is governed by the quality of the optical system and the radial deviation caused by the encoder bearings. The elasticity of the encoder shaft, its coupling to the motor and its mounting also affect the accuracy under transient conditions. It is possible to have a 4 096-line optical encoder where the accuracy is comparable to its resolution, giving much better accuracy for example than a resolver.

In summary, an incremental encoder is a relatively low-cost device giving good performance and good noise immunity (which can be important when the encoder is a long distance from the drive). Apart from physical limitations, high-speed operation can be a problem because the output frequency from the encoder becomes excessively high. Care must be taken when using an optical encoder as it can be damaged by shock loads and will not operate at as high a temperature as a resolver. Other technologies based on inductive or capacitive principles are starting to replace optical devices, as they are lower cost and physically more robust. Their main disadvantage is that they can only provide a relatively low number of pulses per revolution, giving 16 or 32 periods from the rotating disc. This information cannot be used directly as the resolution is too low for motor control; however, interpolation techniques can be used to increase the resolution, as discussed in A5.3.5.

A5.3.3 Incremental encoder with commutation signals

Incremental encoders can be used to give a relative position, which is useful for many applications including induction motor control and speed measurement, but they cannot be readily used for permanent-magnet motor control as the absolute position must be uniquely defined within one electrical revolution. As an alternative to an absolute encoder for permanent-magnet motor control, an incremental encoder can be used with additional outputs referred to as commutation signals (Figure A5.8). These are generated by three additional tracks on the encoder that produce three 120° displaced signals with a period equal to one electrical revolution.

The absolute position is required to locate the motor flux and to derive a reference frame so that the motor currents are applied at the correct angle. After the drive powers up, the commutation signals are used to determine the flux position with a maximum error of $\pm 30°$ electrical. This error could result in up to 13 per cent reduction in the torque per ampere from the motor. Once the motor has moved and there has been a transition on one of the commutation signals, the motor position is known precisely within one electrical revolution, and after this point the incremental signals are used to track the absolute position. There is no further loss of torque once the first commutation signal has changed state. Only the incremental position is required for speed feedback, and so the commutation signals are not used for this as their poor resolution would result in a high level of speed feedback ripple.

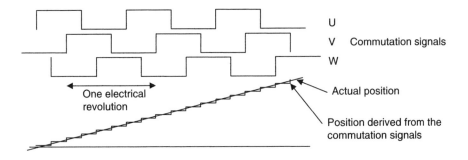

Figure A5.8   Commutation signals

The incremental encoder with commutation signals is a useful extension to the standard incremental encoder, giving absolute position for permanent-magnet motor control. It has all the benefits of the standard encoder, but the disadvantages of increased cost and the extra connections required for the commutation signals.

A5.3.4 Incremental encoder with commutation signals only

For some permanent-magnet motor applications where precise control is not required, but low cost is important, commutation signals are used alone without incremental signals. These can be derived from a simple mechanical sensor or by using Hall effect sensors embedded in the motor; these detect the flux from the magnets. In this case, the commutation signals are used all the time, and not just at power up, to locate the motor flux. If the flux angle is derived directly from the signals for a permanent-magnet motor with sinusoidal flux distribution this would result in 13 per cent torque ripple. These signals cannot be used directly to produce speed feedback by measuring the position change over a fixed sample period as the speed feedback ripple would be unacceptably large. One solution to this problem is to apply a phase-locked loop to the commutation signals to derive higher resolution position feedback. Although this makes the feedback acceptable to control the motor, the dynamic performance is limited by the lag associated with the phase-locked loop.

A5.3.5 SINCOS encoder

The encoders discussed so far only provide the zero crossing information generated from the photosensor signals. However, these signals, which are approximately sinusoidal, can give additional position information using a method called interpolation. A SINCOS encoder produces analogue sinusoidal signals instead of square-wave outputs. The de facto standard for SINCOS encoder outputs is a differential 1 V p–p sine wave. To remove the need for a negative power supply within the encoder, the differential signals often have a 2.5 V offset.

The signals are processed by the drive to give the interpolated position. The differential sinusoidal voltages are converted to quadrature square waves containing the zero crossing information, and a position counter is used in the same way as for the incremental encoder to give a position with a resolution equal to four times the number of sine waves per revolution (Figure A5.9). The sinusoidal signals are also fed into A to D converters so that their magnitudes are available in addition to the zero crossing information. Normally, further processing is carried out by a microprocessor within the drive to give the interpolated position (Figure A5.10).

The position within one period of the sinusoidal waveforms can be obtained with a resolution that is approximately equal to the data resolution after the A to D converters. The resolution of the interpolated position is given by the sum of the number of sine waves per revolution plus the interpolation resolution. Therefore, if the encoder provides 2 048 (2^{11}) sine waves per revolution and 10-bit resolution data is available after the A to D converters, the final interpolated position is a binary value with 21-bit resolution. This shows one of the advantages of using SINCOS technology

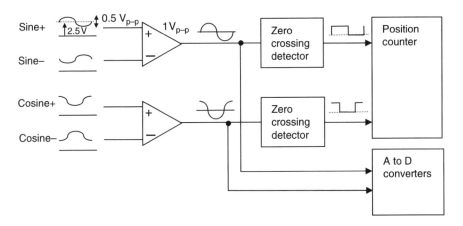

Figure A5.9   SINCOS signal processing

in that the position resolution is much higher than is possible with other methods. SINCOS encoders with a lower number of sine waves per revolution may be used for high-speed applications to limit the frequency of the signals from the encoder while maintaining a reasonable position resolution. Interpolation may also be used with inductive or capacitive encoders as mentioned previously, which can only have a low number of sine waves per revolution, to give position resolution comparable to a digital incremental encoder or a resolver.

Despite the obvious improvement in resolution possible with SINCOS encoders, these devices use small analogue signals and can be affected by electrical noise if care is not taken with the encoder wiring. Filters are necessary in the drive to remove electrical noise and as the frequency of the sine waves increases these filters reduce the magnitude, which in turn reduces the resolution, of the position feedback. The magnitude of the sine waves, and hence the resolution, can also be reduced by voltage drops in long cables between the encoder and the drive. The accuracy of SINCOS encoders is usually substantially less than the resolution. The accuracy is normally specified as the combination of two effects: the position error within a revolution, which is affected by eccentricity of the optical disc and so on, and

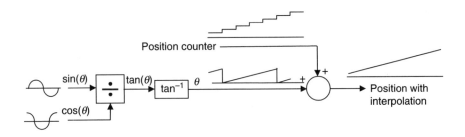

Figure A5.10   SINCOS interpolation

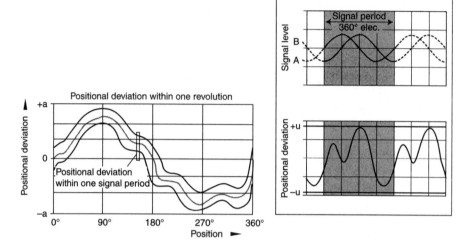

Figure A5.11 SINCOS encoder accuracy

position error within one signal period caused by the deviation of the waveforms from a sinusoidal shape (Figure A5.11).

It is interesting to note that high resolution is required to reduce ripple on the speed feedback, but the higher frequency inaccuracy caused by the error within each signal period can itself introduce speed feedback ripple that gives a reduction in the effective resolution of the encoder.

A5.3.6 Absolute SINCOS encoder

SINCOS technology can be used to improve the resolution of the position feedback, but it still only provides incremental position information. It is possible to use additional commutation signals in the same way as with a digital incremental encoder to give the absolute position within one electrical revolution to facilitate permanent-magnet motor control. An alternative method based on SINCOS technology, providing additional sine and cosine waveforms with a period of one mechanical revolution, can be used to derive an estimate of the absolute position within one turn. This is primarily intended for starting permanent-magnet motors instead of using commutation signals. Interpolation based on the additional waveforms gives an estimate of the initial position and then the normal sine-wave signals are used to track the absolute position. The accuracy of the single period per turn signals is normally quite poor, and so a once per turn marker signal is provided to correct the absolute position. Although this type of encoder is still available, it has largely been replaced by SINCOS encoders with additional serial communications.

A5.3.7 Absolute encoders

So far, two types of absolute encoder have been discussed that provide the absolute position within one electrical or mechanical revolution using additional commutation or sine-wave tracks. A more accurate absolute position within one mechanical turn that is available from power-up can be obtained by adding a number of additional tracks that encode the position as Gray code. The higher the required resolution, the more tracks are required. The Gray code information can be used by itself or it can be used to give the absolute position within a turn for a SINCOS encoder. A disc designed to give the absolute position for an encoder with 2 048 sine waves from its incremental tracks would require a 13-bit Gray code and this requires 13 additional tracks (Figure A5.12a). It would not be practical to connect an additional 13 signals between the encoder and the drive, and so most encoder manufacturers now use a serial communications channel to pass the absolute position information to the drive. More recently the absolute information has been encoded in one track with a serial code structure (Figure A5.12b) that is unique over one complete revolution.

Traditionally, multi-turn absolute position encoders used batteries to store the position at the point when the drive is powered down. These had the obvious disadvantages of using batteries, and the measurement circuits of the encoder would

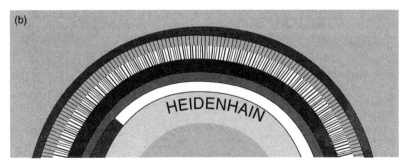

Figure A5.12 Optical encoder discs with absolute information: (a) 13-bit Gray code; (b) serial code

need to remain active to measure any movement while the drive was in the power-down state. This type of encoder has largely been replaced with multi-turn absolute encoders using a small gear box, where the combined position of each of the gears is used to derive the multi-turn absolute position. This has the advantage of not relying on batteries and any movement that occurs while the encoder is powered down can be detected when it powers up again.

A5.3.8 SINCOS encoders with serial communications

SINCOS encoders with additional serial communications are becoming more popular as these devices offer the high-resolution feedback of a SINCOS encoder with the possibility of obtaining the absolute position within one or more mechanical revolutions via the serial communications channel. All the protocols discussed here are based on an EIA-485 physical layer.

A5.3.8.1 EnDat

EnDat is a bidirectional synchronous serial protocol with separate data and clock signals (Figure A5.13). The communications channel is used to obtain the absolute position after the drive has powered up, and then the absolute position is tracked using the sine and cosine signals. If required, the absolute position can be re-read at

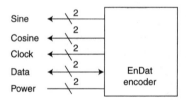

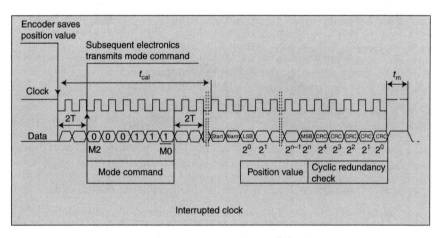

Figure A5.13 SINCOS encoder with EnDAT serial communications

regular intervals to verify the tracked absolute position. The position is sampled at the start of the communications message sent from the drive to request the position, and so, provided the drive tracks any position change from that precise point in time, the absolute position can be determined even if the encoder is rotating. The time taken to obtain the position from the encoder is not critical, and so a low clock rate can be used. This avoids significant skew between the clock and the returning data associated with synchronous protocols due to the transmission delay in long cables. It is important that errors in communications between the drive and encoder are detected, and so a cyclic redundancy check (CRC) is included in all messages. (EnDat is a registered trademark of Heidenhain.)

A5.3.8.2 Hiperface

Hiperface is similar to EnDat and offers similar facilities, except that it uses asynchronous serial communications (Figure A5.14). Hiperface also provides error detection in the form of a checksum that is included in all communications between the drive and the encoder. (Hiperface is a registered trademark of Sick Stegmann.)

A5.3.8.3 SSI

SSI was one of the earlier serial protocols available with encoders. It is a unidirectional synchronous protocol that only allows the transfer of the position from the encoder to the drive without error checking. It is available with some SINCOS encoders, but it does not include error detection and is not as secure as either EnDat or Hiperface.

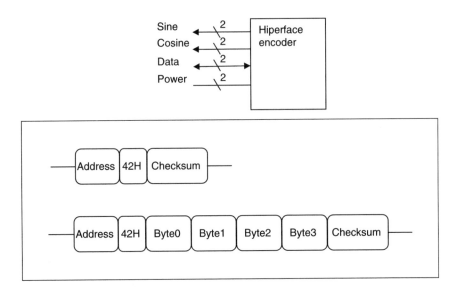

Figure A5.14   SINCOS with Hiperface serial communications

A5.3.8.4 Summary

Both EnDat and Hiperface provide bidirectional communications and the following features are possible in addition to obtaining the absolute position from the encoder.

- The encoder can be identified automatically by a drive so that the drive can set itself up to operate correctly with the encoder without any intervention from the user. This is a useful feature as encoders are becoming more complex with many settings required to make them operate correctly.
- Information can be stored in the encoder, such as the drive setup parameters required for the motor on which the encoder is mounted.
- The encoder can indicate that it has an internal error.
- The zero position of the encoder can be changed electronically from the drive.
- It is possible in some applications to use the independent sine wave and communications channels of a SINCOS encoder to implement safety functions.

A5.3.9 Serial communications encoders

Most modern encoders are based on custom combined optical and electronic integrated circuits. As the level of integration increases it is possible to include more functionality within the encoder. As already mentioned, the sine-wave signals from a SINCOS encoder are susceptible to electrical noise. This problem can be avoided if the interpolation is carried out inside the encoder and the interpolated position is sent to the drive via the serial communications channel.

A5.3.9.1 BiSS

BiSS is a bidirectional form of SSI that includes error detection. The interpolated position can be read from the encoder and the protocol supports additional features that are similar to those provided with EnDat and Hiperface. If the position information is to be used to control a motor, the position must be obtained at a high sample rate, and so a high data rate is required. As BiSS is a synchronous protocol; either the distance between the drive and the encoder must be limited or some form of line length measurement must be carried out and the drive must compensate for skew between the outgoing clock and the data received from the encoder. (BiSS is a registered trademark of IC-Haus.)

A5.3.9.2 EnDat

The EnDat protocol can be used in a similar way to BiSS to obtain interpolated information from an encoder. (EnDat is a registered trademark of Heidenhain.)

In the future, serial communications encoders are likely to replace many other types of feedback device. Optical encoders using serial communications can offer very-high-resolution position feedback with better noise immunity than a SINCOS encoder with analogue sine-wave signals. Inductive or capacitive encoders with a relatively low number of sine waves per revolution using serial communications can offer accuracy and resolution equal to, or better than, digital incremental encoders or resolvers. By using serial communications, it is possible to include absolute position information and other advanced features.

A5.3.10 Wireless encoders

Wireless encoders, or rather radio systems designed to connect to quadrature signal encoders, now exist on the market. The transmission speed and determinism/synchronisation is, however, a significant hurdle to their practical implementation as part of a commutation and/or quality closed-loop speed feedback system. At the time of writing typical update rates for a wireless encoder are of the order of 0.6 ms. This corresponds to the equivalent of approximately 64 data updates per revolution for a motor operating at $1\ 500\ \text{min}^{-1}$.

Chapter A6

Motion control

A6.1 General

The aim of many applications using variable-speed drives is to control motion. This may be simple rotary speed control of a single motor or the movement of an object through a complex profile in three dimensions involving the coordinated movement of several motors. Figure A6.1 shows a generalised motion controller for a single motor, often referred to as a single axis controller.

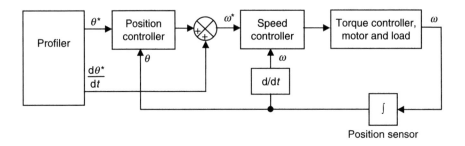

Figure A6.1   Single-axis motion controller

The profiler provides a position reference θ^* and speed feed-forward reference $d\theta^*/dt$. The control of position, speed and torque are covered in detail in Chapter A4. This section focuses on the profiler and its use in single- and multi-axis systems. Rotary motion control as shown in Figure A6.1 will be used as an example unless otherwise stated, but all the principles discussed apply equally to linear motion.

The function of the profiler is to take the required motion and provide appropriate speed and position references for the axis. The following are the three possible functions of the profiler.

- The input to the profiler is a command to move from one position to another using a time-based profile that ensures that the move is made in the shortest time without exceeding certain limits.

- The input to the profiler is the continuously changing position of a master system. The input motion is modified with a position-based profile referred to as a CAM function to give the required movement of the axis.
- The required motion may be the continuously changing position of a master system and the profiler may modify the master position with a simple ratio. This system is usually referred to as an electronic gearbox.

These functions are described in detail later, but it is useful at this point to look at the different control quantities used in the profiler (position, speed, acceleration and jerk) and also the different possible configurations.

A6.1.1 Position, speed, acceleration and jerk

Position, speed, acceleration and jerk are related to each other as shown in the following equations (A6.1), (A6.2) and (A6.3), where t is time:

$$\text{Position} = \int \text{Speed } dt \quad (A6.1)$$

$$\text{Speed} = \int \text{Acceleration } dt \quad (A6.2)$$

$$\text{Acceleration} = \int \text{Jerk } dt \quad (A6.3)$$

Figure A6.2 gives an example profile based on linear speed ramps, and shows position, speed and acceleration across the profile. The speed, acceleration and jerk may be limited as follows.

A6.1.1.1 Speed

If the aim is to move between two positions as quickly as possible, the maximum speed is usually limited by the mechanical design of the system. In other cases the maximum speed for safe operation may be used. The example profile has a maximum speed of 100 rad s^{-1}.

A6.1.1.2 Acceleration

The example profile uses a maximum acceleration of 100 rad s^{-2} throughout the whole of the acceleration and deceleration periods, which gives the linear speed ramp characteristic. As well as overcoming a static load and friction, the torque produced by the motor must accelerate the load inertia. Therefore the maximum possible acceleration is defined by the rating of the drive and the motor. The maximum acceleration may also be limited by the mechanical design of the system.

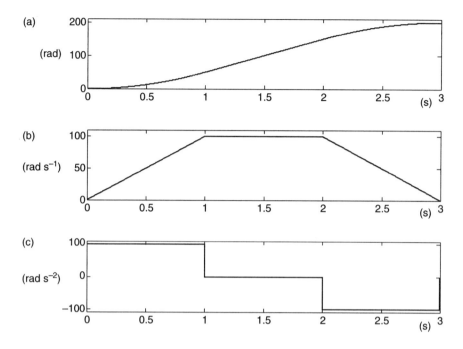

Figure A6.2 Motion profile based on linear ramps: (a) position; (b) speed; (c) acceleration

A6.1.1.3 Jerk

Jerk is the rate of change of acceleration. The example profile has zero jerk throughout the profile except at the start and end of both the acceleration and deceleration periods. At these points there is a step change in the acceleration, producing infinite jerk. The effect of jerk is not obvious, but can be demonstrated by applying a motion profile that is similar to the example to the linear motion of a vehicle and considering its effect on the occupants (see Figure A6.3).

Assuming that the accelerator and brake can be moved instantly to their required settings, the occupants will be forced back into their seats at point A by the infinite positive jerk of the vehicle as the vehicle begins to accelerate. Unless the acceleration is excessive, the occupants become accustomed to the constant force during acceleration, and are then thrown forwards at point B by the infinite negative jerk of the vehicle. As expected, the occupants are thrown forwards again at point C as the brakes are applied. If the decelerating force applied by the brake is constant, the occupants become accustomed to the constant force that tries to force them forwards and out of their seats. It might not be obvious, but when the vehicle stops they are thrown back into their seats, in the same way as when the vehicle began to accelerate, by the infinite positive jerk of the vehicle. This type of motion would not be acceptable to the passengers, and so the driver limits the jerk by avoiding sudden changes in

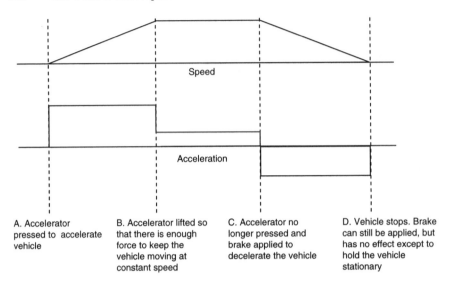

Figure A6.3   Vehicle motion profile

accelerator position and by applying and releasing the brakes smoothly at the start and end of the deceleration period. This modified motion profile is simulated by limiting the maximum jerk and is demonstrated in the S ramp functions described later. One important application for a motion profile with a maximum jerk limit is in the elevator industry, where sudden jerk would be unacceptable to the occupants of the elevator. It is clear that the level of jerk is proportional to the torque or force transients that are being applied to the mechanical system by the drive and motor. Therefore a jerk limit may also be required to limit vibration and stress.

In some applications the rate of change of jerk is also controlled.

A6.1.2   Possible configurations

There are four important configurations that could be used to provide a demand for the profiler in a motion control system. Each case is shown as a multi-axis system, but all except option (1) could be implemented with a single axis. It is assumed that the demand to the profiler is digital.

1. *Target position or speed as reference.* The target position or the target speed (Figure A6.4) is used by the profiler with a time-based profile to move the axis to the target position or speed in the shortest time within the limits imposed on the profile (i.e. maximum speed, maximum acceleration and maximum jerk as required). The profiler(s) and target generator may be part of the same control system; however, if this is not the case the target reference is normally provided over a suitable fieldbus. This configuration is generally used when the changes in the target are not synchronous with the sample time of the profiler, such as

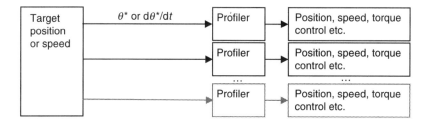

Figure A6.4   Target position or target speed as a reference

in a pick-and-place application, so the fieldbus does not need to provide time synchronisation between the target generator and the profiler.

2. *Position as reference.* Where coordinated movement between several axes is required, for example in a machine tool application, a central motion controller may be used that generates the position profile for each axis (Figure A6.5). The profiler for each axis is included in the central motion controller, which generates a continuously changing position reference for each axis. Usually a speed feed-forward reference is also generated for each axis by the motion controller, and the position reference is delayed by one sample. By doing this the speed feed-forward generates the motion and the position controller only corrects the motion for any deviation from the profile caused by the load. A pulse train can be used as the reference with a separate direction signal [frequency (F) and direction (D)], but this is being replaced by fieldbus systems that provide synchronisation between the motion controller and drive system such as SERCOS, EtherCAT or CTSync. If a fieldbus is used without synchronisation there will be a beating effect as shown in Figure A6.6, which results in no change of position at some samples followed by a change of twice the expected position at the next sample.

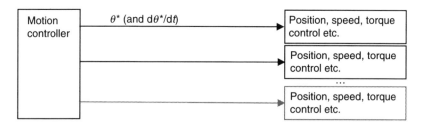

Figure A6.5   Position as reference

3. *Master speed as reference.* A master speed may be used as the reference for one or more axes where movement is required that is relative to a master system (Figure A6.7). Usually the reference is provided as a set of quadrature signals from an encoder on the master system and the profiler uses a CAM or electronic gearbox function to move the axis through a profile that is synchronised to the movement of the master.

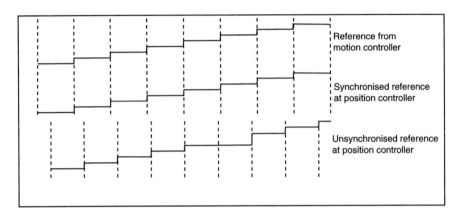

Figure A6.6 Position reference with and without synchronisation

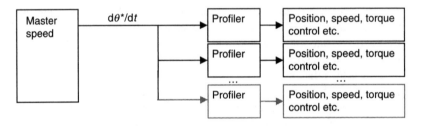

Figure A6.7 Master speed as reference

4. *Master speed with ratio as reference.* In some applications a difference in speed is required between each axis, or sets of axes (Figure A6.8). This method is used to provide tension between different sections of a paper-making machine by setting the speed of each section slightly faster than the previous section.

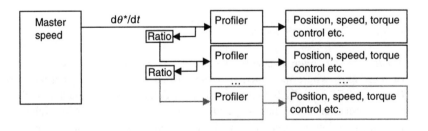

Figure A6.8 Master speed with ratio as a reference

A6.2 Time-based profile

A time-based profile generator provides a profile that moves the position reference to the target position in the minimum amount of time within the limits of the system. For a linear speed ramp profile these limits are maximum speed and maximum acceleration, as demonstrated in the profile shown in Figure A6.2. This type of profile can be generated in a number of ways and an example of one method is shown in Figure A6.9.

The speed reference ω can be set to the maximum speed limit ω_{max} or the target speed ω^*. In the example shown in Figure A6.9, the speed target is zero ($\omega^* = 0$) and the position target is θ_{target}. The ramp controller allows the output speed ($d\theta^*/dt$) to change within the constraints of the profile and it cannot increase or decrease faster than the maximum acceleration a_{max}. The appropriate speed ω is selected at any time so that the profile output achieves its required target speed and position. The profile generator constantly calculates the position change that would occur for the profile output speed to change from its current value to the target value and uses this to make a decision on which speed reference to select. In this case the target speed is zero, so the position change that would occur to reach the target speed is given by

$$\Delta\theta = (d\theta^*/dt)^2 / 2a_{max} \tag{A6.4}$$

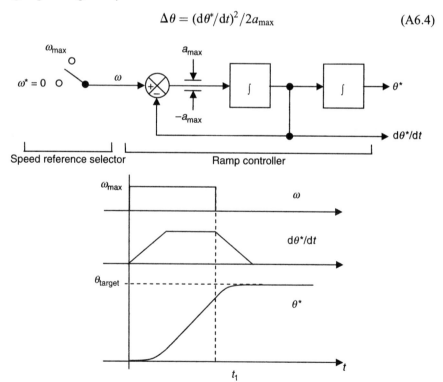

Figure A6.9 *Linear speed ramp profile generator*

The initial speed and position are zero, so the system will not move or reach its target, and so the speed reference selector changes to ω_{max}. The profile output speed increases up to ω_{max} with an acceleration of a_{max}. Eventually, at time t_1, the position that will be moved before the target speed of zero is reached is equal to the distance to the target position, and so the speed reference selector changes to the target speed of zero. The speed then ramps to zero and the position reaches its target.

There may be occasions where the axis is required to stop before reaching its target. If the axis is required to stop within the profile limits, the simplest way to do this is to change the target position to the current position plus the change of position defined by equation (A6.4). The profile generator will then automatically set the speed reference selector to the target speed and the axis will ramp to a stop. For an emergency stop it may be necessary to stop as quickly as possible and ignore the required limits. This is referred to as a shear pin stop and can be achieved by setting $d\theta^*/dt$ to zero and θ^* to the present feedback position θ.

As mentioned earlier, the linear profile includes infinite jerk each time the acceleration changes. If a limit is applied to the jerk the linear speed profile used as an example in Figure A6.2 is converted into the speed S ramp shown in Figure A6.10.

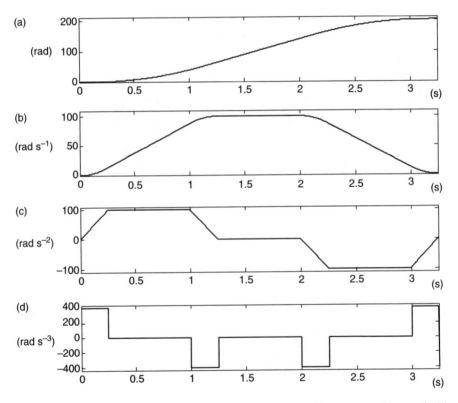

Figure A6.10 Motion profile based on S speed ramps: (a) position; (b) speed; (c) acceleration; (d) jerk

The abrupt changes in acceleration that gave the infinite jerk at each end of the acceleration and deceleration periods are now linear ramps and the acceleration and deceleration sections of the speed profile have the characteristic 'S' shapes that give this profile its name. The disadvantage of using S ramps is that the time taken to reach the required position is longer than with linear speed ramps. In this particular example, the time is extended from 3.0 s for the linear ramp to 3.25 s for the S ramp. The only way to overcome this is to increase the maximum acceleration, which may require a larger drive and motor.

The analysis of S ramps can be very complex, but the following gives some basic equations that may be used to predict the acceleration times and the position moved. First, the time taken to accelerate (or decelerate) and the position moved with a linear speed ramp are calculated to be used later for comparison. The time taken for acceleration with a linear ramp is quite straightforward and is given by

$$t_{acc} = \omega_{max}/a_{max} \tag{A6.5}$$

The total position moved is a combination of the position moved during the acceleration and deceleration, and the position moved when running at constant speed:

$$\Delta\theta = \Delta\theta_{acc} + \Delta\theta_{constant\ speed} + \Delta\theta_{dec}$$
$$= (\omega_{max}^2/2a_{max}) + (\omega_{max} \times t_{constant\ speed}) + (\omega_{max}^2/2a_{max})$$
$$= (\omega_{max}^2/a_{max}) + (\omega_{max} \times t_{constant\ speed}) \tag{A6.6}$$

where $t_{constant\ speed}$ is the time spent running at constant speed.

Figure A6.11 shows acceleration with a speed S ramp in more detail. The time to accelerate is the sum of the times for the S parts and the linear part of the speed profile. Before proceeding any further it is necessary to determine the speed change ω_B that

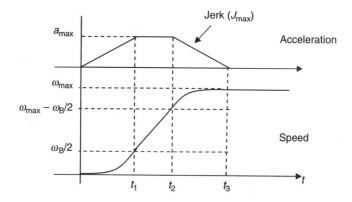

Figure A6.11   Acceleration with speed S ramp

is just large enough to allow the acceleration to reach a_{max}. This speed is the sum of the speed changes during each of the S parts of the speed profile.

$\omega_B/2$ is given by the area under the acceleration characteristic between 0 and t_1, so

$$\omega_B/2 = a_{max} \times t_1/2 = a_{max} \times (a_{max}/J_{max})/2 = a_{max}^2/2J_{max}$$

and therefore

$$\omega_B = a_{max}^2/J_{max} \tag{A6.7}$$

Assuming that the acceleration reaches a_{max} (i.e. $\omega_{max} \geq \omega_B$), the acceleration time is given by the sum of the times for each segment of the profile:

$$\begin{aligned}t_{acc} &= t_1 + (t_2 - t_1) + (t_2 - t_3) \\ &= (a_{max}/J_{max}) + [(\omega_{max} - \omega_B)/a_{max}] + (a_{max}/J_{max})\end{aligned} \tag{A6.8}$$

Substituting for ω_B from equation (A6.7) gives

$$t_{acc} = (\omega_{max}/a_{max}) + (a_{max}/J_{max}) \tag{A6.9}$$

Comparing equation (A6.9), which gives the acceleration time for an S speed ramp with equation (A6.5), which gives the acceleration time for a linear speed ramp, shows that the time for acceleration or deceleration is simply extended by the term a_{max}/J_{max}. This can be confirmed with the profile in Figure A6.10 where $a_{max} = 100$ rad s^{-2} and $J_{max} = 400$ rad s^{-3}. The acceleration time has been extended by $a_{max}/J_{max} = 100/400 = 0.25$ s to 1.25 s when compared to Figure A6.2.

The position moved during the acceleration shown in Figure A6.11 is given by

$$\Delta\theta_{acc} = (\omega_{max}^2/2a_{max}) + [(\omega_{max} \times a_{max})/2J_{max}] \tag{A6.10}$$

The total movement is derived in the same way as for the linear ramp by adding the movement during acceleration, running at constant speed and deceleration:

$$\Delta\theta = (\omega_{max}^2/a_{max}) + [(\omega_{max} \times a_{max})/J_{max}] + (\omega_{max} \times t_{constant\ speed}) \tag{A6.11}$$

Comparing equation (A6.11) with the position moved with a linear speed ramp given in (A6.6) shows that there is an additional position movement during acceleration and deceleration given by the additional term in (A6.11), i.e. $[(\omega_{max} \times a_{max})/J_{max}]$. To ensure that the total position moved over the whole profile is the same as for the original linear speed ramp profile, the time and hence the position moved during constant running must be reduced. During this period the speed is ω_{max}, and so the time reduction is given by dividing the additional position movement by ω_{max}, giving (a_{max}/J_{max}). This can be confirmed with the example profile in Figure A6.10 where the time for running at constant speed has been reduced by

Motion control 243

Table A6.1 Summary of acceleration time and change of position

	t_{acc}	$\Delta\theta$
Linear speed ramp	(ω_{max}/a_{max})	$(\omega_{max}^2/a_{max}) + (\omega_{max} \times t_{constant\ speed})$
S speed ramp where $\omega_{max} \geq \omega_B$ $(\omega_B = a_{max}^2/J_{max})$	$(\omega_{max}/a_{max}) + (a_{max}/J_{max})$	$(\omega_{max}^2/a_{max}) + [(\omega_{max}\,a_{max})/J_{max}] + (\omega_{max} \times t_{constant\ speed})$
S speed ramp where $\omega_{max} < \omega_B$	$2\sqrt{(\omega_{max}/J_{max})}$	$\sqrt{(\omega_{max}^3/J_{max})}$

$(a_{max}/J_{max}) = 100$ rad s^{-2}/400 rad s^{-3} = 0.25 s when compared to the linear speed profile in Figure A6.2.

Table A6.1 summarises the results obtained for a linear ramp and S ramp. Additional results are included for the case where the speed change for the S ramp is less than ω_B, so that the speed change is too small for the acceleration to reach a_{max}, and the ramp does not include a section with constant acceleration.

Linear speed and S speed ramp time-based profiles have been described that can be used to move the axis from one position to another. Other profile types can be used, but these are not covered here.

A6.3 CAM profile

The profiler output may be required to follow a characteristic based on the movement of another system, giving a profile based on an input position instead of using a time base as covered in the previous section. Historically this type of function has been provided by a mechanical CAM as shown in Figure A6.12.

A CAM can be implemented electronically by using the master position as a reference and a look-up table in the profiler to produce the output position in relation to the input position. Normally the change of position reference is used to produce a corresponding change in output position as shown below. The look-up table contains a change of input position and a change of output position for each segment in the CAM profile.

In this example the input speed is constant at 100 units s^{-1}. The first segment ends after 1 s when the input has changed by 100 units, and the output should change by 100 units at this point. The second segment ends after another 1 s as the input has changed by a further 100 units and the output has changed by 300 units. If the CAM table is used without any form of interpolation, a large CAM table is likely to be required to give smooth movement of the output. The change of the reference position $d\theta^*/dt$ (the master speed) is used to drive the CAM table, and so this only provides relative movement between the input and output. However, the position

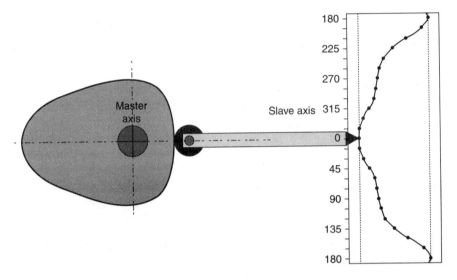

Figure A6.12   Mechanical CAM

reference θ^* can be used to initialise the start position in the CAM table to give absolute position locking between the input and output.

Interpolation can be used to provide smoother movement of the output. The CAM profile from Figure A6.13 has been modified to use linear interpolation as shown in Figure A6.14a.

The sample rate of the profiler is such that many samples occur during each of the segments of the CAM profile. At each sample point the output is changed by a proportion of the required output movement for the whole segment based on the

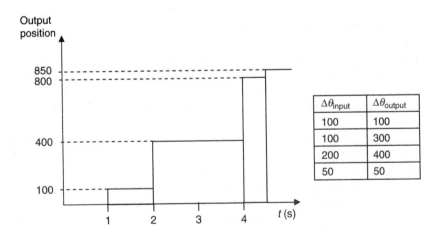

Figure A6.13   CAM profile without interpolation

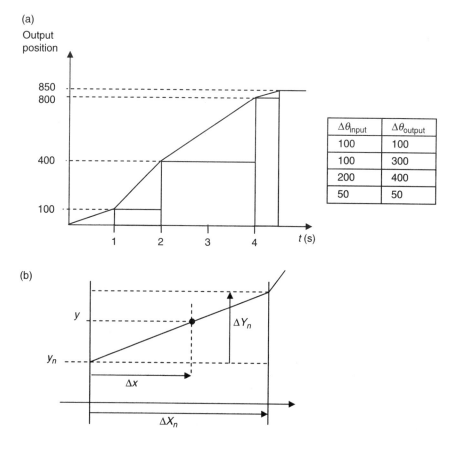

Figure A6.14 CAM profile with interpolation: (a) complete profile; (b) one segment

change of input position as a proportion of the input position change for the segment. Therefore the output position y during the nth segment is given by

$$y = y_n + \Delta Y_n(\Delta x/\Delta X_n) \qquad (A6.12)$$

where y_n is the output position at the start of the segment, ΔX_n and ΔY_n are the required change of input and output position, respectively, for the whole segment from the look-up table, and Δx is the change of input position so far since the start of the segment. Now the movement of the output is likely to be smoother, but there are still abrupt changes of speed at the boundaries between segments as shown in Figure A6.15.

If x represents the input as a proportion of the movement through a segment, i.e. $x = (\Delta x/\Delta X)$, then equation (A6.12) can be rewritten as

$$y = y_n + \Delta Y_n\, f(x) \qquad (A6.13)$$

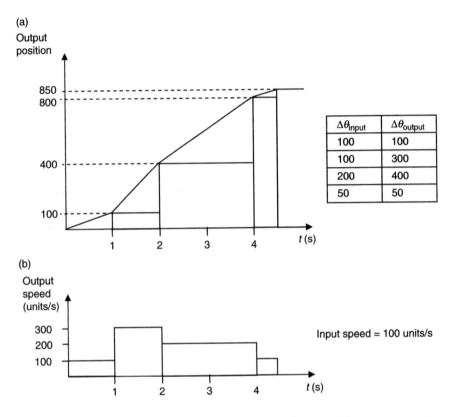

Figure A6.15 CAM profile with linear interpolation: (a) output position; (b) output speed

For linear interpolation, the function $f(x) = x$. Alternative functions may be used to give different types of interpolation. The position and speed for one segment are shown in Figure A6.16 for 'cosine' interpolation where $f(x) = x - [\sin(2\pi x)]/2\pi$.

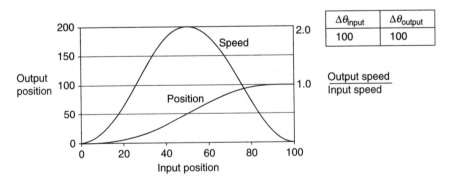

Figure A6.16 CAM profile with cosine interpolation

As the name suggests, the speed follows a cosine function $(1 - \cos \theta)$ and is zero at the start and end of the segment. The acceleration, which is the differential of the speed, is $\sin \theta$. This gives a much better profile than linear interpolation as there are no discontinuities in either the speed or the acceleration, although there is still a discontinuity in the jerk. The only disadvantages are that the maximum speed of the output is now twice the speed that would be produced by linear interpolation, and the output returns to zero speed at the end of each segment.

It is possible to provide continuous motion with speed changes, but without discontinuities in the speed profile, by combining the outputs of two CAM tables. To do this it is necessary to create an acceleration ramp segment with the function $f(x) = x^2$ and a deceleration ramp segment with the function $f(x) = 2x - x^2$. The resulting component profiles are shown in Figure A6.17.

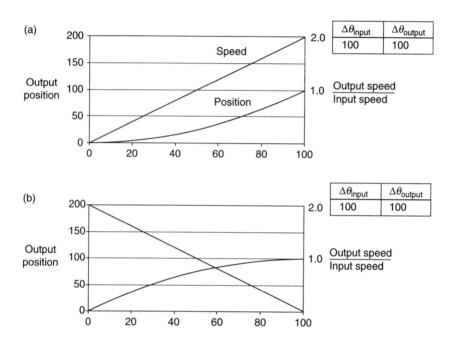

Figure A6.17 CAM profiles to give linear speed changes: (a) acceleration $f(x) = x^2$; (b) deceleration $f(x) = 2x - x^2$

These can then be combined with a CAM table using linear interpolation to give a speed profile without discontinuities, as shown in Figure A6.18. In the first segment, the speed rises to 200 units s^{-1} using an acceleration ramp profile. In the second segment, linear interpolation is used to maintain the speed at 200 units s^{-1}, and then in the third segment this is combined with another acceleration ramp to accelerate to 400 units s^{-1}. The example uses linear ramp segments, but it is possible to create other profile shapes by using alternative functions. For example

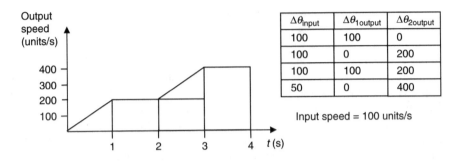

Figure A6.18   CAM profile using two CAM tables

$f(x) = x - [(\sin(\pi x))/\pi]$ and $f(x) = x + [\sin(\pi x)/\pi]$ can be used as the acceleration and deceleration ramps, respectively, giving cosine-shaped speed ramps.

A6.4   Electronic gearbox

An electronic gearbox function is used where the position and speed of two or more systems are linked by a simple ratio. There are three basic forms of electronic gearbox.

- *Direct positional lock.* In some applications such as screw tapping, it is important that two axes are locked in position at all times including periods of acceleration and deceleration. This is the electronic equivalent of a mechanical gearbox.
- *Ramped non-rigid speed lock.* Sometimes it is only important to maintain lock once the slave system is running at the master speed with the required ratio applied. In this case, a time-based profile is used to accelerate the slave axis up to the speed of the master system and then the electronic gearbox function is engaged as shown in Figure A6.19.
  This operates in a similar way to a mechanical system that uses a clutch to engage the two systems, but has the advantage of avoiding the possible mechanical impact when the systems are first locked.

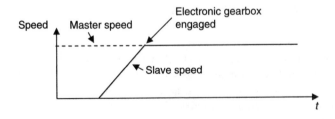

Figure A6.19   Electronic gearbox with ramped, non-rigid speed lock

Motion control   249

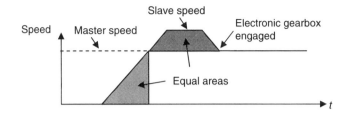

Figure A6.20   Electronic gearbox with ramped rigid speed lock

- *Ramped rigid speed lock.* Another alternative is ramped rigid speed lock. Here, it is important for the position of the master and slave to be locked before the electronic gearbox function becomes active. In the example shown in Figure A6.20 the slave axis is started at a point where it is synchronised correctly with the master axis. It does not maintain its synchronism during acceleration, but it is made to exceed the master speed to 'catch up' until it is once again synchronised and then the electronic gearbox function can be engaged. A rotary knife is a good example of a system requiring this type of control. Only after it is synchronised can the knife be used to cut.

A6.5   Practical systems

Traditionally, the motion controller has been a separate system providing references to the drive. With higher levels of integration, it is now possible to get option modules for many drives that include motion control functions. An example of such a product is the Control Techniques Advanced Position Controller, a software function included in an option module to be mounted inside a drive. It is clear that setting up a motion control system can be complex, and so PC applications are normally provided to assist the user. One example of this is the Control Techniques Indexer software, which provides an easy-to-use graphical interface allowing the user to quickly define a time-based profile.

A6.5.1   Control Techniques' Advanced Position Controller

The position controller shown in Figure A6.21 includes the motion control functions described in this chapter. The user can select any one of the motion functions and the controller will provide a suitable speed reference to the drive.

- *Position reference.* The position reference is used by the time-based profile to give a position movement with limits placed on maximum speed, maximum acceleration and maximum jerk.
- *Speed reference.* The speed reference is used by the time-based profile to set the drive speed reference with the same limits as applied to the profile with a position reference.

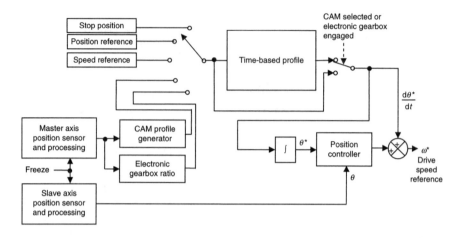

Figure A6.21   Control Techniques advanced position controller

- *CAM.* Two CAM tables are provided with all the CAM profile functions described in Section A6.3. When the CAM function is selected the timed-based profile is not required, and so it is bypassed.
- *Electronic gearbox.* The electronic gearbox uses the time-based profile up to the point where it is engaged, and after that point the time-based profile is bypassed until the electronic gearbox is disengaged again.
- *Stop position.* The stop position is constantly updated so that the controller can perform a profiled or shear pin stop when the stop position is selected.
- *Freeze.* A freeze input is a hardware trigger that can be used to store the master and/or slave position at a precise instant defined by either edge of the freeze signal. The following examples show how the freeze data can be used:
  - As material passes a sensor some feature on the material can cause a freeze event, which is then used to start the CAM function or electronic gearbox. This allows the slave system to be precisely synchronised to the feature on the material.
  - A registration mark can be printed on paper in one colour. This mark is then detected using a freeze function so that the printing of other colours can be automatically aligned.
  - In high-speed systems it is not possible to use a simple sampled system to capture the position, because there is significant position movement between samples. A freeze function can capture the exact position at the required instant in time.

A6.5.2   Control Techniques' Indexer

The Control Techniques Indexer application is intended to help the user to set up an application that consists of the number of time-based profile movements. Trigger events can be placed between each move, so for example a move may not start until a digital input becomes active. In a similar way it is possible for the user, while commissioning, to step through a particular motion sequence. Figure A6.22 shows a screenshot of an example project.

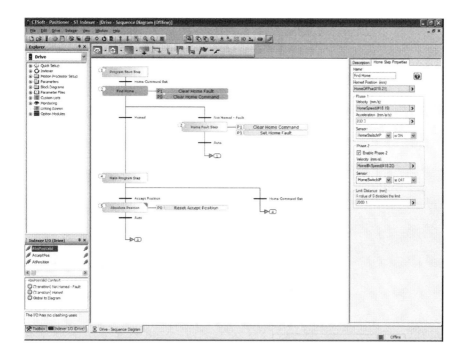

Figure A6.22   Control Techniques indexer

Chapter A7
Voltage source inverter: four-quadrant operation

A7.1 General

The thyristor-based armature converter used in the d.c. motor drive described in Section A2.2.4 allows power to flow in either direction between the mains supply and the motor, and so the drive can be used for full four-quadrant operation. The inverter that forms the output stage of the popular voltage-source a.c. motor drives described in Section A2.4.2 also allows power to flow in either direction between its d.c. connections and the motor (Figure A7.1). A simple rectifier is often used between the mains supply and the d.c. terminals of the inverter, but this does not allow power to flow back into the supply. Therefore, an a.c. motor drive based on this configuration cannot be used where power is required to flow from the motor to the mains supply. This section describes two ways in which this form of a.c. motor drive can decelerate a motor and its load with a simple rectifier input stage. Also described is an alternative active rectifier stage that can be used in conjunction with a voltage-source inverter to facilitate full four-quadrant operation.

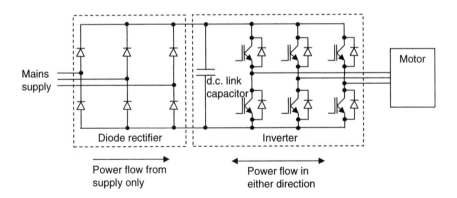

Figure A7.1   Possible power flow in a typical voltage source a.c. drive with input diode rectifier

A7.2 Controlled deceleration

During deceleration with a motor and load that has sufficient inertia, the power flowing from the inverter terminals can cause the voltage across the d.c. link capacitor to rise significantly. The more rapid the rate of deceleration, the faster the voltage rises. To prevent the power circuit from being damaged, d.c. link over-voltage protection is included in most industrial drive control systems; this inhibits the inverter if this voltage exceeds a trip threshold. The power flowing from the inverter depends on the inverter output voltage and current, as shown in Figure A7.2.

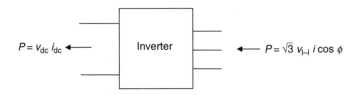

Figure A7.2 Power flow through the output inverter stage

The rms line voltage on the a.c. side of the inverter is given by

$$v_{l-l} = v_{dc} m/\sqrt{2} \tag{A7.1}$$

where m is the modulation index. Neglecting the losses in the power circuit, the power flowing from the d.c. side of the inverter is

$$P = v_{dc} i_{dc} = \sqrt{(3/2)} v_{dc} m i \cos \phi \tag{A7.2}$$

Where the decelerating torque is constant, the power taken to decelerate a motor and load reduces as the motor slows down. This is also indicated by equation (A7.2), which shows that the power is proportional to the modulation index, which is approximately proportional to the motor speed when this is below the point where field weakening begins. The voltage across the d.c. link capacitor can be controlled by regulating i_{dc} to dynamically adjust the deceleration of the motor and prevent this voltage from reaching the trip threshold. This forms the basis of a control method that can be used in a drive with a simple diode rectifier input circuit to prevent over-voltage trips during deceleration. The motor phase current combined with the power factor ($i \cos \phi$) is approximately equal to the motor torque-producing current. Therefore, from equation (A7.2)

$$i_{dc} \approx \sqrt{(3/2)} m i_T \tag{A7.3}$$

Figure A7.3 shows a control scheme that can be included in a drive with an outer speed controller to regulate the d.c. link voltage. When the d.c. link voltage exceeds v_{dcmax}, the output of the PI controller is used to provide the torque-producing current reference. This system remains active until the motor is accelerated again or the speed falls to the point where the losses in the power circuit and motor exceed the power

Voltage source inverter: four-quadrant operation 255

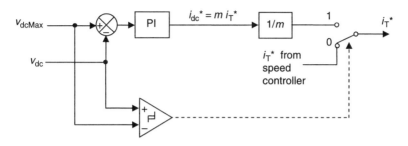

Figure A7.3 D.C. link voltage controller

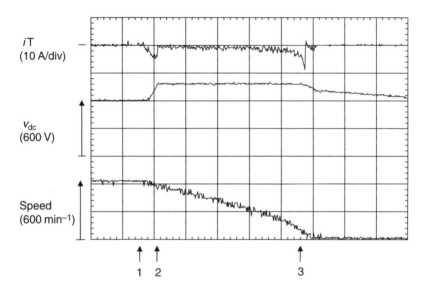

Figure A7.4 D.C. link voltage regulation during deceleration: (1) deceleration begins and v_{dc} begins to rise; (2) v_{dc} reaches v_{dcMax} and torque-producing current is regulated to prevent v_{dc} from rising further; (3) power flow has reduced so that i_T no longer needs to be regulated

from the decelerating the load. Figure A7.4 shows how the d.c. link voltage is regulated during deceleration in a real system.

This method can be used with the permanent-magnet or induction motor drives based on closed-loop current control as described in Chapter A4. A variation of this method can be included in the open-loop induction motor control method also covered in Chapter A4.

A7.2.1 Performance and applications

Regulation of the d.c. link voltage during deceleration can be used in applications where relatively slow deceleration is acceptable. This includes many fan and pump applications, centrifuges, and so on. The advantages and disadvantages are summarised in the following.

A7.2.1.1 Advantages

- A diode rectifier can be used between the mains supply and the inverter and this is usually the lowest cost configuration.

A7.2.1.2 Disadvantages

- Fast deceleration is not possible, because the energy stored in the load is dissipated in the power circuit and the motor. However, it is possible to boost the motor voltage during deceleration to over-flux the motor and increase the losses to shorten the stopping time.
- As a diode rectifier is used, the supply currents include significant harmonic components whether the system is motoring or braking.

A7.3 Braking resistor

To overcome the limited deceleration possible with the scheme described above, the diode rectifier can be supplemented with a d.c. link braking resistor circuit as shown in Figure A7.5.

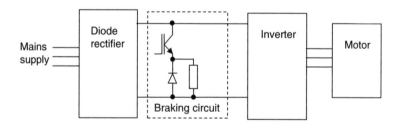

Figure A7.5   A.C. motor drive with a braking resistor circuit

The braking circuit switching device is controlled by the simple system shown in Figure A7.6. When the d.c. link voltage exceeds the braking threshold (v_{dcmax}),

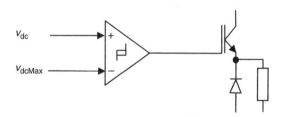

Figure A7.6   Braking circuit control

the switching device is turned on. Provided the braking resistor has a low enough resistance to absorb more power than the power flowing from the inverter, the d.c. link voltage will fall and the switching device will be turned off again. In this way the on and off times are automatically set depending on the power from the inverter, and the d.c. link voltage is limited to the braking threshold. To limit the switching frequency of the braking circuit, some voltage hysteresis may be included as indicated in Figure A7.6. Alternatively, the switching frequency can be limited by enforcing a minimum on time for the switching device.

A7.3.1 Performance and applications

A braking resistor circuit is used in many applications where it is practical and acceptable to dissipate the energy stored in the load in a resistor. Unlike the controlled deceleration method there is no longer a limit on the dynamic performance of the system.

A7.3.1.1 Advantages
- The deceleration rate is no longer limited provided the braking resistor can absorb all the power from the inverter.
- This is a relatively cost-effective solution.

A7.3.1.2 Disadvantages
- For applications with high-inertia loads, the energy that must be taken from the load during deceleration can be significant, and so the required energy rating of the braking resistor may be unacceptably large.
- During deceleration, the energy stored in the load is dissipated in the braking resistor. This may not be acceptable either because it is wasteful or because of the heat generated by the resistor.
- As a diode rectifier is used, the supply currents include significant harmonic components whether the system is motoring or braking.

A7.4 Active rectifier

The diode rectifier shown in Figure A7.1 does not allow power flow from the d.c. link to the supply. Various alternative circuits can be used to recover the load energy and return it to the supply. One such scheme is the active rectifier shown in Figure A7.7, in which the diode rectifier is replaced with an IGBT inverter. Unless the reactance of the supply is high, additional input inductors are required to limit the unwanted currents generated by the switching action of the inverter. It is possible to control the inverter to give good quality, near-sinusoidal, current waveforms with a power factor close to unity.

The input inductors and the supply could be considered as an infinitely large synchronous motor where the 'torque-producing current' controls the real power flow and the 'flux-producing current' controls the reactive power flow. A system based on a

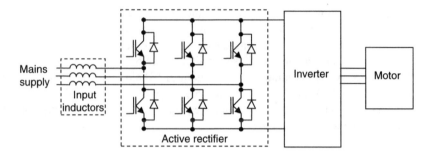

Figure A7.7 A.C. motor drive with active rectifier

reference frame may be used to control the active rectifier in a similar way to the permanent-magnet motor control scheme described in Chapter A4. The reference frame must be defined with the y-axis aligned with the supply voltage vector, so that the y-axis current will represent the real current components and the x-axis current will represent the reactive components. The control circuit for an a.c. motor drive (Figure A4.27) can be adapted as shown in Figure A7.8 for the active rectifier. The torque axis components denoted by the subscript T are now replaced by the real components with subscript y, and the flux axis components denoted with subscript F are replaced by the reactive components with subscript x.

The reactive current reference i_x^* is normally set to zero, so that the currents at the inverter terminals contain only real components giving a displacement factor of unity. In some applications, however, some reactive current can be provided to improve the overall power factor of the system local to the active rectifier. The d.c. link voltage controller produces a current reference that should define the flow of current i_{dc} into the d.c. link capacitors and the inverter connected to the motor, in order to regulate the d.c. link voltage. As shown earlier in equation (A7.3), the d.c. link current reference i_{dc}^* can be converted to the real current reference i_y^* by dividing by the modulation index.

It is possible to define the angle for the reference frame from voltage feedback on the mains supply side of the input inductors, but this requires voltage isolation between

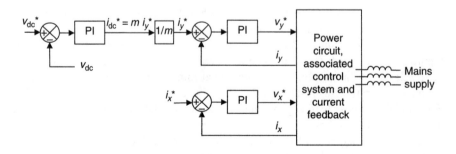

Figure A7.8 Active rectifier control system

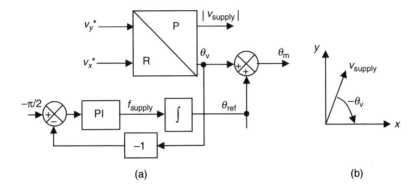

Figure A7.9 *Reference frame angle generation: (a) reference frame generator; (b) supply voltage vector*

the supply and the control system. A much more cost-effective solution is to use the output of the current controllers to define the reference frame angle, because, provided the current controllers are operating correctly, their steady-state output voltages should be a representation of the supply voltages. Figure A7.9 shows a part of the reference frame translation system from Figure A4.33 with the additional elements required to generate the reference frame angle.

Once the rectangular components of voltage from the current controllers have been converted to polar coordinates, the angle between the required reference frame and the voltage vector is $-\theta_v$ as shown. If the reference frame is correctly aligned, then the angle between the voltage vector and the x-axis is $-\pi/2$, so the error is given by $\theta_v - \pi/2$. This error is fed into a PI controller giving the supply frequency, which is then integrated to give the reference frame angle (θ_{ref}). This is a phase-locked loop and if relatively low gains are used in the PI controller it can give very good noise rejection. It should be noted that this system aligns the reference frame with the voltages on the inverter side of the input inductors, which means that the displacement factor on the mains supply side is not unity. However, the input inductors are normally between 8 and 10 per cent reactance, so this phase shift does not significantly affect the displacement factor at the supply connection.

A7.4.1 Performance and applications

An active rectifier can be used along with an inverter where full four-quadrant operation and good quality input waveforms are required. Cranes and elevators, engine test rigs and cable-laying ships are some applications where an active rectifier may be used. The performance characteristics of this configuration are summarised below.

A7.4.1.1 Advantages
- Power flow between the motor and the mains supply is possible in both directions, and so this makes the drive more efficient than when a braking resistor is used. As

well as recovering energy during deceleration, it is possible to recover energy from an overhauling load such as a crane when the load is being lowered.
- Good-quality input current waveforms are possible with good displacement and distortions factors.
- It is possible to boost the d.c. link voltage to a level that is higher than would be possible with a simple diode rectifier.

A7.4.1.2 Disadvantages
- An active rectifier is more expensive than a simple diode rectifier.
- Other input components are required, in addition to the input inductors, to control d.c. link capacitor charging during power up and to prevent inverter switching frequency components from affecting other equipment connected to the same supply.

Chapter A8
Switched reluctance and stepper motor drives

A8.1 General

All the motors described earlier in this part of the book have the characteristic that they can be operated, albeit at essentially fixed speed, when fed by an appropriate d.c. or a.c. supply. Similarly, the converter topologies are in fundamental terms either independent of motor type or variations exist for different motor types. In this section we consider a group of drive systems where the motor can only be used in conjunction with its specific power converter and control, and consequently only overall characteristics are relevant.

Switched reluctance drive systems are of importance in some applications where high, low-speed torque is required, and less importance is placed on smoothness of rotation. Although considerable advances have been made in improving the noise characteristics of this drive, it can still be a limiting factor where a broad operating speed range is required.

Stepper motors systems are somewhat in decline and would rarely be described as a general industrial drive. Their operating characteristic of being controlled by 'a computer' pulse train is now a common feature of many modern servo drives. Also, where rapid settling times are required, stepper drives, which are inherently 'open-loop', are not ideal in both fundamental performance and in the respect that varying mechanical friction has a significant impact. This variability can make stepper drives unacceptable in applications where the transient performance is important.

A8.2 Switched reluctance motors and controllers

A8.2.1 Basic principle of the switched reluctance motor

Like the stepper motor, the switched reluctance motor (SRM) produces torque through the magnetic attraction that occurs between stator electromagnets (formed by winding coils on salient poles) and a corresponding set of salient poles formed on a simple rotor made only of electrical steel (or other ferromagnetic material). The stepper and SRM share the same basic principle of energy conversion, and both are members of the family of 'variable-reluctance' motors.

The intuitively straightforward principle of torque production is easily visualised in the very simple reluctance motor illustrated, in cross-section, in Figure A8.1. The

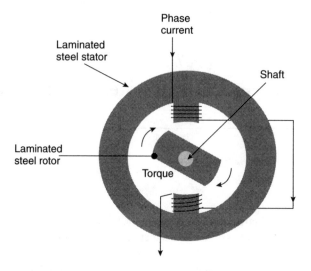

Figure A8.1 '2–2' reluctance motor

motor illustrated would commonly be referred to as having a '2–2' pole structure, the two numbers referring to the numbers of stator and rotor poles, respectively. Intuitively, it can be seen that if current is passed through the stator windings, with the rotor position as shown above, then the rotor will experience a clockwise torque (as indicated by the arrows). Note that, because no permanent magnets are involved, the polarity of the phase current is immaterial. This permits the use of a wide range of power electronic drive circuits, as will be shown later.

If the rotor (and any associated mechanical load) is free to move, this torque will cause the rotor to accelerate clockwise. Torque will continue to be produced in a clockwise sense until the rotor reaches the fully aligned position shown in Figure A8.2. The fully aligned rotor position is sometimes referred to as 'top dead centre' (TDC) by analogy with the internal combustion engine and the following characteristics apply.

- Zero torque is produced.
- Magnetic circuit 'reluctance' is at a minimum.
- Electrical inductance at a maximum.

At TDC, the magnetic circuit – completed by the rotor – offers minimum opposition to magnetic flux. This opposition is known as the magnetic circuit's reluctance, and is analogous to resistance in an electrical circuit. Hence, at TDC, the phase reluctance is at a minimum. This means that, for a given value of phase current, the magnetic flux linked by the windings is maximised, and therefore the electrical inductance of the phase – defined as flux linkage per unit current – is at its maximum value.

Assuming mechanical self-inertia carries the rotor past TDC, it can be seen that the polarity of torque produced by the motor will reverse if we continue to energise the

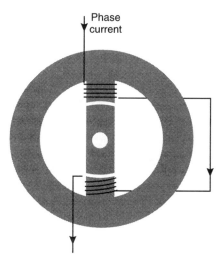

Figure A8.2 Rotor at 'top dead centre' (TDC) (zero torque produced, magnetic circuit reluctance at a minimum, electrical inductance at a maximum)

phase windings beyond the fully aligned position (Figure A8.3). Although the rotor is still turning clockwise under its angular momentum, the torque is now applied in an anti-clockwise sense, and will first reduce the rotor's clockwise angular velocity, and eventually may, depending on the initial rotor speed and the total moment of inertia, cause it to reverse.

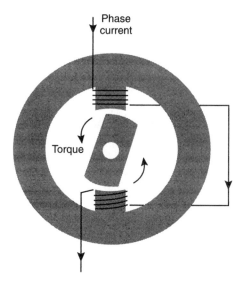

Figure A8.3 Torque reversal beyond TDC

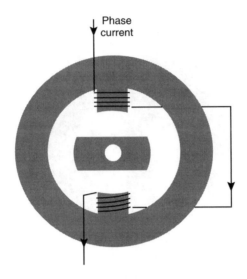

Figure A8.4   *Rotor at 'bottom dead centre' (BDC) (zero torque produced, magnetic circuit reluctance at a maximum, electrical inductance at a minimum)*

Thus the polarity of torque can be reversed, and braking (i.e. generating) can be accomplished without reversing the phase current, in spite of the fact that the machine has no magnets or windings on its rotor. When braking or generating, mechanical work performed on the rotor is converted into energy in the magnetic circuit, which can then be recovered as electrical energy to the power supply by means of the phase winding.

If the rotor turns still further clockwise, it will eventually reach a second position of zero torque, this time when its poles are fully unaligned with respect to the stator poles. Qualitatively, we may say the clockwise and counterclockwise forces now balance each other out, and the net torque is zero. The fully unaligned position is commonly referred to as 'bottom dead centre' (BDC), as illustrated in Figure A8.4.

In contrast with Figure A8.2, the magnetic circuit's reluctance is now clearly at its maximum possible value, and the electrical inductance of the phase is correspondingly at a minimum. The following characteristics apply.

- Zero torque is produced.
- Magnetic circuit 'reluctance' is at a maximum.
- Electrical inductance is at a minimum.

Now we consider how to use the torque-productive intervals illustrated above to operate the switched reluctance machine as a motor or generator.

A8.2.1.1   Operation as a motor

Motoring operation of the SRM requires that the torque generated by the machine acts in the same sense as the *actual* direction of rotation. In other words, the torque should be of a polarity (i.e. direction) such that it reinforces the present direction of rotation.

Through study of Figures A8.1 to A8. 4, we can see that, to operate the machine as a motor, we should energise the winding of the machine only when the rotor and stator poles are approaching each other. In other words, ideally, we should apply phase current only when the magnetic circuit reluctance is decreasing (or equivalently, when the electrical inductance of the phase is increasing) with respect to time.

Assume we want rotation in a clockwise direction, starting with the rotor positioned as shown in Figure A8.1. We should energise the phase winding until the rotor reaches TDC (Figure A8.2) when we must switch the phase current off. The rotor will then coast beyond TDC under its own momentum (Figure A8.3), until it reaches BDC (Figure A8.4). The phase can then be switched on again and the whole cycle repeated. If we fail to switch the phase current off at TDC, then, while the current, or to be strictly correct the magnetic flux associated with it, persists, anticlockwise torque will be produced as the rotor turns further clockwise. This anticlockwise torque will reduce the average motoring torque produced by the machine over a cycle of operation.

Note that the machine will operate equally well as a motor in either direction of rotation, the stator coils just need to be energised over the appropriate range of angular position.

A8.2.1.2 Operation as a brake or generator

So far as basic control of the machine is concerned, there is no fundamental distinction between the terms 'generating' and 'braking'. As an aside, it could be argued that generating is concerned with the efficient conversion of mechanical work to electrical energy, whereas braking is simply the removal of mechanical energy from the load regardless of where that energy ends up. For the purposes of our present simple explanation, the two terms will be regarded as equivalent.

It is fairly easy to see that generating with the SRM is a mirror image of motoring operation. Generating requires that the torque be of a polarity such that it opposes the present direction of rotation. Following on from the previous example, again assume the present direction of rotation is clockwise. We therefore require anticlockwise torque to brake or generate.

This time, considering Figures A8.1 to A8. 4 reveals that we should energise the phase winding when the rotor poles are moving away from the stator poles, as shown in Figure A8.3. In other words, current should be applied when the phase inductance is decreasing (or reluctance increasing). The current should ideally be switched on at TDC, and off again at BDC. Again, if magnetic flux, and current, persist when the rotor has turned beyond BDC, then motoring torque will be produced over part of the machine's electrical cycle, and the average braking (or generating) torque will be reduced.

A8.2.1.3 Summary so far

Motoring torque is produced when a phase is energised while its inductance is increasing with respect to time. Conversely, braking (or generating) torque is produced if the phase is energised while its inductance is falling with respect to time. We can therefore choose motoring or generating operation of the SRM simply by timing the excitation of the phases with respect to the rotor position. Hence, it is a reluctance machine in which the phases are controlled, or 'switched', in accordance with rotor position, giving rise to the name 'switched reluctance motor'. Note that it is a 'self-synchronous' machine, i.e. the stator excitation frequency is locked to the rotor speed. This is in contrast to, say, the induction motor, where rotor speed is a *result* of the applied frequency.

Figure A8.5 shows the variation of phase inductance as a function of angle for a typical SR machine. If we apply current over the intervals discussed above, then the polarity of torque can be seen to depend on whether the inductance is rising or falling with angle.

If the machine is only lightly magnetically loaded (when it will be producing only a modest amount of torque for its size), then the steel from which the rotor and stator are made will behave magnetically in an approximately linear fashion. That is, for a given number of turns on the windings, the phase's magnetic flux will vary approximately in proportion to the phase current. If linearity is assumed, then it can be shown that the torque produced as a function of angle is

$$T = [i^2 \cdot (dL/d\theta)]/2 \tag{A8.1}$$

This relationship shows independence of polarity of phase current (due to the i^2 term) and that the torque polarity depends on the slope of the inductance curve with angle. Given that the angular spacing of L_{max} and L_{min} is fixed by the machine's rotor pole pitch, the magnitude of the inductance's gradient with angle will depend on the difference between L_{max} and L_{min}. Thus the output of the machine, for a given current, depends on this difference, which should ideally be as large as possible. In a practical machine, the ratio of L_{max} to L_{min} will typically lie in the range 4:1 to 10:1.

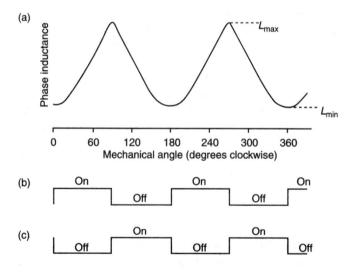

Figure A8.5 *(a) Typical inductance with angle; (b) phase current for clockwise torque; (c) phase current for anti-clockwise torque*

A8.2.1.4 Relationship between torque polarity and motoring and generating

Reversing the polarity of torque, for a given direction of rotation, switches the machine between motoring and braking operation. To do this, it is necessary to simply alter the excitation pattern so that each phase is energised either when the inductance is rising or when it is falling. Note also that braking torque becomes motoring torque – and vice versa – if the direction of rotation is reversed, provided the excitation pattern with respect to rotor *angle* remains unchanged.

The choice of rotational direction defined as positive (increasing) angle, as shown in Figure A8.5, is an arbitrary one. Once this is decided, we can define 'positive' or 'forwards' torque and speed in the same sense. Motoring operation then occurs when the torque sign and speed sign are the same; braking occurs when they are dissimilar.

A8.2.2 Control of the machine in practice

A8.2.2.1 Low-speed operation

The simple control methodology discussed so far assumes that (for motoring operation) a controlled phase current is switched on when the rotor reaches BDC, and off again at TDC. By varying the magnitude of the phase current, the average torque produced by the motor can be regulated.

This approach is indeed used in practice to control the motor at relatively low rotational speeds. The power electronics applies the full d.c. supply voltage to the phase winding(s), thereby causing the magnetic flux – and hence phase current – to rise at the maximum possible rate. When the phase current has reached its working value, which at low speeds will occur within a relatively small rotor angle, the converter must limit the current, which it does by reducing and controlling the average voltage at the winding. This is usually accomplished by switched-mode action, or so-called 'chopping' of the phase current. When the rotor reaches TDC, the power converter applies the full d.c. bus voltage *in reverse* across the winding. This forces the flux linkage, and hence the phase current, to fall at the maximum possible rate, until both the flux and current are zero. Again, at low rotor speeds, this happens in a small mechanical angle.

A8.2.2.2 What happens as speed is increased?

An important point, not considered earlier, is the recognition that magnetic flux and phase current cannot, in practice, rise or fall instantaneously. Faraday's Law states that the rate of change of flux linkage (equal to magnetic flux ϕ multiplied by the number of winding turns N_{ph}) is proportional to the applied voltage V_{ph}; hence,

$$d\phi/dt = (V_{ph}/N_{ph}) \quad (A8.2)$$

Inductance is defined as flux linkage per unit current, i.e.

$$L = (N_{ph} \cdot \phi)/i \quad (A8.3)$$

We can re-write equation (A8.3) to yield an expression for current:

$$i = (N_{ph} \cdot \phi)/L \tag{A8.4}$$

The relationship of equation (A8.4) is key to understanding the current waveforms in the SRM phases. The current at any instant is determined by the ratio of magnetic flux linkage to inductance. Remember that magnetic flux is a function of the applied winding voltage and time (A8.2), whereas the phase inductance is a function of rotor angle and hence has both speed and time dependency.

A8.2.2.3 Medium-speed operation

Equation (A8.2) shows that the rate of change of magnetic flux is constrained by the available d.c. supply voltage. Consequently, from equation (A8.4), the current will take a finite time to reach its working value when switched on at L_{min} and a second (longer) time to fall back to zero when we switch off at L_{max}. At low rotational speeds, these time intervals will occupy negligible rotor angle, and it is possible for the phase current waveforms to closely approximate the ideal square-wave functions shown in Figure A8.5b and c.

However, as the rotational speed increases, the time occupied by an electrical cycle of the machine falls with the reciprocal of speed, and the rise and fall times of the current become significant. This yields the sort of phase current waveforms shown in Figure A8.6.

The finite rise time of the phase current means that a little torque will be lost because the current has not reached its working value when the inductance starts to rise. However, the persistence of current beyond L_{max} (TDC) is more significant,

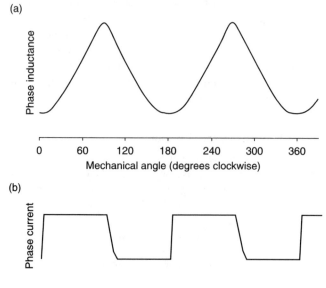

Figure A8.6 (a) Phase inductance and (b) realistic current waveform at 'medium' speeds

because it results in a short period of braking torque. Not only does this reduce the average motoring torque, but the extended phase current also brings a small increase in winding losses. This means that the output has fallen while the losses have increased, and therefore this 'tail' current can noticeably reduce the efficiency of the motor.

The non-linear shape of the current at turn-off is due to the fact that not only is the magnetic flux falling linearly with time, but the inductance also falls. This means that, from equation (A8.4), the current decreases less rapidly than might otherwise be expected. The effect is less marked at L_{min} because here the inductance changes less rapidly with angle (hence, for a given speed, with time).

It is the rate of change of magnetic flux relative to the rate of change of inductance with time that determines whether the rise and fall times of the current are significant or not. Note that the rate of change of magnetic flux is determined by the number of turns on the winding as well as the supply voltage (Equation A8.2). This means that the same effect will be noticed at a lower rotor speed if the number of turns on the phase winding is increased or if the d.c. supply voltage is reduced.

A8.2.2.4 How is performance maintained as speed increases?

The effect of 'tail' current could be minimised by using relatively few turns on the motor windings, thereby ensuring that the rate of change of flux was always large enough to force the flux and current up and down within a negligibly small rotor angle, regardless of speed. However, this would mean that, for a given working value of magnetic flux in the motor, very large phase currents would be needed. This would increase the cost of the power electronics required to control the motor, and this approach is not used in practical drive systems.

A preferred approach is normally adopted to mitigate this effect – 'angle advance'. The switch-on point of the phase current is advanced with respect to the rotor angle, so that the magnetic flux and current have already reached their working values by the time the phase inductance starts to rise. Similarly, the turn-off angle is advanced, so that by the time the inductance starts to fall, the magnetic flux and current have already been forced to relatively low values, and little braking torque results.

If the angle advance is too great, however, torque will again be lost. Too early a switch-on angle will place significant flux in the machine when the inductance is still falling from the previous cycle. The resultant braking torque will reduce the average output of the machine while adding to the phase current, hence worsening efficiency, too. Switching off too early will avoid the braking torque otherwise incurred as the inductance falls beyond TDC, but it will unnecessarily reduce the torque. Increasing the peak phase current can restore the output, but this may worsen the rms phase current, hence increasing the copper losses in the winding and again worsening efficiency. Therefore an optimum degree of angle advance must be chosen for both switch-on and switch-off, so as to maximise efficiency and output for the torque and speed level in question.

A8.2.2.5 High-speed operation

As the rotor speed rises still further, the rate of change of inductance with respect to time increases along with it and the effects illustrated in Figure A8.6 are exacerbated.

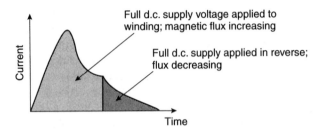

Figure A8.7 Effect of large rate of change of inductance on phase current at high speed

Eventually, a speed is reached where the phase current can be naturally limited by the phase inductance, while still maintaining a sensibly broad pulse of current and flux. Under these conditions, there is no need to limit the phase current by reducing the phase voltage, and the torque can be efficiently controlled simply by adjusting the switching angles with respect to rotor angle. This is sometimes referred to as the 'single pulse' control mode, so-called because the phase voltage is applied as one continuous pulse, rather than being 'chopped' by repeated switching of the power electronic control circuit.

If the speed is high enough, it is possible for the current to inflect and actually decrease, despite the full positive supply voltage still being applied to the winding. This is shown in Figure A8.7. Note the inflection of the waveform at peak current. This occurs in spite of the fact that the full supply voltage is applied to the winding, and that the phase magnetic flux is therefore increasing. The current 'rolls over' because time rate of rise of inductance exceeds the time rate of rise of magnetic flux, and therefore, from (A8.4), the current must fall.

Eventually, if the speed increases still further, a point is reached where it is only just possible to reach the chosen working flux level, and then drive the flux back down to zero, within one electrical cycle. The peak flux, and hence current, is then limited by the available d.c. bus voltage, and any further increase in speed will necessitate a reduction in peak flux and hence in output torque. This defines the base speed of the SRM.

It is possible to operate the machine with a component of continuous (standing) current and flux. That is, the current and magnetic flux associated with a given phase do not fall to zero before the next cycle of operation begins. This so-called continuous current mode has some further implications for the control of the machine, but is useful in increasing the available power output above base speed. It is especially useful in systems where a wide constant power range is needed or where a large transient overload capability is needed at high speeds. This technique has been patented by Switched Reluctance Drives Ltd and Emerson.

A8.2.2.6 Summary of typical/practical control

The phase currents are always switched synchronously with the rotor's mechanical position.

At low speeds, the phases are energised over the entire region of rising inductance, and active current limiting is required from the controller. Torque is controlled by adjusting the magnitude of the phase current.

As speed increases, the rise and in particular the fall times of the phase current occupy significant rotor angle, and it is usual to advance the turn-on and turn-off angles with respect to rotor position. The torque is now controlled by both the current limit level and by the switching angles, although current is usually used as the primary control variable.

At high speeds, the rise and fall times occupy still greater rotor angles. The current naturally self-limits and it is possible and indeed usual to control the torque using only the switching angles. The shape of the current waveform is greatly influenced by the high rate of change of inductance with respect to time.

The use of angle advance, with the switch-on and switch-off angle both independently adjustable, means that it is possible to maintain a high level of energy conversion efficiency as operating conditions are varied. The SR machine, when controlled in this way, is capable of producing high efficiency over a very wide range of torque and speed.

By choosing appropriate switching angles and current levels, together with an appropriate electromagnetic design, the torque–speed characteristic of the switched reluctance drive can be tailored to suit the end application. Furthermore, simply by changing the control parameter selection with torque and speed, a given machine design can be made to offer a choice of different characteristics. It is usual to store the control parameter variations within the motor control system, e.g. as mathematical functions (of torque demand and motor speed), or as look-up tables, either of which can readily be embedded into the controlling microprocessor's software code.

A8.2.2.7 Control of speed and position

When controlled as described, the switched reluctance machine and its basic control system form essentially a *torque*-controlled drive. This can be compared in performance to a separately excited d.c. machine with controlled armature *current*. This is sometimes what we want, e.g. for tension control in winding machinery or for web handling.

However, more commonly, controlled speed (or even position) is what the end user requires. If torque is still produced as the motor speed increases, the speed of the SRM will increase without limit if additional measures are not taken to control it. To do this, we simply include the basic torque controlled SR machine within an outer speed regulating loop, as is done with the traditional d.c. drive. Feedback of speed can be readily derived from the rotor position sensor, if one is used, and otherwise from sensorless control position data.

Figure A8.8 shows a typical control structure for a basic speed-controlled switched reluctance system. Position control of the SRM can be achieved (in the conventional manner) by adding a further position control loop around that shown in Figure A8.8. Note that, under closed-loop position control, the SR machine is not limited to shaft positions corresponding to its natural detents. The shaft position can be controlled

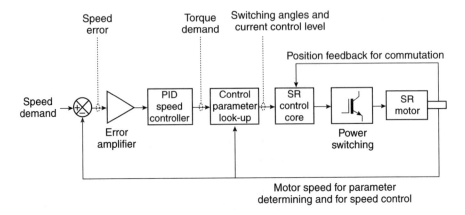

Figure A8.8 Basic speed-controlled switched reluctance system

to any desired resolution using the position control loop. This will in turn demand speed, and hence torque, which will ultimately be translated into the appropriate currents in each phase required to maintain the set position.

A8.2.3 Polyphase switched reluctance machines

The simple single-phase machine we have discussed so far is capable of producing torque over only half of its electrical cycle (which, for the 2–2 pole structure, repeats twice per revolution). Motoring or braking torque from such a machine will necessarily be discontinuous, and hence starting in the desired direction is not possible from all rotor positions. Single-phase motor starting can be ensured by including a small 'parking magnet' within the stator, positioned such that the rotor always comes to rest in a torque-productive position. These limitations are acceptable for some applications, and the single-phase SR motor is especially useful for low-cost high-speed applications such as vacuum cleaner fans.

More demanding applications use higher pole numbers on the machine's rotor and stator, with the stator poles wound and connected into multiple identical phases. Figure A8.9 illustrates the cross-section of a three-phase '6–4' machine. Here, diametrically opposed coils are connected together to form three-phase circuits, denoted as phases A, B and C.

The excitation of the phases (in this case, three in number) is interleaved equally throughout the electrical period of the machine. This means that torque of the desired polarity can be produced continuously, thus greatly reducing the variation in output torque with respect to angle ('the torque ripple'). Furthermore, machines with more than two phases are able to start in either direction without requiring special measures. The number of phases can in theory be increased without limit, but phase counts from 1–4 inclusive are the most common for commercial and industrial applications.

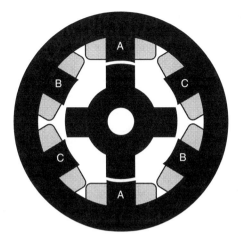

Figure A8.9   Cross-section of three-phase 6:4 SRM

Many different combinations of pole count are possible. It is sometimes beneficial to use more than one stator pole pair per phase; for example, the 12:8 pole structure is commonly used for three-phase applications. Each phase circuit then comprises four stator coils connected and energised together.

Increasing the phase number brings the advantages of smoother torque and self-starting in either direction. (Self-starting is possible with two phases, but only in a pre-determined direction, generally using a profiled or stepped rotor design. Self-starting in either direction requires at least three phases.)

A higher phase count does increase the complexity of the associated power electronics and, to some extent, the signal level controls. However, increased power electronics costs are mitigated by the power throughput being equally distributed between the phases of the controller, and its total volt–ampere rating may therefore be no greater (and in some cases may actually be less) than that required for a machine of low phase count.

A8.2.4   Losses in the switched reluctance motor

Like any other electric motor, the losses in an electric machine may be categorised as follows.

- *Copper losses* (*in the phase winding*). These comprise resistive heating from the main phase current, plus any losses due to circulating or eddy current heating in the copper (sometimes called 'proximity effect' and an important consideration at high frequencies). Skin effect, which effectively raises the winding resistance by concentrating the current around the outside of the conductor, must be taken into account when calculating the effective phase resistance and the copper losses in a high-frequency machine.

- *Iron losses (in the rotor and stator steel).* These are caused by the changing magnetic flux within the steel. Eddy current losses, caused by induced circulating currents in the steel, generally increase with the square of frequency, whereas hysteresis losses follow the frequency of excitation directly. The overall rate of increase of iron loss with frequency therefore lies somewhere between the two cases. The iron losses are heavily dependent on the electromagnetic properties of the steel used.
- *Friction and windage losses.* These are mechanical losses resulting from the rotational speed, such as bearing losses, air friction, turbulence and shearing of air layers at rotor edges.

At relatively low rotational speeds, and especially in smaller machines, the iron losses are usually small, and the copper losses dominate. At higher speeds, the losses in the iron become important, and they are a key factor in the design of large high-speed SR machines. Relatively high windage losses at high speeds result from the rotor structure.

For a given grade of steel and lamination thickness, the iron losses depend not only on the frequency of excitation 'seen' by the steel, but also on the magnitude of the magnetic flux and the flux excursions. The magnetic loading of different sections of the rotor and stator steel varies throughout the cycle of operation as the phases are switched on and off, and as the rotor position changes. Furthermore, in a polyphase machine, many sections of the rotor and stator steel carry flux from more than one phase. The problem therefore has both time and spatial dependency. Modelling and calculating iron losses in the SR machine is a highly complex matter, and obtaining accurate results is notoriously difficult. Good results depend on sophisticated finite-element analysis combined with a detailed knowledge of the machine's operation and the steel type in use.

A8.2.5 Excitation frequency

Each stator pole must complete one full cycle of energisation for each and every rotor pole that passes it, and hence the frequency of excitation per phase is determined by the product of the number of rotor poles and the speed of the rotor in revolutions per second:

$$f_{ph} = (n_r \cdot S)/60 \quad (A8.5)$$

where n_r is the rotor pole count and S is the rotor speed in revolutions per minute.

Because there may be multiple interleaved phases, the total excitation frequency will be given by

$$f_{tot} = (n_{ph} \cdot f_{ph} \cdot S)/60 \quad (A8.6)$$

These figures are important in the calculation of motor and power electronics losses, and also in determining the time periods with which the control electronics will have to work.

A8.2.6 Power electronics for the switched reluctance motor

A8.2.6.1 Power supply and 'front end' bridge

The switched reluctance drive, like a conventional a.c. inverter, controls the motor by using power-switching devices, which connect the various phases of the motor to a d.c. supply of relatively constant voltage. Typically, this d.c. supply will be derived from standard single- or three-phase a.c. power lines by means of a bridge rectifier, filter capacitors and, optionally, a filter choke.

If dynamic braking is required, the d.c. supply will receive an average current flow *back* from the SR machine as mechanical energy is converted back to electrical energy. The d.c. supply must be receptive to the generated power, which will initially cause the voltage across the filter capacitors to rise. The capacitors may be sufficient to accommodate transient braking energy, but thereafter, it must either be dissipated in a braking resistor or returned to the a.c. power supply through an active bridge. In this respect, the switched reluctance system is again no different to a standard a.c. drive.

A8.2.6.2 Power switching stage

It was shown earlier that the rate of change of flux linkage is equal to the applied winding voltage, and that it is necessary to get the flux, and hence current, to the correct value in the shortest possible time (hence angle). It then follows that, at the switch-on and switch-off angles, we want to force the flux up and down, respectively by, applying the maximum possible winding voltage. When the phase is excited, we further need to be able to control the current to any desired level.

Unipolar excitation is normal for the switched reluctance machine, and as such the associated power circuits apply winding current in one direction only. There are many different power converter topologies that can be used with the switched reluctance machine. Discussion of all these variants is beyond the scope here, but a brief summary may be helpful to establish a choice of preferred power circuit.

Most power circuit topologies used for the SR motor have either one or two solid-state switches per phase, each of which requires an actuating circuit to open and close the switch.

In spite of the number of possible power circuit variations, it will be found that the total VA rating (volt–ampere product) of the switching devices used never falls below a certain minimum value. For example, power switching topologies using only one power switching device per phase may be superficially more attractive in cost terms than a two-switch-per-phase circuit. But, the single power switch will have to handle at least twice the voltage seen by each device in the two-switch circuit, and so the total VA rating is the same as, or possibly greater than, the two-switch case. This 'rule' will be seen in the following examples.

A8.2.6.3 Single-switch-per-phase circuits

Figure A8.10 shows one phase of a possible single-switch power circuit.

The power switch S is here shown as a MOSFET, but could equally be an IGBT, GTO, bipolar transistor or other switching device. Two parallel rails are required, the

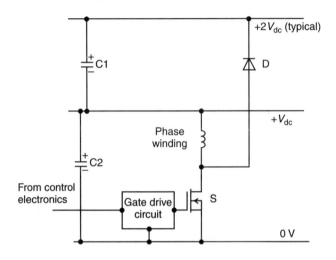

Figure A8.10   A single-switch-per-phase topology

mains d.c. supply ($+V_{dc}$) being connected across C2. When S is closed, V_{dc} is applied to the phase and flux increases accordingly. When S is opened, the phase current commutates to diode D, and the winding 'sees' a negative voltage equal to the potential across C1. Flux and current are then driven down by this negative voltage, which is typically equal to V_{dc} in magnitude, as shown, but need not necessarily be so.

Only a single-gate drive circuit is needed per phase, referenced to the zero-volt rail (which may be connected directly to the signal level controls, and in a low-voltage system may be at ground potential). This can offer savings in low-voltage or non-isolated systems. Simple single-switch-per-phase circuits are occasionally used in low-cost appliance and automotive applications where the saving in the number and/or complexity of gate drive circuits is attractive. Their use in high-power industrial systems has become less common, however, partly because of the availability of MOS-gated power switches such as MOSFETs and IGBTs, which can be operated from simple low-cost gate drive circuits. Additionally, the circuit above has only two switching states and as a result offers less control flexibility than the two-switch circuit, which is discussed later. Note that the switch sees the full winding current and (when returning energy to the supply via D) *twice* the supply voltage V_{dc}.

The circuit of Figure A8.10 requires a centre-tapped power supply, which can conveniently be obtained by splitting the supply reservoir/filter capacitance into C1 and C2, as shown. The motor will draw energy from C2, some of which will be returned to C1 at the end of the electrical cycle via the diode D. Some means must therefore be provided either to dissipate energy from C1, or to return energy from C1 to C2. If this is not done, the respective capacitor voltages will drift towards an unbalanced condition, and ultimately most of the total supply voltage will appear across C1. This circuit is sometimes referred to as the 'C-dump', presumably because a capacitor (here C1) is used as a 'dumping ground' for unconverted energy returned from the motor. An additional switching device is often used to return this energy from C1 to C2, so that an *n*-phase motor will require $(n + 1)$ switches.

A8.2.6.4 Multiple-phase operation

If the motor has two – or any even number of – phases, then the drift of capacitor voltage may be eliminated by connecting the second phase as shown in Figure A8.11. The d.c. supply can then be conveniently connected across C1 and C2 so that the capacitors form a 'centre-tap' or midpoint on the power supply. The idea can be extended to four phases, six phases and so on in the obvious manner, by connecting equal numbers of the phases as 'high-side' and 'low-side' switches. By drawing the energy of the second phase from C1 and returning it to C2, the average current flow in the two capacitors can be equalised and the average (d.c.) voltage balance will be maintained if the phase currents are identical. The circuit is in fact self-stabilising if the phase energisation is not current-limited, because an increase in capacitor voltage will lead to an increase in phase current. If current control is used, however (as it usually will be at low and medium speeds) then additional measures are needed to ensure voltage balance is maintained. Note that the gate drive circuit for the second phase (B) is not referenced to the 0 V rail, and in fact its common-mode potential swings rapidly between V_{dc} and 0 V as the switch S is opened and closed. The gate driver must be carefully designed in order to deal with these conditions. (The common-mode voltage may swing at a rate of many thousands of volts per microsecond using modern IGBT or MOSFET switches.)

Each switch in Figure A8.11 experiences the full supply voltage V_{dc}, but the windings only see half this voltage during flux-up and de-flux. This means that, for the same rate of change of flux as before, the number of turns on the windings must be halved, necessitating twice the phase current for the same working flux level. The total switch VA per phase is therefore (once again) $2 \times V_{dc} \times I_{PH}$.

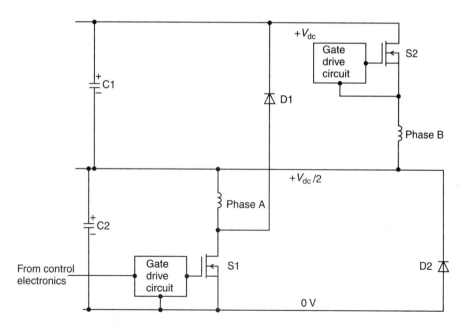

Figure A8.11 Single-switch-per-phase topology with two phases ('H' circuit)

A8.2.6.5 Single-switch circuit using bifilar winding

Another well-known single-switch topology uses a 'bifilar' (two strand) winding in the motor, in which the phase energy is supplied via one winding and returned by the other (Figure A8.12). This circuit has the advantage that all the power switches can be referenced to the zero-volt supply rail, which is attractive for low-voltage systems such as automotive applications.

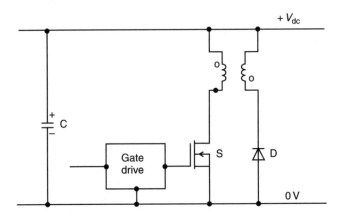

Figure A8.12   Bifilar SR power electronics topology

Because the two halves of the winding are connected in the opposite sense to each other, transformer action means that each switch sees theoretically twice the d.c. supply voltage during energy return (de-flux). However, the electromagnetic coupling between the two strands of the winding is inevitably imperfect, and this results in sometimes large additional voltage overshoots across the switches when they open. This means that the switches must be of significantly higher voltage capability than the theoretical $2 \times V_{dc}$, while still carrying the full winding current. The VA rating therefore in practice always exceeds $2 \times V_{dc} \times I_{PH}$ per phase.

Furthermore, the machine winding is more complex and requires additional connections to the power electronics, both of which increase cost. The efficiency of a bifilar machine is marred because, at any instant, the phase current is flowing through only a part of the available conductor area in the winding. The copper 'fill factor' (i.e. the proportion of winding space actually containing copper) is also a little worse, because of the additional, separately insulated, winding required.

A8.2.6.6 Two-switch asymmetrical bridge

This topology is the most common used for SRM control. Two switches and two diodes are used, as shown in Figure A8.13. Here, the gate drive of S1 is referenced to the zero-volt rail (the 'low-side' switch), while S2 has a 'floating' gate drive (the 'high-side' switch). The drive circuit for S2 must, like the phase B gate driver

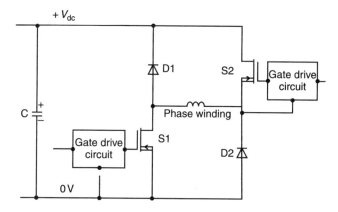

Figure A8.13 Asymmetric bridge power circuit

of Figure A8.11, withstand rapid swings in common-mode potential as S2 opens and closes.

Each power switch is exposed to the full d.c. supply voltage V_{dc}, and switches the full winding current I_{PH}. The total VA rating per phase is therefore again $2 \times V_{dc} \times I_{PH}$. The power electronics rating is therefore no worse than the single-switch cases (and better than bifilar). Although additional power switches are required, the total switch heat losses are distributed among two packages per phase, which eases thermal management, especially in high-power drive systems.

More important in many applications is the fact that the asymmetrical bridge circuit offers additional control flexibility, which is useful in managing the phase current and acoustic noise of the machine.

Note that, because the motor phase is connected in series with the two switches across the power supply rails, this power circuit is not susceptible to the 'shoot-through' fault condition, which can arise in conventional inverters. Indeed, closing both switches at the same time is a prerequisite for normal operation of the machine. This can simplify protection of the power electronics, although it should be noted that a shoot-through fault would occur if the terminals of the motor phase were inadvertently short-circuited, e.g. by incorrect user installation.

Figure A8.14 shows an outline schematic for a complete three-phase SR power converter, with no braking.

A8.2.7 Advantages of the switched reluctance system

The SRM combines the advantages of good low-speed torque, relatively high power density with great simplicity and therefore potentially good robustness.

A8.2.7.1 Rotor construction

Because the rotor carries no windings, magnets or conductors of any kind, the machine is in principle well suited to high-speed operation, and is a good choice for use in harsh environments and high temperatures. Heating of the rotor is confined to eddy

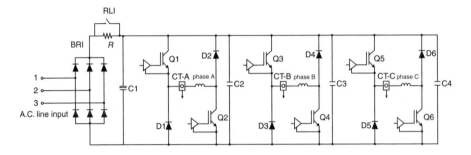

Figure A8.14 Basic three-phase power converter circuit

current and hysteresis losses in the steel, and as a result, the rotor runs relatively cool in the majority of applications, enhancing bearing life. The geometry of the rotor does mean that the windage losses in the motor can be high unless great care in the design is taken.

A8.2.7.2 Stator construction

The stator is also very simple and robust, requiring only short-pitched coils, which are placed over the salient stator poles, and which can easily be pre-wound on a former or bobbin. The stator windings – unlike those of an induction motor – are not distributed over many slots, and the phases do not cross each other in the end-winding region. This largely eliminates the risk of a phase-to-phase insulation failure. The simplicity of the coils allows the end windings to be much shorter than those typically found in induction motors, and the losses associated with the end windings (which do not contribute to the output of the motor) are reduced. This improves efficiency and allows the construction of relatively short stack ('pancake') motors with minimal penalty on specific output. The winding construction tends to yield a lower capacitance to the frame than a conventional a.c. motor, typically 20–30 per cent less. This improves electromagnetic compatibility and reduces radio-frequency interference, because coupling of high-frequency currents to the stator is somewhat reduced.

As mentioned above, with the possible exception of machines operating at very high speed and at high power, the machine's losses are concentrated in the stator, which is relatively easily cooled. Thermal management of the machine is therefore relatively simple. Furthermore, the fact that the rotor heating is minimal, especially during stall, means that the stall endurance of the machine is limited by the thermal time constants associated with the *stator*. These are generally long, due to the large stator mass, and the SRM performs well under conditions of prolonged stall.

A8.2.7.3 Electronics and system-level benefits

The SRM requires the use of power electronic controls for commutation and control, in many ways similar to the inverter used to vary the speed of an induction motor. However, in contrast to the a.c. drive, the motor does not require sinusoidal supplies in order to operate efficiently. As a result, the power converter used with the SRM need not switch at high frequencies. This reduces switching losses in the power semiconductors at low motor speeds, and is especially useful in medium- and high-power drives (e.g. >10 kW) where switching losses can otherwise be significant. To avoid tonal components in the drive's acoustic noise, it is common to use current control schemes with randomised or 'spread-spectrum' switching frequencies.

At higher speeds, the power semiconductors switch on and off only *once* per electrical cycle of the machine (Figure A8.7). The switch turn-on occurs at zero current (which implies zero loss), while turn-off may occur at a current less than the peak phase current due to natural 'roll-over' of the current. The switching losses at high speeds are therefore negligible.

The relatively low electronics switching losses, combined with high torque per amp of phase current, mean that the power semiconductor ratings in the switched reluctance drive may be lower than conventional systems.

The phases of the switched reluctance drive system operate independently of each other, and in the event of a fault developing in one phase, the others are able to continue to produce torque as normal. This gives the machine an inherent fault tolerance with the ability to 'limp-home' in the event of a partial failure.

The SRM is capable of yielding very high overload torque; its ability to do so is really limited only by the thermal time constants associated with the stator windings. This high peak overload torque capability, combined with relatively low rotor mechanical inertia (due to material being removed at the outer diameter to form the salient poles), means that very high rates of angular acceleration are possible.

Careful design and optimisation of control parameters can yield good system efficiency over a wide range of torque and speed (Figure A8.15). This shows the true system efficiency (measured as mechanical power output divided by the raw a.c.

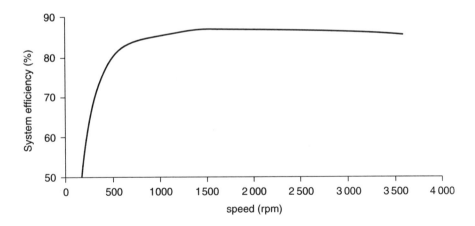

Figure A8.15   *System efficiency (mains to mechanical output) of 7.5 kW SR Drive^{TM} system at rated output (SR Drive is a registered trademark of Switch&d Reluctance Drives Ltd)*

electrical power input to the power electronics) of an IEC 132-frame 7.5 kW SR Drive® system. The drive is delivering a constant 50 Nm torque from stall to 1 500 min^{-1}, and a constant mechanical output power of 7.5 kW from 1 500 to 4 000 min^{-1}.

The switched reluctance system is capable of operating over a wide speed range at constant power, without the efficiency or power electronics cost penalties associated with permanent-magnet and induction motor technologies under these conditions. This is proving an attractive benefit for automotive and traction applications, where significant interest is being shown in the machine, especially when these merits are considered along with its robust and simple construction.

A8.2.8 Disadvantages of the switched reluctance system

Although the machine is very simple in both construction and concept, in practice saturation of the steel means the machine is highly non-linear. As a result, it does not lend itself to the relatively straightforward and well-understood design methodologies used for more traditional machines. The analysis, design and control of high-performance SRMs is a complex matter, and without the necessary tools and expertise, it is easy to design a system of poor or indifferent performance.

The motor and controller must be designed as a system and operated together. The motor cannot be connected directly to a mains power supply if its associated electronic controller fails.

Because each phase operates independently of the others, two cables per phase are usually required, although this is a function of the power circuit used. With the most frequently employed asymmetrical bridge power converter, a three-phase system requires six motor leads.

It has been said of the SRM that it requires a short air gap length to perform well. It is true that the SRM can benefit from a short air gap, which increases L_{max} and hence the difference between L_{max} and L_{min}. It is also true that excellent results can be achieved using air gap lengths equal to, or even greater than, those normally employed in the production of standard induction motors.

A8.2.8.1 Torque ripple

The torque produced by the SRM, when excited by constant phase currents, is not inherently smooth even if three or more phases are used. The SRM produces some residual torque 'ripple'. For a three-phase system this may typically amount to 30–40 per cent (expressed as the peak-to-peak excursion divided by the mean torque). Contrary to popular wisdom, however, there are surprisingly few applications where the presence of such torque ripple causes noticeable process or control problems. Indeed, SRMs have been successfully used in traditionally demanding applications such as positioning drives and for web tensioning, without recourse to torque smoothing by exceptional electronic means. Care must, however, be taken that higher-frequency components (harmonics) of torque ripple do not cause undue acoustic noise through exciting resonances in the load or other associated mechanical arrangements.

If the natural torque ripple really is unacceptable, the torque produced by the SRM can be smoothed using techniques such as 'current profiling'. This involves modulating the phase current (or flux) with respect to the rotor angle, so as to cancel out the torque variations that otherwise occur. Torque ripple can be reduced to a few per cent using such techniques. Further smoothing is possible, but is probably best accomplished using adaptive or self-tuning techniques that are able to take into account the inevitable minor variations from one motor to the next that will occur in mass production. Adaptive techniques can also minimise the effect of rotor position measurement error due to encoder imperfections or misalignment.

A8.2.8.2 Acoustic noise

Acoustic noise was a well-known problem with early SR systems. Some of this may have been due to torsionally induced vibration (from torque ripple, as discussed above), but it is important to distinguish this source of noise from that due to the large 'normal' forces experienced by the stator poles in any electrical machine. These tend to distort the stator from its desired round shape (Figure A8.16).

Because the normal forces exerted on each stator tooth pulsate (as the rotor turns and as each phase is switched on and off), the stator will tend to vibrate in various modes at the phase frequency and harmonics, hence radiating acoustic noise. Noise due to stator distortion and vibration can be minimised by careful attention to mechanical design (including the choice of pole structure and considerations of back-iron thickness), as well as by electronic means. The latter can make a significant impact by controlling the spectral content of the normal forces 'seen' by the stator poles, and hence that of the resulting surface acceleration that generates the noise.

Care must also be taken with general mechanical design and stiffness. Good concentricity must be maintained if low noise performance is to be attained.

Sufficient progress has, however, been made in the control of acoustic noise to permit the use of SRMs in some high-volume domestic appliances such as washing machines, where noise is a major concern (Figure A8.17).

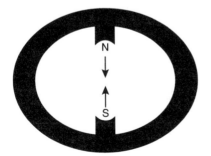
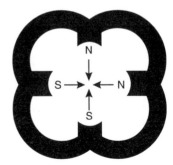

Figure A8.16   Exaggerated normal-force bending of stators

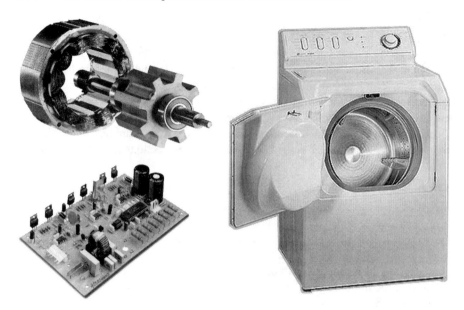

Figure A8.17   Low-noise SR Drive^{TM} system for domestic washing machines [photograph courtesy of Maytag Inc/Emerson] (SR Drive is a registered trademark of Switched Reluctance Drives Ltd)

A8.3   Stepper motor drives

Stepping motors are a group of motors characterised by the fact that the shaft rotates in angular steps corresponding to discrete signals fed into a controller. The signals are converted into current pulses switched to the motor coils in a specific sequence. The motor acts as an incremental actuator, which converts digital pulses into analogue output shaft rotation. The speed of rotation is dependent on the pulse rate and the incremental step angle, whereas the angle of rotation is dependent on the number of pulses fed to the motor and the incremental step angle.

This dependence means that the motor is eminently suited to open-loop position and speed control, within limitations.

A8.3.1   Stepping motor principles

There are three basic types of stepping motor:

- permanent magnet (PM),
- variable reluctance (VR), and
- hybrid.

A8.3.1.1   The permanent-magnet motor

The PM motor has a laminated, slot-wound stator, usually with two, three or four phases. The rotor, mounted on a bearing in each end frame, is usually a solid cylinder

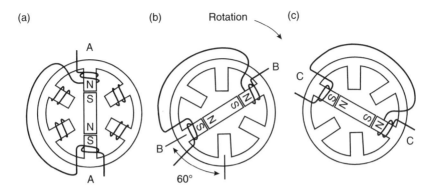

Figure A8.18   Step sequence for a PM rotor stepping motor

magnetised in two-, four-, six- or eight-pole configurations. Rotation of the motor shaft is achieved by switching currents between coils to produce a change of the electromagnetic field alignment. By controlling the sequence of switching, the field can be made to rotate within the stator bore and the rotor will rotate in synchronism (Figure A8.18).

When the electromagnetic field is in a fixed position, the rotor will be aligned with it and the torque exerted will be zero. By switching current to the next set of coils, the electromagnetic field will move to align with those coils and out of alignment with the rotor. This will exert a restoring force on the rotor to bring it into alignment again, which results in a rotational torque being developed at the shaft.

Compared to the VR motor (described below) the PM version develops a higher torque due to the magnet flux strength, and it has a preferred axis of alignment because of the polarised rotor. It should be mentioned that when the rotor is 180° out of alignment the torque is zero, but this is a very unstable position and any small movement to either side of 180° results in the rotor returning to its correct alignment position.

When the stator is not energised the PM rotor still tends to align itself with a pair of poles, normally remaining in the same position as when last energised; this is known as the detent torque. Care must be taken that the current rating of PM steppers is not exceeded, otherwise demagnetisation of the rotor can occur.

When stepping in synchronism with the rotating stator field a back-emf is generated in the field coils due to the PM rotor. The amplitude is proportional to the stepping rate, which reduces current input as speed increases. Further, as a consequence of the inductive nature of the windings, the current input reduces as the switching frequency increases. There are a number of techniques, generally known as 'current forcing', to overcome this problem, which are described in A8.3.2.

A8.3.1.2   The VR motor

The VR motor is similar to the PM type, except that the rotor is not magnetised; it is formed of soft iron material with a number of equally spaced poles, which form paths of minimum reluctance in the overall magnetic circuit. Because the rotor is not

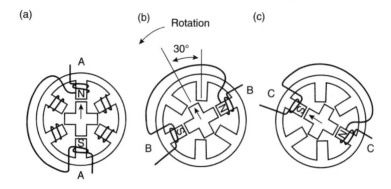

Figure A8.19 Step sequence for a VR rotor stepping motor

magnetised, polarisation is determined only by the stator excitation, and the step angle will be a function of the number of rotor poles compared to stator poles (Figure A8.19). These are often not the same number.

Although the VR motor has a lower static torque rating compared to that of the PM motor, the absence of the permanent magnet in the rotor allows a higher speed range to be achieved for similar input. Furthermore, the detent torque is almost zero, (a small amount may be present depending on the type of rotor and its remnant magnetic flux), so the motor can be moved freely when not energised. There is no problem of demagnetising the rotor, so torque output can be up-rated for short duty, although magnetic saturation of the stator and winding temperature will determine the limits.

A8.3.1.3 The hybrid motor

The hybrid motor is a combination of the PM and the VR motors in that the rotor has a permanent magnet core with soft iron end pieces (Figure A8.20). The magnet is polarised axially and the magnetic path flows out radially through the soft iron end pieces and back via the stator yoke. The principle of operation is therefore quite different from either the PM or the VR motor.

Nevertheless, the generation of torque is due to the forces involved in the alignment of rotor teeth with stator pole teeth and rotation is controlled by switching of the current to the coils in a prescribed sequence.

The most common configuration of the hybrid motor is a four-phase wound stator, each pole having teeth spaced equally at 1/48th of a revolution, and the rotor with 50 teeth equally spaced. By switching currents to each pole pair in sequence, the rotor moves round to align itself with the nearest set of teeth and the angle of rotation is given by

$$\text{Step angle} = 360°/(\text{rotor teeth} \times \text{stator phases})$$
$$= 360°/(50 \times 4)$$
$$= 1.8°$$

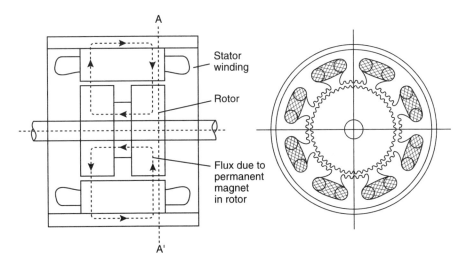

Figure A8.20   Hybrid stepper, sectioned A–A' to show magnetic circuits

The above assumes unipolar switching, which uses the simplest circuits, but by half-step switching, the angle can be reduced to 0.9° and by 'mini-stepping' the angle can be reduced by a factor of ten or more.

A8.3.2   Stepping motor drive circuits and logic modes

A8.3.2.1   General

The performance characteristics of stepping motors are significantly affected by the type of drive used, and the way in which the phases of the motor are energised. Typical logic modes are shown in Figure A8.21 for three- and four-phase VR and PM steppers. The number preceding the mode letter indicates the number of motor phases to which the mode applies.

- *Mode A* (*unipolar, single coil*). Only one phase at a time is energised; in other words, one phase is switched off at the same time as the next phase is switched on. The motor will execute one basic step for each input pulse.
- *Mode B* (*unipolar, two coils*). Two phases are energised at any one time. This mode will produce higher holding and dynamic torque and will reduce rotor oscillations, but power input to the motor is double compared to mode A. The motor will also execute one basic step for each input pulse.
- *Mode AB* (*unipolar, half step*). This is a combination of modes A and B in which the motor phases are energised sequentially in modes A, B, A, B, A, B, A and so on. The motor will execute half the basic step for each input pulse. This mode has the advantage of a smaller step angle and this enables the motor to operate at higher pulse rates.

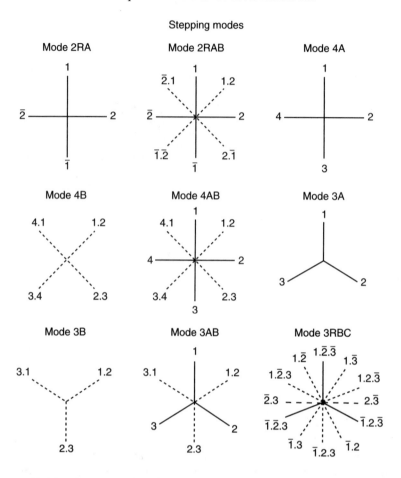

Figure A8.21 Logic modes for three-phase and four-phase VR and PM stepping motors

- *Mode RBC (partial bipolar).* This is a more complex mode requiring reversal of coil polarity as indicated in the diagram by a bar placed above the phase number. This mode is applicable only to a three-phase PM motor and enables it to operate at a quarter of the basic step angle.

A8.3.2.2 Unipolar switching

The arrangement for unipolar switching is shown in Figure A8.22, and results in the pattern of currents shown in Figure A8.23.

Reversing the sequence will cause a reversal in the direction of rotation. With the above switching pattern, the motor steps through its basic or 'full' step angle of 1.8° for

Switched reluctance and stepper motor drives 289

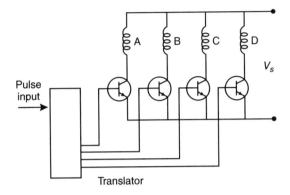

Figure A8.22 Unipolar switching

Step number	Phase			
	A	B	C	D
1	1	1	0	0
2	0	1	1	0
3	0	0	1	1
4	1	0	0	1

Clockwise ↓   Anti-clockwise ↑

Figure A8.23 Unipolar full-step switching

the hybrid stepper. The pattern can be modified as shown in Figure A8.24, introducing half-stepping, i.e. a step angle of 0.9°. This has the following advantages:

- higher resolution,
- smoother drive rotation,
- step resonance minimised, and
- reduced settling time.

The sequence of the switching pattern is also reversible, resulting in reversed motor shaft rotation.

Step number	Phase			
	A	B	C	D
1	1	1	0	0
2	0	1	0	0
3	0	1	1	0
4	0	0	1	0
5	0	0	1	1
6	0	0	0	1
7	1	0	0	1
8	1	0	0	0

Clockwise ↓   Anti-clockwise ↑

Figure A8.24 Unipolar half-step switching

Unipolar switching results in current flowing in one direction only through the winding. Although the direction of flow is not important, at best, only two of the four motor coils are energised at any one time. Improved performance can be obtained by energising all four coils for each full step. This can be achieved with bipolar switching.

A8.3.2.3 Bipolar switching

By connecting pairs of coils in series or in parallel and reversing the flow of current in coils, use can be made of all four coils to give a higher torque (Figs. A8.25, A8.26 and Table A8.1).

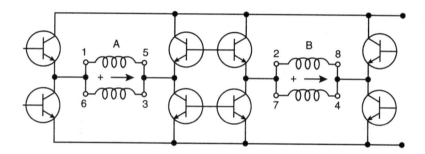

Figure A8.25   Parallel connection

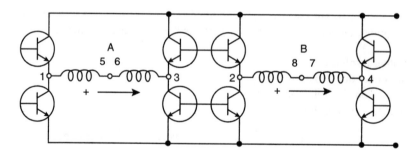

Figure A8.26   Series connection

It should be noted that with coils connected in series the inductance is four times that of coils connected in parallel. So, although the low-speed performance is similar, the current and hence the torque will fall off earlier at higher stepping rates for series-connected configuration. The choice of coil connection should take this into account in addition to the current and voltage capacity of the controller.

A8.3.2.4 High-speed stepping: L/R drives

As stepping increase, the coil inductance and back-emf limit the current and the torque falls off with increasing speed. Various techniques are used to overcome this problem. In the simpler drives, the L/R type, voltage forcing is employed by using a supply

Table A8.1 Step positions with different phase excitation

Full step			Half step		
Step number	Phase A	Phase B	Step number	Phase A	Phase B
1	+1	+1	1	+1	+1
2	−1	+1	2	0	+1
3	−1	−1	3	−1	+1
4	+1	−1	4	−1	0
			5	−1	−1
			6	0	−1
			7	+1	−1
			8	+1	0

higher than the coil voltage and limiting the standstill current by a resistor in series. This reduces the time constant of the coil/resistor circuit and allows a faster rise time at each switch sequence. Owing to the IR loss in the resistor, the use of L/R drives is limited to relatively low power motors.

When operating a unipolar drive in the four-step mode (full stepping, Figure A8.27), the use of two resistors rather than a single will result in a better damped drive capable of operating at higher stepping rates, yet the overall standstill efficiency is the same.

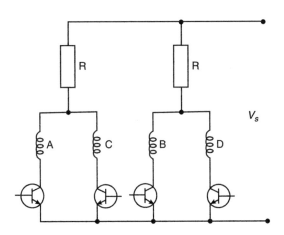

Figure A8.27 Four-step L/R drive configuration

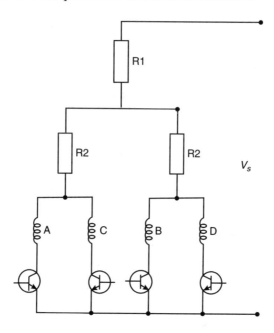

Figure A8.28   Eight-step L/R drive configuration

With the eight-step mode (half-stepping, Figure A8.28) it is necessary to achieve a balance of torque between adjacent steps, i.e. when one coil is on and two coils are on, so a three-resistor network is used.

Voltage forcing by means of series resistors may also be used with the bipolar drives.

A8.3.2.5   Chopper drives

To overcome the losses in voltage-forcing resistors and produce a more efficient drive, essential for the larger motors, a chopper-regulated technique is adopted. A high-voltage d.c. source is used to obtain a fast build-up of current in the motor coils. When the nominal level is reached, the switching device is turned off, but the current tends to be maintained by the motor coil inductance. When the current has decayed below the nominal value, the switching device is turned on again. This process of 'chopping' is repeated with increasing frequency as the motor step rate increases, until the time constant of the motor coil circuit does not allow the current to reach nominal value, when the supply reverts to a constant-voltage source.

The 'chopping' principle can be applied to both unipolar and bipolar drives.

A8.3.2.6   Bilevel drives

A further variation of design is the bilevel drive in which the supply voltage is maintained at a high level for acceleration and deceleration, but it is reduced to a lower level once the motor is running at a constant load/speed. This permits higher peak ratings to be applied to motor and drive combinations.

A8.3.3 Application notes

The optimum ramping rates and stepping rates are determined from motor performance data, after allowing for load friction and inertia, in order to maintain synchronism with the input pulses, over the total traverse (Figure A8.29). The ramping down time can be less than for ramping up, because load friction will assist in the deceleration.

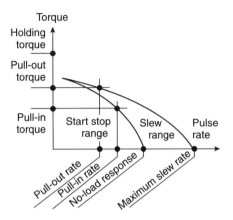

Figure A8.29   Dynamic characteristics

A8.3.3.1 Effect of inertia

The inertia of the load connected to the motor shaft can have a significant effect on dynamic performance (Figure A8.30). Best performance is obtained with a load inertia equal to that of the motor rotor and higher values of load inertia will result in an increase in the mechanical time constant. As a consequence of this, the motor will exhibit lower 'pull-in' rate and a general decrease in dynamic response.

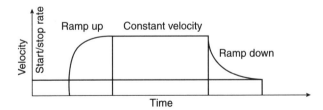

Figure A8.30   Effect of inertia

A8.3.3.2 Resonance

An important point to mention is the phenomenon of 'resonance' suffered by all stepping motors, to some degree or other. This can occur under two particular circumstances. The first is when, at the moment the drive pulse is switched on, the rotor

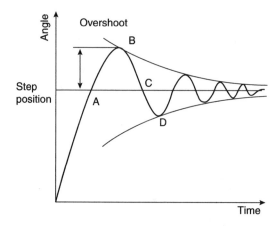

Figure A8.31   Resonance effect

lags so far behind the new step position that the torque developed is not sufficient to accelerate the rotor to the new position. This corresponds to point D in Figure A8.31. The second is when the rotor swings so far forward that the new step position is actually behind the rotor, represented by point B. This will produce a reverse torque, which will slow down the rotor so that it will not be able to respond to the subsequent pulse, and may even result in a reversed direction of rotation.

There are ways to overcome this phenomenon; increasing the load friction can damp it out, but at considerable expense to motor performance. Viscous friction dampers can be used, which affect the transient response but not the steady-state performance. Various electronic circuit modifications have been devised to reduce oscillation, and even changes to the motor magnetic field geometry can be economic for large-quantity applications.

In practice, by using 'half-stepping' drive circuits and arranging to operate motors away from the resonant speed regions, combined with the inherent damping of load friction, resonance does not usually present too much of a problem. If the application permits, altering the acceleration rate will often avoid resonance effects.

A8.3.3.3   Stepper/encoders

A significant advantage of the stepper motor is its ability to operate in open loop mode in position control and velocity control systems. However, as control systems became more sophisticated, the advantage of closing the loop became evident. This has resulted in a major movement in the market towards PM servo drives, but closed-loop stepper motors are available. Because the stepping motor is a digital actuator it is usual to employ a digital feedback device, such as an encoder, or resolver with R/D conversion.

Part B
The drive in its environment

Introduction
B1 The a.c. supply
B2 Interaction between drives and motors
B3 Physical environment
B4 Thermal management
B5 Drive system power management: common d.c. bus topologies
B6 Electromagnetic compatibility (EMC)
B7 Protection
B8 Mechanical vibration, critical speed and torsional dynamics
B9 Installation and maintenance of standard motors and drives

Introduction

This chapter considers the most important aspects of operating a drive in the real world where the effects of an imperfect environment on the drive, and of the drive on its environment, must be considered. In many cases, the issues are quite similar to those for a motor alone, but the use of electronic circuits offers both opportunities for alleviating some effects, and risks that new issues might arise.

It is not possible to consider every aspect and so focus has been given to the key areas which may impact the user or give the opportunity to improve drive system design though a good understanding of opportunities and limitations.

Chapter B1

The a.c. supply

B1.1 General

In this section the impact drives have on the a.c. supply is considered as well as how imperfections in the supply affect the operation of drives. We shall consider power factor and explore how the use of drives requires greater precision in some of the definitions and language that have historically been used in power systems.

B1.2 Supply harmonics and other low-frequency disturbances

B1.2.1 Overview

Supply harmonics are caused when the a.c. input current to the load departs from the ideal sinusoidal wave shape. They are produced by any non-linear circuit, but most commonly by rectifiers.

The supply current waveform is generally measured in terms of the harmonics of the supply frequency it contains. The harmonic current flowing through the impedance of the supply causes harmonic voltage to be experienced by other equipment connected to the same supply.

Because harmonic voltages can cause disturbance or stress to other electrical equipment, there are regulations that apply to public supply systems. If installations contain a high proportion of power electronic equipment such as uninterruptible power supplies or variable-speed drives, then they may have to be shown to satisfy the supply authorities' harmonic guidelines before permission to connect is granted. As well as obeying regulations, users of drives need to ensure that the harmonic levels within their own plant are not excessive.

In the general realm of electronic equipment design and regulation, harmonics are considered to be just one of the many aspects of the discipline of electromagnetic compatibility (EMC). For variable-speed drives, because of the high power levels involved and the intimate connection between the basic design principles and the harmonic behaviour, the subject of harmonics is normally considered independently from other EMC aspects, which are predominantly concerned with high-frequency effects.

Some of the practical problems that may arise from excessive harmonic levels are as follows:

- poor power factor, i.e. high current for a given power,
- interference to equipment which is sensitive to voltage waveform,
- excessive heating of neutral conductors (single-phase loads only),
- excessive heating of induction motors,
- high acoustic noise from transformers, busbars, switchgear and so on,
- excessive heating of transformers and associated equipment, and
- damage to power factor correction capacitors.

An important property of harmonics is that they tend to be cumulative on a power system. That is, the contributions of the various harmonic sources add up to some degree because they are synchronised, and only their phase angles differ. It is worth emphasising this difference from high-frequency EMC effects, which may cause interference in sensitive data and measuring circuits through unintended coupling paths. High-frequency effects tend to be localised and not significantly cumulative, because the various sources are usually uncorrelated. It is important to be clear that with few exceptions, if harmonics cause disturbance it is through direct electrical connection and not through stray paths. Screening is rarely a remedial measure for harmonic problems.

It is necessary to consider the effect of supply harmonics both from the point of view of the possible effect of harmonic emissions from drives on other equipment, and also their possible effect on the drive. However, because most a.c. drives use a simple rectifier at their input, their immunity to harmonics is inherently good and requires no special attention here.

In addition to harmonics, consideration is also given to other possible low-frequency effects on the mains supply, such as interharmonics, voltage notching and lighting flicker.

B1.2.2 Regulations

Regulations may exist to protect the public power network from excessive harmonics, or as part of wider EMC regulations. Although the category of 'low frequency' for EMC standards extends officially up to 9 kHz, in most cases only harmonics up to order 50 are considered, which is 2.5 kHz on a 50 Hz supply and 3 kHz on a 60 Hz supply. There are currently no limits to emission in the range from 2.5 kHz/3 kHz to 9 kHz.

Measurements should be made using equipment that conforms to the current IEC standard for harmonic measuring instruments, which at the time of writing is IEC 61000-4-7:2002. The use of a correctly specified instrument is particularly important in the presence of fluctuating quantities.

There are two kinds of regulations that may be relevant; these are discussed in the next sections.

B1.2.2.1 Regulations for installations

These are imposed by the electricity supply authority to protect other electricity consumers from the effects of excessive harmonics. They are usually based on an agreed level of voltage distortion that can be tolerated by correctly designed equipment. This is specified in terms of a total harmonic distortion (THD), which is the ratio of the harmonic voltage to the fundamental expressed as a percentage. (Where there are a number of harmonic voltages present it is usual to calculate the total harmonic voltage as the square root of the sum of the squares. Alternatively if the rms and fundamental voltages are known, then the harmonic voltage is calculated as the square root of the difference between the squares of these values.)

The internationally accepted maximum THD 'compatibility level' in a low-voltage system is 8 per cent, and to achieve this with a high degree of confidence it is usual to aim for a rather lower level as the 'planning level', typically 5 per cent. Individual harmonics are also subject to limits.

From the point of view of the supply authority, the relevant harmonic voltage is at the point of common coupling (PCC) with other power consumers. The harmonic voltage levels within the consumer's premises may be higher because of the impedance of cables and transformers. In large installations measures may be necessary to prevent harmonic problems within a site. As there are no statutory requirements, a relaxed version of the authority limits can be applied internally. It is not advisable to allow the 8 per cent THD compatibility level to be exceeded, because the majority of equipment will have been designed to be immune only up to this level.

Predicting the voltage distortion for a proposed installation by a calculation can be an expensive undertaking, because it requires existing harmonics to be measured over a period of time, the system parameters such as source impedances to be derived, and the effect of the planned new load to be estimated. For a large installation with a high proportion of the load comprising electronic equipment, it may be cost-effective to complete this exercise in order to avoid either initial overdesign or the application of unnecessary remedial measures. For simpler cases a full analysis would be burdensome.

Regulations such as the United Kingdom Energy Networks Association recommendation G5/4-1 provide simplified 'staged' procedures to permit connection based only on harmonic current data, which can be obtained quite readily from the manufacturers' technical data. This does involves making some simplifying assumptions biased in a cautious direction. If the simplified stage does not permit connection, the full calculation procedure has to be applied.

B1.2.2.2 Regulations and standards for equipment

A further simplification of the guidelines can be made if a product conforms to a relevant harmonic standard, when it can be connected without reference to the supply authority. The international standard for equipment rated at less than 16 A is IEC 61000-3-2, the corresponding CENELEC standard being EN 61000-3-2. These are applied to consumer products and similar equipment used in very large numbers, where individual permission to connect would not be practical. In the European Union, EN 61000-3-2 is mandatory for equipment within its scope. Small variable-speed drives rated at less than about 650 W shaft power fall within the scope of this standard, and can be made to conform to it by the application of suitable measures. However, where they are used in large quantities in a single installation it may be more cost-effective to assess their total current and obtain permission to connect from the supply authority.

A further more recent standard IEC 61000-3-12 (EN 61000-3-12) covers equipment rated up to 75 A and is mandatory in the EU for equipment within its scope.

B1.2.3 Harmonic generation within variable-speed drives

B1.2.3.1 A.C. drives

Harmonic current is generated by the input rectifier of an a.c. drive. The only exception is for an active input stage ('active front end', AFE), where PWM is used to create a sinusoidal back-emf, and there is, in principle, no harmonic current. The only unwanted current is at the PWM carrier frequency, which is high enough to be relatively easy to filter. This arrangement is discussed later (B1.2.6.6).

The essential circuit for a typical a.c. variable speed drive is shown in Figure B1.1. The input is rectified by the diode bridge, and the resulting d.c. voltage is smoothed by the capacitor and, for drives rated typically at over 2.2 kW, the supply current is smoothed by an inductor. It is then chopped up in the inverter stage, which uses PWM to create a sinusoidal output voltage of adjustable voltage and frequency. Supply harmonics do not, however, originate in the inverter stage or its controller, but in the input rectifier.

The input can be single or three phase. For simplicity the single-phase case is covered first. Current flows into the rectifier in pulses at the peaks of the supply voltage as shown in Figure B1.2.

Figure B1.3 shows the Fourier analysis of the waveform in Figure B1.2. Note that all currents shown in the spectra are peak values, i.e. $\sqrt{2}$ times their rms values. It

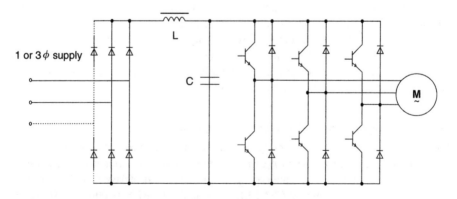

Figure B1.1   Essential features of an a.c. variable-speed drive

The a.c. supply 303

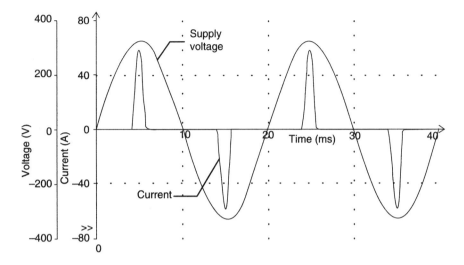

Figure B1.2   Typical input current waveform for a 1.5 kW single-phase drive (with supply voltage)

comprises lines at multiples of 50 Hz. Because the waveform is symmetrical in the positive and negative half-cycles, apart from imperfections, even-order harmonics are present only at a very low level. The odd-order harmonics are quite high, but they diminish with increasing harmonic number. By the 25th harmonic the level is negligible. The frequency of this harmonic for a 50 Hz supply is 1 250 Hz, which is in the audio-frequency part of the electromagnetic spectrum and well below the

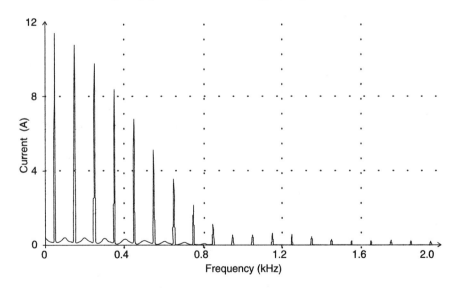

Figure B1.3   Corresponding harmonic spectrum for Figure B1.2

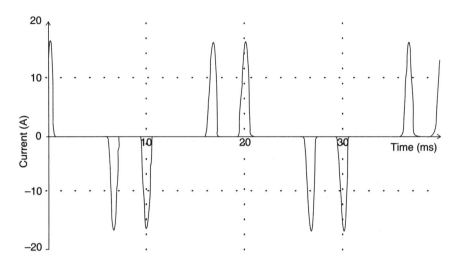

Figure B1.4   Typical input current waveform for a 1.5 kW three-phase drive

radio-frequency part, which is generally considered to begin at 150 kHz. This is important, because it shows that supply harmonics are low-frequency effects, which are quite different from radio-frequency EMC effects. They are not sensitive to fine details of layout and screening of circuits, and any remedial measures that are required use conventional electrical power techniques such as tuned power factor capacitors and phase-shifting transformers. This should not be confused with the various techniques used to control electrical interference from fast switching devices, sparking electrical contacts, and so on.

Three-phase drives cause less harmonic current for a given power than single-phase drives. Figure B1.4 shows the input current waveform for a 1.5 kW three-phase drive. The line current is less in any case, and there are two peaks in each mains cycle, each of about 20 per cent of the peaks in the single-phase drive.

Figure B1.5 shows the corresponding harmonic spectrum for the current waveform in Figure B1.4. Compared with the single-phase case the levels are generally lower, and the triplen harmonics (multiples of three times the supply frequency) are absent.

The actual magnitudes of the current harmonics depend on the detailed design of the drive, specifically the values of d.c. link capacitance and inductance. Therefore the supplier must be relied upon to provide harmonic data.

B1.2.3.2   D.C. drives

There is no difference in principle between the harmonic behaviour of a.c. and d.c. drives, but the following aspects of d.c. drives are relevant.

- The current waveform is not affected by the choice of design parameters (inductance and capacitance) in the drive. It does not therefore vary between

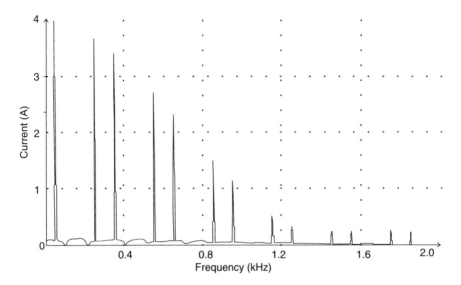

Figure B1.5 Corresponding harmonic spectrum for Figure B1.4

drive manufacturers. It can be calculated from knowledge of the motor armature inductance, source inductance and pulse number.
- The phase angles of all harmonics change with the rectifier firing angle, which, during the constant torque region of operation up to the base speed of the motor, is approximately proportional to speed. For multiple drives, unless their speeds are co-ordinated, the phase angles are effectively random and the harmonic amplitudes do not add up arithmetically.
- Today d.c. drives tend to be most often used at relatively high power levels, and often a dedicated transformer is provided, so 12- and higher-pulse numbers are more readily provided.

Effect of loading
In the case of an a.c. drive, the input current is proportional to the load power, i.e. the product of motor shaft torque and speed. As the load power falls, all of the main harmonics also fall, but not as rapidly as the fundamental. In other words, the THD deteriorates as the load falls. This applies whether the power reduction is through reduced speed or torque or both.

In the case of a d.c. drive, the above applies for variations in output current, and hence motor shaft torque. However, for a given torque the current does not fall significantly as the speed falls. At light load the waveform may improve somewhat at low speed if the d.c. current becomes discontinuous, but at full torque, at low speed the harmonic structure is much the same as at maximum speed.

The highest harmonic current for a given drive invariably occurs at maximum load, but in a system with multiple drives it may be necessary to look in detail at the effect of various possible load combinations.

B1.2.4 The effects of harmonics

Some of the effects of harmonics were summarised in Section B1.2.1. Figure B1.6 shows a voltage waveform where a distribution transformer is loaded to 50 per cent of its capacity with single-phase rectifiers. It shows the characteristic flat-top effect. Although this waveform looks alarming, most modern electronic equipment is undisturbed by it. However, the harmonic content can cause excessive stress in components, especially capacitors, connected directly to the supply.

The diode bridge input circuit in a single-phase a.c. drive is the same as used in a very wide range of electronic equipment such as personal computers and domestic appliances. All of these cause similar current harmonics. Their effect is cumulative if they are all connected at the same low-voltage (e.g. 400 V) supply system. This means that to estimate the total harmonic current in an installation of single-phase units, the harmonics have to be added arithmetically.

Phase-controlled equipment such as lamp dimmers and regulated battery chargers cause phase-shifted harmonics that can be added by root-sum-squares to allow for their diverse phase angles.

In a mixture of single- and three-phase loads, some of the important harmonics such as the 5th and 7th are 180° out of phase and mutually cancel. Sometimes this information can be very helpful, even if there is no certainty that the loads will be operated simultaneously; for example, in an office building that is near to its limit for 5th and 7th harmonics because of the large number of single-phase computer loads, the installation of three-phase variable-speed drives will certainly not worsen the 5th and 7th harmonics and may well reduce them.

Over-loading of neutral conductors is a serious concern in buildings containing a high density of personal computers and similar IT equipment. It is caused by the summation of triplen harmonics in the neutral conductor; the neutral current can

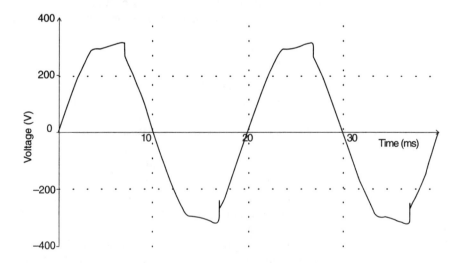

Figure B1.6   Supply voltage waveform with single-phase load of 50 per cent supply capacity

equal or even exceed the individual phase currents, whereas it is common for the conductor to be of reduced cross-section. Single-phase a.c. drives would have a similar effect, but it is unusual for them to be used at such a high density.

B1.2.5 Calculation of harmonics

B1.2.5.1 Individual drives: d.c.

The calculation of the input current for a controlled rectifier is covered in most of the standard textbooks of power electronics. A particularly clear and comprehensive account is given in IEEE Std 519-1992. An account is given in Chapter A2, but for clarity is summarised here.

The basic analysis for the controlled rectifier assumes an infinite inductance load. Then for a p-pulse rectifier the input current has a stepped wave shape with p regularly spaced steps in each cycle. This can readily be shown to contain no even harmonics, and only odd harmonics of the order $n = kp \pm 1$ where k is any integer.

The amplitudes of the harmonics follow the simple rule for a rectangular wave being inversely proportional to the harmonic number:

$$I_n = \frac{I_1}{n} \tag{B1.1}$$

For some purposes this simple calculation is sufficient, but the influence of finite inductance on the d.c. side should be taken into account, and the a.c. side inductance may also have a significant effect.

Figure B1.7 shows the effect of d.c. current ripple on the four dominant harmonics of a six-pulse drive. The 5th harmonic increases steadily with increasing ripple, whereas for moderate ripple levels the other harmonics fall.

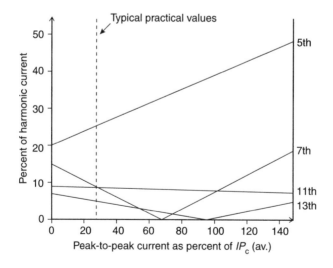

Figure B1.7   Six-pulse converter: variation of line current harmonic content with ripple current

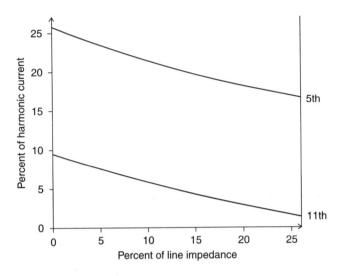

Figure B1.8 Variation of line harmonic content with line impedance $\alpha = 0°$

Figure B1.8 shows the effect of supply inductance on these harmonics for a firing angle of zero. All of them fall with increasing inductance, particularly the 11th and 13th. However, the benefit is reduced at large firing angles because of the more rapid commutation. So for operation at high torque and low speed (low voltage) the benefit of supply inductance may be minimal. Further information is given in IEEE 519.

B1.2.5.2 Individual drives: a.c.

Whereas for a d.c. drive the harmonics are determined largely by external circuit parameters, for an a.c. drive they are determined mainly by the internal inductance and capacitance. It is therefore not usually practical for the user to calculate the harmonic current for an a.c. drive, and it is the responsibility of the drive manufacturer to provide harmonic current data. This should be provided at least for full load, and preferably at part load also. Linear interpolation of the harmonic currents as a proportion of the fundamental can then be used to estimate other loadings.

For small a.c. drives with no internal inductance, the supply impedance has a considerable influence. IEC 61800-3 recommends that the fault level be assumed to be 250 times the drive input current rating. Data should also be provided for where line reactors are fitted.

With large drives the IEC 61800-3 fault level is unrealistic, and a fault level such as 16 kA, which corresponds with the capability of widely available switchgear, may be used.

B1.2.5.3 Systems

The impact of a harmonic current on the power system can be estimated by calculating the resulting harmonic voltage at a point in the supply system shared with other

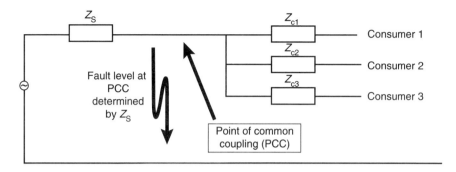

Figure B1.9   Supply system, showing point of common coupling with fault level

equipment. The power supply companies have a duty to control the quality of the power delivered to consumers, so their interest is at the point where the supply is shared with another consumer – the PCC. The basic equivalent circuit for this calculation is shown in Figure B1.9.

The impedance Z_S is usually dominated by the reactance of the distribution transformer, but cable and line inductance may also be a significant contributor.

For the study of harmonics, the principle of superposition is used, which means that the mains source is turned off and the consumer being studied is considered as a source of harmonic current, as shown in Figure B1.10. Each harmonic is considered in turn. The voltage is simply the product of the current and the impedance of the supply system upstream of the PCC. The impedance at 50 Hz (or other mains frequency) can be found from the declared fault level of the supply, which should be available from the supply company. If it is expressed in MVA then the impedance in ohms at mains frequency can be calculated as follows:

$$Z_S = V^2/(\text{MVA} \times 10^6) \tag{B1.2}$$

where V is the voltage between lines and Z_S is the positive-sequence fault impedance.

The impedance is assumed to be predominantly inductive, as is the case with high-power circuits, so that for a harmonic of order n the impedance is nZ_S.

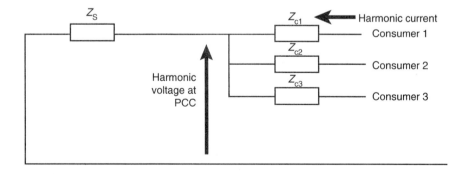

Figure B1.10   Supply system arranged for harmonic analysis

This calculation is required for assessment against Stage 3 of the UK Energy Networks Association recommendation G5/4-1. It is widely accepted as giving a valid basis for assessment of harmonic penetration. The presence of substantial capacitance from high-voltage cables or power factor correction capacitors causes a more complex situation where resonance causes the impedance to rise at certain frequencies. If these coincide with odd harmonics where substantial currents exist, a higher harmonic voltage than estimated can occur. Fortunately this is an unusual situation, as it can be expensive to cure.

Note that in Figure B1.10 the harmonic voltage within the premises of consumer 1 will be higher than that at the PCC, because of the voltage drop in Z_{C1}. Meeting G5/4-1 at the PCC is therefore no guarantee of tolerable harmonic levels within the system of the consumer generating the harmonics.

In order to analyse a practical system, the known harmonic data for all the rectifiers and other distorting loads must be combined to predict a total current. In general, each harmonic from each unit is a vector quantity that can only be added to the others through vector addition. Usually the phase angle is unknown, and in the case of phase angle controllers, such as those used in d.c. drives, it varies with the operating condition.

For uncontrolled rectifiers, the phase angles of the dominant harmonics will be similar, and the amplitudes add directly. For controlled rectifiers, the phase angles can be treated as random and the amplitudes add as $\sqrt{\sum I_n^2}$.

Diversity of loading is also an important issue. In some installations only a small part of the possible load on each drive can occur simultaneously. This must be considered to avoid an over-estimate of the harmonic loading.

B1.2.5.4 Isolated generators

If the system is supplied by isolated generators not connected to a grid, the impedance of the generators must be determined. The relevant parameter is the direct axis sub-transient reactance, x_d''. (Strictly the quadrature axis impedance should also be considered, depending on the load angle. In practice they are usually similar.) Typical values are between 14 and 20 per cent, compared with the 5 per cent of a typical distribution transformer, so generators are less able to tolerate harmonic current than the public supply network.

B1.2.6 Remedial techniques

The first point to make is that harmonics problems are unusual, although with the steady increase in the use of electronic equipment, problems may be more common in future. The situations where problems have occurred most frequently are in office buildings with a very high density of personal computers, and in cases where most of the supply capacity is used by electronic equipment such as drives, converters and uninterruptible power supplies (UPS).

As a general rule, if the total rectifier loading (i.e. variable-speed drives, UPS, PC etc.) on a power system comprises less than 20 per cent of its current capacity then harmonics are unlikely to be a limiting factor. In many industrial installations the capacity of the supply considerably exceeds the installed load, and a large proportion of the load is not a significant generator of harmonics; uncontrolled (direct on line) a.c. induction motors and resistive heating elements generate minimal harmonics.

If rectifier loading exceeds 20 per cent then a harmonic control plan should be in place. This requires that existing levels be assessed, and a harmonic budget allocated to new equipment.

Calculations using the techniques described in Section B1.2.5 may be required to predict the effect on harmonic voltage from connecting additional equipment.

If necessary, the following measures, which are discussed with reference to a.c. drives unless otherwise specified, can be used to reduce the harmonic level. Most techniques are applicable to d.c. drives but consideration of the speed-dependent phase of the harmonics is necessary.

B1.2.6.1 Connect the equipment to a point with a high fault level (low impedance)

When planning a new installation, there is often a choice of connection point. The harmonic voltage caused by a given harmonic current is proportional to the system source impedance (inversely proportional to fault level). For example, distorting loads can be connected to main busbars rather than downstream of long cables shared with other equipment.

B1.2.6.2 Use three-phase drives where possible

As shown above, harmonic current for a three-phase drive of given power rating is about 30 per cent of that for a single-phase drive, and there is no neutral current. If the existing harmonics are primarily caused by single-phase loads, the dominant 5th and 7th harmonics are also reduced by three-phase drives.

B1.2.6.3 Use additional inductance

Series inductance at the drive input gives a useful reduction in harmonic current. The benefit is greatest for small drives where there is no d.c. inductance internally, but useful reductions can also be obtained with large drives.

Additional a.c. supply line inductance
The addition of a.c. input inductance to the single-phase drive improves the current waveform and spectrum from those shown in Figures B1.2 and B1.3 to those shown in Figures B1.11 and B1.12, respectively. It is particularly beneficial for the higher-order harmonics, but the 5th and 7th are reduced by a useful degree. Only the 3rd harmonic is little improved.

Because the three-phase rectifier has no 3rd-harmonic current, the a.c. inductor is even more beneficial, as shown in Figures B1.13 and B1.14. In these examples the value of the a.c. inductor is 2 per cent (i.e. 0.02 p.u.). Values higher than this need to be applied with caution, because the drive output voltage at full load begins to be reduced significantly.

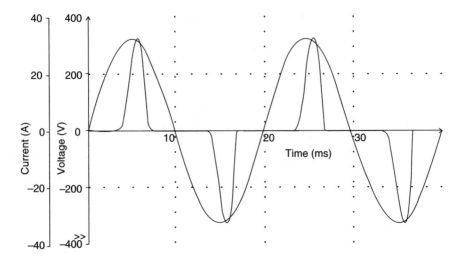

Figure B1.11   Input current waveform for single-phase drive as in Figure B1.2 but with 2 per cent input inductor

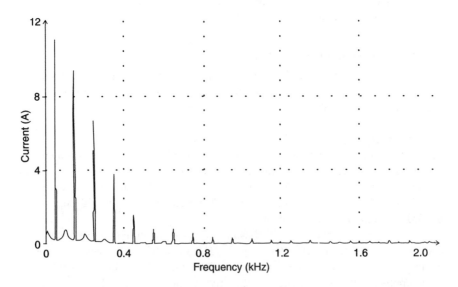

Figure B1.12   Corresponding harmonic spectrum for Figure B1.11

The a.c. supply 313

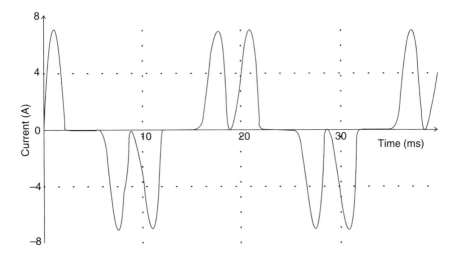

Figure B1.13   Input current waveform for three-phase drive as for Figure B1.4 but with 2 per cent input inductors

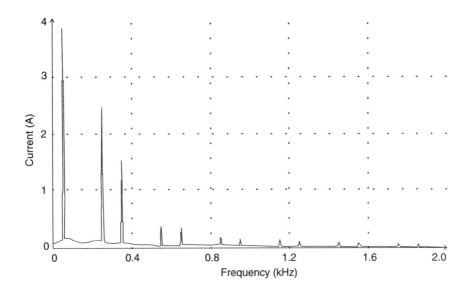

Figure B1.14   Corresponding harmonic spectrum for Figure B1.13

Additional d.c. inductance

A.C. drives rated at 4 kW or more usually have three-phase input and include inductance built in to the d.c. link or the a.c. input circuit. This gives the improved waveform and spectrum shown in Figures B1.15 and B1.16 respectively, which are for a hypothetical 1.5 kW drive for ease of comparison with the previous illustrations.

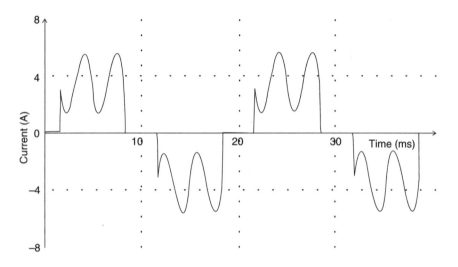

Figure B1.15 Input current waveform for a three-phase 1.5 kW drive with d.c. inductance

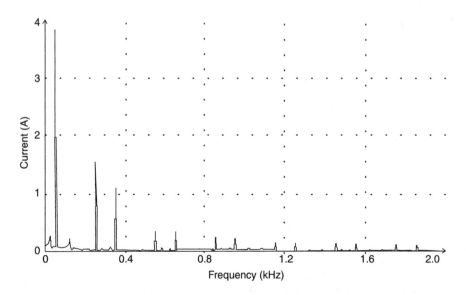

Figure B1.16 Corresponding harmonic spectrum for Figure B1.15

Further improvement is possible by adding a.c. inductance as well as d.c., as shown in Figures B1.17 and B1.18. This represents the limit of what can be practically achieved by very simple low-cost measures.

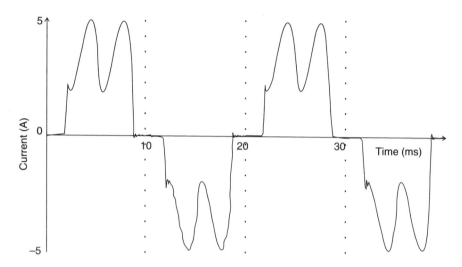

Figure B1.17   *Input current waveform for a three-phase 1.5 kW drive with d.c. and 2 per cent a.c. inductors*

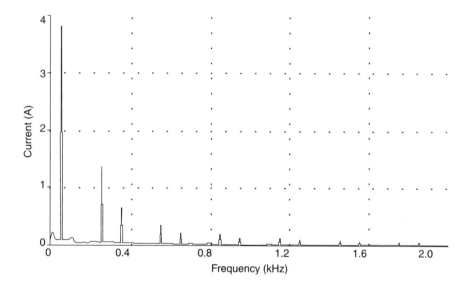

Figure B1.18   *Corresponding harmonic spectrum for Figure B1.17*

B1.2.6.4   Use a lower value of d.c. smoothing capacitance

For a three-phase rectifier, the capacitance value can be much reduced provided that the inverter is adapted to compensate for the resulting voltage ripple. The input current waveform is then improved and tends towards the 'ideal' case with a large d.c. inductance, where the current is approximately constant during the 120° conduction period.

This is a useful technique for a cost-sensitive application where harmonic current is a critical factor. There are disadvantages resulting from the reduced capacitance, in that the d.c. link voltage becomes more sensitive to transient conditions, both from rapid variations in load and also from supply disturbances. This approach is therefore most attractive in applications where the load does not exhibit highly dynamic behaviour. Higher ripple voltages also effectively reduce the maximum possible output voltage.

Another practical factor is that the capacitor now has a high ripple current relative to its capacitance value, so that a conventional aluminium electrolytic capacitor cannot be used and a relatively large and expensive plastic dielectric is required.

B1.2.6.5 Use a higher pulse number (12 pulse or higher)

Standard three-phase drives rated up to about 200 kW use six-pulse rectifiers. A 12-pulse rectifier eliminates the crucial 5th and 7th harmonics (except for a small residue caused by imperfect balance of the rectifier groups). Higher pulse numbers are possible if necessary, the lowest harmonic for a pulse number p being $(p-1)$.

Individual a.c. drives may be supplied with d.c. from a single bulk 12-pulse rectifier, or where the loading on drives is known to be reasonably well balanced, individual six-pulse drives may be supplied from the two phase-shifted supplies.

If the transformer rating matches the total drive rating reasonably closely then its inductance gives a very useful additional reduction of the higher-order harmonics. For ratings up to about 1 MW it is unusual to require pulse numbers greater than 12.

A 12-pulse system is illustrated in Figure B1.19. The star and delta windings (or zig-zag windings) have a relative 30° phase shift, which translates to 180° at the

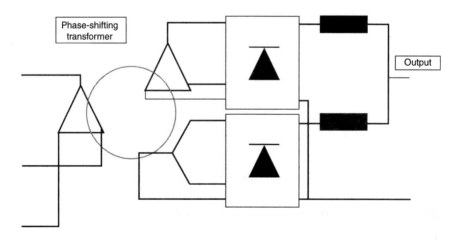

Figure B1.19 Basic 12-pulse rectifier arrangement

5th and 7th harmonics (as well as 17, 19, 29, 31, and so on), so that flux and hence primary current at these harmonics cancels in the transformer.

The transformer input current waveform and spectrum are shown in Figures B1.20 and B1.21, respectively.

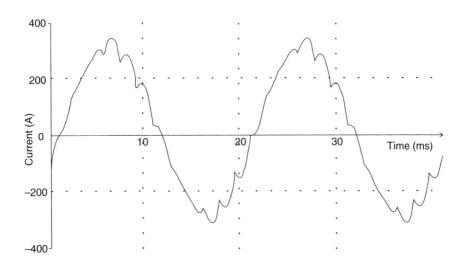

Figure B1.20   Input current waveform for 150 kW drive with 12-pulse rectifier

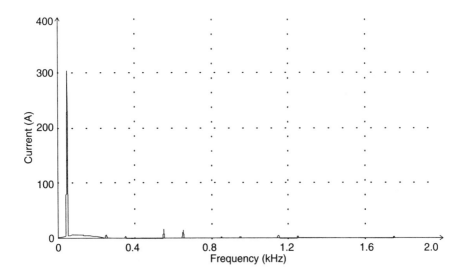

Figure B1.21   Corresponding harmonic spectrum for Figure B1.20

B1.2.6.6 Use a drive with an active input converter

An active input converter using PWM generates negligible harmonic current, as well as permitting the return of power from the load to the supply.

The input current for an active input converter contains negligible harmonic current if the supply voltage is sinusoidal. There are two side effects that must be considered.

- The input-stage PWM frequency causes input current, which will usually have to be filtered. This is *in addition to the radio frequency filter*.
- Existing voltage harmonics in the supply will cause some harmonic current to flow into the drive. This should not be mistaken for harmonic emission.

B1.2.6.7 Use a harmonic filter

Three types of harmonic filter can be distinguished.

- *Passive parallel filter.* The traditional form of filter is typically based upon a power factor correction capacitor. By the addition of series inductances the filter can be made to have a very low impedance at key harmonics such as 5 and 7 or 11 and 13. This kind of filter is a natural choice for a d.c. drives because it provides the necessary power factor correction as well as attenuation of key harmonics. It is less suitable for a.c. drives, which inherently have a displacement factor close to unity. The parallel filter is applied close to the origin of the plant power supply, and attenuates the harmonic voltage arising from all sources of harmonic current. As it presents a low impedance at these harmonics, it sinks harmonic current from existing harmonic voltages at the supply source. In extreme cases this can cause over-loading of the filter, resulting in operation of the necessary over-current protection device. The selection and application of this kind of filter requires a careful survey of existing harmonic voltage and an estimate of the resulting loading on the filter.
- *Passive series filter.* This type of filter is designed specifically for use with drives or similar distorting loads. The filter is connected in series with one or more defined loads, and reduces their harmonic current emission by offering a low-pass characteristic, possibly with additional attenuation at the 5th and 7th harmonics by a notch in the frequency response. The benefit of the series filter is that it is largely unaffected by existing supply harmonics. It may also offer improved performance with an unbalanced three-phase supply, when compared with high-pulse-number rectifiers.

    Some possible disadvantages include the following:
    - a tendency to excessive voltage drop in the d.c. link with increasing load,
    - a significant standing capacitive current, which may exceed the stability limit for a local synchronous generator, and
    - possible disturbance to the control systems for thyristor input stages, as used in some drive designs for input inrush current control.

    These disadvantages have been overcome by ingenious design. The remaining disadvantage is cost, because the series filter has to carry the entire supply current, whereas the parallel filter carries only the harmonics. At the time of

writing, series filters have been found to be cost effective for drive power ratings up to about 400 kW.
- *Active parallel filter.* The active filter uses power electronic techniques to generate harmonic current in opposition to that generated by the load. It can be arranged to target only the load current, and it has current limiting so that in the presence of excessive harmonic current it will operate up to the limits of its capability and not trip due to over-loading. Currently the cost of such filters is high and they are only used in critical situations.

B1.2.7 Typical harmonic current levels for a.c. drive arrangements

Table B1.1 shows typical harmonic current levels for a.c. drive arrangements.

Table B1.1 Harmonic current levels for a.c. drive arrangements

	Harmonic current as percentage of fundamental					
	I_3	I_5	I_7	I_{11}	I_{13}	I_{THD}
Single-phase, no inductance	97	91	83	62	51	206
Single-phase, 2% inductance	90	72	50	13	6	130
Three-phase, no inductance	0[a]	49.6	28.2	6.6	6.0	58
Three-phase, 3% inductance	0[a]	35.0	12.2	7.4	3.9	38
12-pulse	0[a]	1.8	0.6	4.5	3.1	5.8
Active input converter	0[a]	1.4	0.3	0.5	0.2	3.3

[a] For a balanced supply.

B1.2.8 Additional notes on the application of harmonic standards

The requirements of individual standards are beyond the scope of this book. However, some issues recur frequently and are worth highlighting. In particular, questions relating to harmonic behaviour and testing at a variety of loads often occur, as do questions relating to the correct base or reference current by which to refer to harmonic currents.

B1.2.8.1 The effect of load

For most equipment, the harmonic current becomes larger as a proportion of the overall current as the load is reduced. This is particularly so with rectifiers, because at reduced load the inductances become proportionally smaller and the capacitances larger relative to the load. This is illustrated in Figure B1.22 for an a.c. drive.

With a simple rectifier the key harmonic currents (3, 5 and 7) fall in absolute terms as the load is reduced. Only some of the higher harmonics may increase over some parts of the load range. This is illustrated in Figure B1.23 for the same arrangement as for Figure B1.22.

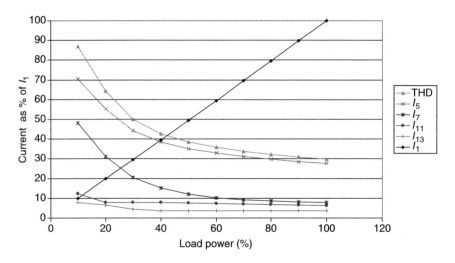

Figure B1.22 Variation of input current harmonics and THD as a percentage of fundamental current with variation of load power for an a.c. drive

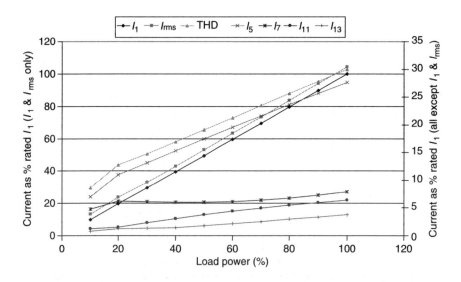

Figure B1.23 Variation of input current harmonics and THD as a percentage of rated fundamental current with variation of load power, for an a.c. drive

Note that for an a.c. drive it is the load power that determines the input current, including its harmonics. With d.c. thyristor drives it is the load current.

Product standards such as IEC 61000-3-12 make it clear that the equipment must be tested at the rated load, and that harmonics are measured as a proportion of the rated input current. However, with drives used in machinery it is often difficult to operate the machine at rated load in an EMC test environment. Frequently the rated load can only be achieved briefly. It is common for measuring instruments to indicate harmonics and THD as a proportion of the fundamental current at the same instant as the measurement, in other words not taking account of the equipment rated current. This can result in a wrong conclusion (apparent failure to pass the test). Only the data at rated load must be used, and if rated load cannot be achieved then the results must be corrected to refer to the rated fundamental current of the equipment.

B1.2.8.2 Choice of reference current: application of IEEE Std 519-1992

For a product test there is a clearly defined reference current, which is the manufacturer's declared rated current. For an installation assessment, a reference current must be established. In IEEE Std 519-1992 this is referred to as the 'maximum fundamental load current' for a particular electricity consumer at the PCC. This is calculated as the average current of the maximum demand for the preceding 12 months. The intention is to choose a reference current that is representative of the relative size of the consumer, so that the harmonic absorption capability of the power network can be fairly apportioned. Harmonic limits are then set by reference to this maximum current, and the term 'total demand distortion' (TDD) is used to express the ratio of the rms harmonic current to the maximum current.

IEEE Std 519 is often applied as a contractual condition for the purchase of power equipment, and it is not uncommon for the reference current to be taken as the rated current for a specific item of equipment. This results in a very stringent limit on harmonic current, which can only be achieved by using a special harmonic reduction technique such as a 12-pulse or 18-pulse rectifier. Harmonic emission is minimised, but at the cost of many unnecessary phase-shifting transformers and filters.

B1.2.9 Interharmonics and emissions up to 9 kHz

Low-frequency emission may occur with a frequency that is not an integer multiple of the supply frequency. This is referred to as an interharmonic if it lies within the range from zero to the 50th harmonic. There is no agreed terminology for higher frequencies.

Excessive interharmonics can result in undesirable modulation effects, leading to lighting flicker and modulation of d.c. supplies derived by rectification of the a.c. supply. Excessive higher frequencies can result in telephone noise on analogue telephones, and disturbance to waveform-sensitive equipment such as UPS systems.

A conventional drive is usually not a significant source of interharmonics or of the higher frequencies.

An active input converter will be a source of emission at frequencies related to both the PWM switching frequency and the supply frequency. The same applies to an active filter. Although there are no agreed limits for such emissions, a sensible practical limit would be the same as is applied to the highest-order harmonics: for example, adapting the limits from IEC 61000-2-2 would give

$$V_f = \frac{1930}{f} - 0.27 \ (\%) \tag{B1.3}$$

where f is the frequency of emission considered. For example, emission at 2.9 kHz, which is a dominant modulation product from a 3 kHz switching frequency and a 50 Hz line frequency, would be limited to 0.4 per cent. The cost of a filter to achieve this is much less than for a harmonic filter, although not insignificant.

B1.2.10 Voltage notching

Notching is an effect specifically associated with naturally commutated thyristor rectifiers, as used in d.c. drives. During the overlap interval where commutation between the semiconductor switches takes place, the voltage between lines at the drive terminals is forced to zero, and it recovers abruptly at the end of overlap, so that the line-to-line voltage has a notch. Although this phenomenon also occurs with a diode rectifier, it happens at the zero crossing point so there is no initial step change of voltage and the recovery is slow.

Notching can disturb equipment sharing the supply. The rapid voltage steps can induce high pulse currents and voltage overshoots. Oscillatory effects in the power system can result in multiple zero-crossings, which may affect equipment that detects them. Although theoretically the notches can be represented by harmonic series, it is the abrupt change in the time function that causes the disturbance, and this is the reason for distinguishing this phenomenon.

Typical limits to notch depths are 40 per cent for industrial supplies, as given in IEC 60146-1-1 and IEC 61800-3, and 20 per cent for residential supplies as give in G5/4-1.

For heavy plant with large d.c. drives a local depth of up to 80 per cent might be tolerated, but the thermal effects of notches are cumulative so care must be taken where many such drives are present.

Notch depth is controlled by using input inductances with the drive. The notch depth at the supply is given by

$$D = Z_S/(Z_S + Z_C) \ \% \tag{B1.4}$$

where Z_S is the supply impedance and Z_C is the impedance of the added inductance.

To estimate the notch depth at the supply PCC the value of Z_S would be the same as is used for harmonic voltage calculations. Within a given plant, allowance must be made for the inductance of cables and any transformers.

In practice it can be difficult to obtain the relevant impedance data, and many d.c. drives are operated with a 'rule of thumb' 2 per cent line inductance without any practical difficulties. Good practice dictates that some inductance must always be incorporated, because 100 per cent notches can cause widespread malfunctions, and the thyristor switches within the drive itself might be subjected to excessive di/dt stress if there are capacitors connected to the same supply.

Voltage notching is discussed further in Section B1.4.9.

B1.2.11 Voltage dips and flicker

It is often not appreciated by drive users how great an advantage the use of variable frequency gives in starting an induction motor. Direct on line (DOL) starting of a standard cage induction motor draws a current of typically five times its rated full-load current. Furthermore, the power factor of this current is very low because the actual power delivered is small (high torque but low speed). The voltage drop caused in the mainly inductive supply source impedance is therefore at a maximum. This can affect other loads, causing disturbance to electronic systems as well as lighting flicker, and it can cause difficulty in starting the motor if the load has significant static friction. Starting a motor by a correctly adjusted variable-speed drive causes a current in the motor not exceeding its short-term rated operating current, giving torque equal to the short-term rated value, while the drive input current begins at nearly zero and rises with the actual delivered shaft power. The drive itself supplies the reactive current for the motor, so the voltage drop caused in the inductive supply impedance is minimal.

'Flicker' refers to the effect of supply voltage fluctuations on lighting levels. Whereas occasional lighting fluctuations are tolerable, the human eye and brain are very sensitive to periodic flicker, especially with frequency in the region of 5–20 Hz. Supply companies operate guidelines to assist them in case of complaints, and standards such as IEC 61000-3-3 and IEC 61000-3-11 lay down limits on permissible voltage fluctuations when equipment is connected to a supply with a defined source impedance. There is a curve that defines the permissible flicker limits over a range of frequencies up to 25 Hz. There is also a much less stringent limit for single voltage changes, such as that occurring at switch on.

The effect of a drive on flicker is largely neutral, in that a fluctuating load will generally result in a fluctuating drive input power because the stored energy in the drive is insufficient to provide smoothing. The improved displacement factor of an a.c. drive when compared with a directly fed induction motor gives some reduction in the resulting voltage fluctuation, and the controlled motor starting and controlled inrush current of the drive reduce single voltage changes.

Direct a.c. to a.c. converters (see Section A2.5) can cause flicker problems and this needs to be considered when considering such converters.

There are some applications such as low-speed reciprocating pumps where the load torque inherently pulsates at a frequency within the sensitive range. It may be possible to use an a.c. drive to reduce the resulting flicker by deliberately allowing the speed to vary by more than the usual speed regulator would allow. Better use is then made of the system inertia to smooth the fluctuations in power flow. The details depend upon the drive design, but one example would be to use a negative value for the slip compensation parameter.

B1.3 Power factor

The power factor of an a.c. load is a measure of the ratio of the actual power to the product of rms current and voltage. The latter is a measure of the loading on the power distribution infrastructure, and therefore reflects the cost of the electrical hardware associated with supplying the load. Power factor can be expressed in the following form:

$$\text{Power factor} = \frac{\text{Power (kW)}}{\text{rms voltage (V)} \times \text{rms current (A)}} \quad (B1.5)$$

If a sinusoidal supply voltage and a linear load is considered then the power factor would be a measure of the phase shift between the voltage and the current, and is frequently expressed as $\cos \phi$, where ϕ is the angle between the current and voltage. $\cos \phi$ is frequently referred to as the fundamental power factor or the displacement factor.

Displacement factor

$$= \frac{\text{Power (kW)}}{\text{Fundamental rms voltage (V)} \times \text{Fundamental rms current (A)}} \quad (B1.6)$$

The power factor is dimensionless and always less than unity.

If the load is inductive where the current lags the voltage the power factor may be referred to as lagging. If the load is capacitive where the current leads the voltage, the power factor may be referred to as leading.

Power factor indicators, which work from the phase angles of voltage and current, will indicate this value. Most industrial users pay for their electricity based not simply on the actual power used, but there will be some penalty for drawing power at a poor power factor, by being charged for kVA rather than kW, by having a minimum allowable power factor or a combination of these. It is in industrial users' interests to operate their facilities at a high power factor.

Power factor in sinusoidal, linear systems is well understood and is subject to a number of methods of description. Apparent power, the product of the rms voltage (V) and rms current (A), is frequently used and has a useful conceptual value in the sinusoidal and linear domain. The in-phase component is the 'real' power. Unfortunately, many people extend this definition to call the out-of-phase component, $\sin \phi$, the 'reactive' power. The apparent justification for such a concept, in the sinusoidal and linear domain, is that there is energy flow associated with the reactive kVA into as well as out of the inductive and/or capacitive components. There is no net power flow associated with this energy and the term reactive power should not be used. Reactive kVA or kVAr is acceptable.

In systems where the supply voltage contains harmonics and/or the load is non-linear, harmonic effects impact the overall power factor. The distortion factor describes these harmonic effects as follows:

$$\text{Distortion factor} = \frac{\text{Fundamental voltage (V)} \times \text{Fundamental current (A)}}{\text{rms voltage (V)} \times \text{rms current (A)}} \quad \text{(B1.7)}$$

Considering equations (B1.5), (B1.6) and (B1.7), it can be seen that

$$\text{Power factor} = \text{Displacement factor} \times \text{Distortion factor} \quad \text{(B1.8)}$$

For a public low-voltage supply, the total harmonic distortion is limited to 5 per cent, so it is common to assume that the supply voltage is sinusoidal. In this case the distortion factor defined in equation (B1.7) simplifies to

$$\text{Distortion factor} = \frac{\text{Fundamental current (A)}}{\text{rms current (A)}} \quad \text{(B1.9)}$$

The distortion factor is a measure of the effect of harmonic current. It can be related to the more familiar current I_{THD} by the relation

$$\text{Distortion factor} = \frac{1}{\sqrt{1 + \text{THD}^2}} \quad \text{(B1.10)}$$

where THD, the total harmonic distortion of the current, is expressed as a fraction.

An example will clarify the relevance of these factors. Consider a typical 11 kW motor operating at full rated load, connected either directly to the mains supply or through an a.c. variable-speed drive. In this example the drive uses a d.c. link choke of about 3.5 per cent based on input power. Table B1.2 shows comparative supply data.

Table B1.2 Comparing supply power factor for a DOL motor and a motor fed from an a.c. drive

At supply terminals	DOL motor	Motor via a.c. drive	Notes
Voltage (V)	400	400	
rms current (A)	21.1	21.4	No significant change
Fundamental current (A)	21.1	18.8	Reduced due to no need to provide magnetising current
Displacement factor/ cos ϕ	0.85	0.993	Improved with drive, which has negligible phase angle
Distribution factor	1	0.877	Reduced with drive due to harmonic current
Power factor	0.85	0.871	Slight improvement with the drive
Power (W)	12 440	12 700	Slight increase at full load due to drive losses

We can summarise by saying that the typical PWM voltage source a.c. drive improves the power factor as compared with a direct on line motor; it reduces the requirement of the supply to provide the magnetising current for the motor, but in return generates harmonics. The harmonics are not charged for, so there is a saving in demand charges. Power consumption is increased by the approximately 2 per cent additional losses of the drive at the full load point. If the drive were to run at rated speed continuously then there would probably be an increase in net running cost, but of course the purpose of the drive is to control the speed, torque or position to suit the process, and in many applications the energy savings and so cost saving gained are very considerable.

B1.4 Supply imperfections

B1.4.1 General

The mains supply is subject to imperfections for which the impact must be considered. The main imperfections that are recognised and controlled by the supply utilities are given in EN 50160, which also gives limits applicable within the EU. International limits are given in IEC 61000-2-2. In addition, when thyristor converters such as those in d.c. drives are involved, voltage notching as described earlier in Section B1.2.10 must also be considered.

Generally the effect of an a.c. variable-speed drive is to 'buffer' the motor from the effects of supply imperfections, but for some effects such as temporary power loss, care is required in selecting the appropriate drive settings.

B1.4.2 Frequency variation

This is caused by the normal power regulating functions of the power grid, and for public supplies the permitted range of frequency is very small. When operating from local generation much greater changes of frequency may occur, and high rates of change may also be experienced. Deviations greater than 15 per cent from nominal would not normally be tolerated in any case.

The induction motor speed varies with supply frequency, although for the practical range of variation of public supplies this effect is negligible.

The basic voltage source a.c. variable-speed drive that is fed from the supply through a diode rectifier bridge is unaffected by any likely frequency variation. D.C. thyristor drives and active input converter a.c. drives, which need to maintain synchronisation with the supply, have restricted ranges of operating frequency and of rate of change of frequency. These will be specified by the drive manufacturer.

B1.4.3 Voltage variation

This is caused by the normal voltage regulating functions of the power grid. The induction motor speed varies slightly with voltage, in a way that depends upon the torque/speed characteristic of the load.

The a.c. variable-speed drive provides regulation against supply voltage variations. However, from base speed upwards the available torque is usually restricted by the available voltage (see Section A1.3.4).

B1.4.4 Temporary and transient over-voltages between live conductors and earth

Temporary power-frequency over-voltages can result from earth faults, which do not result in the power system protection operating. Transient over-voltages usually result from lightning activity. These voltages affect the design of the insulation system. Whereas a motor will normally have an insulation system designed to withstand the highest expected voltage impulses, the drive may make use of over-voltage suppression devices to protect thin insulation materials as are used in power semi-conductors to allow efficient heat transfer. Such devices must be correctly selected to ensure that they limit transient over-voltages while not being at risk of damage by temporary power-frequency over-voltages.

B1.4.5 Voltage unbalance

Voltage unbalance (in percent) may be defined as

$$\text{Voltage unbalance (\%)} = 100 \times \frac{\text{Maximum voltage deviation from average}}{\text{Average voltage}} \quad (B1.11)$$

For example, a system with line-to-line voltages of 400, 408 and 392 V, the average is 400 V, the maximum deviation from the average is 8 V, and the percentage unbalance is $[100 \times (8/400)] = 2$ per cent.

Voltage unbalance results primarily from unbalanced loads on the power system. Faults in tap-changers can also cause temporary unbalance.

A.C. motors are sensitive to supply unbalance (negative sequence components) because it results in a counter-rotating magnetic field inducing unwanted current into the rotor and generating reverse torque. The standard IEC 60034-1 gives a limit of 1 per cent for the long-term tolerance of motors to negative phase-sequence voltage, whereas EN 50160 gives a limit for public supplies of 2 per cent, but also states that in some places up to 3 per cent may occur. Large machines will have negative phase-sequence protection fitted, whereas smaller machines will generally operate satisfactorily in the presence of unbalance except that the losses and internal temperatures will be abnormally high. Extreme unbalance will result in operation of the motor thermal protection device.

A de-rating curve (Figure B1.24) is given by NEMA in MG-1, which should be applied to induction motors operated on an unbalanced supply. The standard also

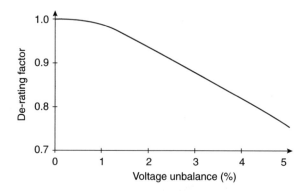

Figure B1.24   Induction motor de-rating factor due to unbalanced supply voltage

recommends that motors should not be operated on supplies with a voltage unbalance in excess of 5 per cent.

The a.c. drive naturally protects the motor from the effect of supply unbalance. Unbalance does, however, increase the ripple current in the d.c. smoothing circuit, and the current in individual rectifier diodes. Figure B1.25 shows the rectified three-phase supply voltage waveforms. Each positive device in the bridge conducts when the voltage is higher than in the other two phases, and each negative device in the bridge conducts when the voltage is lower than the other two phases.

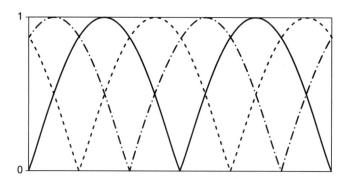

Figure B1.25   Three-phase rectified voltage: balanced supply

Consideration of the voltage waveform shows that the conduction periods for each phase are even and symmetrical positive and negative.

Figure B1.26 shows the case of a 5 per cent unbalanced case. The increased d.c. voltage ripple is readily seen, but consideration of the waveforms shows that the conduction periods for each phase will be very different between phases and unsymmetrical.

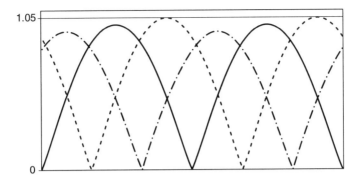

Figure B1.26   Three-phase rectified voltage: 5 per cent unbalanced supply

Drives designed according to the IEC 60146 family of standards will operate at full load with a 2 per cent negative phase sequence. Beyond that, they will usually have some form of protection using thermal or thermal modelling techniques that trips the drive or reduces its load to avoid damage.

B1.4.6   Harmonic voltage

Harmonic voltage results from the operation of non-linear loads such as electronic equipment of most kinds. A.C. motors are sensitive to harmonics other than co-phasal ones, because these result in rotating magnetic fields with high slip frequencies, causing added rotor losses (see Section A1.3). IEC 60034-1 gives a limit defined in terms of a 'harmonic voltage factor', which takes account of the harmonic order. This corresponds approximately to a limit of 4.5 per cent for the 5th harmonic. This is compatible with widely applied practical limits for public LV supplies of 3 or 4 per cent, but not to the limits given in EN 50160 and IEC 61000-2-2, which are 6 per cent.

A.C. and d.c. drives are generally not significantly affected by realistic levels of harmonic voltage. In small a.c. drives that use no inductance, the d.c. link ripple current could in principle be increased quite substantially by the presence of harmonic voltages, but the degree of this effect depends entirely upon their phase angles. This is better understood by considering the supply waveform. If the harmonics result in the typical 'flat topped' distorted sine wave, as shown in Figure B1.6, then the ripple current is actually less than with an ideal supply and the effect is beneficial. If the waveform has a 'peaky top' then the ripple current is increased. This would result in a reduction in life expectancy of the d.c. link capacitors.

B1.4.7   Supply voltage dips and short interruptions

These can result from faults in the power system, and dips can also result from the starting of loads such as induction motors that have a high starting current with low power factor. Faults usually affect a single phase initially, and if cleared locally by the single-phase protection then the dip or interruption is restricted to the affected phase. If the upstream three-phase protection operates then all phases are interrupted. One common kind of short interruption occurs when a power system fault is cleared successfully by the protection system, which then auto-recloses after a delay in the region of 200 ms.

Note that it is the supply voltage that is interrupted, and not the connection; in other words, the supply impedance remains low during the interruption. This needs to be considered if carrying out simulation tests. Depending on the drive design, its reaction to an open-circuit supply may be different from its reaction to a low-impedance supply.

Usually neither motors nor motor–drive systems carry sufficient electrical or magnetic stored energy to continue to operate through a supply dip or interruption lasting longer than about one-quarter of a supply cycle. The load may, however, have sufficient inertia to maintain operation at reducing speed, or inertia can be added, if the application requires this.

Careful consideration is required in deciding the correct actions to take during a power dip or interruption, because in many applications it is unacceptable for a machine to re-start without warning after a power interruption. However ride-through during brief interruptions is clearly desirable.

Whether or not a drive is used, if ride-through operation is required then attention must be paid to the reaction of the system both to the initial loss of power and to the return. For example, a single directly started motor will re-start if required, but a system containing many motors might trip if the total starting current is excessive.

Drives have a popular reputation for being rather too likely to trip during supply disturbances, particularly during dips and interruptions. There is usually a range of options for accommodating such disturbances, which may have to be selected from a drive function menu, and these options are not always well understood. The default behaviour may be to trip on supply loss, but there will usually be alternative behaviours that can use the mechanical stored energy in the load to maintain a supply to the drive by controlled deceleration. Attention may be required to the load behaviour in this situation, and if the motor could continue to turn after the drive has lost control then a special 'spinning motor' algorithm may be required for it to regain control when the supply returns.

Note that for a single-phase dip or interruption a three-phase drive will usually continue to run with an unbalanced input current until its thermal input protection device operates.

After a successful 'ride-through' operation, when the supply returns, the drive d.c. bus voltage will usually be low, but not sufficiently low for the inrush control system to operate. A high inrush current will then result. The drive should be designed to withstand this, but the combined inrush current from a number of drives has been known to cause a complete machine or system input protection device to operate.

The account above applies to a.c. drives. Thyristor d.c. drives rely upon the supply voltage for commutation, so that a loss of just one phase of the supply for more than one half-cycle is very likely to result in the drive tripping, and possibly operation of the protection devices (typically semiconductor fuses).

B1.4.8 Interharmonics and mains signalling

Interharmonics result from asynchronous processes in loads. One source is wound-rotor induction motors, and a more recent source is active rectifiers using asynchronous PWM. Mains signalling is used in some countries for load control, when it is often referred to as 'ripple control'. EN 50160 gives limits for the signal, for example at 1 kHz the limit is 5 per cent of nominal supply voltage.

Usually neither motors nor drives are sensitive to these effects. A few cases have been experienced where the ripple control signal at a particular site has been enhanced to an abnormal level, possibly by some local resonance mechanism, and has affected the synchronisation system on a d.c. thyristor drive. In that case additional filtering of the synchronisation signal was required.

B1.4.9 Voltage notching

Voltage notching is an effect peculiar to line-commutated converters, and is discussed in Section B1.2.10. It is not identified in EN 50160 or IEC 61000-2-2, but it is referred to in the UK harmonics guideline G5/4-1.

At the converter terminals the notch depth is 100 per cent twice per cycle on each line pair, and there are also four lesser notches where the other line pairs have 100 per cent notches. This is generally considered intolerable for associated circuits, and line reactors are incorporated as good practice to restrict the maximum notch depth to a less disturbing level such as 40 per cent (as required by IEC 60146-1-1 class B) or 20 per cent (as required by G5/4-1 at a low voltage PCC).

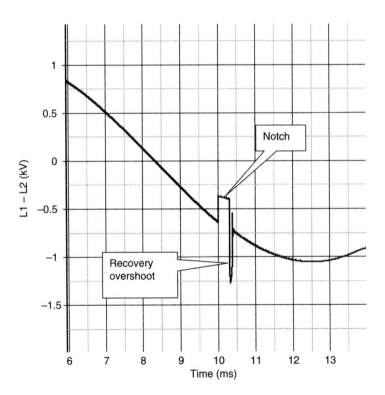

Figure B1.27   Example of a line voltage notch

Voltage notching can be represented by a combination of harmonics, and its effect on linear circuits is correctly predicted in this way, but to predict the effect on non-linear circuits it must be represented as a time function.

Figure B1.27 illustrates a line notch, and Figure B1.28 shows the notch in detail. Voltage notching has no significant effect on directly connected motors.

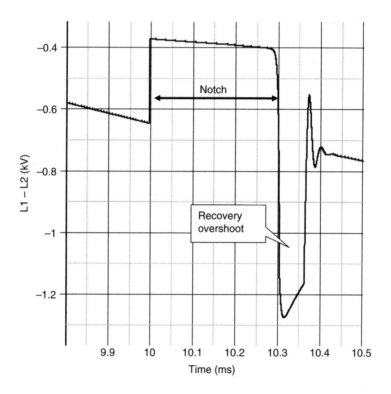

Figure B1.28   Example of a line voltage notch: expanded view

For drives, the effect of notching depends upon design details. An a.c. voltage source drive using a diode input rectifier with a.c. or d.c. inductance is unlikely to be affected. Any drive that uses thyristors, such as a d.c. drive or an a.c. drive with thyristor input rectifier, might be disturbed by notching if it causes thyristors to be switched off prematurely. This results in unexpected heating of damped snubber circuits, and if the thyristor gate drive is not sustained throughout the required conduction period it might result in serious anomalies in the current waveform, with a risk of over-stress to rectifier components. The drive should be designed to be immune at least to the level specified in IEC 60146-1-1.

Smaller a.c. drives that use no internal inductance can be upset by the overshoot following the notches, because the rectifier charges the capacitor towards the peaks.

At light load the d.c. bus voltage can rise up towards the peak of the overshoot, and where the notch is close to the peak of the supply voltage cycle this can lead to over-voltage tripping. In extreme cases the drive may be destroyed because after tripping the bus voltage may continue to rise. The solution here is to reduce the notch depth by adding line inductance, or to improve the d.c. drive snubber, or in some cases it may be more cost-effective to fit an input inductance and a braking resistor to the a.c. drive.

B1.4.10 EMC standards

Immunity standards exist in the IEC 61000-4 family for many of the effects listed above (Table B1.3). In particular, tests for immunity to surges (transient over-voltages) and supply dips and interruptions are required by the generic immunity standards IEC 61000-6-1 and IEC 61000-6-2, which are effectively mandatory in the EU for many types of equipment.

Table B1.3   EMC immunity standards

Standard IEC 61000-4	Area covered	Required by generic standards?
5	Surge	Yes
11	Voltage dips, short interruptions and voltage variations	Yes
13	Harmonics and interharmonics including mains signalling	No
14	Voltage fluctuation	No
27	Unbalance	No
28	Variation of power frequency	No
34	Voltage dips, short interruptions and voltage variations (input current >16 A)	No

Chapter B2
Interaction between drives and motors

B2.1 General

The control of electric motors by means of power electronic converters has a number of significant side-effects. These are primarily due to the introduction of additional frequency components into the voltage and current waveforms applied to the motor. In the case of a.c. machines, which are originally intended to operate at fixed speed, there are additional implications that need to be considered, including mechanical speed limits and the possible presence of critical speeds within the operating speed range.

The additional frequency components may or may not be harmonically related to the fundamental operating frequency, depending upon whether modulation processes in the converter are synchronised with the fundamental. This has little effect on the losses, so in this section we will follow the custom that 'harmonics' refers to all higher-frequency components.

B2.2 Drive converter effects upon d.c. machines

The effects due to deviation from a smooth d.c. supply are, in general, well understood by drive and motor manufacturers. Ripple in the d.c. current clearly increases the rms current, which leads to increased losses and hence reduced torque capacity. The harmonics associated with the current ripple lead to the now universal practice of using laminated magnetic circuits, which are designed to minimise eddy currents. With chopper converters, which are used in servo amplifiers and traction drives, frequencies in excess of 2 kHz can be impressed on the motor. Special care is needed to select a motor with sufficiently thin laminations and laminations made with high-quality magnetic grade steels.

The ripple content of the d.c. currents significantly affects commutation within a d.c. machine. The provision of a smoothing choke can be extremely important in this respect, and recommendation should be made by the motor manufacturer depending upon the supply converter used.

Beside the thermal and commutation impacts, the ripple current also results in pulsating torque, which can cause resonance in the drive train. Laminating the armature not only reduces the losses in the motor but also its dynamic behaviour by decreasing the motor time constant.

B2.3 Drive converter effects upon a.c. machines

B2.3.1 Introduction

It is often stated that standard 'off-the-shelf' a.c. motors can be used without problem on modern PWM inverters. Although such claims may be largely justified, switching converters do have an impact and certain limitations do exist.

NEMA MG1-2006, Part 31, gives guidance on the operation of squirrel-cage induction motors with adjustable-voltage and adjustable-frequency controls. IEC 60034-17 and IEC 60034-25 give guidance on the operation of cage induction motors with converter supplies, and design of motors specifically intended for converter supplies, respectively.

B2.3.2 Machine rating: thermal effects

Operation of a.c. machines on a non-sinusoidal supply inevitably results in additional losses in the machine. These losses fall into three main categories.

- *Stator copper loss*. This is proportional to the square of the rms current. Additional losses due to skin effect must also be considered.
- *Rotor copper loss*. The rotor resistance is different for each harmonic current present in the rotor. This is due to the skin effect and is particularly pronounced in deep bar rotors. Because the rotor resistance is a function of frequency, the rotor copper loss must be calculated independently for each harmonic. Although these additional losses used to be significant in the early days of PWM inverters, in modern drives with switching frequencies above 3 kHz the additional losses are minimal.
- *Iron loss*. This is increased by the harmonic components in the supply voltage.

The total increase in losses does not directly relate to a de-rating factor for standard machines because the harmonic losses are not evenly distributed through the machine. The harmonic losses mostly occur in the rotor and have the effect of raising the rotor temperature. Whether or not the machine was designed to be stator critical (stator temperature defining the thermal limit) or rotor critical clearly has a significant impact on the need for, or magnitude of, any de-rating.

For PWM voltage source inverters using sinusoidal modulation and switching frequencies of 3 kHz or higher, for a 50/60 Hz supply, the additional losses are primarily core losses and are generally small. It is also observed that motors designed for enhanced efficiency, e.g. to meet the EC Eff1 requirements, also experience a proportionately lower increase in losses with inverter supplies because of the use of reduced-loss cores.

Many fixed-speed motors have shaft-mounted cooling fans. Operation below the rated speed of the motor therefore results in reduced cooling. Operation above the rated speed results in increased cooling. This needs to be taken into account by the motor manufacturer when specifying a motor for variable-speed duty.

B2.3.3 Machine insulation

B2.3.3.1 Current source inverters

Current source inverters feeding induction motors have motor terminal voltages characterised as a sine wave with the superposition of voltage spikes caused by the rise or fall of the machine current at each commutation. The rate of rise and fall of these voltage spikes is relatively slow and only the peak magnitude of the voltage is of practical importance in considering the impact on machine insulation. The supply voltage never exceeds twice the crest voltage of the sinusoidal waveform, and is consequently below almost all recognised insulation test levels for standard machines.

Current source inverters feeding synchronous machines are even gentler on insulation systems, as the sinusoidal terminal voltage is reduced during commutation, producing the same effect as notching on the supply associated with supply converters.

B2.3.3.2 Voltage source inverters

PWM inverter drives are used with standard induction motors in very large numbers throughout the world, and their advantages are well known in terms of improved energy efficiency and flexibility of control. Occasionally drive users are advised to take special precautions over the motor terminal voltage because of an effect sometimes referred to as 'spikes' or 'dv/dt', which could possibly damage the motor insulation. This section explains the effect and prescribes what steps should be taken to ensure that the motor insulation system gives a long reliable life when used with a PWM drive.

Overview

The main effects of PWM drive waveforms on motor insulation are as follows.

- Motor winding insulation experiences higher voltages between turns on a coil and between coils, when used with a voltage-source PWM inverter drive than when driven directly from the a.c. mains supply. This effect is caused by the fast-rising PWM voltage pulse edges, which result in a transiently uneven voltage distribution across the winding, as well as short-duration voltage overshoots because of reflection effects in the motor cable. It is a system effect, which is caused by the behaviour of the drive, cable and motor together.
- For supply voltages up to 500 V a.c., the voltage imposed by a correctly designed inverter is well within the capability of a standard motor of reputable manufacture.
- For supply voltages over 500 V a.c., an improved ('reinforced') winding insulation system is generally required to ensure that the intended working life of the motor is achieved.
- When the motor used is of uncertain quality or capability, additional circuit components can be added to protect it.

338 *The Control Techniques Drives and Controls Handbook*

Guidance to avoiding problems and explanation of the phenomena involved

1. The voltage at the drive terminals is limited within tight bounds by the drive circuit, it is clamped by the returned-energy diodes and has negligible overshoot. The motor cable increases the peak motor voltage. In applications with short motor cables (i.e. 10 m or less) no special considerations of any kind are required.
2. Output inductors (chokes) or output filters are sometimes used with drives for reasons such as long-cable driving capability or radio-frequency suppression. In such cases no further precautions are required because these devices also reduce the peak motor voltage and increase its rise time.
3. In all other cases the following guidance should be followed:
   (a) Preferred approach: select a suitable motor
      (i) *For supply voltages less than 500 V a.c.* Check that the motor has the capability to operate with a PWM drive. Most reputable motor manufacturers have assessed their products for drive applications and can give an assurance of compatibility. Alternatively, Figure B2.1 shows the peak voltage/rise-time withstand profile, which is required for reliable operation. The motor supplier should be asked to confirm this capability. Figure B2.1 also shows the capability of a typical good-quality motor, which comfortably exceeds the requirement. However, note that conformity of the motor with IEC 60034-17 alone is not sufficient.
      (ii) *For supply voltages in the range 500–690 V a.c.* Select an inverter-rated motor. An enhanced insulation system is required. The permitted voltage/rise-time curve should equal or exceed that shown in Figure B2.2. Figure B2.2 also shows the capability of a typical inverter-rated motor for use up to 690 V, which comfortably exceeds the requirement. Note, however, that conformity of the motor with NEMA MG31 alone is not sufficient.

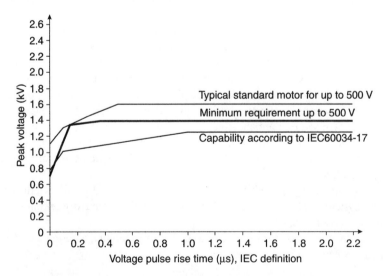

Figure B2.1 *Peak voltage/rise-time profile requirements for supplies up to 500 V a.c.*

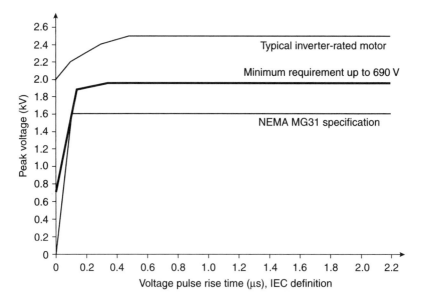

Figure B2.2 Peak voltage/rise-time profile requirements for supplies up to 690 V a.c.

(b) Alternative approach – use additional preventative methods

It may not be possible to follow the above recommendations, for example because the drive is to be retrofitted with an existing motor or data are not available for the motor concerned.

In this case additional preventative measures are recommended. The most cost-effective measures are usually drive output line chokes for lower power systems, and motor cable termination networks for higher powers. More details are given later.

(c) Factors affecting motor selection
  (i) Star windings are preferable to delta windings.
  (ii) Windings with single conductors are preferable to those containing parallel paths.
  (iii) Motor loading and duty should be carefully assessed to ensure that the motor does not over-heat; the insulation system is degraded by excessive temperature.

Special cases

1. *High braking duty.* Where the drive spends a large part of its operating time in braking mode, the effect is similar to increasing the supply voltage by about 20 per cent and the relevant precautions must be taken for the higher voltage.
2. *Active front end (regenerative/sinusoidal/unity power factor input drives).* For drives with active front ends (regenerative and/or unity power factor), the effective supply voltage is increased by about 20 per cent and the relevant precautions must be taken for the higher voltage.

3. *Special control schemes.* Some drive designs using flux vector control with fast-acting flux orientation frequently called direct torque control (see Section A4.4.2) can generate continuous 'double pulses' where the output voltage changes by twice the d.c. link voltage in a single step. This can result in four times the d.c. link voltage appearing at the motor terminals, causing increased stress and possibly premature motor failure. The stress is so extreme that a combination of inverter-rated motor and additional measures such as line chokes may be required to prevent motor damage. The drive supplier should be consulted for detailed guidance in this case.

Additional preventive measures

The two most cost-effective techniques are as follows:

1. *Output inductors (chokes) and output filters.* These are all connected at the drive in series with its output. They all work by forming a low-pass filter in conjunction with the motor and motor cable impedance, thus reducing the rate of rise of the drive output voltage. Some overshoot still occurs, which is controlled by damping or clamping. This results in some power loss, which must be allowed for in sizing the inductors or selecting the filter. The loss is roughly proportional to the motor cable length and the drive switching frequency.

    The use of an inductor with 2 per cent impedance at the maximum output frequency is sufficient to lengthen the rise-time to a point where it is no longer a consideration: 5 μs is easily attainable. The natural high-frequency loss in a standard iron-cored inductor gives sufficient damping, and this is more cost-effective than using a low-loss inductor with separate damping components.

    Commercially available dv/dt and sinusoidal filters should not normally be considered purely for motor protection, as their cost is excessive. They may, however, be specified for other reasons such as EMC or motor acoustic noise.

    The inductance should be chosen so that the impedance does not exceed 3 per cent pu at maximum frequency, otherwise the voltage drop will cause significant loss of torque at high speed.

    Conventional iron-cored inductors are suitable. Allowance should be made for additional core loss because of the presence of high frequencies. Special low-loss high-frequency inductors should not be used because severe resonance problems can occur.

    Individual phase inductors or three-phase inductors are equally effective. Other benefits include the following:

    - reduced loading effect on the drive from the cable capacitance and
    - reduced radio-frequency emission from the motor cable (EMC).

    Disadvantages include the following:

    - voltage drop,
    - power loss,
    - modest cost at low current ratings but increasing rapidly with increasing rating, and
    - for power levels above 100 kW the inductance at 3 per cent pu may be insufficient for the purpose.

2. *Motor cable terminating unit.* With increasing drive rating the above methods become increasingly expensive because they have to pass the entire drive output current. For powers exceeding about 70 kW it may be more cost-effective to use a terminating unit. This is a resistor–capacitor network, which is connected at the motor terminals in parallel with the power connections, and presents an impedance approximately matching the characteristic impedance (surge impedance) of the cable, during the pulse edges. This suppresses the reflection. It does not change the rise time, but it virtually eliminates the overshoot. It has the advantage of not carrying the drive output current, but power loss tends to be greater than for an inductor, and mounting at the motor terminals may be inconvenient and require a special sealed construction to match that of the motor.

Figures B2.3 and B2.4 show typical waveforms produced by these methods.

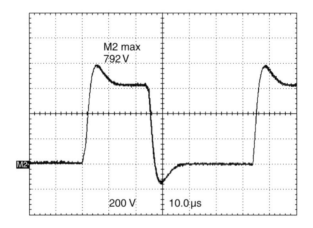

Figure B2.3   Motor terminal voltage with inductor

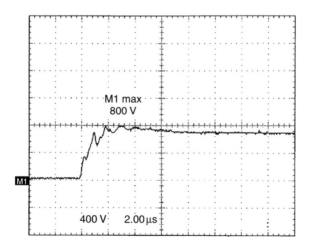

Figure B2.4   Motor terminal voltage with terminating unit (note different timescale)

Table B2.1 Relative costs of alternative techniques

Motor rating (kW)	Drive (%)	Motor (%)	Output inductor (%)	dv/dt filter (%)	Sinusoidal filter (%)	Terminator (%)
(400 V supply)						
2.2	350	100	74	443	334	170
75	220	100	14	99	146	9
250	120	100	5	65	Not practical	3

A precaution should be taken in that the unit must have an enclosure rating (e.g. IP number or NEMA category) suitable for the motor application.

There are no other benefits to using this system. There are, however, the following disadvantages:

- the control of some kinds of flux vector or other closed-loop controllers may be affected;
- there may be power loss;
- there may be additional cost and inconvenience of motor terminal mounting.

3. *Output filters.* More advanced output devices are available, in the form of dv/dt filters and sinusoidal filters. They have similar benefits for motor terminal voltage, but as they are relatively expensive they are unlikely to be cost-effective unless they are also needed for other reasons.

Table B2.1 shows some relative costs of typical examples of these alternative techniques. From this it may be concluded that output inductors are the most economic measure for systems rated up to about 70 kW, beyond which terminators become more attractive.

Technical explanation of the phenomena

1. *Review of PWM principles.* The output voltage of the drive is a series of pulses with a magnitude of either $+V_{dc}$ or $-V_{dc}$, where V_{dc} is the drive d.c. link voltage, with pulse width modulation (PWM). Because the motor load has inductance, the current flowing, and the magnetic flux in the motor, comprise mainly the underlying fundamental of the PWM with a small ripple component at the switching frequency. Figure B2.5 illustrates in greatly simplified form a part of the output voltage waveform, with the associated motor magnetic flux.

    V_{dc} is typically about 1.35 times the rms supply voltage, for example 540 V with a 400 V three-phase a.c. supply.

    Typical frequencies and times would be as shown in Table B2.2.

    Note the timescales. The rise time is five orders of magnitude shorter than the output period.

    Drive designers generally aim to use the highest practical switching frequency, because this has a variety of benefits including reducing the audible noise from

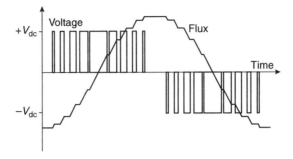

Figure B2.5 PWM inverter output voltage and current waveforms

the motor. This means they are constantly seeking to use faster power switching devices, which give lower switching losses through shorter rise times.

All of the pulse edges in Figure B2.5 have amplitude equal to the d.c. link voltage. Standard PWM controllers only generate these unipolar pulses. Some special control schemes without PWM modulators can generate bipolar pulses, which change from $+V_{dc}$ to $-V_{dc}$ in one transition.

2. *Motor voltage.* The PWM pulse rise times are so short that the time for the pulse to travel down the motor cable can easily exceed the rise time. For example, the velocity of the pulse is typically 1.7×10^8 m s^{-1}, so in 100 ns it has travelled only 17 m. When this happens, analysis needs to rely on transmission line theory. Full details are beyond the scope of this book, but the essential mechanism is as follows:
   - At each pulse edge the drive has to charge the inductance and capacitance of the cable, so a pulse of energy is delivered into the cable. The pulse travels at a velocity which is characteristic of the cable and is typically 1.7×10^8 m s^{-1}.
   - When the edge reaches the motor terminals, a reflection occurs because the motor surge impedance is higher than that of the cable (this is true for most low-voltage motors, although the impedance does fall as the motor rating increases). The voltage tends towards double the step magnitude; i.e. there is an overshoot approaching 100 per cent.
   - The reflection returns to the drive where it is again reflected, but in the negative sense because of the low impedance of the drive.
   - When this second reflection returns to the motor terminals, it cancels the over-voltage. Therefore the overshoot lasts for about twice the time of flight in the cable. If the rise time of the pulse is longer than twice the time of flight in the cable, then the overshoot is cancelled before it reaches 100 per cent.

Table B2.2 *Typical switching frequency/rise-time relationships*

	Frequency (Hz)	Period/time
Power output	50	20 ms
Switching	3 000	333 μs
Pulse rise time	–	100 ns

For a single pulse of magnitude V_{dc}, regardless of the motor cable length the overshoot can never exceed 100 per cent of V_{dc}. However, the duration of the overshoot does increase with increasing cable length.

For an ideal lossless cable, the rise time of the pulse is maintained along the cable so that the rate of change of voltage at the motor terminals (dv/dt) approaches twice that at the drive. However, in practice the cable exhibits high-frequency loss, which causes an increase in the rise time. This also means that the rise time at the motor terminals is fixed mainly by the high-frequency behaviour of the cable, so that contrary to statements sometimes made it is not the case that the introduction of new faster-switching power semiconductor devices increases the stress on the motor.

The account given above relates to a simple pulse travelling in a single-phase cable. The real situation typically has three-phase cores in a cable with surrounding screen, and multiple propagation modes must be considered. The key principles are unchanged, however.

Note that bipolar pulses, produced when two phase outputs change state in opposite directions simultaneously, have pulse edge magnitudes of $2V_{dc}$. These are also increased by 100 per cent so that the total voltage during the reflection is then $4V_{dc}$. These can be generated by some kinds of drives with direct torque control, although in most implementations of that system restrictions on the modulation are imposed to avoid such double transitions and the resulting very high over-voltage.

Figure B2.6 shows some typical measured voltage waveforms, which show the effect in practice. Even with 4 m of cable some overshoot is apparent. With 42 m the overshoot is virtually 100 per cent.

3. *Motor winding voltage.* The voltage overshoot has little effect on the main motor insulation systems between phases and from phase to earth, which are designed to withstand large overvoltage pulses. Typical dielectric strengths for motors of reputable origin are about 10 kV. However, some small low-cost motors may have had economies made in the interphase insulation, which can lead to premature insulation failure.

Owing to its short rise time, the pulse also affects the insulation between turns, and especially between coil ends. The voltage pulse travels around the motor winding as it does along the motor cable. Figure B2.7 illustrates in a simplified lossless model how this results in a large part of the pulse appearing across the ends of a coil during the time between it entering one end and leaving the other.

In practice even in the largest low-voltage motors the voltage between electrically adjacent turns is insignificant, but between the ends of the coil it may briefly reach a substantial part of the pulse magnitude. In this simplified illustration the entire pulse voltage appears across the coil. In practice, magnetic coupling between turns reduces this. Figure B2.8 gives a summary of the results of measurements made with a range of rise times on a variety of motors.

With a sinusoidal supply voltage the coil ends only experience a fraction of the phase voltage, as determined by the number of series coils. With a drive therefore there is a considerable increase in the voltage stress between the coil ends.

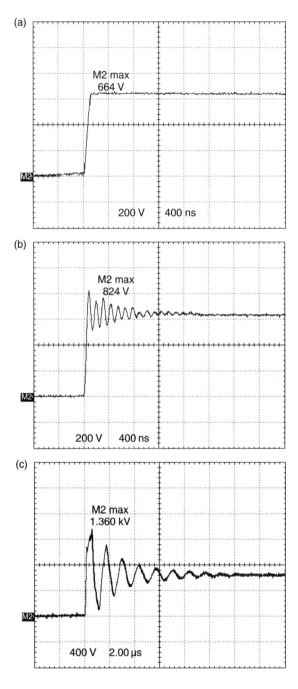

Figure B2.6 *Motor terminal voltage waveforms for varying cable lengths (note scale changes): (a) cable length = 0.5 m, (b) cable length = 4 m, (c) cable length = 42 m*

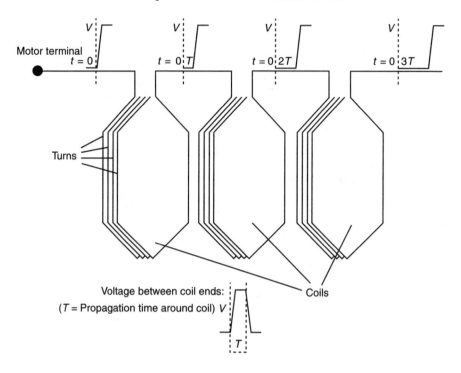

Figure B2.7   Propagation of pulse through motor windings

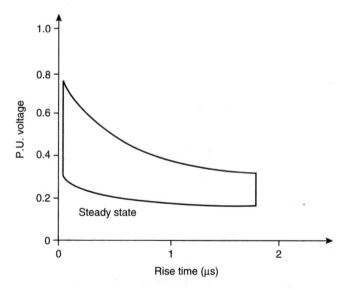

Figure B2.8   First coil voltage distribution against incident-voltage rise time

The effect of this depends on the motor construction. Large motors using form winding are constructed so that the coil ends are not in contact. The inter-turn insulation does not then experience the high-voltage pulse. Smaller random-wound motors may however have coil-end wires in contact, so special attention is then required regarding the quality of the inter-turn insulation.

4. *Motor inter-turn insulation design.* Modern motors of good-quality manufacture use advanced winding wire, which has a multilayer insulation system, and is easily capable of withstanding repetitive peak voltages of the order of 1 400 V, as could be generated by a drive with a 500 V a.c. supply. In the United States, where supplies of this level are common, many motor manufacturers routinely use an inverter-grade wire with further enhanced insulation, withstanding at least 1 600 V.

There is a possibility of a low-energy electrical discharge effect called partial discharge, which can occur in voids between wires. This is because of the electric field concentration in such voids where the permittivity of the gas or air is lower than that of the insulation material. At every pulse edge a small discharge of energy occurs, which may gradually degrade the insulation system. If the effect is excessive, the motor fails prematurely with an inter-turn fault. Resin impregnation, particularly under vacuum in order to minimise voids, suppresses this effect, as well as contributing to the physical stability of the winding under high mechanical stress or vibration.

For a.c. supply voltages higher than 500 V, further measures are required to prevent partial discharge. Inverter-rated motors use inverter-grade winding wire, which is resistant to partial discharge, as well as multiple impregnation regimes to minimise voids, and enhanced inter-phase insulation.

Motor standards
The international standard IEC 60034-17 gives a profile for the minimum withstand capability of a standard motor. It is in the form of a graph of peak terminal voltage against voltage rise time. This replaced the old IEC34-17 standard, which gave a rather arbitrary 500 V μs^{-1} limit without a clear rationale. The new standard is based on research into the behaviour of motors constructed with the minimum acceptable level of insulation within the IEC motors standard family. There is a great deal of published technical information on this subject. One of the best descriptions is contained in a paper written by workers at Dresden University who carried out a major research exercise on the subject (Kaufhold et al. 1996).

Tests show that standard PWM drives with cable lengths of 20 m or more produce voltages outside the IEC 60034-17 profile. However, most motor manufacturers produce standard motors with a capability substantially exceeding the requirements of the standard. Figures B2.9 and B2.10 give the actual requirements for supply voltages up to and exceeding 500 V, respectively. Standard motors are widely available to meet the requirements of Figure B2.9. Usually a special 'inverter-rated' motor is required to meet the requirements of Figure B2.10. Such motors carry a price premium of typically 3 to 10 per cent depending on the rating.

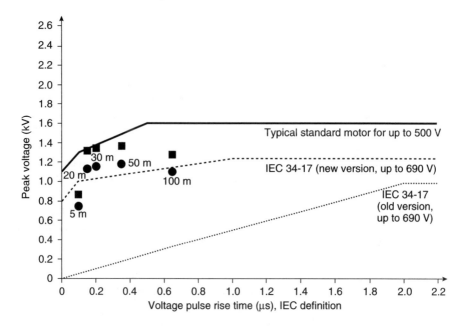

Figure B2.9   IEC limits, manufacturers' limits and measurements

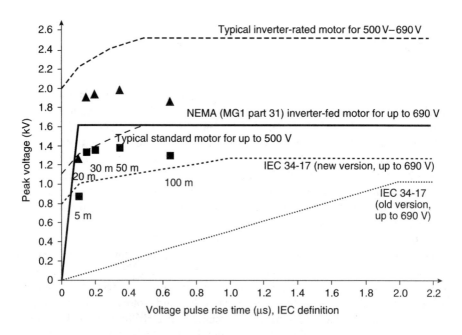

Figure B2.10   Limits and measurements for motors rated over 500 V, and NEMA MG1

Figure B2.9 also gives some measured voltages for a typical system, showing that they exceed the IEC60034-17 limits but they do not exceed the capability profile of a typical standard motor from a well-known manufacturer.

This graph illustrates clearly the effect of lengthening the motor cable. The rise time increases steadily with increasing length, while the overshoot falls off after a peak at about 50 m. The voltage stress on the motor therefore falls above quite moderate cable lengths.

NEMA publishes similar limits in the United States in MG1 part 31, shown in Figure B2.10. The practical measurements suggest that these limits are insufficient for drives operating much above 500 V. However, inverter-rated motors are readily available with much improved capability, as shown.

B2.3.4 Bearing currents

With a balanced three-phase sinusoidal supply, the sum of the three stator currents in an a.c. motor is zero and there is no further current flow outside the motor. In practice, however, there are conditions that may result in currents flowing through the bearings of a.c. motors even when fed with a sinusoidal 50 or 60 Hz supply, and the risk is further increased when using an inverter supply.

B2.3.4.1 Root causes of bearing currents

Magnetic asymmetry

It is well understood that an asymmetric flux distribution within an electrical machine can result in an induced voltage from one end of the rotor shaft to the other. If the bearing 'breakover voltage' is exceeded (the electrical strength of the lubricant film being of the order of 50 V) or if electrical contact is made between the moving and fixed parts of the bearing, this will result in a current flowing through both bearings. The current is of low frequency and its amplitude is limited only by the resistance of the shaft and bearings, so it can be destructive. In some large machines it is common and good practice to fit an insulated bearing, usually on the non-drive end, to stop such currents flowing.

This mains-frequency issue is well understood and with modern motors such problems are rare.

Supply asymmetry

An ideal power supply is balanced and symmetrical. Furthermore, the 'neutral' is at zero potential with respect to the system earth. With all modern PWM inverter supplies, although it can be assumed that the wanted component of the supply feeding the motor is indeed balanced and symmetrical, it is instantaneously unbalanced when pulses of different widths are produced. The resulting 'neutral voltage' is not zero with respect to earth, and its presence equates to that of a common-mode voltage source. This is sometimes referred to as a zero sequence voltage. It is proportional in magnitude to the d.c. link voltage (itself proportional to the supply voltage), and has a frequency equal to the switching frequency of the inverter.

This common-mode voltage will lead to the flow of currents through stray impedances between the inverter phase connections and earth. It can result in high-frequency current in the bearings. Generally such current is limited in amplitude and duration by its relatively high source impedance, so that it does not result in rapid severe damage, but it can cause premature bearing wear out by continual spark erosion of the bearing surfaces. There are three related but distinct mechanisms:

1. *Direct capacitive coupling from the stator winding to the rotor body.* The winding is largely surrounded by stator core, but there is some capacitance from the winding at the stator slot ends to the rotor body, so the rotor acquires a charge at each common-mode voltage pulse edge. Some designs for screening arrangements that suppress this effect have been published. It should be noted that the level of charge involved is small and there is no evidence of this mechanism causing bearing wear out in standard motors.
2. *Motor earthing current flowing through the shaft.* The motor earth current comprises relatively large current pulses because the capacitance from the winding to the frame is relatively high. Normally this current flows harmlessly into the earth connection of the motor. Under some abnormal conditions these current pulses may flow through the motor shaft via the bearings into the driven machinery. In this case, damage to the motor bearings can result, as well as to the bearings of the driven machinery and possibly other sliding surfaces such as gear teeth. Normally this does not happen, even if the earthing arrangement for the motor is poor, because the structure of the driven machinery is naturally and intentionally in electrical contact with the motor frame. If this is not the case, for example if the motor is mounted on a concrete base and not fixed to the driven machine, and if the earthing arrangement is not particularly successful in returning high-frequency current from the motor frame to the inverter, then bearing current may result. The current in this case is quite high; pulses of the order of 2 A or more are quite common, and serious damage can occur rapidly, e.g. in a period of a few weeks.
3. *Circulating current.* The high-frequency current flowing from the motor winding stray capacitance to the stator core is not uniformly distributed, and has a component that is circulating in an axial plane. This results in a magnetically induced emf in the loop comprising the shaft, bearings and stator frame. The effect is strongly dependent on physical dimensions, and for larger machines, having a shaft height over 280 mm, the induced emf may be sufficient to break down the insulation of the lubricating film and result in a circulating current.

Note that mechanisms 2 and 3 are functions of the current caused by the charging of stray capacitance between the winding and the stator core, and are therefore sensitive to the rate of change of the winding voltage. They can therefore be influenced by filtering the voltage. Mechanism 1 is simply a function of the winding voltage and the relative capacitances from the winding to the rotor and the rotor to the stator and frame, and is not amenable to straightforward filtering, although some form of common-mode filtering is possible.

B2.3.4.2 Good practices to reduce the risk of bearing currents

The first thing to remember is that although a great deal has been discussed and published on the subject of bearing currents associated with inverter-fed motors, damage from this cause is in practice a rare event. However, any motor may be subject to bearing currents if its shaft is connected to machinery at a different ground potential from the motor frame.

The first priority is to avoid the risk of shaft grounding current, by ensuring that the motor frame is connected through a low-inductance route to the structure of the driven machinery. If this connection is not made naturally though the construction of the machine, then a short direct electrical bond connection must be made between the motor frame and the machine.

The general precautions that are recommended for good EMC behaviour also have the benefit of minimising the tendency of high-frequency current pulses to propagate around the earthing system, including through the bearings. As a minimum, there must be a dedicated earth conductor that follows the same route as the phase conductors between the inverter and the motor. The magnetic coupling between the common-mode current and the earth conductor then tends to encourage the current to return through the desired route. A screened cable should preferably be used, which gives virtually 100 per cent magnetic coupling between the phase conductors and the earth (screen) provided that the screen is connected directly to the relevant metal parts at both ends.

For motor frame sizes exceeding 280 mm the recommended method for avoiding circulating bearing current is to use an insulated bearing at the non-drive end. Motors specified for inverter duty will normally be supplied with the appropriate precautions incorporated. In this case any shaft transducer at the non drive end (NDE) will also need to have insulation to avoid creating an alternative bearing earth path.

In all cases care must be taken with the terminations of the motor cable screen. All terminations must be low resistance and low inductance or the benefit will be reduced.

It has been suggested that the motor cable should preferably be symmetrical; either it should not contain an earth core or alternatively it should contain three cores in a symmetrical arrangement relative to the power cores. Tests carried out at Control Techniques have not verified any benefit from this, but there may be some benefit in reducing the short-term motor terminal voltage unbalance, and hence reducing the tendency to circulating bearing current.

In situations where these precautions cannot be taken, for example where the motor or its installation are inaccessible, the risk of excessive bearing current can also be reduced by filtering the inverter output. A low-pass filter must be used to increase the rise time of the pulses arriving at the motor, and this filter must be effective in the common mode. It cannot be assumed that a 'dv/dt filter' is effective in the common mode, because simpler designs that are intended to reduce motor winding voltage stress primarily operate on the voltage between phase lines. A simple form of filter using a ferrite ring common-mode inductance at the inverter, and taking advantage of the motor cable capacitance, gives a useful reduction in high-frequency common-mode current. Specialist filter manufacturers offer common-mode filters that are effective for radio-frequency interference and bearing current.

B2.3.5 Overspeed

Most standard industrial motors may be capable of operating at speeds above their 50/60 Hz rated speed, but it is important to have the manufacturer's assurance on the suitability of any motor for operation above base speed. The bearings and type of balancing of the standard rotor dictate a maximum mechanical speed, which cannot be exceeded without endangering the motor or its expected life. The following issues should be considered by the motor manufacturer when sanctioning use in the 'overspeed' range.

- Mechanical stress at the rotor bore and assurance that the shaft to core fit is secure should be considered.
- Bearing life, which is a function of the speed for anti-friction bearings, should be examined. Each bearing has overspeed and temperature limits that need to be reviewed.
- Bearing lubrication, which is also a function of speed and operating temperature, should be assessed. Grease may not adhere properly and oil may churn or froth.
- Vibration, which is a function of the square of the angular velocity, should be considered. Care must be given to ensure not to operate near system-critical or natural frequencies.
- Airborne noise can be dramatically increased at higher speeds.
- Winding stress caused by vibration of the windings at high frequency may require additional winding bracing and treatment.
- Balancing weights affixed to the rotor or fan assembly should be attach properly. At high speeds shear stress levels may be exceeded.

Table B2.3 Maximum motor speeds and balancing for L-S MV (2/4/6 pole) motors

Motor type	Maximum speed (min^{-1})	Balancing level
80	15 000	S
90	12 000	S
100	10 000	S
112	10 000	S
132	7 500	S
160	6 000	R
160 LU	5 600	R
180	5 600	R
200	4 500	R
225 ST/MT/MR	4 100	R
225 M/MK	4 100	R
250	4 100	R
280 SP	3 600	R
315	3 000	R

- Methods of shaft coupling should be reviewed. This would apply to any other auxiliary devices attached to the motor shaft, notably including speed and position transducers.
- The speed rating and energy absorption capability of brakes should be reviewed.
- Acceptable internal stress levels and fits of cooling fans should be maintained.
- Decreased motor efficiency caused by increased losses should be assessed.
- Motor torque capability at increased speed should be considered.

As an example, Table B2.3 shows the maximum speeds that can be tolerated by Leroy-Somer MV motors in horizontal and vertical operation, directly coupled to the load and with no radial or axial loading.

B2.4 Motors for hazardous (potentially flammable or explosive) locations

Motors are generally available with approvals for use in hazardous locations, complying with, for example, the European ATEX or North American Hazloc regulations. Because of the factors described above, any approval of a motor for use in a hazardous atmosphere cannot be taken to apply when it is inverter fed unless this is specifically stated on the safety name-plate. The main risk to be considered is of excessive temperature, because of the increased losses and the reduced effectiveness of the cooling fan. However, the adequacy of the insulation for the high peak voltages must also be considered. The motor safety data should specify any special requirements for the installation; for example, there will normally need to be measures with demonstrable high reliability to be incorporated for protection against excessive motor temperature.

The regulations for hazardous locations are complex. Independent guides are available, for example, from the UK trade association GAMBICA.

Chapter B3
Physical environment

B3.1 Introduction

This section looks at the environmental requirements, conditions and impacts that motors and drives have to operate within. Particular focus will be placed upon

- enclosure and the protection it affords,
- mounting arrangements and standards,
- terminal markings and direction of rotation,
- operating temperature,
- humidity and condensation,
- noise,
- vibration,
- altitude,
- storage, and
- corrosive gases.

B3.2 Enclosure degree of protection

B3.2.1 General

Both motor and drive enclosures are usually described with reference to the Ingress protection classifications laid out in IEC 60034-5. The designation used as defined in IEC 60034-5/EN 60034-5 consists of the letters IP (International Protection) followed by two numerals signifying conformance with specific conditions. Additional information may be included by a supplementary letter following the second numeral. (Interestingly this system is contained within NEMA MG 1 but is not widely adopted by the industry in the United States.)

IEC 60034-5 stipulates the format shown in Figure B3.1 for the definition of protection classifications. The first characteristic numeral indicates the degree of protection provided by the enclosure with respect to persons and also to the parts of the machine inside the enclosure (see Table B3.1).

The second characteristic numeral indicates the degree of protection provided by the enclosure with respect to harmful effects due to ingress of water (see Table B3.2).

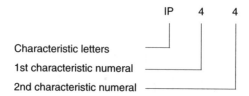

Figure B3.1 Format for the definition of protection classifications

B3.2.2 Motor

B3.2.2.1 General

Electrical machines are available in a wide number of different enclosures and classifications. Large users of motors can specify customised and application-specific enclosures, but for most users they will be selecting from established standard designs.

For open internally air-cooled machines suitable for use under specific weather conditions and provided with additional protective features or processes, the letter 'W' may be used. This additional letter is placed immediately after the IP e.g. IPW 54. Similarly the letter 'R' is used to indicate duct-ventilated machines (in such cases the air discharge must be located outside the room where the motor is installed).

The most frequently used degrees of protection for electrical machines are summarised in Table B3.3.

Brushless servomotors are normally IP65 or IP66. This single feature is often the key reason for a user to select brushless servomotors for specific applications such as in the food industry where washdown is a common requirement.

B3.2.2.2 US practice

Although the IP classifications are known in the United States, it is common practice for manufacturers of electrical machines there to adopt less formal designations. Typical designations are summarised in Table B3.4.

B3.2.3 Drive

The use of air for cooling in the majority of industrial drives means that most drives are of ventilated design. They are susceptible therefore to degradation by solid contaminants. Dust and fibres, which may be released by industrial processes, must be prevented from contaminating cooling fans and surfaces, and an excessive accumulation will typically cause an over-temperature trip. Electrically conductive contaminants such as metal particles or carbon are clearly hazardous for drive reliability, and a combination of hygroscopic contaminants with damp conditions can have a similar effect.

Access to operating drives must be considered in the context of electrical safety. The system of ingress rating ('IP code') covers protection against ingress both of human-generated materials such as the finger or metal rod, or of environmentally generated materials such as dust or water. This blending of safety and environmental

Table B3.1 The first characteristic numeral

First characteristic numeral	Brief description	Definition
0	Non-protected	No special protection.
1	Protected against solid objects greater than 50 mm diameter	No accidental or inadvertent contact with or approach to live and moving parts inside the enclosure by a large surface of the human body, such as a hand (but no protection against deliberate access). Ingress of solid objects exceeding 50 mm in diameter.
2	Protected against solid objects greater than 12 mm in diameter	No contact by fingers or similar objects not exceeding 80 mm in length with or approaching live or moving parts inside the enclosure. Ingress of solid objects exceeding 12 mm in diameter.
3	Protected against solid objects greater than 2.5 mm in diameter	No contact with or approaching live or moving parts inside the enclosure by tools or wires exceeding 2.5 mm in diameter. Ingress of solid objects exceeding 2.5 mm in diameter.
4	Protected against solid objects greater than 1 mm in diameter	No contact with or approaching live or moving parts inside the enclosure by wires or strips of thickness greater than 1 mm in diameter.
5	Dust protected	No contact with or approaching live or moving parts within the enclosure. Ingress of dust is not totally prevented but dust does not enter in sufficient quantity to interfere with the satisfactory operation of the equipment.
6	Dust-tight	No contact with or approach to live or moving parts inside the enclosure. No ingress of dust.

Table B3.2 The second characteristic numeral

Second characteristic numeral	Brief description	Definition
0	Non protected	No special protection.
1	Protected against dripping water	Dripping water (vertically falling drops) shall have no harmful effects.
2	Protected against dripping water when tilted up to 15° from the vertical	Vertically dripping water shall have no harmful effect when the equipment is tilted at any angle up to 15° from the vertical, from its normal position.
3	Protected against spraying water	Water falling as a spray at an angle up to 60° from the vertical shall have no harmful effect.
4	Protected against splashing water	Water splashing against the equipment from any direction shall have no harmful effect.
5	Protected against water jets	Water projected by a nozzle against the equipment from any direction shall have no harmful effect.
6	Protected against heavy seas	Water from heavy seas or water projected in powerful jets shall not enter the equipment in harmful quantities.
7	Protected against the effects of immersion in water to depths of between 0.15 and 1 m	
8	Protected against the effects of prolonged immersion at depth	

data does cause confusion, because although the IEC standard makes no claim to be a product safety standard there is a tendency to use the code as an indicator for the kind of installation expected. For example, many modular drives that are intended for location within a closed panel or in a closed electrical operating area are designed to IP20, whereas those designed for exposure to a more aggressive industrial

Table B3.3   Degrees of protection for electrical machines

First numeral	Second numeral								
	0	1	2	3	4	5	6	7	8
0	IP00		*IP02*						
1		*IP11*	*IP12*	*IP13*				*IP17*	*IP18*
2		IP21	IP22	IP23					
3									
4					IP44				
5					IP54	IP55	*IP56*		
6						IP65	IP66		

Note: Italics indicate machines are available with this degree of protection but are somewhat unusual.

Table B3.4   Degree of protection designations for electrical machines in the United States

Open drip proof (ODP)	A machine in which the ventilating openings are so constructed that successful operation is not interfered with when drops of liquid or solid particles strike or enter the enclosure at any angle from 0° to 15° downward from the vertical. These are motors with ventilating openings, which permit passage of external cooling air over and around the windings.
Totally enclosed fan cooled (TEFC)	The TEFC-type enclosure prevents free air exchange but still breaths air. A fan is attached to the shaft, which pushes air over the frame during operation to help the cooling process.
Totally enclosed air over (TEAO)	The TEAO enclosure does not utilise a fan for cooling, but is used in situations where air is being blown over the motor shell for cooling such as in a fan application. In such cases the external air characteristics must be specified.
Totally enclosed non-ventilated (TENV)	A TENV type enclosure does not have a fan.
Washdown duty (W)	An enclosure designed for use in the food-processing industry and other applications that are routinely exposed to washdown, chemicals, humidity and other severe environments.

environment and to industrial operators are designed to IP54. The IP codes become a kind of de facto specification, whereas for a proper assessment of the equipment safety a full knowledge of the risk factors in a particular environment is needed.

IP20 cabinets are readily achievable using standard cubicle design techniques and following the drive manufacturer's installation guidelines.

IP54 cubicles for small drives, where the losses and so heat to be dissipated in the panel are small, are also straightforward to design, but large drives where a lot of energy needs to be dissipated are more of a challenge. Some manufacturers have designed their drive product such that the drive can be mounted in the panel with the heat sink protruding out of the back of the enclosure. In this way the bulk of the heat is not dissipated inside the panel. As an example, a 55 kW Control Techniques Unidrive SP at rated output and 3 kHz switching frequency dissipates a total of 1 060 W, of which less than 20 per cent is dissipated into the panel when through-panel mounted. Alternatively air–air or water–air heat exchangers are readily available.

B3.3 Mounting arrangements

B3.3.1 Motor

B3.3.1.1 General

Internationally agreed coding applies to a range of standard mountings for electric motors, d.c. and a.c., and covers all the commonly required commercial arrangements.

Incorrect mounting of a motor can cause premature failure and loss of production. All motor manufacturers will provide advice on the suitability of a particular build for a specific application.

Within IEC 60034-7/EN 60034-7 there are examples of all practical methods of mounting motors. NEMA publishes alternative standards within NEMA Standards publication No. MG1 – *Motors and Generators*.

B3.3.1.2 IEC 60034-7 standard enclosures

Electrical machines have been categorised within this standard by the prefix IM (international mounting), a letter and one or two subsequent digits. It is unusual for the prefix IM to be used and it is more usual to see only a letter followed by one or two digits; e.g. B3 = foot mounting.

The most usual types of construction for small- and medium-sized motors are summarised in Figure B3.2.

B3.3.1.3 NEMA standard enclosures

Motor mounting and location of the terminal box location is designated in accordance with the arrangements summarised in Figure B3.3.

B3.3.2 Drive

The mounting arrangements for drives are clearly laid out by the drive manufacturer. Although the mechanical mounting of the drive itself merits little comment here, it

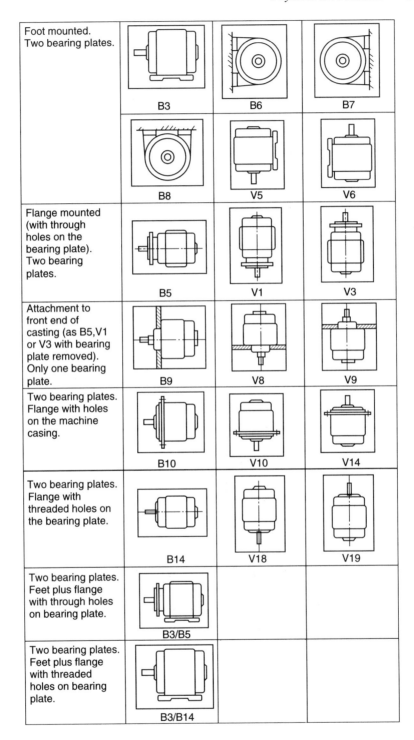

Figure B3.2   The most common types of construction for small and medium motors

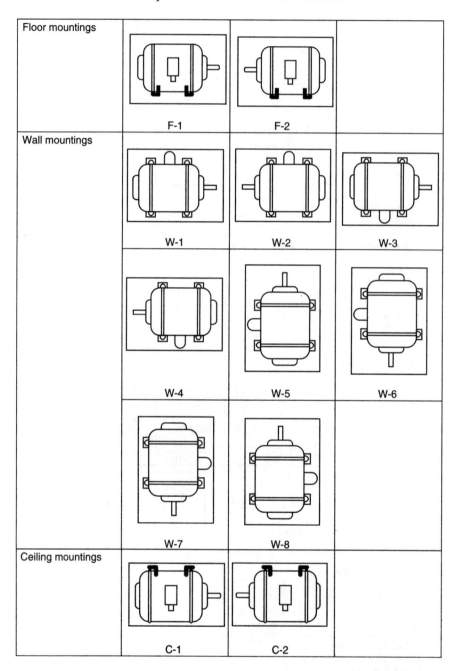

Figure B3.3   Motor mountings and terminal box location

should be recognised that although a drive is mounted in an enclosure in much the same way as other electrical control equipment, it does have a relatively high mass itself and there are a number of heavy components within the drive. Vibration is a consideration and can cause problems, not only in the application but also during transit of the enclosure.

B3.3.3 Integrated motor drive

In some applications it is convenient to mount the drive very close to the motor. Indeed it can be ideal to mount the drive on the motor itself. The design of such a product presents many challenges and opportunities, and there are a number of products on the market today, such as that shown in Figure B3.4. Mounting of these products should carefully follow the manufacturer's recommendations.

Figure B3.4   Integrated motor drive [photograph courtesy of Leroy–Somer]

B3.4   Terminal markings and direction of rotation

B3.4.1   Motor

B3.4.1.1   General

Terminal markings and directions of rotation are set out in the international standard IEC 60034-8/EN 60034-8. NEMA also defines the terminal markings in NEMA Standards Publication No. MG-1. For clarity, the two standards are discussed separately.

B3.4.1.2 IEC 60034-8/EN 60034-8

General

IEC 60034-8 describes the terminal windings and direction of rotation of rotating machines. A number of broad conventions are followed.

- Windings are distinguished by a CAPITAL letter (e.g. U, V, W).
- End points or intermediate points of a winding are distinguished by adding a numeral to the winding (e.g. U1, U2, U3).

Direction of rotation

IEC 60034-8 defines the direction of rotation as an observer facing the shaft of the motor (viewing from the drive end). In a.c. polyphase machines (without a commutator), the direction of rotation will be clockwise when the alphabetical sequence of the terminal letters of a phase group corresponds with the time sequence of the supply/terminal voltages.

Terminal markings for a.c. machines

The terminal markings for three-phase stator windings of synchronous and asynchronous machines are marked as shown in Figures B3.5 to B3.8.

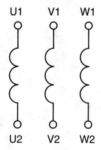

Figure B3.5   Single winding with six terminals

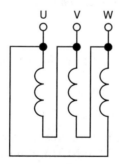

Figure B3.6   Delta connection with three terminals

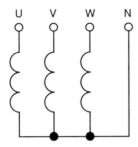

Figure B3.7 Star connection with four terminals

Figure B3.8 D.C. excitation winding of a synchronous machine with two terminals

Terminal markings for d.c. machines
The terminal markings for individual windings and specific d.c. machine types are detailed as follows.

- *Individual windings.* The terminal markings for d.c. commutator machines is marked as shown in Figures B3.9 to B3.13.
- *Compensated d.c. motor with compensating and commutating windings for clockwise rotation.* The terminal markings are as shown in Figure B3.14.
- *Compound d.c. motor with commutating windings for clockwise rotation.* The terminal markings are as shown in Figure B3.15.
- *Shunt wound d.c. motor: connections for clockwise rotation.* The terminal markings are as shown in Figure B3.16. Note that the direction of rotation will be clockwise, regardless of voltage polarity, if connections are made as in the figure.
- *Series d.c. motor: connections for clockwise rotation.* The terminal markings are as shown in Figure B3.17. Note that the direction of rotation will be clockwise, regardless of voltage polarity, if connections are made as in the figure.
- *Series d.c. generator: connections for clockwise rotation.* The terminal markings are as shown in Figure B3.18. Note that the direction of rotation will be clockwise, regardless of voltage polarity, if connections are made as in the figure.

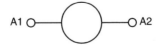

Figure B3.9 Armature winding with two terminals

Figure B3.10   Commutating winding with two terminals

Figure B3.11   Compensating winding with two terminals

Figure B3.12   Series excitation winding with two terminals

Figure B3.13   Separately excited field winding with two terminals

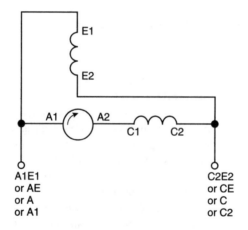

Figure B3.14   Compensated d.c. motor

B3.4.1.3   NEMA

General
NEMA MG-1 provides for the general guidance for terminal markings for motors, generators and their auxiliary devices as detailed in Table B3.5.

A.C. motor direction of rotation
NEMA MG-1 states quite clearly that terminal markings of polyphase induction machines are not related to the direction of rotation.

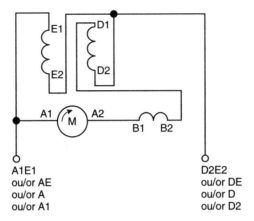

Figure B3.15   Compound d.c. motor

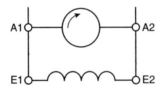

Figure B3.16   Shunt wound d.c. motor

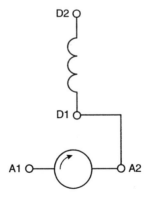

Figure B3.17   Series d.c. motor

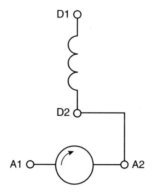

Figure B3.18 Series d.c. generator

Table B3.5 NEMA MG-1 general terminal markings

Armature	A1, A2, A3, A4, etc.
Brake	B1, B2, B3, B4, etc.
A.C. rotor windings (slip rings)	M1, M2, M3, M4, etc.
Capacitor	J1, J2, J3, J4, etc.
Control signal lead attached to commutating winding	C
Dynamic braking resistor	BR1, BR2, BR3, BR4, etc.
Field (series)	S1, S2, S3, S4, etc.
Field (shunt)	F1, F2, F3, F4, etc.
Line	L1, L2, L3, L4, etc.
Magnetising winding (for initial and maintenance magnetisation and demagnetisation of permanent magnet fields)	E1, E2, E3, E4, etc. Note: E1, E3 and other odd-numbered terminals should be attached to the positive terminal of the magnetising supply for magnetisation
Resistance (armature and miscellaneous)	R1, R2, R3, R4, etc.
Resistance (shunt field adjusting)	V1, V2, V3, V4, etc.
Shunt braking resistor	DR1, DR2, DR3, DR4, etc.
Space (anti-condensation) heaters	H1, H2, H3, H4, etc.
Stator	T1, T2, T3, T4, etc.
Starting switch	K
Thermal protector (e.g. thermistor)	P1, P2, P3, P4, etc.
Equalising lead	= (equals sign)
Neutral connection	Terminal letter with numeral 0

For synchronous machines numerals 1, 2, 3 and so on indicate the order in which the terminals reach the maximum positive values (phase sequence) with clockwise shaft rotation.

Again, it is best to consult with the manufacturer.

D.C. motor direction of rotation

The standard direction of rotation of the shaft for d.c. motors is anti-clockwise (counter-clockwise) as viewed from the end opposite the motor shaft.

The direction of rotation depends upon the relative polarities of the field and armature, so if the polarities are both reversed then the direction of rotation will be unchanged. Reversal can be obtained by transposing the two armature leads or the two field leads.

The standard direction or rotation for generators will of course be the opposite of the case for motors.

Terminal markings for d.c. machines

The connections for the three main types of d.c. motor are as follows.

- *Shunt motors.* See Figures B3.19 and B3.20.
- *Series motors.* See Figures B3.21 and B3.22.
- *Compound motors.* See Figures B3.23 and B3.24.

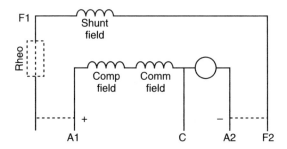

Figure B3.19   Shunt wound motor anti-clockwise rotation facing non-drive end

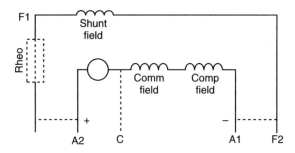

Figure B3.20   Shunt wound motor clockwise rotation facing non-drive end

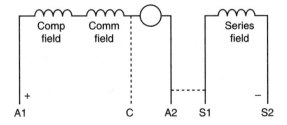

Figure B3.21   Series wound motor anti-clockwise rotation facing non-drive end

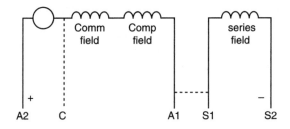

Figure B3.22   Series wound motor clockwise rotation facing non-drive end

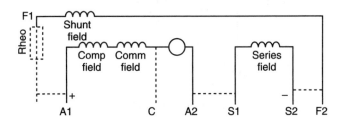

Figure B3.23   Compound wound motor anti-clockwise rotation facing non-drive end

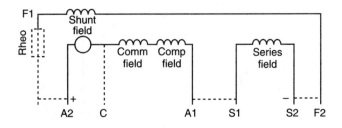

Figure B3.24   Compound wound motor clockwise rotation facing non-drive end

B3.4.2 Drive

The normal convention for the input or supply terminals for a drive are L1, L2, L3, and the output terminals will normally be identified to match the motor terminals to which there are intended to be connected, which for a typical a.c. motor would be U, V, W.

B3.5 Ambient temperature

B3.5.1 Motor

One of the key factors that determines the rating of an electric motor is the temperature at full load of its active materials. The permissible temperature rises for various classes of insulation materials are specified in the standards. Both NEMA and IEC use 40 °C as the base ambient temperature.

The life of a motor is equal to the life of its winding insulation (disregarding the wear of bearings, brushes, slip rings or commutator, which can be replaced at relatively low cost). Any service conditions influencing temperature rise and thus the condition of the insulation must be given particular attention.

If a motor is installed in an ambient temperature above its rated value, the permissible temperature rise will need to be reduced to keep the absolute value of the maximum temperature at its design level. This is a key consideration for all users, and high ambient temperatures must be discussed with the supplier in order to ensure high availability and long life.

The temperature rise in the motor results from the losses caused by the conversion of energy (electrical to mechanical), which can be expresses in the following equation:

$$P_{loss} = P_{elec} - P_{shaft}$$

In practice it is not the losses of a motor but its efficiency η that is quoted; this is calculated as

$$\eta = (P_{shaft} \times 100)/P_{elec}$$

The resulting energy losses are stored in the motor and the greater part is dissipated to the surrounding atmosphere by ventilation, the condition depending upon the heat storage capacity of the motor and the temperature rise. With constant load, the steady-state condition is reached when the amount of heat produced by the losses in the machine is equal to the heat dissipated. In continuous duty, this state of equilibrium is typically reached for industrial motors after about 3 to 5 hours. The resulting temperature rise of the winding and other parts of the motor is the difference between the temperature of the particular motor part and the coolant temperature. It may be determined from the increase in resistance of the winding:

$$\Theta = [(R_W - R_K)/R_K] \times (235 + t_{cold} - t_{coolant})$$

where Θ is the temperature rise of the winding (°C), t_{cold} is the temperature of the cold winding (°C), $t_{coolant}$ is the temperature of the coolant (°C), R_W is the motor winding resistance at operating temperature (Ω), and R_K is the motor winding resistance when cold (Ω).

As a rule of thumb, the life of typical winding insulation decreases by about 50 per cent for each 10 °C.

It should be noted that the frame temperature is neither a criterion for the quality of the motor nor for the temperature rise of the winding. An extremely 'cold' motor may have higher losses and higher winding temperatures than an extremely 'warm' motor.

B3.5.2 Drive

B3.5.2.1 Maximum operating temperature

The limited operating temperature capability of silicon power semiconductors (see Chapter A3) means that drives are always temperature sensitive. The industry convention is for the drive to be rated for continuous operation at an ambient temperature of 40 °C, often with a defined de-rating for 50 °C. This de-rating can be significant, and drive manufacturers will be able to supply specific information on this. The de-rating is usually switching frequency dependent.

The ambient temperature in which a drive operates is defined as the temperature in which the drive itself sits. If the drive for which an ambient temperature is stated is mounted inside an enclosure, then the ambient temperature referred to is the temperature at the point within the enclosure where the drive is mounted. Typical general practice for enclosure design is to have the internal temperature no more than 10 °C higher than the external temperature.

The drive itself may significantly affect the ambient temperature. Typical efficiency for a PWM drive at full load is 98 per cent, so allowance must be made for the effect of the 2 per cent power loss on the ambient temperature. The cost of providing for this loss, for example in ventilation or air-conditioning capability, may be high, so for multi-drive systems it may be important to use a realistic operating profile. The single largest component of loss in the drive is the inverter stage losses (switching and conduction losses), so the loss depends on the motor current, not on the motor power, as is the case with supply harmonic current.

B3.5.2.2 Minimum operating temperature

The minimum operating temperature is determined by a number of factors. This is usually the minimum operating temperature of the electronic components. It should be noted that although some components appear to work at very low temperatures, their performance is significantly impaired, and so the manufacturer's recommendations should always be followed. A good example of this is the Hall effect current sensors, which have characteristics that are very non-linear at low temperatures. This means that although a drive may well start at low temperatures the auto-calibration of the current sensors frequently undertaken at power up would be inaccurate and would result in poor performance when the drive temperature increased.

It is usual for drives to be specified for operation down to 0 °C. A second, start-up minimum temperature is quoted by some manufacturers, but in this case a recommendation is usually made to re-start the drive once it has warmed up to the operating minimum temperature.

For drives mounted in an enclosure the use of small panel heaters is a simple solution to overcome minimum temperature issues.

B3.6 Humidity and condensation

B3.6.1 Motor

Condensation, which occurs when moist air temperature is reduced to the dew point, is a problem that can seriously affects electrical machines. Where bare terminals are exposed to condensation, tracking can occur, leading to flashover between terminals. Hygroscopic insulations suffered badly under moist conditions.

Non-hygroscopic insulation systems have largely eliminated this problem; however, condensation can still cause significant problems. In some cases bearing seals act to trap moisture into the motor housings and quite a substantial amount of water can accumulate there. Drain holes at the lowest point of the enclosure are frequently included in the design to allow the water to drain away. These drain holes are usually sealed with plugs that must be removed and then refitted as part of the routine maintenance procedure. Different types of plug are used, including a screw, a siphon, and a breather. For certain applications it is advisable to leave the drain hole open, though this does have an impact on the IP rating of the motor.

B3.6.2 Drive

Whereas modern insulation materials maintain good performance over the entire range of air humidity, a complex electronic circuit cannot be made to operate safely and reliably in the presence of water, so internal condensation must be avoided. The risk of condensation occurs primarily when the drive power has been removed, and the temperature fluctuates because of external factors such as building heating or solar heating. Most drives must be protected from condensation, typically by providing some low-level heating when the drive is disconnected from the power supply. A sealed enclosure can prevent moisture ingress, but the difficulty of managing the operating temperature, and also of ensuring that the internal air is dry so that no condensation can occur at low temperature, results in fully enclosed and moisture-resistant designs carrying a considerable cost premium.

It is typical for industrial drives to be specified for a maximum humidity of 95 per cent non-condensing at 40 °C.

B3.7 Noise

B3.7.1 Motor

It is inevitable that even an economically designed and efficient electric motor will produce audible noise, due for example to magnetic torsions and distortions, bearings or air flow. The latter type of noise is most predominant in two-pole and four-pole machines with shaft-mounted fans, and most d.c. machines where a separately mounted fan is most common.

Procedures for testing for motor noise are clearly laid out in the standards. Although there are detailed differences between them, the principles are the same.

Referring to Figure B3.25, a series of background sound pressure readings are taken at the prescribed points. The motor will then be run on no load and at full speed. A.C. motors will be supplied at rated voltage and frequency. Synchronous machines will be run at unity power factor. Certain correction factors may be applied where the test reading is close to the background reading.

In most industrial applications the motor is not the predominant source of noise and the overall situation must be taken into account when making a noise analysis.

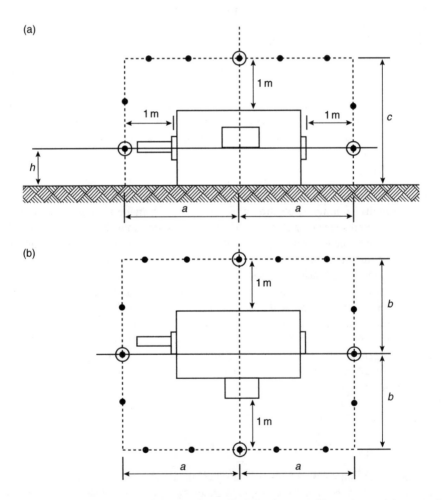

Figure B3.25   Location of measuring points for horizontal machine: (a) vertical plane; (b) horizontal plane

Many steps can be taken to reduce the noise of electrical machines:

- the use of oil lubricated sleeve bearings, which are much quieter than most other bearing types,
- careful choice of bearing lubricant, which can affect the noise of the machine,
- careful design of the machine air circuits, which will minimise ventilation noise.

Although the d.c. motor is generally less troublesome than the induction motor, there are applications where special action is needed. The following measures can be made to reduce noise:

- use of a reduced magnetic loading (lower flux densities),
- increasing the number of armature slots,
- skewing the armature slots (or, less commonly, the pole shoes),
- using continuously graded main pole gaps or flaring the gaps at the edges of the main pole,
- increasing the air gap,
- bracing the commutating poles against the main poles,
- using semi-closed or closed slots for the compensating winding,
- selecting the pitch of the compensating winding slots to give minimum variation in air gap permeance,
- using a 12-pulse rather than a six-pulse d.c. drive or fitting a choke/reactor in series with the machine to reduce the current ripple.

After the designer has taken whatever steps to minimise the noise generation at source, it may still be higher than is acceptable. To achieve the specification it may now be necessary to apply external silencing. This may be in the form of inlet or outlet air duct silencers or even the fitting of a complete enclosure.

The use of acoustic partitioning requires a good knowledge of at least the octave band sound pressure levels present in order that good silencing can be achieved. Airborne noise striking an acoustic partition will, like other forms of energy, be dissipated in various ways as shown in Figure B3.26.

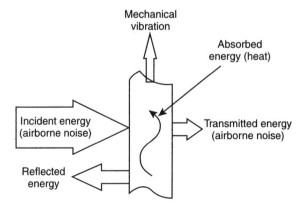

Figure B3.26   Energy flow in an acoustically excited partition

B3.7.2 Drive

Compared with the motor, the drive is a relatively unimportant source of noise and is usually easily managed. However, in certain installations it may be critical. The main noise source is the cooling fan, and the tendency for drive power density to be progressively increased has resulted in the increasing use of high-speed fans, which are relatively noisy. It is helpful for the fan speed to be controlled by the drive, because even a small reduction in speed results in a useful reduction in noise, as well as reduced power consumption and increased fan lifetime.

The other potential source of noise from a drive is wound components such as chokes. Correct detail design of these parts prevents excessive noise, but the possibility should not be overlooked. Chokes that operate with substantial ripple current over the frequency range where human hearing is most sensitive, e.g. about 1 to 5 kHz, can be difficult to design cost-effectively with low acoustic emission.

To give some scale to the issue, a 22 kW Control Techniques Unidrive SP gives worst-case acoustic noise levels at 1 m distance of 43–56 dBA.

B3.7.3 Motor noise when fed from a drive converter

As has already been discussed, electric motors produce noise when operated without variable-speed drives. The purpose of this section is to discuss the new factors that arise when a drive is applied.

The general trend when using variable speed is for the mechanically generated noise and vibration to fall as the speed falls and rise as it increases, because most of the driving forces are proportional to the square of speed, and the efficiency of coupling into the outside environment also falls as the frequency falls. The complicating factor is the possible existence of resonances. If the machine was originally designed for operation at fixed speed then some resonant frequencies may be unknown, or have been removed or adjusted to prevent their excitation under normal conditions. When the speed is varied then unexpected resonances may come to light.

Table B3.6 Principle driving frequencies

Drive type	Source of noise or vibration	Frequency	Behaviour
D.C. thyristor	D.C. current ripple	$np f_i$	Increases as voltage falls, due to increasing ripple current
A.C. quasi-square	Voltage harmonics	$n f_o$	Depends on voltage control technique
A.C. PWM	Switching frequency	$n_s f_s \pm n_o f_o$	Virtually constant if f_s is constant

p is the phase number of converter, f_i the drive supply frequency, n the harmonic order (evens and triplens absent), f_o the output frequency, f_s the PWM switching frequency, n_s the harmonic order of switching frequency, n_o the harmonic order of output frequency (for symmetrical PWM n_o is odd when n_s is even and vice versa).

New frequencies of magnetic excitation occur when using a drive, because of the switching action of the power semiconductors. For operation at a given speed and power, increases in acoustic noise between 1 and 5 dB have been observed with inverter supply according to IEC 60034-25:2007 (clause 13.3). When using PWM with a switching frequency of more than about 2 kHz, the weighted sound pressure or sound power emission from a motor at rated speed is affected little by the converter, although the sound can be subjectively more irritating than its weighted value suggests if it occupies the sensitive range of human hearing (about 1 to 5 kHz). Table B3.6 shows the principal driving frequencies of the additional magnetic flux that can be expected. The resulting noise and vibration has frequencies that are the results of modulation between the working magnetic flux and the additional flux.

Small machines generally have resonant frequencies well above the main exciting frequencies. For large machines consideration must be given to the following possible effects:

- Vibration
  - Shaft critical speed (including complete drive train)
  - Main structural resonance excited by mechanical unbalance (once per revolution) or magnetic excitation (twice per electrical cycle)
- Noise
  - Higher-order resonance excited by inverter switching frequency and related frequencies (or inverter harmonics for non-sinusoidal inverter outputs)
  - Higher-order resonance excited by rotor slot-passing frequency

Table B3.7   Methods of reducing the additional noise

Drive type	Source of noise or vibration	Technique	Comment
D.C. thyristor	D.C. current ripple	Additional d.c. inductance	Reduce ripple current
		Twelve-pulse or higher	Increase ripple frequency and reduce current; high cost
A.C. quasi-square	Voltage harmonics	Higher pulse number, multi-level	Increase ripple frequency and reduce current; high cost
A.C. PWM	Switching frequency	Increase switching frequency	May require drive de-rating
		Filter ('sinusoidal filter')	May not work with closed-loop control; high cost; drive performance may be reduced

In some applications the drive can be programmed to avoid undesirable speed or frequency ranges, sometimes referred to as 'skip frequencies'.

If the new noise frequencies introduced by the drive are unacceptable, improvements can be made either by changing the frequencies or by filtering. Table B3.7 summarises the main possibilities.

B3.8 Vibration

B3.8.1 Motor

Magnetic, mechanical and air-flow inaccuracies due to construction lead to sinusoidal and pseudo-sinusoidal vibrations over a wide range of frequencies. Other sources of vibration can also affect motor operation, such as poor mounting, incorrect drive coupling, end shield misalignment and so on.

The vibrations emitted at the operating frequency, correspond to an unbalanced load which swamps all other frequencies and on which the dynamic balancing of the mass in rotation has a decisive effect.

In accordance with ISO 8821, rotating machines can be balanced with or without a key or a half key on the shaft extension. ISO 8821 requires the balancing method to be marked on the shaft extension as follows:

- 'H' half key balancing
- 'F' full key balancing
- 'N' no key balancing

The testing of the vibration levels is undertaken with either the motor suspended (Figure B3.27) or mounted on flexible mountings (Figure B3.28).

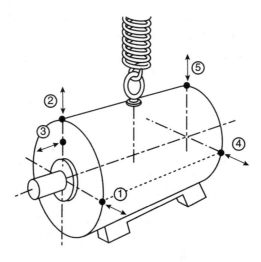

Figure B3.27 System for suspended machines (measuring points as indicated)

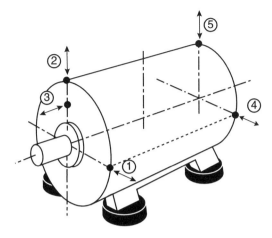

Figure B3.28 System for machines with flexible mountings (measuring points as indicated)

The vibration speed can be chosen as the variable to be measured (in mm s^{-1}). This is the speed at which the machine moves either side of its static position.

As the vibratory movements are complex and non-harmonic, it is the rms value of the speed of vibration that is used to express the vibration level.

Other variables that could be measured include the vibratory displacement amplitude (in micrometres) or vibratory acceleration (in m s^{-2}).

If the vibratory displacement is measured against frequency, the measured value decreases with frequency. High-frequency vibrations are not taken into account.

If the vibratory acceleration is measured against frequency, the measured value increases with frequency. Low-frequency vibrations (unbalanced loads) cannot be measured.

The maximum value of rms speed of vibration is the variable chosen by the standards and is generally classified as detailed in Table B3.8 for medium-sized machines.

Table B3.8   Maximum rms speed of vibration

Class	Speed n (min^{-1})	Frame size H (mm)		
		$80 \leq H \leq 132$	$132 < H \leq 225$	$225 < H \leq 315M$
N (normal)	$600 < N \leq 3\,600$	1.76	2.83	4.45
R (reduced)	$600 < N \leq 1\,800$	0.70	1.13	1.76
	$1\,800 < N \leq 3\,600$	1.13	1.76	2.83
S (special)	$600 < N \leq 1\,800$	0.44	0.70	1.13
	$1\,800 < N \leq 3\,600$	0.70	1.13	1.76

B3.8.2 Drive

Because drives are intended to control machines, it is to be expected that they might be subjected to some vibration from machinery when operating. A drive intended for industrial operation can reasonably be expected to have undergone some form of vibration test or verification. Typically a vibration withstand test is planned to occupy a realistic time such as a single working day. The test levels are set high enough to represent a situation at the severe end of the likely range, such as transportation in a standard road vehicle with no special precautions. There is a natural desire to correlate this test result with an estimate of the continuous vibration level that could be withstood throughout an expected working life of 20 years. However, considering the complex structure and multiple wearout processes involved, it is probably unrealistic to expect to make such an estimate with any useful degree of confidence.

Drives intended for permanent installation in vehicles are designed to a more severe vibration specification, which would not be cost-effective for an industrial product.

Another form of vibration, which might be caused or enhanced by drive operation, is torsional vibration of the drive train. This is a function of the tuning of the drive and its control loops, and is discussed in detail in Chapter B8.

B3.9 Altitude

Owing to the fact that air density reduces with increasing altitude, it is necessary to allow for the resulting reduction in cooling capacity of the air when motors and drives are operated at altitudes in excess of their rating. It is normal for motors and drives to be rated for a maximum altitude of 1 000 m.

Some manufacturers rate their machine for combinations of ambient temperature and altitude without quoting a de-rating factor. For example 40 °C/1 000 m or 30 °C/2 000 m or 20 °C/3 000 m.

It should be remembered that although the outdoor temperature at higher altitudes is usually low, the motors and drives will probably be installed indoors at higher ambient temperatures.

The IEC recommendation is to reduce the permissible temperature rise by 1 per cent per 100 m above 1 000 m.

The operating altitude is important and should be specified when purchasing a drive or a motor.

B3.10 Corrosive gases

B3.10.1 Motors

Motors are available for operation in a variety of environments including the presence of many corrosive gasses. The specific gases and the worst-case concentrations should be clearly specified to the motor manufacturer when the motor is being specified.

More details can be found in Section A1.8.

B3.10.2 Drives

Drives are usually designed to meet the requirements of Class 3C1 of IEC 60721-3-3. This corresponds to the levels typical of urban areas with industrial activities and/or heavy traffic, but not in the immediate neighbourhood of industrial sources with chemical emissions.

Hydrogen sulphide is a corrosive gas that has been seen to cause problems in relation to the silver, tin, lead and copper content of many electronic components including the base printed circuit board. The problem is that in the presence of an applied voltage and some moisture, dendrites (a crystalline structure illustrated in Figure B3.29) grow between conductors and eventually result in a short circuit that can cause the product to fail. The time to failure can be as short as 30 minutes or as long as several years.

Many manufacturers have moved to gold finishes in order to avoid this type of problem, which can occur at quite small concentrations.

Figure B3.29 Example of dendrite growth bridging tracks on a printed circuit board

Chapter B4
Thermal management

B4.1 Introduction

Motor and drives are both, in general, highly efficient power conversion devices, but when dealing with high powers, even with high efficiency there are significant losses to be dealt with. Motor and drive designers know that thermal management is one of their most significant challenges, and significant technology is brought to bear in this area. Motor and drive designers rely on advanced thermal simulation packages in order to optimise their designs. These designs are, however, based upon defined conditions in which the equipment is expected to function. This section describes the typically available cooling arrangements for motors and the design criteria for designing drive modules into cubicles.

Note that thermal protection is described in Chapter B8. The enclosure ingress protection classifications (IP) are described in Chapter B3.

B4.2 Motor cooling

B4.2.1 General

Very much related to the enclosure of the machine, but not synonymous with it, is the method of cooling. All rotating electrical machines designed for economy of materials and dimensions require an effective form of cooling to ensure that internal losses are dissipated within the limits of the maximum temperature rise for the class of winding insulation employed, and so that bearing and surface temperature rise figures are kept within safe limits.

There are three distinct differences between what is considered a standard enclosure for d.c, a.c. industrial (induction) and servo motors:

- d.c. motors tend to be of drip-proof construction (IP23) and the cooling is generally provided by a fixed-speed fan mounted on the exterior of the motor, blowing air directly over the motor windings, as illustrated in Figure B4.1.
- a.c. induction motors tend to be totally enclosed (IP44 or IP54) with a shaft mounted fan at the non drive end running within a cowl to duct the cooling air over a finned motor body as shown in Figure B4.2. Note also the cast 'paddles' on the rotor 'endshields', which provide internal air circulation and turbulence

Figure B4.1   Typical forced ventilating fan on a d.c. motor [photograph courtesy of Leroy–Somer]

to assist with transmitting the heat from the motor windings to the stator housing and from there to the atmosphere.
- Servo motors tend to be totally enclosed to a very high degree of protection, IP65. They tend not to have a cooling fan and rely significantly on conduction from their mounting flange for cooling. In this respect, servo motors are somewhat unusual.

It should be stressed at this stage that these are typical for standard industrial motors and that a wide range of cooling arrangements are available as standard from different

Figure B4.2   Typical shaft-mounted external cooling fan on an a.c. induction motor [photograph courtesy of Leroy–Somer]

manufacturers as well as custom solutions for OEM customers. The more common forms of motor cooling are specified by IEC 60034-6/EN 60034-6.

B4.2.2 D.C. motors

D.C. motors (and wound field synchronous motors) are generally of IP23 enclosure and therefore suitable for constant-velocity forced-ventilation cooling, either from a frame-mounted fan as already mentioned or, where the working environmental conditions are so difficult as to require an IP54 enclosure, from a remote fan mounted in a clean air position, with ventilating ductwork between motor and cooling fan unit.

The d.c. IP23 machine is restricted to use in a clean air environment or to one in which an air filter on the ventilation fan inlet gives sufficient protection to the winding and commutator. Where the working environment is so difficult as to require a totally enclosed machine, an attractive form of cooling is 'single-pipe' motor ventilation. This makes use of a remote fan, drawing air from a clean source and delivering it through a pressurised duct system to an adapter on the commutator-end end shield (IC17 of IEC 60034-6/EN 60034-6). Alternatively, double-pipe ventilation supplies cooling air from a remote fan as described, and in addition a discharge duct taking the used and warmed air away from the motor (IC37).

Ducted ventilation has the advantage that injurious gases and contaminants are unlikely to invade the motor winding space from the working environment. Additionally, the use of double-pipe ventilation and the introduction of (say) a 10 min purging period before main motor starting may sometimes allow the IP55 enclosed d.c. motor to be used in an atmosphere where there is an explosive gas risk, for example in the printing industry, where the volatiles from some inks constitute a hazard.

Although ductwork costs have to be taken into account, such single- or double-pipe ventilated d.c. motors can prove an attractive alternative to totally enclosed d.c. motors, with either closed-air-circuit air cooling (CACA) or closed-air-circuit water cooling (CACW).

For the screen protected IP23 d.c. machine, it is relatively easy to arrange the machine with a frame-mounted fan, blowing constant-velocity air through the winding space. For the much less popular IP54 totally enclosed fan-cooled machine, the fan being motor shaft mounted, the mechanical arrangement is less satisfactory, requiring a separate cooling fan motor to be mounted external to the motor but inside the cooling fan cowl.

CACW and CACA machines are available as both a.c. and d.c. designs, and can be very efficient, leading to potentially reduced motor size. Although a relatively expensive solution, this is worth investigating where difficult ambient conditions preclude lower-cost alternative enclosures. It is also useful to consider where a supply of cooling water is available and/or where a very compact motor design is required.

Flameproof d.c. motors are of restricted availability, particularly above 20 kW, and on account of the small demand are very expensive.

B4.2.2.1 Air filters

The fitting of an air filter to a forced-ventilation fan can provide useful protection against internal motor contamination. However, heavy contamination of the filtering element can reduce cooling airflow markedly. Although thermal devices protect against this circumstance, closer, more direct protection is possible by using an air flow, or air proving switch. This is generally arranged to monitor the air pressure driving the air through the motor winding space. As filter contamination gradually builds up, pressure falls and the air switch will indicate an alarm condition when the pre-set limit is detected. The drive system can be arranged either to shut down immediately or after a pre-set time interval to allow the driven machine to be cleared of its product or to give a warning to the machine or process operator of imminent shutdown.

B4.2.3 A.C. industrial motors

The external shaft-mounted cooling fan of the standard a.c. induction motor offers a simple and effective solution. It is important, however, that the motor is mounted correctly so as to ensure the correct operation of the fan. A typical recommendation is that a minimum distance of 1/3 of the frame size is allowed between the end of the fan cover and any possible obstacle such as a wall or part of a machine.

It should be noted that for variable-speed a.c. applications, the cooling performance of the shaft-driven internal and/or external fans of the standard motor varies inversely as the square of the shaft speed. At half speed they provide only 25 of the full-speed cooling effect.

On a typical constant torque load requiring constant motor current, the motor has approximately constant losses over its speed range. Consequently, a standard motor will require significant de-rating of its output. The precise de-rating is very dependent upon the particular design of the motor and the manufacturer should be consulted.

An alternative to de-rating is to use a motor with cooling air provided by a fan driven at constant speed independently of the motor shaft speed. Such motors are readily available but they tend to be significantly more expensive than the standard motor. Furthermore, the presence of an external fan can pose a problem when the application requires a speed or position feedback signal, because the encoder is normally mounted co-axially at the rear of the motor. With close coupled bearingless encoders of short axial depth, however, it is possible to arrange satisfactory mounting that is compatible with the forced-cooling arrangement.

B4.2.4 High-performance/servo motors

The thermal management of high-performance permanent-magnet (PM) servo motors is critically important if reliable operation and long life is to be obtained. In order to correctly apply a PM servo motor it is important to understand how heat is dissipated from a typical motor. In a standard air-cooled motor, heat is removed from the motor by convection and radiation heat transfer from three main areas of the motor:

- the motor housing (typically 50 per cent of the losses),
- the motor mounting plate (typically 40 per cent of the losses), and
- the non-drive end and encoder housing (typically 10 per cent of the losses).

It is clear that the heat transfer through the motor mounting plate is of crucial importance, and this is an area users frequently misunderstand. It is greatly complicated by the lack of standardisation in the industry in relation to the conditions under which motor ratings are quoted. For this reason it is critical that the user seeks guidance from the motor manufacturer.

A typical approach to motor rating, which leads to the protection strategy, is as follows. The motor is mounted to an aluminium plate heat sink of differing dimensions according to the frame size of the motor under test, and run in a test rig as shown in Figure B4.3. The motor is run throughout its rating envelope and the motor winding temperature is continuously measured using an embedded thermocouple within the winding. The difference between motor ambient temperature and motor winding temperature is recorded during the test. This temperature difference is called the Δt.

Generally around six thermocouples are installed within the winding to detect the hottest point within the winding (inside slot and winding overhang), and they differ due to stator orientation (when mounted as shown in Figure B4.3, the bottom of motor is hotter than the top due to convection). Thereafter the hottest winding temperature is used as the reference winding temperature.

At the set motor speed and torque, the motor temperature is allowed to saturate. The criterion for saturation is that the winding temperature is within ± 2 K of Δt of 100 K and the winding temperature rise is less than 1.2 K per hour. The motor 'rating' curve is established under fixed winding temperature criterion. Note that IEC60034-1 defines a temperature gradient of 2 K per hour, but some manufacturers reduce this

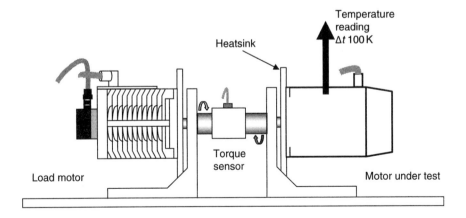

Figure B4.3   Typical test arrangement for motor rating confirmation

level to improve the repeatability of test data and to allow the rotor temperature to stabilise. For example, Control Techniques apply a temperature gradient of 1.2 K per hour.

The motor 'rated' torque is measured at motor 'rated' speed at Δt of 100 K. Motor 'stall' torque is measured at near-zero motor speed at Δt of 100 K.

B4.2.4.1 Intermittent/peak torque limit

Above the continuous zone is an intermittent zone where the motor may be safely operated for short periods of time. Operation within the intermittent zone is permissible provided that $\Delta t < 100$ K, in accordance with the motor type rating.

Clearly the mounting arrangement of the motor is critical to both the lifetime and the possible rating of the motor.

Protection of the motor is through advanced thermal models within the drive, and such a model requires good information about the motor. Ultimately, however, it is good practice to install thermistors in critical points of the motor to provide a defined trip temperature.

B4.2.4.2 Forced-air (fan) cooling

Forced-air cooling can be used to enhance the motor output power. Figure B4.4 shows the motor rated torque performance with natural cooling and with forced-air (fan) cooling. Fan cooling reduces the thermal convection resistance between the motor housing, the mounting plate, the feedback device housing and the surrounding ambient.

Reduction of the thermal resistance between the motor surface and the ambient means that the motor can dissipate more heat, and so can withstand higher losses. Therefore, higher current can flow in the motor while still keeping the critical temperature within safe limits.

As an example, at air velocity of 10 m s^{-1}, the housing convection resistance will be lowered by approximately 60 per cent compared with natural convection resistance

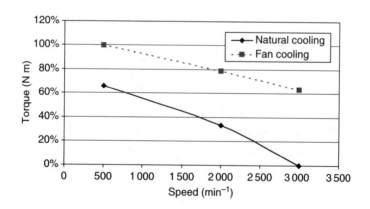

Figure B4.4   Example of rated torque for natural and fan-assisted cooling

under rated conditions. This illustrates the effectiveness of air cooling. The results shown in the figure confirms this.

In the case of PM servo motors, care must be taken to ensure that the peak current (or peak torque) is appropriately defined when enhancing the torque performance using forced air. The peak current capability is also a function of the rotor magnet temperature and with forced cooling the rotor magnets do not enjoy the same heat transfer improvement.

B4.3 Drive cooling: the thermal design of enclosures

B4.3.1 General

The reliable trouble-free operation of all industrial equipment is dependent upon operation in an environment for which that product was designed. The single most significant reason for the premature failure of a variable-speed drive controller is operation in excessive ambient temperature. The design of the enclosure in which the drive is housed is therefore of critical importance.

In order to design the smallest possible enclosure, a finite-element thermal simulation is needed. This is expensive and very time consuming and is only generally practicable for original equipment manufacturers (OEMs) who are producing a significant number of the same panel design.

The following guidance covers the basic calculations necessary to ensure that heat generated by a drive can be satisfactorily transferred to the air surrounding the cubicle. When making the calculation, remember to take account of all power dissipated in the cubicle, not simply that generated by the drive. Further, in the internal layout of the cubicle, where possible, avoid placing electronic components at the top of the cubicle (hot air rises!), and where possible provide fans to circulate internal air. Remember, as a rule of thumb, an electronic product lifetime halves for every 7 °C temperature rise!

B4.3.2 Calculating the size of a sealed enclosure

The enclosure itself transfers the internally generated heat into the surrounding air by natural convection, or external forced airflow. The greater the surface area of the enclosure walls, the better is the dissipation capability. Remember also that only walls that are not obstructed (not in contact with walls, floor or another 'hot' enclosure) can dissipate heat to the air.

Calculate the minimum required unobstructed surface area A_e for the enclosure as follows:

$$A_e = P/[k(T_i - T_{amb})]$$

where A_e is the unobstructed surface area (m^2), P the power dissipated by *all* heat sources in the enclosure (W), T_{amb} the maximum expected ambient temperature outside the enclosure (°C), T_i the maximum permissible ambient temperature inside the enclosure (°C) and k the heat transmission coefficient of the enclosure material (W m^{-2} °C^{-1}).

Example
Calculate the size of an enclosure to accommodate the following:

- two Control Techniques 4 kW Unidrive SP drives (3 kHz switching frequency) operating at full load, full speed continuously,
- EMC filter for each drive,
- braking resistors mounted outside the enclosure,
- a maximum ambient temperature inside the enclosure of 40 °C, and
- a maximum ambient temperature outside the enclosure of 30 °C.

$$\text{Maximum dissipation of each drive} = 83 \text{ W}$$
$$\text{Maximum dissipation of each EMC filter} = 9 \text{ W}$$
$$\text{Total dissipation} = 2 \times (83 + 9) = 184 \text{ W}$$

The enclosure is to be made from painted 2 mm (3/32 in) sheet steel having a heat transmission coefficient of 5.5 W m^{-2} °C^{-1}. Only the top, the front and two sides of the enclosure are free to dissipate heat, as shown in Figure B4.5.

The minimum required unobstructed surface area A_e for the enclosure is

$$A_e = P/[k(T_i - T_{amb})]$$
$$= 184/[5.5(40 - 30)]$$
$$= 3.35 \text{ m}^2$$

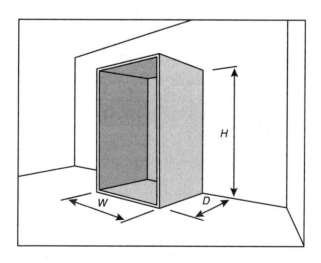

Figure B4.5   Enclosure having front, sides and top panel free to dissipate heat

If we select an enclosure with a height H of 2 m, a depth D of 0.6 m, and minimum width W_{min}

$$\text{Dissipating surfaces} > 3.35 \text{ m}^2$$
$$\text{Top} + \text{front} + (2 \times \text{sides}) > 3.35 \text{ m}^2$$
$$(W_{min} \times 0.6) + (W_{min} \times 2) + (2 \times 0.6 \times 2) > 3.35 \text{ m}^2$$
$$W_{min} > (3.35 - 2.4)/2.6$$
$$> 365 \text{ mm}$$

If the enclosure is too large for the available space it can be made smaller only by

- reducing the power dissipation in the enclosure,
- reducing the ambient temperature outside the enclosure,
- increasing the permissible ambient temperature inside the cubicle if possible by de-rating equipment in line with the manufacturer's recommendations,
- increasing the number of unobstructed surfaces of the cubicle, and
- ventilating the cubicle.

B4.3.3 Calculating the air-flow in a ventilated enclosure

In this case the dimensions of the enclosure are determined only by the requirements to accommodate the equipment, *making sure to provide any recommended clearances*. The equipment is cooled by forced air flow. This being the case it is important in such an arrangement to ensure that the air flows over the heat-generating components to avoid localised hot spots.

The minimum required volume of ventilating air is given by

$$V = 3kP/(T_i - T_{amb}) \text{ m}^3 \text{ hr}^{-1}$$

where V is the cooling air flow (m³ hr⁻¹), P the power dissipated by *all* heat sources in the enclosure (W), T_{amb} the maximum expected ambient temperature outside the enclosure (°C), T_i the maximum permissible ambient temperature inside the enclosure (°C), and k the ratio of p_o/p_i (p_o and p_i are the air pressures at sea level and the installation, respectively). Typically a factor of 1.2 to 1.3 can be used to allow for pressure drops in dirty air filters.

Example
Calculate the size of an enclosure to accommodate the following:

- three Control Techniques 15 kW Unidrive SP drives (3 kHz switching frequency) operating at full load, full speed continuously,
- EMC filter for each drive,
- braking resistors mounted outside the enclosure,

- a maximum ambient temperature inside the enclosure of 40 °C,
- a maximum ambient temperature outside the enclosure of 30 °C.

$$\text{Maximum dissipation of each drive} = 311 \text{ W}$$
$$\text{Maximum dissipation of each EMC filter} = 11 \text{ W}$$
$$\text{Total dissipation} = 3 \times (311 + 11) = 966 \text{ W}$$

Then the minimum required volume of ventilating air is given by

$$V = 3kP/(T_i - T_{amb})$$
$$= (3 \times 1.3 \times 966)/(40 - 30)$$
$$= 377 \text{ m}^3 \text{ hr}^{-1}$$

B4.3.4 Through-panel mounting of drives

Some manufacturers have designed their drive product such that the drive can be mounted in the panel with the heat sink protruding out of the back of the enclosure. In this way the bulk of the heat is not dissipated inside the panel. As an example, a 55 kW Control Techniques Unidrive SP at rated output and 3 kHz switching frequency dissipates a total of 1 060 W, of which less than 20 per cent is dissipated into the panel when through-panel mounted. The implications of this on enclosure design are obvious.

Chapter B5
Drive system power management: common d.c. bus topologies

B5.1 Introduction

It is usual to consider a drive and its associated motor as a single power train. For an a.c. voltage source PWM drive this comprises an a.c. to d.c. supply side converter, a d.c. to a.c. motor side converter and a motor each performing a power conversion function. Figure B5.1 shows a typical topology.

In systems comprising more than one motor, it may be advantageous to have a power configuration in which there is a single or bulk a.c. to d.c. converter feeding a d.c. bus, which then feeds multiple a.c. to d.c. converters, as shown in Figure B5.2.

Such an arrangement has a number of benefits including a reduction in the number of supply contactors and protection equipment, which can be a significant cost consideration.

In applications where regenerative energy is present for significant periods of time on one or more drives it is possible to recycle this energy via the d.c. bus to another

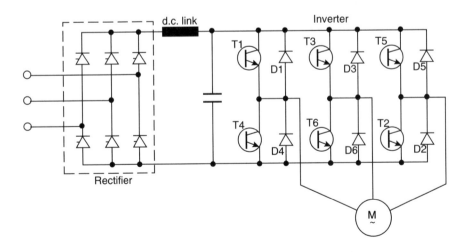

Figure B5.1   Typical PWM inverter drive

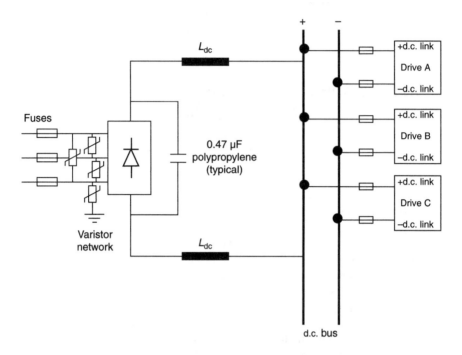

Figure B5.2   Common d.c. bus fed from a bulk a.c. to d.c. converter for multiple drives

drive that is in the motoring condition. Using a common d.c. bus arrangement as shown in Figure B5.2, the regenerative energy will flow through the d.c. bus to a motoring drive without the need for any further conversion process.

Compared to the conventional arrangement of individual power conversion with a non-regenerative a.c. to d.c. converter where regenerative energy is usually dissipated in braking resistors, the system efficiency is significantly improved. This can be particularly advantageous where one or more drives may be 'holding back' a line to provide tension. It is often applied in high-performance drive applications where substantial amounts of energy are used in accelerating and braking drives.

Another common example is a winder/unwind application as shown in Figure B5.3. In this example the input rectifier in a standard drive is adequately rated for the worst-case supply current that the total system will draw. It is important to note that the input rectifier has to be adequately rated for not only the worst-case operational duty but must be able to supply the inrush current to the total d.c. link capacitance. Drive manufacturers will be able to advise of the maximum d.c. link capacitance that can be added to the capacitance built into the drive.

Most common d.c. bus systems such as that illustrated in Figure B5.3 can be operated in such a way that there is never any net regenerative energy and so a unidirectional supply side converter can be used. In an application where, for example,

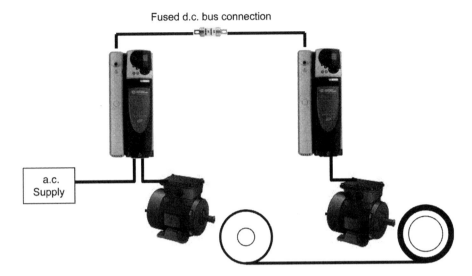

Figure B5.3   Winder/unwind application using a common d.c. connection

on shut down there was to be net regenerative energy, a common d.c. bus system would allow a complete system to feed a single braking resistor.

In applications such as a textile knitting machine where under all conditions including loss of the a.c. supply the machine must slow down in a controlled and synchronised manner, a common d.c. bus can be used to control the deceleration of all elements in an optimal way, with the net energy from the machine inertia feeding the d.c. bus.

In summary the advantages of a common d.c. bus system are as follows:

- a reduction of energy losses,
- reduced input stage systems cost (reduced a.c. input cabling, a.c. input fuses, contactors etc.),
- system controlled power down on mains loss,
- common braking solutions, and
- reduction in cubicle size because of reduced converter losses.

There are disadvantages however, and care needs to be taken in the implementation of such as system. The direct connection of the d.c. links of a.c. drives usually entails the direct connection of the d.c. link capacitor banks of all inverters. These capacitors store substantial amounts of energy. In the event of a fault, the combined stored energy of all the d.c. link capacitors in the group of drives will be fed into the fault, causing substantial damage to the original failed drive as well as to others. Protection against this type of fault must be implemented in the form of fuses between the d.c. bus and the individual drives. The fusing strategy is critical and will be discussed later.

B5.2 Power circuit topology variations

B5.2.1 General

Having decided on a common d.c. bus system, it is necessary to decide upon the form of power supply connection. This is dependent upon many factors, including the following:

- a requirement to regenerate energy back into the mains supply,
- fail-safe braking requirements,
- supply harmonic limitations,
- use of standard/commercial components,
- whether all drives to be connected to the bus are, or can be, of the same rating,
- the peak current to be drawn from the system in relation to the individual ratings of the drives, and
- the location of the d.c. bus soft start/charging circuit within the individual drives.

There are a number of different ways of connecting the drives together and paralleling the d.c. buses; these are detailed in the following sections.

B5.2.2 Simple bulk uncontrolled external rectifier

The use of a bulk input converter, as illustrated in Figure B5.2, is strongly preferred if the installation requires drives of different ratings to be connected together. The d.c. link chokes of some standard drives tend not to be in circuit if the drive is supplied via the d.c. terminals, and hence it is necessary to supply an external choke with a specification depending on the total rating of the drives connected to the d.c. link. Splitting the inductance equally between the positive and negative d.c. link can provide some impedance to limit fault current if an earth fault occurs in either the positive or negative d.c. link.

It is also necessary to provide an external rectifier module. Inrush current limiting is not required when the individual drives have their soft start circuits (inrush resistors and relays/contactors) in circuit until the d.c. link is at the correct level.

A high-voltage polypropylene capacitor should be fitted across the d.c. terminals of the rectifier module. This helps reduce the reverse recovery voltage spikes, and can also help provide a path for radio frequency interference (RFI) currents in applications where long motor cables are used.

The varistor network on the input phases provides protection against line–line and line–earth voltage surges. This is required because the a.c. input stages of the drives are not being used.

The d.c. link inductance should be selected to keep the resonant frequency reasonably constant. As the d.c. link capacitance is the arithmetic sum of the capacitance of all connected drives, the required d.c. link inductance can be calculated as follows:

$$1/L_{dc} = 1/L_{drive1} + 1/L_{drive2} + \cdots$$

where L_{drive1}, L_{drive2} etc. are the design values of the chokes fitted in the standard drives. This data is available from the drive manufacturer.

This arrangement has the following advantages.

- It allows drives of different ratings to be connected together.
- It reduces a.c. supply side component sizes.
- It reduces energy losses (heat loss from braking resistors).
- It is possible to connect in a 12-pulse configuration to reduce supply harmonics.

There is the following disadvantage.

- It cannot regenerate, so dynamic braking may be required.

B5.2.3 A.C. input and d.c. bus paralleled

This configuration, illustrated in Figure B5.4, is where the a.c. supply connections of multiple drives are fed from a common supply as they would be if they were configured as individual drives and the d.c. links are connected. This topology seeks to take on the majority of the efficiency and operational benefit without the simplification of the supply connection.

If a.c. supply and d.c. link connections of drives with different ratings are connected, unequal and disproportionate current sharing in the input diode bridges will result, due mainly to the different choke sizes presenting different impedances. If

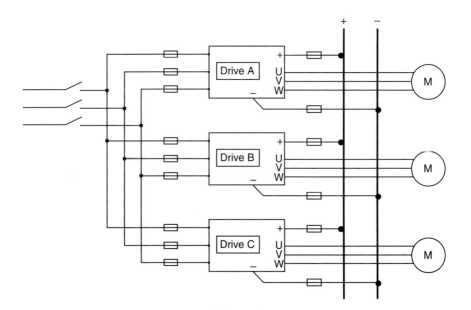

Figure B5.4 Hard paralleling of all drive input rectifiers

drives without a d.c. link choke are to be connected in this way, then individual a.c. supply reactors are required.

If the drives of equal rating are similarly connected, the impedances seen on the input stages are very similar, and current sharing is good. However, a 10 per cent de-rating is strongly recommended to allow for the small imbalances that will exist.

In the design of such systems a number of further issues need to be considered.

- A.C. and d.c. power wiring should be of a star form and not daisy chain, and efforts should be made to equalise cable length.
- Drives should be located close to each other.
- The resistance of the a.c. fuses may help in sharing current, so it is important to use the same fuse types in corresponding positions.
- Deliberately adding some external resistance could help with sharing.
- A.C. line chokes can be used to help sharing.
- As the brake threshold voltage will be slightly different for each drive, there should be only one main brake resistor for all the paralleled drives. It is possible to design thermal protection in such a way as to protect multiple resistors; however, care should be taken and the drive supplier consulted for full details.

The arrangement has the following advantage.

- No additional rectifier circuit is required.

There are the following disadvantages.

- The arrangement cannot regenerate, so dynamic braking may be required.
- It is not possible to connect in a 12-pulse configuration to reduce supply harmonics.
- It only allows drives with the same power stage to be connected together.

B5.2.4 One host drive supplying d.c. bus to slave drives

This arrangement, illustrated in Figure B5.5, is possible only where the peak current drawn from the system/group of drives is lower than the rated current of the largest drive. This can be the case in applications such an unwinder–winder, or in a machine tool where there is a large spindle and small axis drives. In applications where this condition is only marginally satisfied, an over-rated drive may be used to facilitate this solution. In this case no additional d.c. inductance is required, because the inductance within the large drive is being utilised. Care needs to be taken to confirm that the charging circuit of the large drive can carry the full charging current for the increased d.c. link capacitance.

Advantages of the system include the following.

- The number of a.c. supply side components are reduced.
- The host d.c. bus supply drive controls dynamic braking (if required).
- Drives of different ratings may be connected together.

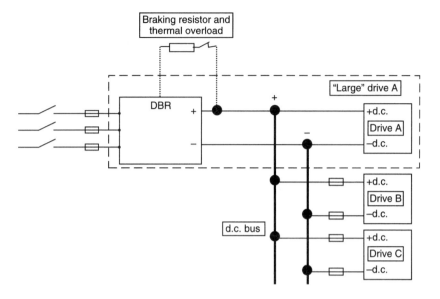

Figure B5.5   The mains converter in one drive (A) supplying all drives (B, C etc.) from its d.c. link

There are the following disadvantages.

- It cannot regenerate, so dynamic braking may be required.
- It is not possible to connect in a 12-pulse configuration to reduce supply harmonics.

B5.2.5 A bulk four-quadrant controlled rectifier feeding the d.c. bus

The attraction of such a system is that it can, in principle, be realised using a standard, four-quadrant d.c. drive. There are, however, limitations. First, standard d.c. drives can typically only regenerate up to 1.15 times the rms supply voltage, making this the upper limit for the d.c. link voltage. This link voltage is lower than the standard rectified supply voltage of 1.35 times the rms supply voltage, and consequently the available base speed is reduced to $1.15/1.35 = 85$ per cent of normal. This can be addressed by feeding the converter from a step-up transformer.

The selection of the required d.c. link inductance is important in order to ensure that the resonant frequency of the d.c. link is within the control capability of the drive 'speed' loop and ensures stable operation under light load conditions.

As the d.c. link capacitor current ripple is directly related to the lifetime of that capacitor, it is important to keep it within its design limits.

It is easy to see that what sounded like one of the simplest forms of implementing a common d.c. bus solution requires a significant amount of detailed engineering requiring intimate knowledge of both a.c. and d.c. drive designs. Such a solution must therefore be undertaken in collaboration with the drive manufacturer.

B5.2.6 Active bulk rectifier

This is a similar topology to the simple bulk rectifier described in Section B5.2.2 but with the diode bridge replaced by a four-quadrant PWM converter as illustrated in Figure B5.6. Some standard commercial drives such as the Control Techniques Unidrive SP have been designed to operate in such an arrangement. The mains supply is connected to what would normally be the motor terminals (U, V, W) through the input inductors L_{in}, and filter L_f and C_f, as shown. The d.c. bus connections are made to the +ve and −ve d.c. bus terminals. Another single inverter drive, or multiple inverter drives, can be connected across the d.c. bus as described in other systems to produce a four-quadrant drive system with control of one or more motors. Input inductors must be provided to give a known minimum amount of source impedance, to allow the inverter to operate as a boost converter, and to limit the PWM switching frequency related currents to an acceptable level for the converter. Increasing the value of inductance tends to reduce the PWM ripple current while reducing the current loop bandwidth and the power factor. A supply filter (formed by L_f and C_f) will be required to further attenuate switching frequency-related distortion so as to meet any applicable standards on mains supply harmonics, or to prevent other equipment connected to the same supply from being affected by high-order harmonics.

Some converters allow the user to set the power factor as seen by the supply. It is possible to operate such converters at a leading power factor, therefore acting to improve the overall plant power factor but at the cost of requiring a higher converter rating to handle the higher current.

When a.c. power is first applied, the d.c. link capacitance is charged through the inverter anti-parallel diodes and pre-charge resistor. The resistor prevents excessive charging current that could damage the diodes and potentially trip any input protection. Once the d.c. link capacitor is charged, the contactor is closed to short-circuit the resistor for normal operation. Note that this contactor is for pre-charging and is needed in addition to standard control switchgear.

This form of supply gives excellent regulation of the d.c. bus even in the event of a transient condition such as a high-speed motor reversal, as shown in Figure B5.7.

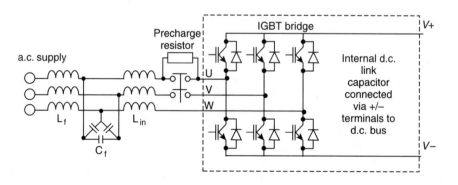

Figure B5.6   *A four-quadrant PWM converter feeding the d.c. bus*

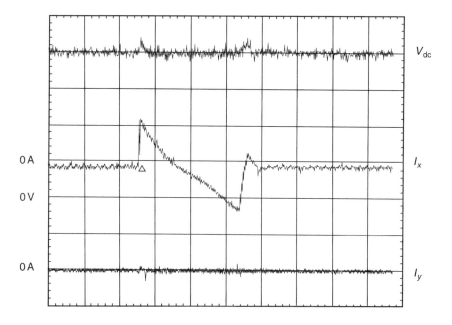

Figure B5.7 D.C. voltage regulation during a high-speed motor reversal for V_{dc} (d.c. link voltage) = 180 V/div and time = 200 ms/div

The control system is seen to limit the change in d.c. link voltage to less than 5 per cent with very rapid changes in power flow. In overload conditions where the d.c. link power exceeds the maximum a.c. power, the PWM rectifier will be forced into current limit. Figure B5.8 shows transient overload operation where the PWM rectifier goes into current limit; however, the system remains stable and the d.c. link voltage deviates from the set point until the load is reduced.

As well as providing a very well regulated d.c. bus, which will usually provide long d.c. bus capacitor life, the PWM solution is an elegant approach for applications requiring regeneration of energy back into the supply. It benefits from low harmonic distortion on the mains, although careful system design is necessary to ensure high (switching) frequency harmonics do not cause interference to other equipment.

There are the following advantages to the arrangement.

- It can regenerate back into the supply.
- Dynamic braking not required.
- The PWM converter (which may be a standard a.c. drive available from some manufacturers) offers a greater reduction in supply harmonics.
- It allows drives of different ratings to be connected together.

There is the following disadvantage.

- There is extra cost and size requirements in adding an extra drive to the system.

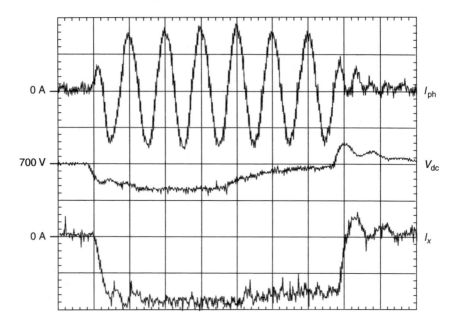

Figure B5.8 *D.C. link voltage during transient overload with V_{dc} (d.c. link voltage) = 100 V/div and time = 20 ms/div*

B5.3 Fusing policy

The subject of fusing for common d.c. bus systems is somewhat confusing, but only with regard to the policy of some drive manufacturers of not recommending fusing. Although in some cases the fuses are integral to the drive and are therefore present but not discussed, in others reliance is clearly placed upon the electronic monitoring and control to protect the drive. The merits of such a philosophy are unclear and it must be assumed that they are based upon a reliance on the d.c. link components, notably the d.c. link capacitors.

It should also be noted that when supplying the drives via their d.c. bus connection, the Underwriters Laboratories Inc. (UL) listing may be invalidated. The supplier of the drive should be contacted to check.

B5.4 Practical systems

B5.4.1 Introduction

The most common applications that take advantage of common d.c. bus systems are dynamometers, notably motor and gearbox test rigs, and machines that unwind and rewind the material being processed. Rather than consider each application, it is helpful to consider the engineering aspects of generic power solutions.

B5.4.2 Variations in standard drive topology

So far in this section, there has been no consideration of the subtle but important variations in power circuit topology that exist not only between different drive manufacturers but also in different drive ranges and powers of individual manufacturers. You cannot start connecting d.c. power circuits without a full understanding of exactly what is being connected.

Variations of topology within the voltage source PWM family generally arise through differences in d.c. link filter design, the method for charging the d.c. link capacitor at start up, and the internal braking device. Figures B5.9 and B5.10 illustrate such variations.

Figure B5.9 is typical for a drive up to about 30 kW, and uses a resistor to limit the capacitor charging current. Small drives below 3 kW would tend not to have a d.c. link inductor. Figure B5.10 is typical for a larger rating of drive in which the charging current is limited by phase control of the supply-side converter. It is clear from the these figures that the d.c. link power connections are not at the same point and great care is needed in relation to the charging circuit if direct connection is to be made.

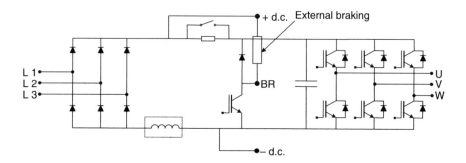

Figure B5.9   Typical low-power drive topology

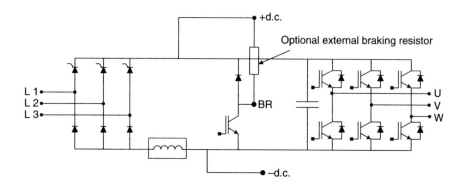

Figure B5.10   Typical higher-power drive topology

B5.4.3 Inrush/charging current

When drives are connected in parallel, the capacitor bank is much larger, which increases the inrush current on power up when the capacitors are initially charged. Inrush current is normally controlled using either a resistor/bypass relay circuit (Figure B5.9) or a controlled thyristor bridge that is gradually phased forward (Figure B5.10). The input configuration (including fuses) must be capable of supplying this inrush current.

If drives designed as in Figure B5.9 are connected to a common d.c. bus then no additional inrush circuit is required, because each drive will limit the inrush current through their own charging resistor, as shown in Figure B5.11.

It is not possible to use the inrush circuit to limit the charging current for any other drive using the charging resistor in such a drive due to its location in the circuit (after the d.c. terminals). If a drive is to be supplied from the d.c. bus of this drive then it must have its own inrush circuit present in the d.c. path.

The peak inrush current must be checked to ensure that it does not exceed the peak rating of the power source device or a.c. supply fuses.

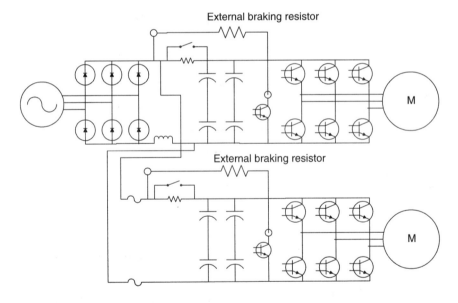

Figure B5.11 Drives fed from a common input bridge with individual charging circuits

B5.4.4 Continuous current

The input converter must be capable of supplying the highest level of continuous current that will be required by the drives at any point in time. In the case of an unwind/rewind application the input could be much smaller than the motoring drives because the losses in the system can be quite low compared to the power flowing around the d.c. circuit. In this case it is common for the a.c. supply to be connected to one or two drives only, with further drives being connected via the d.c. terminals only. The duty cycle of each drive must be considered under all possible operating conditions. If both drives can be motoring at the same time the input must be capable of supplying this current. Starting the machine should be considered, as this can be the worst-case condition.

If the a.c. supply is connected to more than one drive in a common d.c. bus application, the input impedance of the input stage of each drive must be considered. The input current will be shared in inverse proportion to the difference in input impedance. If different input impedances are present in a system, care must be taken to ensure that the input with the lowest input impedance is not overloaded. In general it is easier to ensure that only drives with the same input impedance are connected to both a.c. and d.c. because the current will then be shared equally.

Drives generally have either a d.c. bus choke or a three-phase a.c. line reactor that smoothes the current flowing to the d.c. link capacitors. In the case of a d.c. bus choke the most common solution is to fit the reactor in either the positive or the negative connection. This has significance when connecting both the a.c. and d.c. terminals of drives as shown in Figure B5.12.

It can be seen in Figure B5.12 that the impedance of the d.c. inductance provides impedance that aids current sharing between the bridges for the negative or lower diodes only. In this case three-phase line reactors, as shown, are also required to ensure that both the upper and lower diodes in the rectifier stage share current in the correct proportion.

Diodes have a negative temperature coefficient such that the impedance decreases with temperature. Parallel diodes without any series impedance can therefore thermally run away if tolerances result in more current flowing in one diode compared to another.

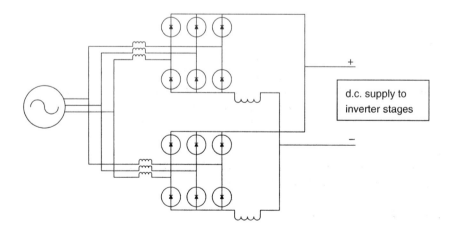

Figure B5.12 *A.C. and d.c. terminals connected: d.c. inductance in negative bus*

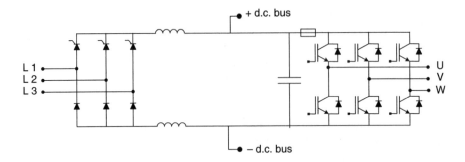

Figure B5.13 Topology with d.c. bus inductance split between both positive and negative connections

If the d.c. bus choke is split such that both positive and negative connections have the same impedance, as shown in Figure B5.13, then the a.c. line reactor is not required. If the system is to consist of many drives of different ratings the best solution is often to use one large common rectifier and supply all drives via d.c. only.

In the case of a drive with a three-phase inductor instead of a d.c. bus choke, as long as the inductances are such that the current will be shared in the correct proportion then no additional components are required. Tolerances in inductance values must be taken into consideration to ensure no single input is overloaded.

B5.4.5 Implementation: essential knowledge

B5.4.5.1 A.C. and d.c. terminals connected: drives of the same current rating only

The drives should be supplied using a common contactor/isolator to ensure that drives power up at the same time. The inrush current is then shared between the drives on power up. If drives are powered up sequentially then the first drive will charge the whole d.c. bus, which could exceed the peak input current rating for the input of that one drive.

Using a common contactor allows the normal a.c. fuses to be fitted to each drive as recommended by the drive manufacturer. If the drives have d.c. bus chokes in one connection only, a.c. line reactors should be fitted to ensure current is shared in both the upper and lower diodes. A.C. line reactor values should be sized such that at least 2 per cent of the phase voltage is dropped across the reactor.

Example
Calculate the appropriate a.c. line reactance for a 400 V, 50 Hz three-phase supply, and a drive with rated input current of 10 A:

Phase voltage is $400/\sqrt{3} = 230$ V
2 per cent voltage drop $= 4.6$ V
Drive input current $= 10$ A

Therefore the required line inductance $= 4.6/(10 \times 2\pi \times 50) = 1.46$ mH per phase.

Typical inductor manufacturing tolerance is ± 5 per cent depending on the specific inductor, so a line reactor of 1.54 mH $\pm$ 5 per cent should be specified.

Fuses should be fitted in both positive and negative d.c. connections between drives. It is also good practice to apply a de-rating of 10 per cent to all drives in the system to take into account tolerances in the line reactor values to ensure no input can be overloaded.

B5.4.5.2 A.C. and d.c. terminals connected: drives of different current ratings

Only drives with the same input stage type as discussed in Section B5.4.2 should be connected with both a.c. and d.c. connections. Let us consider the case for drives with input diode bridges.

A common contactor must be used to power all drives up simultaneously. A.C. chokes must be fitted to each drive in proportion to the input current of that drive. Any values in the correct ratio will cause the current to be shared correctly. Values that are too high will result in excessive voltage drop. Any deviation from the ideal values should be applied in the form of a de-rating to all drives. It is also good practice to apply a nominal de-rating of 10 per cent to all drives in the system to take into account tolerances in the line reactor values to ensure no input can be overloaded.

B5.4.5.3 One host drive supplying d.c. bus to slave drives

The a.c. supply is usually connected to the largest drive in the system. The maximum system loading should be calculated and if this exceeds the rating of the largest host drive then either a larger rating should be selected for the host drive or an alternative topology used. One or more smaller drives are fed from the d.c. bus terminals of the larger drive.

The total inrush current to the secondary drives must not cause the peak current to exceed the maximum capability of the larger drive.

General advice on the detailed design of this type of system is complicated by the variations in topology described earlier. It is essential that the topologies being used are clearly known, and key design data are available, including the following:

- the d.c. link capacitance values of the host drive,
- the maximum possible additional capacitance that can be added to the host d.c. link,
- the I^2t limit for the host rectifier,
- the d.c. link capacitance of each of the other drives to be added to the link, and
- the worst-case inrush current for each drive type.

If the system consists of drives with a topology such as that in Figure B5.9, the inrush current is limited in each drive by its own resistor/relay circuit to its design level. For other drive types where their d.c. link capacitance is added directly to the d.c. link, then the charging current needs to be supplied by the host converter.

Using the manufacturer's data detailed above, the worst-case total inrush current must still be checked to ensure it remains within the I^2t limit of the bridge rectifier on the drive connected to the a.c. supply.

A.C. fuses to the drive connected to the a.c. supply must be chosen to ensure no spurious fuse failures occur. Generally the standard fuses recommended for drives with phase-controlled input stages are likely to be suitable, as they should have been designed for worst-case supply brownout conditions.

D.C. fuses should be connected to both negative and positive connections of each drive.

B5.4.5.4 Simple bulk uncontrolled external rectifier

The rectifier must be chosen such that it can supply the highest motoring current required based on the sum of the duty cycles of all drives on the system. Either an uncontrolled bridge with an appropriate charging circuit, or a controlled bridge ideally with current control, can be used in this arrangement. Again, a clear understanding of the drive topologies to be connected to the d.c. link is necessary, and normal power design rules applied.

D.C. fuses should be connected to both negative and positive connections of each drive.

B5.4.6 Practical examples

B5.4.6.1 Winder/unwinder sharing energy via the d.c. bus

Consider an application where both drives are of the same rating and have a topology as shown in Figure B5.9, where the input arrangement comprises a diode bridge with resistor/relay inrush circuit that is in circuit when supplied via the d.c. bus, and a d.c. inductor in the negative line of the bus, which is on the input converter side of the d.c. terminals.

There are two preferred solutions:

- Connect the a.c. supply to one drive and supply the other from its d.c. link.
- Connect both drives to the a.c. and d.c. supplies.

The most cost-effective solution would be the first, because less cabling and fewer fuses are required.

Inrush current check
Both drives will have their individual inrush resistors in circuit. It is therefore only necessary to check that charging current does not exceed the I^2t value of the bridge rectifier.

Continuous current check
Because in a winder/unwinder application, as illustrated in Figure B5.3, one drive is always regenerating, the majority of the current circulates around the d.c. connections. The a.c. connection is only required to supply current for the losses in the system, which comprise drive losses, motor losses, and mechanical friction and windage

losses. In general these will be quite small in comparison to the drive rating and thus the continuous input current to the one drive is much smaller than the rating of one drive.

If there is any doubt, a measurement should be taken to ensure that the current remains below the typical value recommended by the manufacturer.

D.C. fuses should be connected to both negative and positive connections of each drive.

B5.4.6.2 Four identical drives with a single dynamic braking circuit

We will consider the case of four large drives of say 75 kW, which are designed to operate up to full load at the same time, but with the desire to have a single dynamic braking circuit to share between the drives. Again the drive circuit topology is critical. Let us assume the topology is as shown in Figure B5.10, with a controlled thyristor input stage and a d.c. inductor in the negative connection. Because the drives are all required to run at full load there are two possible solutions.

- Supply all drives using a.c. and d.c. connections.
- Supply all drives with d.c. only from a separate single converter.

The first option is the most cost-effective solution because no extra devices are required.

Inrush current check
If the d.c. capacitance of a single drive is 1 100 μF, then the total system d.c. link capacitance is 4 400 μF. Check the maximum capacitance that can be charged by the input stage of one drive. If it is higher than the total system capacitance no special measures are required to synchronise the start up, as all four drives can be powered up at the same time through a single supply-side converter.

The lowest-cost option of either one single contactor or individual contactors may be used; however, the system must not be allowed to run unless all contactors are closed. If any one contactor fails to close then the other three input converters could be overloaded once the run command is given. It is recommended that auxiliary connections are used to provide feedback to indicate that all contactors are closed.

Continuous current check
Because the drives have a d.c. bus inductor in the negative connection, a.c. line chokes are required to ensure the current is shared in the positive diodes. Two per cent line reactors are required, which can be calculated as described in Section B5.4.5.1.

B5.4.7 Note on EMC filters for common d.c. bus systems

If one EMC filter is used for the complete system it is important to note that it needs to be rated for the total drive motor cable length.

Chapter B6
Electromagnetic compatibility (EMC)

B6.1 Introduction

B6.1.1 General

The purpose of this section is to set out the necessary considerations for system designers and others when incorporating electronic variable-speed drives into complete machines and systems without encountering problems with electromagnetic interference, and in compliance with relevant regulations. Of necessity, only general guidelines have been provided, but because real installations have a wide variety of detailed requirements, explanation of the underlying principles is given, in order to allow the designer to cope with specific situations.

B6.1.2 Principles of EMC

All electrical equipment generates some degree of electromagnetic emission as a side-effect of its operation. It also has the potential to be affected by incident electromagnetic energy. Equipment using radio communication contains intentional emitters and sensitive receivers.

The basic principle of EMC is that electromagnetic emission of electrical equipment, whether intentional or unintentional, must not exceed the immunity of associated equipment. This means that controls must be in place on both emission and immunity. Given the variety and uncertainty of effects and situations, some margin of safety must be provided between these two factors.

Although all equipment exhibits some degree of emission and susceptibility, the limiting factors in most common environments tend to be related to radio equipment, with its powerful transmitters and sensitive receivers. Therefore the majority of EMC standards are related to the requirements of radio communications systems.

In principle, EMC covers phenomena over an unlimited range of frequencies and wavelengths. The EU EMC Directive limits itself to a range of 0–400 GHz. This range is so wide that a perplexing number of different effects can occur, and there is a risk that all electrical phenomena become included in the scope of EMC. This is also the reason for the proliferation of EMC 'rules of thumb', some of which are contradictory. A technique that is effective at a high radio frequency (RF) will probably not be effective at power frequency, and vice versa.

With current industrial electronic techniques, it is unlikely that equipment will exhibit unintentional sensitivity or emission above about 2 GHz. Below 2 GHz, it is convenient to make a first rough separation into 'high-frequency' effects, which correspond to radio frequencies beginning at about 100 kHz, and 'low-frequency' effects. Broadly speaking, low-frequency effects operate only by electrical conduction, whereas high-frequency effects may be induced and operate at a distance without a physical connection. Of course there is no precise dividing line between the two, and the larger the geometry of the system, the lower the frequency at which induction becomes effective. However, this division is helpful in understanding the principles.

B6.1.3 EMC regulations

Regulations exist throughout the world to control intentional and unintentional electromagnetic emission, in order to prevent interference with communications services. The authorities generally have the power to close down any equipment that interferes with such services.

Many countries have regulations requiring consumer and other equipment to be tested or certified to meet emission requirements – for example, the FCC rules in the United States and the C-tick system in Australia. The EMC Directive of the European Union is unusual in requiring immunity as well as emission to be certified.

It is not possible in the limited space available to explain all of these regulations. The EU EMC Directive has been the subject of much written material. Most emission regulations are based on international standards produced by CISPR, and the three basic standards CISPR11, CISPR14 and CISPR22 underlie most other emission standards.

B6.2 Regulations and standards

B6.2.1 Regulations and their application to drive modules

The underlying principle of all EMC regulations is that equipment should not cause interference to other equipment, and especially to communications systems. In addition, in many countries there is a requirement that equipment must be certified in some way to show that it meets specific technical standards, which are generally accepted as being sufficient to show that it is unlikely to cause interference.

Equipment standards are primarily written to specify test methods and emission limits for self-contained products such as electrical consumer goods and office equipment, which are basically free-standing units, even if they have the capability to interconnect with peripherals and networks. The emission levels set in such standards allow to some extent for the possibility that several items of equipment may be co-located. A corresponding approach in the realm of industrial products such as variable-speed drive modules has caused difficulty, because it is clear that some drive modules are used as virtually self-contained units, whereas others are built in to other end-user equipment, sometimes singly but also possibly in considerable numbers. The module cannot meaningfully be tested without its associated motor, cables and other peripherals, and the effect of a number of co-located modules is difficult to predict. Large fixed installations may contain numerous drives and other electronic products, and cannot practically be tested against standards, which were primarily intended for compact free-standing consumer products.

Most drive manufacturers have adopted a practical approach by testing their products in arrangements that are reasonably representative of their final use, and providing installation guidelines.

The original EU EMC Directive was replaced in 2004 (implemented 2006) by a revised version that contained useful clarifications of these topics. In particular, sub-assemblies liable to generate or be affected by electromagnetic disturbance were specifically included in the definition of 'apparatus', and requirements were added for such apparatus to be provided with information for its installation and operation so as to meet the essential requirements of the Directive. This means that the practical approach adopted by many drives manufacturers has now effectively been endorsed, with the added requirement that a CE mark for EMC is now specifically required.

B6.2.2 Standards

Standards with worldwide acceptance are produced by the International Electrotechnical Commission (IEC). Standards for application under the EU EMC Directive are European Harmonised standards (EN) produced by CENELEC. Every effort is made to keep these two families of standards in line, and most of them have the same number and identical technical requirements.

Emission standards work by specifying a limit curve for the emission as a function of frequency. A measuring receiver is used with a coupling unit and antenna to measure voltage (usually up to 30 MHz) or electric field (usually beyond 30 MHz). The receiver is a standardised calibrated device, which simulates a conventional radio receiver.

Immunity standards are rather diverse because of the many different electromagnetic phenomena that can cause interference. The main phenomena tested routinely are the following:

- electrostatic discharge (human body discharge),
- radio-frequency field (radio transmitter),
- fast transient burst (electric spark effect),
- surge (lightning induced), and
- supply dips and short interruptions.

There are very many more tests available; those listed are the required tests under the CENELEC generic standards.

The most important standards for drive applications are the following, all of which have equivalent EN and IEC versions:

- IEC 61800-3 Power Drive Systems (contains emission and immunity requirements);
- IEC 61000-6-4 Generic Emission Standard for the industrial environment;
- IEC 61000-6-2 Generic Immunity Standard for the industrial environment.

The product standard IEC 61800-3 applies in principle to variable-speed drive modules where they are sold as end products. There are, however, many cases where the drive will be incorporated into an end product, which is not in itself a power drive system, and is more likely to fall under the scope of the generic standards. In this case it is the generic standards that are of interest. The permitted levels are generally similar, except that IEC 61800-3 defines a special environment where the low-voltage supply network is dedicated to non-residential power users, in which case relaxed emission limits apply. This can permit useful economies in input filters.

B6.3 EMC behaviour of variable-speed drives

B6.3.1 Immunity

Most drives can be expected to meet the immunity requirements of the IEC generic standard IEC 61000-6-2. Control Techniques drives meet them without any special precautions such as screened signal wires or filters, except for the case of particularly fast-responding inputs such as data links and incremental encoder ports.

The standard sets levels corresponding to a reasonably harsh industrial environment. However, there are some occasions where actual levels exceed the standard levels, and interference may result. Specific situations that have been encountered are shown in Table B6.1.

B6.3.2 Low-frequency emission

Drives generate supply-frequency harmonics in the same way as any equipment with a rectifier input stage. Supply harmonics are discussed in detail in Section B1.2. Harmonics generated by an individual drive are unlikely to cause interference, but

Table B6.1 Methods to improve immunity

Situation	Effect	Cure
Very inductive d.c. loads such as electromagnetic brakes, without suppression, and with wiring running parallel to drive control wiring	Spurious drive trip when brake released or applied	Fit suppression to brake coil, or move wiring away from drive wiring
High RF field from powerful radio transmitter (e.g. in airport facility adjacent to aircraft nose)	Drive malfunction when transmitter operates	Provide RF screening, or move to a location further from the transmitter antenna
Severe lightning surges due to exposed low-voltage power lines	Drive trip or damage from over-voltage	Provide additional high-level surge suppression upstream of drive

they are cumulative so that an installation containing a high proportion of drive loads may cause difficulties.

Apart from supply harmonics, emission also occurs as a result of the switching of the power output stage over a wide range of frequencies, which are harmonics of the basic switching frequency (i.e. size times the supply frequency for a six-pulse d.c. drive), and the PWM carrier frequency for a PWM drive. This covers a range extending from 300 Hz for d.c. drives, up to many megahertz for a.c. drives. Unwanted electromagnetic coupling is relatively unusual at frequencies below about 100 kHz. Few standards set limits in that range, and interference problems are unusual.

B6.3.3 High-frequency emission

The power stage of a variable-speed drive is a potentially powerful source of electromagnetic emission ('noise') because of the high voltage and current, which is subject to rapid switching. Thyristors are relatively slow switching devices, which limits the extent of the emission spectrum to about 1 MHz, whereas with IGBTs it may extend to about 50 MHz. If attention is not paid to installation guidelines then interference is likely to occur in the 100 kHz to 10 MHz range where emission is strongest. Where conformity with an emission standard is required, it is not unusual to find excessive emission anywhere in the 150 kHz to 30 MHz range, where emission is measured by conduction, and occasionally in the 30 to 100 MHz range, where emission is measured by radiation.

This frequency range is lower than that associated with personal computers and other IT equipment, which tend to cause direct radiated emission associated with the internal microprocessor clock and fast digital logic circuits. The drive itself is not an important source of direct emission, because its dimensions are much less than a half wavelength over the relevant frequency range. There may be strong electric and magnetic fields close to the drive housing, but they diminish rapidly, by an inverse cube law, with increased distance from the drive. However, the wiring connected to the drive can be widespread and is likely to be long enough to form an effective antenna for the frequencies generated by the drive.

The power output connections of a drive carry the highest level of high-frequency voltage. They can be a powerful source of electromagnetic emission. Because the cable connecting the drive to the motor is a dedicated part of the installation, its route can usually be controlled to avoid sensitive circuits, and it can be screened. Provided the screen is connected correctly *at both ends*, emission from this route is then minimised.

Output filters can also be used, and are offered by specialist filter suppliers. Their design is quite difficult because they must offer high attenuation in both the common and series modes, while presenting an acceptable impedance to the drive output circuit and avoiding unacceptable voltage drop at the working frequency. They tend to be very costly, and they will usually only work with simple open-loop control because of their complex impedance within the drive closed-loop bandwidth. They may be justifiable in applications where it is not practicable to use a screened motor cable.

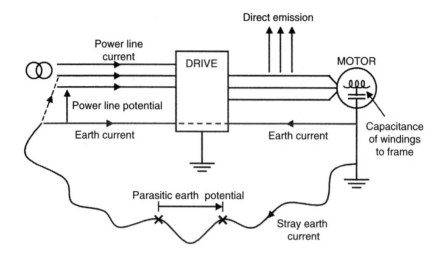

Figure B6.1 High-frequency emission routes

The power input connections of a drive carry a high-frequency potential, which is mainly caused by the current flowing from the drive output terminals to earth through the capacitance of the motor cable and motor windings to earth. Although the voltage level here is rather lower than at the output, control measures may be needed because these terminals are connected to the widespread mains supply network. Most commonly, a radio-frequency filter of some kind is installed here.

The control terminals of the drive carry some high-frequency potential because of stray capacitance coupling within the drive. This is usually of no consequence, but screening of control wires may be required for conformity with some emission standards.

Figure B6.1 summarises the main emission routes for high-frequency emission. Note that the current paths are in the common mode, i.e. the current flows in the power conductors and returns through the earth. Series mode paths are relatively unimportant in high-frequency EMC.

Because the return currents in the common mode all flow in the earth (equipotential bonding) wiring, earthing details are particularly important for good EMC. Much of the installation detail is involved with controlling the earth return paths and minimising common inductances in the earth system, which cause unwanted coupling.

B6.4 Installation rules

B6.4.1 EMC risk assessment

When a drive is to be installed, a cost-effective approach to EMC is to initially assess the risk of interference problems arising. This is in addition to considering any legal constraints on emission levels. Most industrial electronic instrumentation and data-processing equipment have good immunity, and can operate with drives with only modest precautions to control emission. Some specific types of equipment have been found to be susceptible to interference from drives. The following list shows some product families that call for special attention:

- Any equipment that uses radio communication at frequencies up to about 30 MHz. (Note this includes AM broadcast and short-wave radio, but not FM, TV or modern communications services, which operate at much higher frequencies.)
- Analogue instrumentation using very low signal levels, such as thermocouples, resistance sensors, strain gauges and pH sensors.
- Other analogue instrumentation using higher levels (e.g. 0–10 V or 4–20 mA), only if very high resolution is required or cable runs are long.
- Wide-band/fast-responding analogue circuits such as audio or video systems (most industrial control systems are intentionally slow acting and therefore less susceptible to high-frequency disturbance).
- Digital data links, only if the screening is impaired, or not correctly terminated, or if there are unscreened runs such as rail pick-up systems.
- Proximity sensors using high-frequency techniques, such as capacitance proximity sensors.

B6.4.2 Basic rules

For installations where it is known that no particularly sensitive equipment is located nearby, and where no specific emission limits are in force, some simple rules can be applied to minimise the risk of interference caused by a drive. The aspects requiring attention are described in the following sections.

B6.4.2.1 Cable segregation

The drive supply and output cables must be segregated from cables carrying small signals. Crossing at right angles is permitted, but no significant parallel runs should be allowed, and cables should not share cable trays, trunking or conduits unless they are separately screened and the screens correctly terminated.

A practical rule of thumb has been found to be as follows: No parallel run to exceed 1 m in length if spacing is less than 300 mm.

B6.4.2.2 Control of return paths, minimising loop areas

The power cables should include their corresponding earth wire, which should be connected at both ends. This minimises the area of the loop comprising power conductors and earth return, which is primarily responsible for high-frequency emission.

B6.4.2.3 Earthing

By 'earthing' we refer here to the process for connecting together exposed conductive parts in an installation, which is done primarily for electrical safety purposes but can also have the effect of minimising difference voltages between parts of the installation that might result in electrical interference. The more correct term is 'equipotential bonding', which is desirable at power frequencies for safety, and at higher frequencies to prevent interference. At power frequencies the impedance of the bond is dominated by its resistance, but at high frequencies the inductance is dominant. The actual connection to the earth itself is generally not important for either safety or EMC within an installation, but becomes important for connections between installations with separate earth arrangements.

The main drive power circuit earth loop carries a high level of RF current. As well as minimising its area as described above, these earth wires should not be shared with any signal-carrying functions. There are three possible methods for minimising shared earthing problems, depending on the nature of the installation.

1. *Multiple earthing to a 'ground plane'.* If the installation comprises a large mass of metallic structure then this can be used to provide an equipotential 'ground plane'. All circuit items requiring earth are connected immediately to the metal structure by short conductors with large cross-sectional area, preferably flat, or by direct metal-to-metal assembly. Screened cables have their screens clamped directly to the structure at both ends. Safety earth connections are still provided by copper wire where required by safety regulations, but this is in addition to the EMC ground plane.
2. *Dedicated earth points, earth segregation.* If a single earthed metallic structure does not extend throughout the installation, then more attention must be given to the allocation and arrangement of earth connections. The concept of separate 'power earth' and 'signal earth' or 'clean earth' has been discredited in EMC circles recently, but it is valid in widely spread installations where a good equipotential earth structure is not available. It is not necessary to provide separate

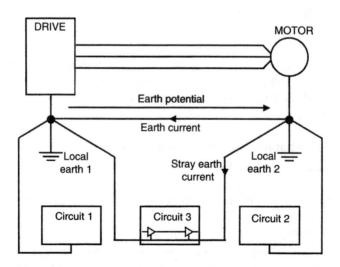

Figure B6.2   Earth potentials and their effect on signal circuits

connections to the earth itself, but to allocate a specific single point in the earth network as the 'reference earth' to which signal and power circuits are separately connected.

Figure B6.2 illustrates how two 'earthed' circuits in a system may have different noise potentials. Local circuits earthed to either point will work correctly (circuits 1 and 2), but if a single circuit is earthed at both points then it will experience a noise potential that might cause disturbance (circuit 3).

The solution is to nominate one earth point as the 'signal earth' and use it as the sole reference point for shared signal circuits, as illustrated in Figure B6.3. This prevents creating loops for the noise current. Circuit 3 must now be provided with signal isolation means if it is to communicate with the motor environment. The disadvantage is that this situation is difficult to manage in a large complex installation, and sneak paths can easily arise that cause problems and are difficult to trace. Alternatively, circuit 3 must be designed to be able to accept a high earth potential difference, for example by using optical isolation between the circuits associated with the motor and the drive.

3. *Common bonded network, correct use of screened cables.* An extension of the 'ground plane' principle for a large installation where a solid ground plane cannot be realised is to use a network of earth or equipotential bonding conductors. These conductors are selected to have sufficiently low resistance to ensure safety during electrical faults and to minimise low-frequency circulating current, but it has to be accepted that their inductance is too high to prevent differential potentials over their area at high frequencies. High-frequency differential potentials are prevented from affecting signal circuits by the correct usage of screened cables. The explanations given in Section B6.6 should assist in understanding it.

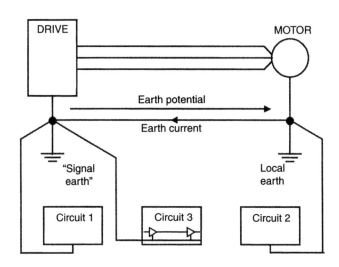

Figure B6.3   Use of single 'signal earth'

B6.4.3 Simple precautions and 'fixes'

There are some simple techniques that can be used to reduce high-frequency emission from a drive at modest cost. These techniques should preferably be applied in conjunction with the basic rules given above, but they may also be useful as a retrospective cure for an interference problem.

The single most effective measure that can be taken is to fit capacitors between the power input lines and earth, as illustrated in Figure B6.4. This forms a simple RFI filter, giving a reduction of typically 30 dB in overall emission into the supply network, sufficient to cure most practical problems unless exceptionally sensitive equipment is involved. Emission from the motor cable is not affected by this measure, so strict cable segregation must still be observed. The capacitors must be safety types with voltage rating suited to the supply voltage with respect to earth. Earth leakage current will be high, so a fixed earth connection must be provided. The values shown represent a compromise between effectiveness at lower frequencies, and earth leakage current. Values in the range 100 nF to 2.2 µF can be used.

The length of the motor cable affects emission into the power line, because of its capacitance to earth. If the motor cable length exceeds about 50 m then it is strongly recommended that these capacitors should be fitted as a minimum precaution.

A further measure, which reduces emission into both supply and motor circuits, is to fit a ferrite ring around the output cable power conductors, also illustrated in Figure B6.4. The ring fits around the power cores but not the earth, and is most effective if the conductors pass through the ring three times (a single pass is shown, for clarity). Section B6.5.4 gives an explanation of the effect of the ferrite core. The ferrite should be a manganese–zinc 'power grade'. Care must be taken to allow for the temperature rise of the ferrite, which is a function of motor cable length; the surface temperature can reach 100 °C.

B6.4.4 Full precautions

If there is known sensitive equipment in the vicinity of the drive or its connections (see list in Section B6.4.1), or if it is necessary to meet specific emission standards, then

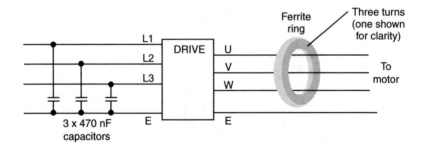

Figure B6.4   Some low-cost emission reduction measures

full precautions must be observed. The drive installation guide should give these precautions in full detail for specific drives. The following outlines the essential principles.

- A suitable input filter must be fitted. The filter specified by the drive manufacturer should be used, and any limits on motor cable length or capacitance and on PWM switching frequency adhered to. Many filters that are not specifically designed for this application have very little benefit when used with a drive.
- The filter must be mounted on the same metal plate as the drive, and make direct metal-to-metal contact, to minimise stray inductance. Any paint or passivation coating must be removed to ensure contact. A back-plate of galvanised steel, or other corrosion-resistant bare metal, is strongly recommended.
- The motor cable must be screened. A copper braid screen with 100 per cent coverage works best, but steel wire armour is also very effective, and steel braid is adequate.
- The motor cable screen must be terminated to the drive heat sink or mounting plate, and to the motor frame, by a very-low-inductance arrangement. A gland giving 360° contact is ideal, a clamp is also effective, and a very short 'pigtail' is usually tolerable, but the drive instructions must be adhered to.
- The input connections to the filter must be segregated from the drive itself, the motor cable, and any other power connections to the drive.
- Interruptions to the motor cable should be avoided if possible. If they are unavoidable then the screen connections should be made with glands or clamps to an earthed metal plate or bar to give a minimum inductance between screens. The unscreened wires should be kept as short as possible, and run close to the earthed plate. Figure B6.5 illustrates an example where an isolator switch has been incorporated.

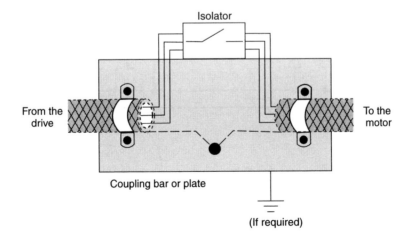

Figure B6.5   Managing interruptions to motor cable

With some drives, the control wiring needs to be screened with the screen clamped to the heat sink or back-plate. The installation instructions should be adhered to in this respect. Omitting this is unlikely to cause interference problems, but may cause standard limits for radiated emission to be exceeded.

B6.5 Theoretical background

B6.5.1 Emission modes

Although the digital control circuits, switch-mode power supplies and other fast-switching circuits in a modern digital drive can all contribute to the RF emission, their suppression is a matter for the drive designer, and suitable internal measures can keep such emission under control. It is the main power stage, especially the inverter of a PWM drive, that is an exceptionally strong source of emission because the fast-changing PWM output is connected directly to the external environment (i.e. the motor and motor cable). This is also the reason for the installation details having a major effect on the overall EMC behaviour.

Figure B6.6 shows the main circuit elements of an a.c. inverter variable-speed drive. The output PWM waveform has fast-changing pulse edges with typical rise times of the order of 50–100 ns, containing significant energy up to about 30 MHz. This voltage is present both between output phases and also as a common-mode voltage between phases and earth. It is the common-mode voltage that is primarily

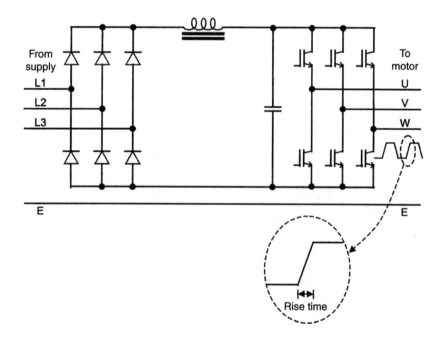

Figure B6.6 Main elements of an a.c. inverter drive

responsible for emission effects, because it results in high-frequency current flowing to earth through the stray capacitances of the motor windings to the motor frame, and the motor cable power cores to the earth core and/or screen.

High-frequency current causes unexpected voltage drops in wiring because of the wiring self-inductance. The significance of this can be illustrated by a simple example. A 1 m length of wire has a typical inductance of about 0.8 μH – the true value of course depends upon the current return path, but this is a typical value. The output current from a drive to charge the stray capacitance of a motor winding would be typically 2 A peak with a rise time of 100 ns. This current would cause a voltage pulse of 16 V with duration 100 ns in the 1 m of wire. Whether this causes interference with associated circuits depends on their design, but certainly a 16 V 100 ns pulse is sufficient to cause a serious error in a digital circuit or a fast-acting analogue one.

Figure B6.1 shows the main emission paths. Owing to the high voltage in the motor cable it is the main potential source of emission. It will be an effective transmitting antenna at a frequency where the motor cable length is an odd number of quarter wavelengths. For example, a 20 m cable will be particularly effective at about 3.75 MHz and also at 11.25 and 18.75 MHz. This will be modified somewhat by the presence of the motor and by the distance of the cable from the surrounding earthed objects. In order to prevent this emission, the cable is usually screened.

Figure B6.1 also shows how the high-frequency voltage in the motor and cable causes current to flow into the earth, because of their capacitance. The capacitance of a motor winding to its frame may be in the range 1 to 100 nF, depending on its rating and insulation design, and the capacitance from the cable power cores to earth is generally between 100 and 500 pF m^{-1}. These values are insignificant in normal sinusoidal supply applications, but cause significant current pulses at the edges of the PWM voltage wave. The current returns through a variety of paths, which are difficult to control. In particular, current may find its way from the motor frame back to the supply through any part of the machinery, and if it passes through earth wires in sensitive measuring circuits it may disturb them. Also, a major return route to the drive is through the supply wiring, so any equipment sharing the supply may be disturbed.

Figure B6.7 shows the effect of using a screened motor cable and an input filter. Fields emitted from the motor cable are suppressed by the screen. It is essential that both ends of the screen are correctly connected to the earthed metal structure at the motor and the drive, in order that the magnetic field cancellation property of the cable can give its benefit. The screened cable also minimises the earth current flowing from the motor frame into the machinery structure, because of its mutual inductance effect. This subject is generally not well understood outside the EMC profession (see Section B6.6.1 and the references for a fuller explanation).

The input filter provides a low impedance path from the earth to the drive input lines, so that the high-frequency current returning from the motor cable screen has an easy local return route and does not flow into the power network. The primary role of the filter is to suppress common-mode high-frequency emission from the drive. There is also some series-mode emission because of the non-zero impedance

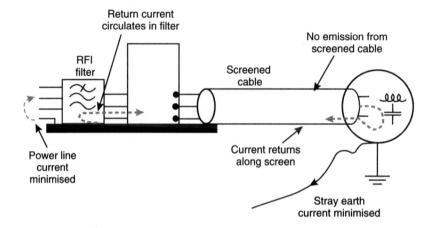

Figure B6.7   The effect of an RFI filter and screened motor cable

of the d.c. smoothing capacitor in the drive. The filter provides some series-mode attenuation to control this.

B6.5.2   Principles of input filters

Figure B6.8 shows the circuit of a typical input filter. The capacitors between lines provide the series-mode attenuation, in conjunction with the leakage inductance of the inductance. The capacitors to earth and the inductance provide the common-mode

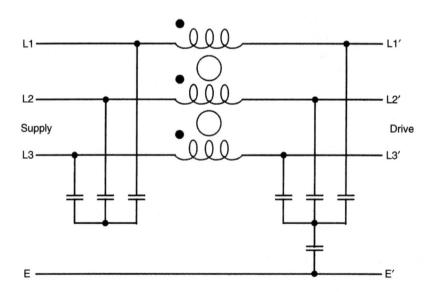

Figure B6.8   Basic input filter

attenuation. The inductance is constructed as a common-mode component, which is not magnetised by the main power current, therefore minimising its physical dimensions. It uses a high-permeability core, which can accept only a very limited unbalance (common-mode) current.

Filters for voltage source drives are carefully optimised for the application. The drive presents an exceptionally low impedance source to the filter, which means that conventional general-purpose filters may have little benefit. The usual method for specifying a filter is in terms of its insertion loss in a test set-up with 50 Ω source and load impedance. An alternative test attempts to be more realistic by using 0.1 Ω source and 100 Ω load. Neither of these tests correctly represents a drive application, and neither can be used as any more than a very rough guide to the suitability of a filter.

B6.5.3 Screened motor cables

The screening capability of screened cable is generally measured by the parameter Z_T, the transfer impedance per unit length. In an ideal cable, any current flowing in the internal circuit produces no voltage between the ends of the cable screen, and conversely current flowing in the screen from an external source produces no voltage in the inner circuit. These two aspects minimise the emission from the cable and the immunity of inner signal circuits to external disturbance, respectively. In practice the resistance of the screen, its imperfect coverage and other details cause a departure from the ideal and a non-zero value of Z_T.

The transfer impedance is not, however, the only factor involved. Because it is not terminated in its characteristic impedance, the cable exhibits strong internal resonances, which cause high currents to flow internally. The current is limited by the natural damping caused by electrical losses in the cable. Steel sheaths have a higher resistance and therefore give better damping than copper sheaths. Steel gives an inferior transfer impedance to copper, but the two factors largely cancel so that a steel wire armoured cable gives no greater emission with a drive than a good-quality copper braided screened cable.

B6.5.4 Ferrite ring suppressors

The use of a ferrite ring as an output suppressor was introduced in Section B6.4.3. The ferrite ring introduces impedance at radio frequencies into the circuit that it surrounds, thereby reducing the current. Because of its high permeability the ring will not work if it surrounds a conductor carrying power current, due to magnetic saturation, but if it surrounds a three-phase set then the magnetic field is only caused by the common-mode current, and saturation is avoided. The manganese–zinc ferrite exhibits high loss in the 1–10 MHz frequency range where motor cable resonance occurs, and this gives useful damping of the resonance and a substantial reduction in the peak current. The loss in the ferrite does cause a temperature rise, and with long motor cables the temperature of the ferrite rises until its losses stabilise, close to the Curie temperature.

The recommendation for using three turns on the ring is based on experience. The number of turns obviously affects the inserted impedance, nominally by a square law relationship, but the inter-turn capacitance limits the benefit and it is rarely effective to exceed three.

B6.5.5 Filter earth leakage current

Because of the low source impedance presented by the drive, suitable filters generally have unusually high values of capacitance between lines and earth. This results in a leakage current (or touch current) to earth at supply frequency exceeding the 3.5 mA that is generally accepted as permissible for equipment that derives its safety earth through a flexible connection and/or plug/socket. Most filters require the provision of a permanent fixed earth connection with sufficient dimensions to make the risk of fracture negligible. Alternative versions of filter with low leakage current may be available, which will have more severe restrictions on the permissible motor cable length.

B6.5.6 Filter magnetic saturation

With long motor cables the common-mode current in the filter may rise to a level where the high-permeability core of the filter inductor becomes magnetically saturated. The filter then becomes largely ineffective. Filters for drive applications therefore have limits on motor cable length.

The capacitance of the cable causes additional current loading on the drive and the filter. Screened cables with an insulating jacket between the inner cores and the screen present a tolerable capacitance. Some cables have the screen directly wrapped around the inner cores. This causes abnormally high capacitance, which reduces the permissible cable length. This also applies to mineral insulated copper clad cables.

B6.6 Additional guidance on cable screening for sensitive circuits

The subject of signal circuit cable screening is often misunderstood. It is quite common for such circuits to be incorrectly installed. This applies particularly to critical signal circuits for drives, such as analogue speed references and position feedback encoders, and also to circuits in other equipment in the same installation as the drive. This section outlines the principles, in order to assist readers in avoiding and troubleshooting EMC problems in complete installations.

B6.6.1 Cable screening action

Correctly used, a cable screen provides protection against both electric and magnetic fields, i.e. against disturbance from both induced current and induced voltage. Electric field screening is relatively easy to understand. The screen forms an equipotential surface connected to earth, which drains away incident charge and prevents current from being induced into the inner conductor. Magnetic field screening is more

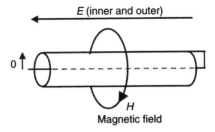

Figure B6.9   Magnetic field induction in screened cable

subtle. An incident alternating magnetic field, which corresponds to a potential difference between the cable ends, causes EMF to be induced in both the screen and the inner conductor. Because the screen totally surrounds the inner conductor, any magnetic field linking the screen also links the inner conductor, so an identical EMF is induced into the inner. The voltage differential between the inner and the screen is then zero. This is illustrated in Figure B6.9.

In order for this benefit to be realised, it is essential that the screen be connected at both ends. Whereas high-frequency engineers routinely observe this practice, it is common in industrial control applications for the screen to be left unconnected at one end. The reason for this to is to prevent the screen from creating an 'earth loop', or an alternative earth path for power-frequency current.

The problem of the 'earth loop' is specifically a low-frequency effect. If the impedance of the cable screen is predominantly resistance, as is the case at low frequencies, then any unwanted current flowing in the screen causes a voltage drop that appears in series with the wanted signal. This is illustrated in Figure B6.10.

At higher frequencies the cable screen impedance is predominantly inductive. Then the mutual inductance effect takes over from resistance, and the voltage induced in the internal circuit falls. A further factor is that the skin effect in the screen causes the external current to flow in the outer surface so that the mutual resistance between inner and outer circuits falls. The net result is that at high frequencies the cable screen is highly effective. A cut-off frequency is defined at the point where the injected voltage is 3 dB less than at d.c., and is typically in the order of 1–10 kHz. Where

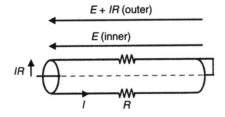

Figure B6.10   Resistance coupling in screened cable

disturbing frequencies exceed the cable cutoff frequency, the screen should be connected at both ends.

B6.6.2 Cable screen connections

The conclusion of this is that for all but low-frequency interference, the screen should be used as the return path for data, as shown in Figure B6.11. Whether the screen is connected to earth at each end, or to the equipment metalwork, is less important than that it be connected to the circuit common terminals. The recommendations of the equipment manufacturer should be followed. It is usual to clamp the screen to a metallic structural part because this gives the least parasitic common inductance in the connection. A 'pigtail' causes a loss of screening benefit, but a short one (up to 20 mm) may be acceptable for drive applications where screening is not critical.

The screened cable should ideally not be interrupted throughout its run. If intermediate terminal arrangements are included with pigtails for the screen connections, every pigtail will contribute additional injection of electrical noise into the signal circuit. If interruptions are inevitable, either a suitable connector with surrounding screening shell should be used, or a low-inductance bar or plate should be used for the screen connection as in Figure B6.5. Suitable hardware is available from suppliers of terminal blocks.

Low-frequency interference associated with earth loops is not important for digital data networks, digital encoder signals or similar arrangements using large, coarsely quantised signals. It is an issue with analogue circuits if the bandwidth is wide enough for errors to be significant at the relevant frequencies, which are primarily the 50/60 Hz power line frequency. Many industrial control systems have much lower bandwidths than this and are not affected by power frequency disturbance. Very-high-performance/servo drives, however, do respond at power line frequency, and can suffer from noise and vibration as a result of power frequency pick-up. The cable screen should not be used as the signal return conductor in this case. The correct solution for wide-band systems is to use a differential input. Analogue differential inputs give very good rejection of moderate levels of common-mode voltage at power line frequency (see Section C3.3.2.3). This rejection falls off with increasing frequency, but then the screening effect of the cable takes over. The combination of

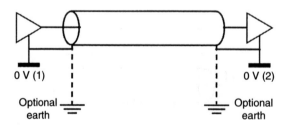

Figure B6.11   Correct screened cable connection for high-frequency screening effect

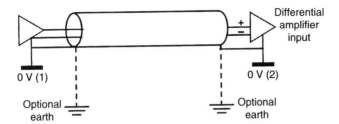

Figure B6.12   Correct screened cable connection for high-frequency screening and low-frequency interference rejection

differential connection and correct cable screening gives good immunity over the entire frequency range. A typical arrangement is illustrated in Figure B6.12.

There are electrical safety issues associated with earthing decisions. A galvanically isolated port with the screen connected only to the isolated common rail prevents low-frequency circulating current, but carries the risk that a fault elsewhere might make it electrically live and a hazard to maintenance staff. Cable screens should be earthed in at least one place for every disconnectable length, to prevent a length becoming isolated and live. This approach is used in the Interbus industrial data network, where each link is earthed at one end and isolated at the other.

Earthing at both ends carries the risk that an electrical fault might cause excessive power current to flow in the cable screen and overheat the cable. This is only a realistic risk in large-scale plant where earth impedances limit power fault current levels. The correct solution is to provide a parallel power earth cable rated for the prospective fault current. An alternative is to provide galvanic isolation, although this carries the risk of a transiently high touch potential at the isolated end during a fault.

Some galvanically isolated inputs include a capacitor to earth, which provides a high-frequency return path but blocks power-frequency fault current. This is actually a requirement of certain serial data bus systems. The capacitor will exhibit parallel resonance with the inductance of the cable screen, and will only be effective for frequencies above resonance. In principle such a capacitor should not be necessary, but it may be required to ensure immunity of the isolated input to very fast transients,

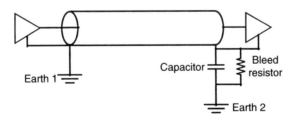

Figure B6.13   Use of capacitor for high-frequency earthing while blocking power fault current

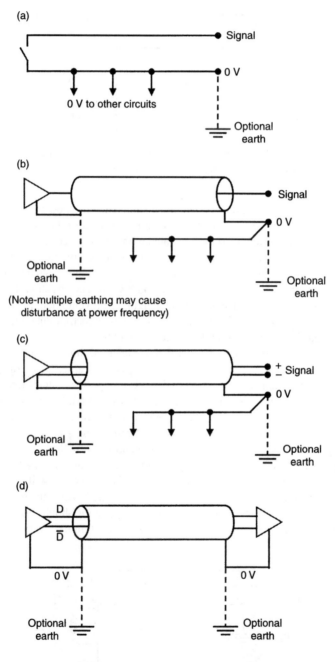

Figure B6.14 Preferred connection methods: (a) low-speed digital circuit with no special precautions and open wiring; (b) low-bandwidth, low-precision analogue circuit, screened; (c) wide-bandwidth and/or high-precision analogue circuit, screened and differential; (d) wide-bandwidth digital data circuit, screened and differential

Electromagnetic compatibility (EMC)   431

or to suppress RF emission from microprocessors, and so on. The capacitor must be rated at the mains voltage. It is usual to provide a parallel bleed resistor to prevent accumulated static charge. Figure B6.13 illustrates this capacitor arrangement.

B6.6.3 Recommended cable arrangements

Figure B6.14 summarises the recommended connection methods for the following cases:

- low-speed digital circuit – no special precautions, open wiring
- low-bandwidth low-precision analogue circuit – screened
- wide-bandwidth and/or high-precision analogue circuit – screened, differential
- wide-bandwidth digital-data circuit – screened, differential.

Chapter B7
Protection

B7.1 Protection of the drive system and power supply infrastructure

B7.1.1 General

This section considers the requirements for protection of the complete drive and motor system from the effects of faults and voltage transients. As in other chapters, much of the underlying philosophy is common to all electrical power installations and it is not the purpose to repeat it here. There are some special considerations in relation to drive systems, and these are addressed.

The most common form of circuit protection is the fuse, followed by various forms of circuit breaker. Standard industrial fuses and circuit breakers are designed to protect the distribution circuits from the effects both of long-term overloads and of short circuits. In the case of a circuit breaker there are separate actuators (thermal and magnetic) for these two functions, whereas the thermal design of the fuse is adjusted to effectively model the thermal behaviour of the cable in both respects. In all cases the device must also carry predictable safe peak currents such as the starting current of direct-on-line (DOL) motors without undue deterioration.

It is difficult to protect semiconductor devices from damage caused by a short circuit using a circuit breaker, as the response time needs to be very rapid, so the following account refers primarily to fuses. It should be understood, however, that in some cases, especially at the lower power levels, a circuit breaker may be able to fulfil the function equally well.

B7.1.2 Fuse types

There are three key fuse types that are relevant for drive systems. Here, we will consider the IEC terminology; other systems are also in use such as that used in the United States and defined in ANSI/UL standards. There are also many commercial variants that offer special features or combinations of features:

- standard industrial fuses for overload and fault protection, e.g. IEC type gG;
- semiconductor fuses for semiconductor short-circuit protection, e.g. IEC type aR; and
- Type gR fuses, which combine the characteristics of a fast-acting fuse with a slower-acting cable protection fuse in one package.

The type gG fuse is used throughout industrial plant for basic protection. It limits the fault current into a short circuit sufficiently to protect standard cables and to allow coordination with an upstream fuse two steps higher in the rating scale. It also provides overload protection for standard cables.

The type aR fuse is intended purely for protection of semiconductor devices against the effects of a short circuit. The fault current is severely limited so that the energy deposited in the circuit is minimised. For semiconductors this is measured in terms of the $i^2 t$ let-through rating of the fuse, which must be coordinated with the survival rating for the semiconductor. This type of fuse is not designed for overload protection, i.e. it will clearly operate if the continuous current is excessive, but not in a fashion that can be used to ensure the safety of the associated circuit.

Type gG fuses are designed for use with an a.c. supply. Fuses for d.c. circuits are more specialised and physically larger, because of the difficulty of interrupting a d.c. arc. Type aR fuses are usually specified for either a.c. or d.c. application, with a limitation of the L/R ratio for a d.c. circuit.

Type gR fuses are for use on the a.c. supply. They are applicable on drives up to approximately 132 kW.

B7.1.3 Application of fuses to drive systems

Virtually all modern drive designs provide internal protection against the effects of overloading, so provided the protection is trustworthy then there is no need to duplicate this protection with fuses. However local wiring regulations always take precedence, and in most installations the local safety code requires there to be 'branch circuit protection', which protects the downstream circuit from the effects of overloads as well as faults.

In many drive systems it will be necessary to use semiconductor fuses. For d.c. thyristor drives this is a requirement to prevent destruction of thyristors from a short circuit in the output, because thyristors cannot interrupt current. For a.c. drives the output is protected by the fast-acting over-current trip arrangement, which protects the output semiconductors, and the impact at the input of a short circuit at the output is negligible. Here, it is the effect of internal faults that must be considered, such as a failed power semiconductor or capacitor. Owing to the high energy density of such devices, it is often only possible to ensure equipment safety in the event of a failure by using semiconductor fuses at the input. When the drive manufacturer stipulates a particular fuse arrangement it is essential to follow this, and in some cases the independent approval status (e.g. UL listing or recognition) is conditional upon the correct protection being used.

A common fuse arrangement to fulfil the above requirements is a standard industrial fuse at the circuit branch point, with semiconductor fuses in the drive panel associated with the relevant drive modules. The question of whether branch overload protection is also required within a panel for power distribution to individual modules needs to considered with reference to the relevant safety standard for the intended place of use.

For the highest power converters it is common practice to provide a semiconductor fuse for every individual semiconductor device, rather than for each supply phase.

This gives the best device utilisation, in that the fuse must be selected to meet the i^2t capability of the device as well as the operating current of the equipment. Often if a single fuse is used for more than one device it will be found that a larger semiconductor than that required for purely thermal reasons has to be provided in order for the protection to be effective. In a d.c. drive there may also be an output fuse to provide protection from the available fault current from the rotating machine.

B7.1.4 Earth faults

In many applications there is little difference in outcome between short circuits between lines and from a line to earth. Wherever there is a protective bonding conductor with continuity to the source of supply (neutral or, occasionally, one phase), an earth fault results in a fault current comparable with the short-circuit current, and the protective device clears the fault current rapidly. The impedance of the earth path has to be coordinated with the prospective fault current, as limited by the protective device, so as to keep the temporary touch potentials of exposed conductive parts within safe limits.

The very fast short-circuit protection and the inherent limiting of fault current within an a.c. PWM drive means that the earth fault current in the output circuit is much reduced when compared with the equivalent DOL motor. However, in the absence of a system for safety certification, the bonding is usually designed on the assumption that the prospective fault current is the same as at the input.

If an earth fault occurs at the output of an a.c. drive when its output voltage happens to be set close to zero, and if the fault occurs into wiring that has considerable inductance, then it is possible that the fault current may not rise quickly to the trip point. This happens if the low-frequency voltage component is insufficient to establish a trip current, while the high-frequency current is limited by the stray inductance of the earth wiring. The drive internal protection system must be designed to ensure that such a fault can be detected, in the presence of a reasonable loop inductance, and the drive tripped before damage can occur to the semiconductors. Alternatively, separate earth fault protection can be used. Such a system must be designed to distinguish between current caused by an earth fault and the normal repetitive charging current flowing into the stray capacitance of the motor cable and motor at the PWM pulse edges.

B7.1.5 IT supplies

IT supplies, which are insulated from earth, may be used in industrial installations where the low-voltage circuit is restricted to that installation. In this kind of installation a single earth fault results in no fault current, but it can be detected by an earth impedance monitor circuit, and action is taken to correct the fault within a reasonable time, during which operation can continue. A similar arrangement is often used in ships and other marine installations.

The effect of an earth fault in an IT system requires special consideration when a.c. drives are used.

An earth fault on the supply system results in the supply-phase voltages becoming unbalanced relative to earth, with two of them approaching the normal line voltage. There is no reason why the drive should not continue to run in this condition, but it must be designed to withstand this temporary abnormal voltage. Safety standards such as IEC 61800-5-1 give rules for insulation design with this kind of system. Generally, any special requirements are quite modest, special attention being required primarily for the design of RF input filters.

An earth fault at the drive output results in an abnormal condition, which might be sustained and has the potential for causing damage. The PWM inverter output voltage is imposed on the entire power system, relative to earth, by such a fault. The consequences are difficult to predict with confidence as they depend on the nature of the other equipment connected to the supply. There are three basic scenarios.

- Operation continues unaffected. This will happen if the other equipment is not particularly sensitive to the voltage appearing between lines and earth.
- The drive trips. This will occur if there is a sufficiently low impedance between power lines and earth for the PWM inverter pulses, caused usually by RFI filter capacitors and/or cable capacitance, for the inverter trip current level to be reached.
- Operation continues but other equipment is disturbed or even damaged by the PWM voltage.

The first scenario is desirable, and if the power circuit is well-defined and limited then it may be entirely acceptable. For a widespread installation it carries a risk that equipment installed at some time in the future might be adversely affected. Any equipment using an RFI filter with capacitors connected to earth is potentially at risk of damage.

The second scenario is safe but may be unacceptable because of lost production. The only certain solution then is to use an isolating transformer at the drive input.

Clearly the last scenario is unacceptable, and it is necessary to decide in advance whether it is a possibility. It can be prevented by using some form of sensitive earth current trip arrangement, and/or by adding capacitance between lines and earth so that the drive does trip. If continued running into a fault is required then an isolating transformer must be used at the drive input.

B7.1.6 Voltage transients

Voltage transients exist in power systems because of natural lightning activity and power system switching and protection operations. Although the nature of transients can vary widely, the most common kind that is considered both for electrical insulation design and electromagnetic immunity is the unidirectional $1.2/50$ μs impulse referred to in a variety of standards such as IEC 61000-4-5, IEEE c62.45 and IEC 61800-5-1. Tests for insulation are referred to as 'impulse tests' and for EMC immunity as 'surge tests'; the same basic effect is covered except that for an insulation test the only consideration is possible breakdown of the insulation, whereas for an EMC test the interest is in whether the equipment operation is upset.

Table B7.1 Over-voltage categories for a system voltage of 300 V

Over-voltage category	Impulse voltage	Category meaning
I	1 500	Measures are taken to reduce over-voltages to a low level
II	2 500	Equipment such as appliances and household equipment
III	4 000	Permanent connection to an industrial installation
IV	6 000	Origin of the installation, e.g. electricity meter

For the design of electrical insulation the magnitude of impulse voltage is determined by the 'over-voltage category', an example of which is given in Table B7.1 for a 300 V system voltage which is defined as the continuous voltage between a phase and earth when operating with the maximum nominal supply voltage.

The over-voltage category is decided by the location of the equipment in the power system. The rationale for this process is rather obscure, because it is not obvious why the over-voltage should reduce between the origin of the installation and the appliance. Only for category I are specific measures used to reduce over-voltages, and the impulses are too slow to be much attenuated by the cable impedance. The answer appears to be that this process is really a coordination process, and that the over-voltage level is actually decided by the electrical spacings used in the distribution infrastructure and connected equipment. This is the reverse of the concept usually applied, which is to choose spacings through air to avoid the risk of breakdown.

Industrial equipment is generally considered to operate in category III. This means that for a standard low-voltage supply we can expect transient voltages between lines and from lines to earth of up to 4 kV, and if we take account the fact that there are no specific measures to reduce the category IV voltages to category III, we might decide that voltages of 6 kV could occur. The implications of this are quite severe – for a mains-connected rectifier the semiconductors would apparently have to be rated for non-recurrent voltages of 4 or even 6 kV in order to be sure of reliable operation, whereas for normal operation only 1.2 kV is required and devices rated at over 1.8 kV are specialised and have higher conduction voltage drops. This would make for expensive and less efficient rectifiers, and the solution in practice is either to make use of the natural voltage-limiting property of the circuit – for capacitor-input rectifiers – or to use voltage surge suppression devices such as metal oxide varistors to reduce the overvoltage.

In order to select the correct device a knowledge of the source impedance for the impulse is also required. The most commonly used model is the 'combination wave generator' defined in standards such as IEC 61000-4-5, which gives a short-circuit current waveform to correspond to the voltage. The peak current into a short circuit is numerically half the voltage, and a cautious approach to matching a varistor to a semiconductor for a 400 V a.c. supply is to match the 3 kA voltage drop of the varistor with the non-repetitive voltage rating of the device. This addresses the voltage stress experienced by the semiconductor junctions.

438 The Control Techniques Drives and Controls Handbook

The stress on the earth insulation of an isolated device must also be considered, and because most devices are not intended to withstand 6 kV impulses, a voltage suppression arrangement between lines and earth may also be required.

B7.2 Motor thermal protection

B7.2.1 General

The reason for motor protection requiring careful consideration can be illustrated by considering a typical large (>100 kW) induction motor, which has an efficiency of the order of 95 per cent. It is designed to dissipate the 5 per cent losses at full load while remaining at a suitable operating temperature, although to be cost-effective some of its materials will be quite close to their limits for reliable long-term operation. The remaining 95 per cent of input power passes out of the motor as mechanical power at the shaft. If a fault were to cause a large part of this 95 per cent of input power to be dissipated in the motor rather than leaving through the shaft, then the heat input to the motor would rise to 20 times its normal value without necessarily being detected from electrical measurements. The temperature could rapidly rise to a destructive and possibly dangerous value. Unlike for a transformer, a differential protection scheme that compares input and output is very difficult to implement.

A conventional induction motor is designed so that the most common causes of abnormal losses, such as excessive load torque and stall, can be detected by an increase in the input current, so that when operating from a fixed supply a simple protection relay is sufficient.

Whether operating direct on the mains supply or under the control of a variable-speed drive, a motor can be thermally overloaded by any of the following:

- increased ambient temperature;
- reduction of the coolant flow;
- increased losses in the machine, caused by
  - overloading, continuously or dynamically (i.e. by repeated overloading or starting),
  - phase loss,
  - excessive voltage and/or magnetic flux density,
  - abnormal supply, e.g. unbalance or excessive harmonic voltage.

B7.2.2 Protection of line-connected motor

For a line-connected induction motor, protection against overloading is usually provided by a current-sensing thermal relay in the starter, with a time response corresponding to that of the motor windings. Electronic relays are also available and offer a wider range of adjustments than the simple thermal kind. The setting is chosen so that continuous or dynamic overloads result in the relay tripping to protect the motor, whereas operation with high peak loads that are within the motor capability remains available. A simple current-sensing relay does not necessarily provide protection against repeated starting beyond the motor capability, because in this situation the rotor power dissipation is disproportionately high due to operation at high slip. An additional timer/counter arrangement may be required to prevent excessive starting.

A simple thermal relay has an inherent 'memory', in that its internal temperature continues to mirror that of the motor winding when the power is removed. It is therefore able to protect the motor within its capabilities when repeated starts are performed.

The motor starter often also provides phase-loss protection. Excessive voltage is unlikely for a direct line-connected motor, except in the unlikely event of a malfunctioning star/delta starting switch. Large machines may also have a negative sequence current relay for protection against excessive supply unbalance.

For protection against increased ambient temperature or reduction of cooling, special provisions would be made if there were considered to be an exceptional risk of such events. A temperature sensor embedded in the motor winding and connected to a suitable relay provides protection against most foreseeable causes of thermal overload.

B7.2.3 Protection of inverter-driven motor

The possible causes of thermal overload are the same when operated from an inverter, but the more flexible range of operating conditions increases the chances of some of the more unusual causes.

The inverter drive normally takes over the function of the motor protection relay. The same thermal protection arrangement is often used for the motor and drive, because the drive is typically designed with short-term overload capability to match the motor. With inverter operation the constraint on frequency of starting does not apply because the stator current is always under control during starting and the slip is low, so the rotor dissipation does not exceed its normal peak value and no additional start protection is needed.

The use of a drive facility for a basic safety protection function requires careful consideration. For example, in order for the safety approvals body UL to accept such a facility to ensure motor overload protection, it is required to undergo 100 per cent testing in production. The function uses drive firmware, which cannot develop a fault in the hardware sense, but the usual precautions for firmware change control should be in place in order to avoid inadvertent alteration of the function, and steps should be taken to ensure that the operation is simple and transparent, as human error in setting up the function is more likely if it is complex or offers multiple options that can readily be confused.

Because the protection is carried out by calculations in the drive processor, power must be present at the drive. This is the case when the motor is controlled through the drive control functions, but in applications where the drive power is removed during operation, the thermal data will usually be lost unless special provision is made to store it in non-volatile memory. To fully simulate the action of a conventional thermal relay a real-time clock is used so that the calculations can re-commence when power is re-applied.

Thermal protection by an embedded temperature sensor is facilitated with a drive, because it usually offers a suitable sensor input with programmable alarm and trip functions. It is recommended that this facility be used wherever practicable, because of the multiple possible causes of excessive motor temperature. In addition to the more obvious risks, such as the reduced cooling at reduced speed when using the motor shaft fan, the intelligent control capability of a drive increases the possibility that under abnormal conditions the drive might operate the motor under a stable controlled condition that could result in excessive motor dissipation, rather than giving a simple over-current trip as would be the case with DOL operation.

B7.2.4 Multiple motors

An inverter may supply more than one motor. In that case the motor protection function cannot normally be performed within the drive as it has no data for the individual motors. Each motor must be provided with a protection relay.

The inverter output circuit is a harsh environment for an electronic sensing circuit, and generally the simple thermal relay is preferred to an electronic type unless the latter is specifically designed for inverter supplies. Even then, attention must be given to the effect of the current pulses that charge the motor cable capacitance at each pulse edge. For small motors with long cables, for example a 400 V, 4 kW motor with cables longer than 50 m, the rms value of the high-frequency impulse current can be comparable with the motor operating current. In such a case, the thermal relay then responds to this current by tripping prematurely. The following are some practical solutions to this difficulty.

- Locate the relays at the motor ends of the cables.
- Use a series choke to reduce the high-frequency current.
- Add a capacitor to bypass the high-frequency current around the circuit containing the relay and choke.

B7.2.5 Servo motors

The above has focused primarily on standard industrial induction motors. The same considerations apply to servo motors, but the emphasis is then much more on the requirement for highly dynamic operation with high transient loading. The drive can provide an advanced thermal model of the motor, allowing the highest winding temperature to be estimated from a knowledge of the current history and temperature measurements such as that of the motor body or stator core. This can be used to control the current delivered to the motor in order to prevent excessive temperature, allowing operation very close to the true limit of capability of the motor. This avoids the need to over-size the motor to provide a rating margin, as well as preventing an unexpected trip if the application turns out to exceed the capability of the motor.

The duty cycle of the load and the cooling are critically important when dimensioning a servo motor and the manufacturer should be consulted to obtain the optimum solution and protection.

Chapter B8
Mechanical vibration, critical speed and torsional dynamics

B8.1 General

Many mechanical drive trains, both fixed-speed and variable-speed, experience vibration. As operating speeds and controller performance continue to increase and motor mass and inertia fall, the danger of resonance problems increases. This subject area is complex and this description will be limited to an overview of the principles and identification of the key sources of excitation of mechanical resonances.

The vibration level of a mechanical drive train (motor, coupling, load, and so on) is the result of imposed cyclic forces either from residual imbalance of the rotor, or from some other cyclic force and the response of the system to these forces.

Problems tend to exist in one of three categories of application:

- high-power, high-speed applications, where operation above the first critical speed is required;
- applications where torque ripple excites a resonance in the mechanical system;
- high-performance closed-loop applications, where the change of motor torque can be very high, and the shaft linking the mechanical parts twists, and the control loop sustains the vibration

Although the torque ripple produced from modern variable-speed drives is small, in comparison with earlier technology, harmonic torques are produced. It is natural therefore to initially conclude that, in a system employing an electrical variable-speed drive, torque ripple is the problem; in practice this is rarely the case.

It is important to recognise that any system in which masses (or inertias) are coupled together via flexible elements is capable of vibrating. Even a simple motor and load can be seen to be a two-mass torsional system.

Consider the following completely general case. For the two inertias shown in Figure B8.1 the equations of motion, if zero damping is assumed, are

$$J_1(d^2\theta_1/dt^2) + K(\theta_1 - \theta_2) = 0$$
$$J_2(d^2\theta_2/dt^2) + K(\theta_2 - \theta_1) = 0$$

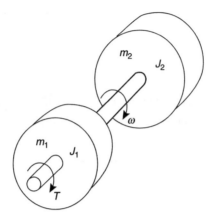

Figure B8.1   Two-mass torsional system

Eliminating θ_2 or θ_1 gives the general solution

$$\theta = A + Bt + C\cos(\omega_{ntf} + \phi)$$

where

$$\omega_{ntf} = \sqrt{[K(J_1 + J_2)/(J_1 \times J_2)]}$$

or

$$f_{ntf} = (1/2\pi)\sqrt{[K(J_1 + J_2)/(J_1 \times J_2)]}$$

where ω_{ntf} is torsional natural frequency (rad^{-1}), f_{ntf} is torsional natural frequency (Hz) and K is torsional stiffness ($= GJ_p/L$, where G is shear modulus of elasticity, J_p is polar moment of inertia of the shaft, which is equal to $r^4/2$ for a circular shaft of radius r, and L is the length of shaft being twisted).

Example
An a.c. motor of inertia 0.5 kg m^2 is coupled to a load of inertia (0.4 kg m^2) via a shaft with a torsional spring constant of 60 × 10^3 N m rad^{-1}.

$$\begin{aligned} f_{ntf} &= (1/2\pi)\sqrt{[k_t(J_1 + J_2)/(J_1 \times J_2)]} \\ &= (1/2\pi)\sqrt{[60 \times 10^3 (0.5 + 0.4)/(0.5 \times 0.4)]} \\ &= 82.7\,\text{Hz} \end{aligned}$$

The system designer needs to be able to calculate the source and expected level of the forces in the system, and then design each component accordingly. Some of the important principles and factors to be considered are detailed in the following sections.

B8.2 Causes of shaft vibrations independent of variable-speed drives

Consider first the causes of vibration, which are entirely independent of a variable-speed drive. It is helpful to consider these in the following categories:

- sub-synchronous vibrations (vibration frequency below the shaft rotational frequency),
- synchronous vibrations (vibration frequency at the shaft rotational frequency),
- super-synchronous vibrations (vibration frequency above the shaft rotational frequency), and
- critical speeds.

B8.2.1 Sub-synchronous vibrations

The most common cause of sub-synchronous vibration is in induction motor systems where beating at slip frequency, or a multiple thereof, can occur. This is due to the fact that all electromagnetic forces in an induction motor occur at frequencies equal to or a multiple of the supply frequency. The rotational speed, however, is slightly less, and mechanical imbalance forces will be cyclic at this reduced frequency. This is a classic example of two cyclic forces at relatively close frequencies combining to give a low-frequency beat.

In the fault condition of an induction motor with a broken rotor bar, the vibration at slip frequency will dramatically increase.

B8.2.2 Synchronous vibrations

The most likely cause of vibrations at the shaft rotational frequency is mechanical imbalance or shaft misalignment.

Mechanical imbalance may be simply a specification/manufacturing quality issue. It is surprising how common it is for confusion as to whether the motor is balanced with the shaft key fit or not or a half key. ISO 8821 has led to a broad acceptance of the half key convention. The definition of a half key recognises that a key profiled to fill the whole volume of the keyway is often impractical and allows a full-length rectangular key of half height or a half length key of full height, the latter to be centred axially in the keyway. Refer to Section B3.8.1 for more details on vibration and balancing of motors.

Imbalance may also be due, in larger motors, to shaft bending owing to unequal cooling or heating of the rotor. Thermal problems are usually time dependent, and are characterised by a gradual change in vibration with time and/or load.

Offset rotor or an elliptical stator bore will both result in cyclic magnetic forces at shaft rotational frequency. If the system inertia is sufficient and it is possible to disconnect the motor supply, this problem will suddenly disappear.

B8.2.3 Super-synchronous vibrations

Super-synchronous vibrations tend to be associated with out-of-roundness bearings or shaft asymmetry along its length due for example to a single large keyway along its length. It is unusual for either of these issues to present a practical problem.

Wear or damage to roller bearings could be a problem and should be considered. Initial bearing problems will result in only a small increase in bearing vibration, but as wear increases, 'sudden' catastrophic failure of the bearing could occur, leading possibly to shaft or even stator core damage.

B8.2.4 Critical speeds

As the physical ratings and, through the use of variable-speed drive systems, operating speeds of motors increase, motors are being designed for application above their first (and in some cases even their second) critical speed. Such applications require careful consideration and detailed analysis to avoid problems.

The critical speed of a shaft is not solely dependent upon the characteristics of the shaft alone, but is greatly affected by the stiffness of the bearing supports.

The actual magnitude of the shaft and bearing housing vibration is dependent upon the resonance curve for the shaft, and the closer it is running to the critical speed the higher the vibration level.

Reference to the motor manufacturer must be made if there is any concern about operating near the critical speed of the motor. It is not common for manufacturers to publish critical speed data. It is common for maximum motor speeds to be published, and they provide comfort for the vast majority of applications. Refer to Section B2.3.5 for more details on maximum speed of operation.

B8.3 Applications where torque ripple excites a resonance in the mechanical system

Torque ripple is inherent in almost all electrical variable-speed drives. The frequency and magnitude is dependent upon the type of converter and control applied.

Consider first the d.c. motor fed from a six-pulse converter. The six-pulse converter is so named because the resultant motor armature current has ripple comprising six peaks for every cycle of mains frequency. On a 50 Hz mains supply this ripple has a fundamental frequency of 300 Hz (60 Hz → 360 Hz). In a separately excited d.c. machine the torque is proportional to armature current, and the torque ripple has a frequency of six times mains frequency. The magnitude of this ripple is typically in the range 10–20 per cent rated torque. The frequency of the torque ripple is a function of the mains frequency and is independent of operating speed, so provided that 300 Hz

(360 Hz) is not close to the natural resonant frequency of the mechanical system, no problems should result. Torque ripple due to the commutation process in the motor is also important where a small number of commutator segments are used. This component of torque ripple has a frequency that is proportional to speed.

Considering now the situation with a PMW inverter system. Although the theory of torque ripple calculation is complex, torque ripple at six and twelve times the output frequency of the drive are of most practical importance. The magnitude of the torque ripple is dependent upon the magnitude of the current harmonics, which is in turn very dependent upon the drive and control type and the demands of the application. It is not possible to give exact calculations of torque ripple, but a general comparison by drive type given in Table B8.1 is helpful.

It is interesting but of limited value to consider the mathematics behind this form of resonance.

Consider again the equations of motion for a two-mass system described in the introduction to this section, but as we are considering torsional vibrations considering the relative displacement of one body to the other, we obtain

$$J(d^2\theta/dt^2) + K\theta = 0$$

where J is the total inertia of the system.

If we have a driving force $r(t)$, i.e. a force that is time variant, we can obtain the differential equation of the new system by adding the driving force to the above equation:

$$J(d^2\theta/dt^2) + k_1\theta = r(t)$$

If we consider the simple case $r(t) = F \cos \omega t$, then

$$\theta(t) = k_2 \cos(\omega_0 t - \delta) + \{F/k_1[1 - (\omega/\omega_0)^2]\} \cos \omega t$$

Table B8.1 Comparison of torque ripple by drive type

Drive/motor type	Torque ripple level	Source of ripple	Frequency of ripple
D.C. drive	(a) High	(a) Armature current ripple	(a) 6 × mains
	(b) Very low	(b) Commutator	(b) Speed × segments rev^{-1}
PWM fed induction/PM motors	Very low	Interaction of flux harmonics with fundamental current	Sub-harmonics of synchronous speed
SR/stepper	High	Torque waveform inherently pulsed	Harmonics of number of 'steps' per revolution

The key issue to note is that this represents a superposition of two harmonic oscillations. The frequency of the first is the natural frequency of the system $\omega_o/2\pi$, the second ω, the frequency of the input.

By inspection, the amplitude ρ of the oscillation at the input frequency depends upon ω and ω_o. As $\omega \rightarrow \omega_o$ the amplitude ρ tends to infinity. This is as expected, because when the forcing frequency is the same as the natural frequency of the system, resonance occurs.

B8.4 High-performance closed-loop applications

B8.4.1 Limits to dynamic performance

A rule of thumb that is contained in EN 61800-2:1998/IEC 61800-2:1998 is that in order to avoid vibration during operation, the speed controller of a drive must be tuned to a value such that

$$f_{ntf} \gg 10/T_R$$

where T_R is the requested response time of the speed controller; i.e. the time required following the initiation of a change in demand for an output going in the correct direction.

This is a very global, somewhat imprecise and certainly conservative rule of thumb, and in that context, it is reasonable to assume T_R is the time to reach 80 per cent of the demanded speed following a step change in demand.

B8.4.2 System control loop instability

Instability of this type is where the change of motor torque can be very high, the shaft linking the mechanical parts twists, and the control loop sustains the vibration; i.e. the control loop acts as a positive feedback to the vibration rather than negative damping feedback.

As a further complication in high-performance closed-loop systems, it needs to be remembered that the speed/position feedback device constitutes a further third mass in the drive train. This is practically important in some applications and can limit torque bandwidth performance to below 1 kHz.

This form of instability is highly complex in nature and, although practically important is beyond the scope of this book.

B8.5 Measures for reducing vibration

Some vibration of a motor drive train is inevitable, and providing resonances, critical speeds and so on are avoided, they can be tolerated. In such systems steps can be taken to reduce vibration and the resulting noise. The primary measures fall into four primary categories.

1. Improve balancing and stiffening to reduce the amount of vibration generated and ensure proper alignment of all rotating parts. Resonance may also occur at certain oscillation frequencies in the surfaces of speed-controlled machines. An example of this may be a tie bar used to link the end frames of a motor. Simple measures of applying an intermediate node would resolve any possible problems.
2. Use isolation to prevent vibrations being transmitted. A wide variety of isolators including simple rubber mats, custom machine shoes through to specialised dampers offering specific stiffness in various directions.
3. Use appropriate non-linear, detuner-type torsional couplings.
4. Build robust foundations. The dimensioning of foundations is critical to ensuring that vibrations from the machine are not transmitted to the structure and that no resulting damage occurs. It is not possible to include the detailed design calculations for foundations here. It is important to recognise the vital nature of foundations to a successful installation.

Chapter B9
Installation and maintenance of standard motors and drives

B9.1 Motors

B9.1.1 General

The installation, commissioning and maintenance of industrial power drive equipment requires careful regard to the relevant safety legislation. Industrial power supply voltages and high-speed high-torque drive systems, unless handled properly, can represent a serious safety hazard.

All equipment must be used in accordance with the duty, rating and conditions for which it is designed, and particularly the power supply must be in accordance with that shown on rating plates, subject to standard tolerances. The loading and speed of the motors must not exceed those of their rating plates or any overload ratings agreed formally with the manufacturer. No attempt should be made to open inspection apertures or similar openings, unless the motor is known to be fully isolated from the power supply and the motor cannot be rotated from the load side. All safety and protection guards and covers should be in place before motors are started.

Power and control cables must be of adequate current-carrying capacity and be voltage grade for the duty, and properly mounted and secured. The power supply system to which the equipment is connected must have an adequate short-circuit fault level.

Motors must not be operated under ambient conditions for which they were not designed. Ambient air temperature should not exceed 40 °C or the temperature agreed, nor should cooling air be contaminated in any way injurious to the machine, nor restricted in flow into ventilating inlet apertures.

B9.1.2 Storage

If motors are stored prior to installation, storage conditions should be such as to avoid deterioration, otherwise the manufacturer's guarantees could be invalidated. Motors, either cased or not, should never be stored out of doors. Packing cases are invariably not weatherproof.

Storage should be clean, dry and free of vibration, which can cause brinelling damage to motor bearings. Extremes of temperature and humidity can cause injurious condensation and should be avoided.

B9.1.3 Installation

Motors should be mounted on rigid, level foundations or flange mountings that are horizontal or vertical or as specified. Mounting rigidity is important, particularly with brush gear machines, because motor frame vibration can cause excessive brush wear. Any deflection of the motor frame by bolting down upon an out-of-true base must be avoided by the use of shim washers.

It is wise to blank off any open motor apertures during installation to avoid foreign material entering the motor. If wide temperature and humidity variations are possible, heat should be carefully applied to ensure that the motor is dry internally before starting up.

A check of the insulation values of the windings to earth should indicate not less than 20 MΩ. If a figure below this value is seen, refer to the supplier who will advise suitable action. It is important that during such insulation tests any electronic apparatus including any drive and protection components associated with the motor should first be disconnected, otherwise damage may occur.

With machines containing brushes, check that all brushes are in position, can be moved freely in their holders, press on to their running surface with equal pressure, are fully bedded onto this running surface, and that all connections are tight. If practicable, rotate the shaft by hand to check freedom and smoothness. If a speed/position feedback device is fitted, check any coupling for security and accuracy of fitting. If the feedback device is stub-shaft mounted, check for concentric running (0.05 mm or 0.002 in eccentricity at the stub shaft end may be regarded as maximum) and for correct axial positioning, with the brushes in the centre of the commutator width in the case of a d.c. tachogenerator.

Ensure that there is no obstruction to the cooling air flow to the motor, and that the body/fin casing of a totally enclosed fan-cooled machine is clean, free from debris, and that the air inlet to the fan cowl is unobstructed. With screen-protected motors ensure that inlet and outlet ventilating apertures are clear. Check that the motor is not in the direct path of hot air flow from other machines or equipment, and that all guards and covers are in position.

If the motor has a forced vent fan unit, ensure that the fan impeller rotates in the correct direction. This needs to be done by careful inspection, because the wrong direction of running does not necessarily reverse the airflow but can reduce the air volume below 60 per cent flow, with a risk of winding burnout. An arrow normally indicates the correct direction. If in doubt, refer to the manufacturer. If this is not convenient, a good indication can be obtained by checking the airflow in both directions of impeller rotation. The largest volume and the highest air velocity usually indicate the correct direction of rotation.

The fitting of pulleys and couplings to motor shafts calls for care if these are of interference fit, because excessive axial pressure or impact force can easily damage modern precision bearings. Heating the pulley or coupling prior to mounting is good practice, as is supporting the motor shaft at the non-drive end before pressing on. The use of a tapped hole in the motor shaft end, now a standard with many manufacturers, to pull on a pulley or coupling is obviously a sound technique.

Installation and maintenance of standard motors and drives 451

For motors fitted with brush gear in particular, it is good practice to specify a balanced pulley. Most manufacturers balance rotors to good standards; a seriously out-of-balance pulley can adversely affect brush gear performance, and in any case imposes excess load on bearings.

Correct belt alignment and tensioning is required for best belt life as for best motor bearing life. Correct coupling alignment is important in avoiding excessive radial and axial loading on the motor bearings.

Perhaps the most common coupling between the motor and its load is the V belt, which often allows the use of higher speed, more efficient, more economically priced driving motors. It also acts as a buffer against mechanical shock loading between the motor and load. The wedge action of the V belt gives increased friction between belt and pulley for a given belt tension. This reduces belt slip, thus extending belt life, and by reducing motor bearing side loading by comparison with the equivalent flat belt, extends bearing life also.

Another advantage of the V belt is that wide drive ratios of up to 4 or 5 to 1 are practical, as are short centre drives when space is limited. The shortest spacing between centres approximates to the diameter of the larger V pulley. Modern belts initially stretch in service, before stabilising. It is therefore most important to check V belt tension after a few days of running and to readjust the tension to recommended values.

B9.1.4 Maintenance guide

Maintenance involves much that is self-evident yet is of increasing importance where emphasis is upon minimum downtime and production loss. Regular inspections ensure that motors remain clean and dry externally and internally. Oil in particular must not be allowed to accumulate on motor brush gear, commutator and slip rings. Ventilating grills and air filters should not be allowed to become obstructed. Compressed air for cleaning should be used with care, so that contaminants are not driven inside the motor or between winding coils. A small centrifugal blower is more appropriate.

The majority of motors below 100 kW have sealed bearings, which are capable of good performance over many years of service. In fact many such bearings are shielded rather than sealed, for true seals at modern motor shaft speeds would produce significant additional bearing heating. Where after several years of running a motor is taken out for a service overhaul, it is customary to replace the bearings, or at least melt out the existing grease, clean up and replace with the correct specification of heated grease, ensuring its penetration onto the bearing running tracks.

Motors with provision for bearing re-greasing should be given four or five strokes from a grease gun filled with the correct grease for the bearing type and speed, at intervals of approximately 2/4 000 hours determined by running speed. Some large machines incorporate grease escape valves on the bearing housings; these bearings should be re-lubricated in accordance with the lubrication plate fitted to such machines. Over-lubrication is the most common cause of bearing failure. An over-greased bearing can overheat, seriously damaging the bearing and changing the lubricating properties of the grease, which in turn does further damage.

Ball and roller bearings require little attention other than a periodic check while running for unusual noise or signs of overheating. Whistling noises are usually caused by defective lubrication, and rumblings by contaminated grease or damaged surfaces. Bearing running temperature depends upon loading, speed, motor and ambient temperatures.

Modern motors run quite warm and it is expected that bearing temperature will be slightly above that of the motor frame for typical speeds. For high-speed motors, bearing temperature can be expected to be somewhat higher and bearings will operate quite satisfactorily at temperatures up to 100 °C, although this would be high for current designs. Any significant changes in running temperature requires investigation, and if there are any signs of bearing damage – heavy black staining or blueing, surface cracking, brinelling, track indentation or excessive wear – the bearing should be replaced.

Detailed lubrication instructions are generally provided by the manufacturer and these should of course be followed.

B9.1.5 Brush gear maintenance

For safety, the drive should be isolated from the electric power supply before attempting any work within the motor. The covers should be removed at regular intervals so that brush gear can be inspected and cleaned. Experience with the installation will suggest the frequency, determined by motor use, speed, vibration level and general cleanliness. After initial commissioning, it is wise to check brush gear at weekly intervals to obtain an appreciation of brush gear behaviour. When this is confirmed to be good, the intervals can be increased with confidence.

In general, brush wear, after an initial bedding period of a month or so, should not be less than 3/4 000 hours per centimetre of brush length. Total permissible wear is to about 50 per cent of the original length. If upon examination the brushes have not worn excessively, have an even glaze (but not a polished shine) on the running face, are not scored or chipped or broken, slide freely in the brush boxes, are not discoloured, and have flexible conductors correctly attached, then they can be returned to service.

Any dust accumulation on the brush gear can be removed by a stiff brush or an air blower, and in replacing the brush it should be ensured that the flexible connection is securely fixed and free of any obstruction.

Brush contact pressure is fixed by the manufacturer in the case of 'tensator' clock spring-type tensioning. Both this type and spring-arm-type tensioning can be checked with a spring balance. As a general rule, brush pressure should be around 0.2 kg cm^{-2} of brush area for speed up to 2 500 rpm, 0.25 kg cm^{-2} for higher speeds, and for traction machines subject to high vibration levels, 0.3 kg cm^{-2}.

With spring-arm brush tensioning, adjustment can be made by using alternative slots on the arms. However, changes from those set by the manufacturer should only be made for good reason.

In fitting new brushes, it is important that brushes of the originally fitted grade are used and carefully bedded in, one brush at a time. Lift each brush in the holder and place a long strip of grade '0' glass paper, abrasive side to the brush, between the brush and the running surface. Draw the glass paper backwards and forwards following the curved surface of the commutator or slip ring to ensure that the whole face of the brush is shaped to the correct curve. Repeat for each brush in turn, ensuring that carbon and glass dust is kept out of the motor. Never use carborudum paper, the particles of which can embed permanently in the brush face and thereafter score the commutator surface.

Any disturbance of the brush box mounting arm holding ring position (rocker ring), requires that the ring be returned to the previous position precisely and clamped. Its position is determined for the best commutation during the manufacturer's testing and marked with paint or a stamped mark on the rocker ring.

Brushes are reaching the end of their useful life when worn down to about 50 per cent of their new length. All brushes should be changed together as a set, even if some are worn less than 50 per cent.

If a commutator or slip-ring surface is marked, blackened or grooved, it can be corrected by the careful application of a commutator stone (glass/epoxy 'commstone'). The stone should rest on a convenient brush arm, be kept in firm contact with the commutator or slip ring surface, and moved axially to ensure that the whole working surface is cleaned as the motor shaft is rotated briskly by hand. Never attempt to clean a commutator with power on the motor.

Serious marking or scoring requires the commutator or slip rings to be skimmed on a lathe, to maintain concentricity. Thereafter the skimmed surface has to be lightly polished. Skimming will sometimes require commutator mica insulation between segments to be undercut, that is, cut back to avoid abrasive contact with the brush face. At the time of undercutting the segment, slot edges should be given a slight chamfer to ensure that carbon does not accumulate in the slot. This is skilled work and requires proper training.

The running surfaces of commutators and slip rings should have a smooth chocolate-brown appearance after a few weeks of use. This 'patina' indicates low friction and good electrical performance and should not be disturbed. Slight pinpoint white sparking at the brushes is generally non-injurious and in fact appears to assist in the early establishment of a good patina.

In the event of marked sparking, blackening, or scoring of the running surface, inspect for rough areas, eccentricity, or flats developing. With commutators, inspect also for protruding or recessed copper segments or high slot micas. Depending upon the severity of the cause, the damaged surface may be cleaned with a strip of glass paper or a commutator stone. Again it is stressed that these operations should only be carried out by a skilled operator. If this correction proves unsuccessful, skimming between lathe centres, and mica undercutting in the case of the commutator, will be necessary.

A common cause of poor commutation, excessive brush wear rates, and commutator scoring or blackening is low brush current density for prolonged periods. Driving a d.c. motor mechanically with no current flow through the brushes would result in very high brush wear rates and damage to the commutator surface. Current flow through the brush/commutator contact area and also the presence of some water vapour in the atmosphere are essential to good brush gear performance. It appears that a state of ionisation in this contact area, with carbon and water molecules present, provides the 'lubrication' necessary and, as previously mentioned, a little pinpoint white sparking helps.

In general, at least 60 per cent of the nameplate current should, on average duty, flow through the brushes. This is easy to assess in terms of current density, knowing that half the total number of brushes are connected to the positive and half to the negative armature terminals. By measuring the cross-section of each brush and multiplying that area by half the brush number, the total brush area per terminal is known.

The current density for standard motor designs should not be less than 77 mA mm^{-2} (50 A in^{-2}). Where the current density is below 60 per cent of full load current continuously, it can be raised by removing one brush per brush arm from the commutator track, leaving an equal number of brushes on each brush arm. If in doubt, the motor manufacturer should be consulted.

B9.2 Electronic equipment

B9.2.1 General

It is essential that electrical equipment is isolated from the incoming supply and that sufficient time is allowed for any internal supplies to discharge fully before any work or internal adjustment is started. In general terms there is little need, if any, for routine maintenance of modern power electronic equipment because the solid-state technology involved simply has no components, beyond power fuse gear, that require planned replacement. However, there are occasions when work is required; for example, where an electronic control fault is suspected it may be useful to monitor important supply voltages within the equipment or to substitute a spare printed circuit board. The following points are pertinent.

B9.2.2 Location of equipment

In both a.c. and d.c. applications, important considerations apply to the location of the drive module. A primary consideration is ventilation, which is essential to allow the drive to perform to its full specification. Other vital considerations are ambient temperatures, humidity and purity of the cooling air. This is especially important in locations where carbon black or flammable solvents are present. Therefore the location of the drive has to be carefully planned and in severe cases ducting may have to be arranged to carry air from outside the area or building. The ambient temperature of the air drawn into the ducting should also be taken into account, to ensure that the heat sink is able to dissipate the heat generated by the output devices.

Normal limits of ambient temperature for electronic drive modules are -10 to $+40$ °C before alternative methods of cooling have to be considered. If ambient temperature falls below -10 °C, controlled heating may be necessary.

Good practice within high-power drive cubicles is to have separate cooling-path ventilation built into the panels. When locating these cubicles, care must be taken

Installation and maintenance of standard motors and drives 455

to ensure that the inlet and outlet vents are not obstructed in any way by other equipment (which in itself may generate heat) or by any other structure.

B9.2.3 Ventilation systems and filters

Power electronic equipment, typically below 20 A, is often cooled by natural convection via an arrangement of electrically isolated heat sinks that themselves form part of the equipment structure. Equipment above the 20 A level is most often housed in wall mounting or freestanding cubicles and frequently has a ventilation system consistent with the environmental rating of the enclosure itself. This may in its simplest form be a totally enclosed cubicle of sufficient volume to ensure that the dissipation over the surface area of the exposed sides is adequate for the heat generation within. Next would be a simple louvre system which, beyond ensuring that the louvres are not restricted in any way, would present no reason for concern.

Forced air arrangements, making use of an air input and exhaust route, are very common and normally include air filtering of a conventional paper or fibre type. It is unusual, however, to have a system that warns of filter contamination beyond those normally guarding the power electronics themselves against over-temperature, and consequently it is important to check such filters regularly. This service period is normally defined by experience of the actual environment.

Filters are most commonly of the disposable or non-reusable kind. However, some may be reused and the recommendations of the supplier should be followed. Do not, ever, risk operating the equipment without the filter element; the atmosphere may carry electrically conductive particles, which will eventually cause malfunctioning if allowed to settle on the electronic equipment where high voltages are nearly always present.

B9.2.4 Condensation and humidity

Generally, equipment intended for use in areas of high humidity or condensation is designed to minimise the possible generation of water vapour and water droplets because the presence of water in any form is definitely undesirable in equipment of this type. Anti-condensation heaters are commonly used and should be regularly checked for correct functioning.

Early in the service of any installation it is advisable to watch for the formation of any unexpected condensation, especially in situations where the equipment is not powered up continuously and where temperature cycling may be a contributory factor. In such cases anti-condensation heaters should be retrofitted.

B9.2.5 Fuses

Fuses have a finite life and may be expected to fail due to fuse-element ageing, especially if the normal current lies towards the upper end of the fuse rating. Many fuse gear manufacturers quote life expectancy under stipulated conditions, but the operating conditions of most installed systems are seldom known and only an average life can be rightfully expected from any fuse.

Consequently, when encountering an open-circuit fuse it is important not to immediately assume that there is a specific reason for failure other than ageing. However, it must be stressed that before a fuse is replaced without further question, it is most important that appropriate electrical tests are made to ensure that there are no obvious short-circuits or overloads in the protected circuit. Extensive supplementary damage can be caused to both electronic equipment and associated electrical equipment by replacing fuses or resetting circuit breakers before conclusive tests.

Modern fuse technology is extremely complex, with many special-purpose fuses being specified, especially so in the protection of semiconductor devices. It is not acceptable to simply fit a replacement fuse of the same current rating. The replacement fuse must be either a direct replacement of the original or an exactly comparable type of approved and listed characteristic demonstrating equivalent current, voltage and rupturing capability. If necessary, the fuse supplier or manufacturer should be asked for verification. Replacement of any fuse by one of a greater value is rarely necessary and any such decision should be carefully considered.

Part C
Practical Applications

Introduction
C1 Application and drive characteristics
C2 Duty cycles
C3 Interfaces, communications and PC tools
C4 Typical drive functions
C5 Common techniques
C6 Industrial application examples

Introduction

The correct and optimised application of a drive in a system requires detailed knowledge and understanding of the application as well as a clear understanding of what if possible to achieve with a drive. Armed with this combined information world class solutions can emerge. This is frequently achieved to best effect when OEM's or end users work together with the motor and drive manufacturer to design the solution. The applications knowledge is primarily in the domain of the user but when the user clearly articulates the problems he has to solve rather than the drive specification he wants, breakthroughs in system design performance and cost can result.

It is interesting to note that in many applications it is difficult to obtain good information about the load torque/speed characteristic and the duty cycle of the machine. Both of these characteristics directly impact the design, cost and performance of the system.

The aims of this section of the book is to illustrate, by example, some of the opportunities that the application of a drive can bring. This ranges from energy saving through to improvements in system performance. Some typical applications will be analysed with a view to showing what functionality and benefits are possible.

Methods of interfacing to a drive are discussed as this is again a critical area of system design, which is changing as Fieldbus systems start to dominate the industrial landscape. The days of a service technician with their trusty soldering iron and passive components to tune a drive system are long gone. Today, commissioning, data logging and fault finding are all done with a PC at the drive, in the control room or in principle on the other side of the world. PC Tools are an important consideration for many drive users, many allowing users to create their own specific functions within the drive.

It has been described in other areas of the book how a modern drive contains many functions beyond a basic speed control loop, and some of these are considered, though and exhaustive review is not attempted.

Control techniques which are found in common and some specific applications are described. Some applications have been selected for further consideration as examples of how drives can be used to provide elegant solutions.

Chapter C1
Application and drive characteristics

C1.1 General

It is not practical to describe all the characteristics for every application and/or every electrical variable-speed drive. This chapter aims to provide an insight into some of the possibilities/opportunities. Typical characteristics are covered. This is an overview only.

C1.2 Typical load characteristics and ratings

In order to successfully select and apply the optimum drive system, it is necessary to understand the essential features of both the alternative drive technologies and the load to be driven. The following tables list common loads and could prove useful when selecting a drive.

Table C1.1 Metals industries

Drive duty	Rating range	Comments on application and usual drive type
Rolling mill	Up to 1 000 s of kW	Some mills involve high-impact torque loading as the ingot is gradually reduced in gauge and converted to plate. A constant power speed range is used as the rolling torque reduces with gauge. Specific steel works' specifications and an arduous environment require careful consideration. Closed-loop induction motor drives and d.c. drives predominate.
Strip mill	Up to 100 s of kW	Normal torque loading (150% maximum). A constant power speed range is used as the rolling torque reduces with gauge. Specific steel works' specifications and an arduous environment require careful consideration. Closed-loop induction motor drives and d.c. drives predominate.
Bar and rod mills	Up to 100 s of kW	Normal torque loading (150% maximum) and constant power speed range. Specific steel works' specifications and an arduous environment require careful consideration. Requires integrated drive control to achieve correct speed ratios between stands. Closed-loop induction motor drives and d.c. drives predominate.
Continuous casting lines	10 s of kW	High reliability required and accurate load sharing. Normal torque loading (150% maximum) and constant power speed range. Specific steel works' specifications and an arduous environment require careful consideration. Closed-loop induction motor drives and d.c. drives predominate.

Application and drive characteristics 463

Levellers, slitters, perforators and sheeters	50 to 150 kW	In metal finishers plant, environment and specification easier than above – normally IP23 enclosure; forced ventilation with filter is acceptable. Drives are often part of an integrated system with coiler and uncoiler (wind/unwind stand) drives. Closed-loop induction motor drives and d.c. drives predominate.
Coiler/uncoiler; winder/unwinder	Up to 200 kW	Constant power rating over diameter build-up range. Regenerative braking with four-quadrant operation. Used in hot and cold mill applications. Steel works' specification and difficult environment need consideration. In metal finishers plant easier conditions apply as above. Closed-loop induction motor drives and d.c. drives predominate. Require special winder control functions.
Cut to length lines	10 to 50 kW	Various types, flying shear, rotary knife and guillotine systems. In metal finishers plant, environment and specification easier – normally IP23 enclosure; forced ventilation with filter is acceptable. Often involves uncoiler and other processes such as leveller. Closed-loop induction motor drives and d.c. drives predominate. May employ permanent-magnet servo drive for precise control of cut length.
Tube mill	Up to 300 kW	Constant power rating over pipe diameter range. Usually four-quadrant with regenerative braking. Environment can be difficult with oil spray present. Closed-loop induction motor drives and d.c. drives predominate.
Cast tube spinner	20 to 50 kW	High values of acceleration and deceleration torque required. Four-quadrant regenerative. Typically 4 to 6 speeds required: clean, spray, fill, spin 1 and spin 2. Difficult environment. Single-pipe vent or box enclosed motor with filtered air supply. Closed-loop induction motor drives and d.c. drives predominate.

Table C1.2   Machine tools

Drive duty	Rating range	Comments on application and usual drive type
Machine tool spindle	Up to 150 kW (5 to 30 kW typical)	High-speed applications up to 80 000 min^{-1}. Mostly flange mounting. Always significant constant power control range, and reversible, four-quadrant drive. Often with encoder for spindle orientation. Forced-vent with filter; often coaxial fan unit to 60 kW. Stator rotor units are supplied to large OEMs for integration into the spindle head. Closed-loop induction motor drives predominate, although at low powers permanent-magnet servo drives are used. Historically, d.c. drives were universally used and some are still applied. Motors tend to be of high inertia to provide very smooth rotation.
Machine tool axis drives	Up to 200 kW (4 to 30 kW typical)	Foot or flange mounting gearbox is usual, otherwise as above. Some linear motors are being applied in this area. Permanent-magnet servo drives. At higher powers closed-loop induction motor drives and d.c. drives are used. On machines requiring rapid point to point motion, motors tend to be of low inertia. On precise contouring machines where very smooth rotation is required higher inertia motors are used.

Table C1.3   Plastics

Drive duty	Rating range	Comments on application and usual drive type
Extruder	5 to 400 kW	Constant torque drive with high torque required to start a stiff extruder screw. Environment can be difficult with plastic particle and fume risk. Single-pipe vent of motor is advisable. Closed-loop induction motor drives and d.c. drives predominate. Open-loop induction motor drives may be used in some applications.

(Continued)

Table C1.3   Continued

Drive duty	Rating range	Comments on application and usual drive type
Process line wind up associated with blown film lines, bubble wrap and many other film production process lines.	1.5 to 15 kW	Constant power drive over diameter build-up ratio but often sized as constant torque drive. Braking usually mechanical. Environment can be difficult. TEFC IP55 used for low power ratings. Closed-loop induction motor drives and d.c. drives predominate. Requires special winder control techniques.

Table C1.4   Rubber

Drive duty	Rating range	Comments on application and usual drive type
Banbury mixer	Up to 1 000 kW	Very heavy peak duty, duty cycle rated typically as 250% full-load torque for 10 s, 150% for 20 s, 100% for 120 s, 10% for 30 s repeating continuously. Difficult environment with particle rubber and carbon black; single- and double-pipe vent is usual, with some CACA and CACW motors used. Check through Banbury manufacturers drive specification – safety environment. Closed-loop induction motor drives have now largely replaced the traditional d.c. drive solution.
Calender	Up to 500 kW	Environment as above, with easy duty, constant torque, but 200% dynamic braking. Check through manufacturer's drive specification, taking particular note of safety aspects. Closed-loop induction motor drives and d.c. drives predominate.

Table C1.5 Chemical applications

Drive duty	Rating range	Comments on application and usual drive type
Mixer	Up to 150 kW	Generally constant torque, but could be rising torque requirement with increasing mix stiffness. Often explosion-proof or hazardous-location enclosure requirement; environment can be difficult. Closed-loop induction motor drives and d.c. drives predominate.
Extruder	Up to 100 kW	Usually constant torque; often explosion-proof or hazardous-location enclosure requirement. Environment can be difficult. Closed-loop induction motor drives and d.c. drives predominate.
Stirrers and agitators	Up to 400 kW	High-reliability requirement to avoid loss of mix through drive shutdown. Often rising torque with mix stiffness. Energy saving of importance as process can occupy many days. Drive often outdoor mounting with CACA weatherproof enclosed motor. In particularly exposed positions motor and gearbox should have additional protection. Closed-loop induction motor drives and d.c. drives predominate.

Table C1.6 Material handling

Drive duty	Rating range	Comments on application and usual drive type
Conveyor	0.5 to 20 kW	Cascading of multiple drives can be a requirement with progressive speed increase or synchronised drives. Constant torque application with dynamic or regenerative braking. Open-loop induction motor drives predominate unless synchronisation/close coordination required, in which case closed-loop induction motor drives are used.
Automated warehousing	0.5 to 20 kW	Motors could be sited outdoors. Usually three-axis systems, constant torque four-quadrant 150% full-load torque at starting to duty cycle rating. Long travel drive must accelerate high inertia. Closed-loop induction motor drives predominate. D.C. drives can be used and in less demanding applications open-loop a.c. is used.

Table C1.7 Lift, hoist and crane applications

Drive duty	Rating range	Comments on application and usual drive type
Lift and hoist	30 to 100 kW	Four-quadrant with up to 300% full-load torque at starting. Starting duty 90–200 starts per hour. Smooth, quiet response very important. Closed-loop induction motor drives are widely used. D.C. drives are used. Open-loop drives are applied to applications and installations where ride quality is not critical.
Crane	3 to 75 kW	Four-quadrant high-torque requirement. Sometimes weatherproof enclosure. Operating duty cycle requires evaluation. As for lift and hoist, closed-loop induction motor drives are widely used. D.C. drives are used. Open-loop drives are applied to applications and installations where ride quality is not critical.

Table C1.8 Concrete pipe manufacture

Drive duty	Rating range	Comments on application and usual drive type
Pipe spinner	10 to 100 kW	Motor needs particular protection against water, cement and vibration. Four-quadrant multi-speed drive requirement with regenerative braking. Duty cycle requires evaluation. Traditionally a d.c. motor has been used but the closed-loop induction motor drive is now the usual choice.

Table C1.9 Fans and blowers

Drive duty	Rating range	Comments on application and usual drive type
Axial flow fan	0.5 to 40 kW	Cage motors specially adapted for air stream use with impeller on motor shaft. Existing motors will be retained under inverter control and may require slight de-rating or slightly reduced maximum speed. Inverse cube law relationship between fan power and speed. Noise falls as the fifth power of fan speed. Open-loop induction motor drives are most widely used.

(*Continued*)

Table C1.9 Continued

Drive duty	Rating range	Comments on application and usual drive type
Centrifugal fan	0.5 to 500 kW	Cube law power–speed relationship. Need to extend acceleration time on large fans to moderate starting current requirement. Power saving important on large fans as is top speed fan noise. Open-loop induction motor drives are most widely used.
Rootes type blowers	3 to 200 kW	Positive displacement blowers are constant torque and kilowatt loading is linear with speed into a fixed system resistance. Load pulsates heavily. Rootes blowers are noisy but easily started. Power saving can be important. D.C. drives and closed-loop induction motor drives predominate. Open-loop induction motor drives can be used with care.

Table C1.10 Pumps

Drive duty	Rating range	Comments on application and usual drive type
Centrifugal pumps	0.5 to >1 000 kW	Power saving on large drives important. Open-loop induction motor drives predominate.

Table C1.11 Paper and tissue

Drive duty	Rating range	Comments on application and usual drive type
Paper machine and pumps	Up to 500 kW	Environment difficult with water, steam and paper pulp present. Pipe vent motors common. Often non-standard a.c. and d.c. motors. Usually closely co-ordinated drives in a paper line. Closed-loop induction motor drives and d.c. drives predominate.
Winders and reelers	5 to 200 kW	Constant power range over diameter build-up range. Four-quadrant operation with regenerative braking. IP23 motor enclosure with filter is common. Closed-loop induction motor drives and d.c. drives predominate. Many motors are specified to operate at speeds that avoid the need for gear boxes.

Table C1.12   Printing

Drive duty	Rating range	Comments on application and usual drive type
Printing press	Up to 200 kW	Some special co-axial motor designs for tandem (series) connection on line shafts. Field weakening for wide speed range. Four-quadrant with slow ramp acceleration and inch/crawl control plus emergency stop. Some safety issues need careful consideration on inching. Pipe vent where ink fumes may be a hazard. Shaftless systems require precise position and registration control from the drive. Closed-loop induction motor drives and d.c. drives predominate, but permanent-magnet servo drives are used in some applications.
Folders, unwind and rewind stands	Up to 100 kW	Often integrated in printing line drive with press drive and unwind stand drive under 'master' control. Otherwise as above. Reel-stands often tape (surface) driven for speed synchronisation for splicing then tension controlled by friction brake. Closed-loop induction motor drives and d.c. drives predominate.

Table C1.13   Packaging

Drive duty	Rating range	Comments on application and usual drive type
Boxing, stamping, folding, wrapping	Up to 75 kW	Mostly four-quadrant with slow ramp acceleration with inch/crawl and emergency stop. Often integrated line control. Permanent-magnet servo drives are widely used in precision packaging machines. Closed-loop induction motor drives (and some d.c. drives) are also used.

Table C1.14   Engineering industries

Drive duty	Rating range	Comments on application and usual drive type
Test rigs of many types	Few kW to >15 MW	Test rig drives require careful engineering. Often high-speed with fast response, accurate speed and torque measurement, and usually four-quadrant with a constant power range. Engine test rigs require special knowledge of throttle control drive/absorb changeover and power measurement. Drive control/monitoring is particularly important. Closed-loop induction motor drives and d.c. drives predominate. Permanent-magnet servo drives are also used for some applications to 300 kW but this is not usual.

Table C1.15   Wire and cable applications

Drive duty	Rating range	Comments on application and usual drive type
Wire drawing	5 to 75 kW	Older systems based on constant-power speed range with individual motor field control on multi-block drives with single controller. Modern installations use individual controllers for each block, often with a.c. drives fed from a common d.c. bus. Progressive speed increase between blocks (heads) as wire diameter reduces and speed increases. Dancer arm tension control between blocks (heads) and tension controlled winder take off. Soap powder lubricants make dusty environment. Forced-vent motors require filter against wire end entry. Four-quadrant drives with precise acceleration/deceleration control. Closed-loop induction motor drives with inverters are the usual choice, but d.c. drives are sometimes still applied.
Bunchers and stranders	10 to 150 kW	Generally multiple drives with cage or bow, capstan plus take-up drives under integrated control. Constant torque, except take-up, with four-quadrant acceleration/deceleration with inch/crawl/E-stop controls. Motors require filter protection against metal dust entry. Accurate speed ratio control necessary to maintain correct cable lay. Closed-loop induction motor drives and d.c. drives predominate.

(Continued)

Table C1.15   Continued

Drive duty	Rating range	Comments on application and usual drive type
Capstan	5 to 100 kW	As for Bunchers and stranders.
Take-up and unwind stands	5 to 50 kW	As for Bunchers and stranders but constant power over diameter build-up ratio.
Extruders	5 to 150 kW	Used to apply insulating materials. See extruders under plastics (Table 3), but control often integrated in cable line drives.
Armourers	10 to 150 kW	As buncher/strander drive above.
Caterpillars	1.5 to 30 kW	Constant-torque duty and low power rating in view of low haul-off cable speeds. Often integrated in cable line drives. Motor protection generally no problem. Closed-loop induction motor drives and d.c. drives predominate.

Table C1.16   Hydraulics applications

Drive duty	Rating range	Comments on application and usual drive type
Pump and motor test rigs	Up to 250 kW	Hydraulic fluid is a contamination risk. Pipe vent often used. Generally constant torque to medium/high speeds with four-quadrant drive. Speed torque and power measurement often required with full drive monitoring on endurance rigs. Closed-loop induction motor drives and d.c. drives predominate.

Table C1.17   Electric motors and alternators

Drive duty	Rating range	Comments on application and usual drive type
A.C. and d.c. motors/ generators/ alternators test-bed rigs	Few kW to >15 MW	All rotating electrical machine manufacturers have elaborate test-bed rigs, supplying their own rotating machines and obtaining control systems from the drives industry to their own requirements. Closed-loop induction motor drives and d.c. drives predominate.

Table C1.18   Textiles

Drive duty	Rating range	Comments on application and usual drive type
Ring frame machines, carding machines, looms	Up to 150 kW	Difficult environment in which IP55 enclosure has become a standard. Open-loop induction motor drive predominates. All drives constant-torque four-quadrant for speed modulation (ring frame) or best speed holding accuracy with slow ramp acceleration/deceleration on carding drives. Special a.c. cage loom motors are high-torque, high-slip designs, although these are being replaced by standard motors with appropriate control characteristics embedded in the drive. Today a.c. induction motor drives predominate with the small braking duty handled with a brake chopper into a resistor.

Table C1.19   Food, biscuit and confection applications

Drive duty	Rating range	Comments on application and usual drive type
Extruder	5 to 400 kW	Hose-proof motors for plant cleaning. Continuous production requiring high levels of reliability, control and monitoring. Otherwise as plastics industry extruders (Table C1.3).
Mixer	5 to 150 kW	As Table C1.5 and see chemical industry mixer drive.
Conveyors	0.5 to 120 kW	As Table C1.6 and see material handling industry conveyor drive.

C1.3   Drive characteristics

C1.3.1   General

The successful integration of electronic variable-speed drives into a system depends upon knowledge of a number of key characteristics of the application and the site where the system will be used. The correct drive has also to be selected. This is not as big an area of concern as it used to be as the voltage source PWM inverter topology has become universally adopted for almost all applications. The control characteristics to meet most loads can be incorporated into this topology with relative ease. It is still important to understand the general characteristics of the various types of drive. In Table C1.20 the key characteristics of many of the most popular drive types are described in a much simplified form. The data on overloads relates to typical industrial equipment, and different figures can be found in the market. Disadvantage and disadvantages of the different drive types are shown in Table C1.21.

Application and drive characteristics 473

Table C1.20 *Characteristics of popular drive types*

	D.C. drives (separately excited motor)		A.C. drives			
	Phase-controlled	Chopper	Open-loop induction motor (V/f)	Open-loop induction motor (vector)	Closed-loop induction motor (flux vector)	Brushless servo drive permanent-magnet motor
Operating speed range	Zero to base speed at constant torque above base speed at constant power Maximum approx. $4 \times$ base speed		10–100% base speed at constant torque	3–100% base speed at constant torque	0–100% base speed at constant torque	0–100% base speed at constant torque
			Above base speed at constant power	Above base speed at constant power	Above base speed at constant power	No operation above base speed
			Maximum approx. $20 \times$ base	Maximum approx. $20 \times$ base	Maximum approx. $20 \times$ base	
Braking capability (%)	150 (4-quadrant drive)	150 (4-quadrant drive)		150		>200
Speed loop response (Hz)	10	50	5	10	100	>100
Speed holding (100% load change)	0.1%, tacho f/b 0.05%, encoder f/b	0.05%, tacho f/b 0.001%, encoder f/b	3%	1%	0.001%	0.0005%
Torque/speed capability	Constant torque + field weakening	Constant torque + field weakening	Constant torque + field weakening	Constant torque + field weakening	Constant torque + field weakening	Constant torque
Starting torque	150% (60 s)	150% (60 s)	150% (60 s)	150% (60 s)	150% (60 s)	200% (4 s)

(*Continued*)

Table C1.20 Continued

	D.C. drives (separately excited motor)		A.C. drives			
	Phase-controlled	Chopper	Open-loop induction motor (V/f)	Open-loop induction motor (vector)	Closed-loop induction motor (flux vector)	Brushless servo drive permanent-magnet motor
Minimum speed with 100% torque	Standstill	Standstill	8% base speed	2% base speed	Standstill	Standstill
Typical motor: IP rating Inertia Size Cooling (NC/FV)	IP23 High Large Forced-vent	IP54/IP23 High (low avail.) Large Forced-vent	IP54 Medium (low avail.) Medium Natural cooling	IP54 Medium (low avail.) Medium Natural cooling	IP54 Medium (low avail.) Medium Natural cooling	IP65 Medium (high avail.) Small Natural cooling
Typical feedback device	Tachogenerator or encoder	Tachogenerator or encoder	N/A	N/A	Encoder	Encoder

Table C1.21 Advantages and disadvantages of popular drive types

	D.C. drives (separately excited motor)		A.C. drives			
	Phase-controlled	Chopper	Open-loop induction motor (V/f)	Open-loop induction motor (vector)	Closed-loop induction motor (flux vector)	Brushless servo drive permanent-magnet motor
Principal advantages	Low-cost controller	Good dynamic performance	Low maintenance	Good dynamic performance	Very good dynamic performance	Excellent dynamic performance
	Relatively simple technology	Relatively simple technology	Utilises standard motors	Full torque down to very low speeds	Full torque available down to standstill	Low-inertia motors
			High motor IP rating	Good starting torque	Standard motor (but with feedback added)	High motor IP ratings
			Simple application and setup		No zero torque dead band	Very smooth rotation possible
			Can supply multiple motors from a single inverter			

(Continued)

Table C1.21 Continued

	D.C. drives (separately excited motor)		A.C. drives			
	Phase-controlled	Chopper	Open-loop induction motor (V/f)	Open-loop induction motor (vector)	Closed-loop induction motor (flux vector)	Brushless servo drive permanent-magnet motor
Principal disadvantages	Expensive motor <100 kW Brush gear maintenance (note: low loads reduce brush life) Four-quadrant version required for regeneration Zero torque dead band Possible failure on supply loss/dip Possible instability on fan/pump type loads Low motor IP rating Chopper available at modest powers only (<10 kW) Additional converter required for regeneration to the supply		Possible motor instability at light loads Additional converter required for regeneration to the supply	Additional converter required for regeneration to the supply Limited torque and speed loop response	Additional converter required for regeneration to the supply	Additional converter required for regeneration to the supply Field weakening range difficult to facilitate

Chapter C2
Duty cycles

C2.1 Introduction

The rating of an electrical machine is usually determined by a temperature limitation, and therefore the duty cycle of the application can significantly affect the rating. There are other limits to motor capacity such as the commutation limit in a d.c. motor, and so duty cycle is not the only determinant of rating; however, if the most cost-effective motor solution is to be obtained it is always good practice to consider the duty cycle with care.

IEC 60034-1 defines duty cycles and these are described in the following with a selective interpretation of their application.

C2.2 Continuous duty: S1

Continuous duty rating S1 relates to a duty where the on-load period is long enough for the motor to attain a steady thermal condition (Figure C2.1). With rated load, this refers to its maximum permitted temperature. Starting and braking are not taken into consideration on the assumption that the duration of these events is too short to have any effect on the temperature rise of the motor. Short time overloads are acceptable. An off-load dwell period is of no significance if it is followed by a load run. The load torque must not exceed rated torque.

Example of motor selection for continuous duty S1
The application calls for power P_1 over a speed range of $n_{a1} \leq n_a \leq n_{a2}$ (min^{-1}). The motor is rated according to the maximum load torque and the maximum speed n_{a2}.

The selected motor must comply with the following requirements:

Rated power	$P_N \geq P_1 \times (n_{a2}/n_{a1})$	kW
Rated speed	$n_N \geq n_{a2}$	min^{-1}
Rated torque	$P_N/n_N \geq P_1/n_{a1}$	kW/min^{-1}

where P_N and n_N are the rated power and speed, respectively, of the motor.

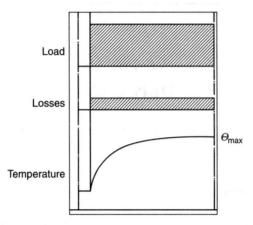

Figure C2.1   Load, losses and temperature for duty type S1

C2.3   Short-time duty: S2

With this form of duty cycle the on load period is too short for the motor to reach a steady thermal condition and the subsequent off-load period is long enough for the temperature of the motor to drop to that of the cooling medium (even with the motor at rest). Starting and braking are not taken into consideration. With S2 duty the load torque may be greater than the rated torque, but only for an appropriately short period (Figure C2.2).

When specifying short-time duty S2 it is also necessary to state the on period, e.g. S2 − 30 min. The standard specifications recognise the following on periods: 10, 30, 60 and 90 min.

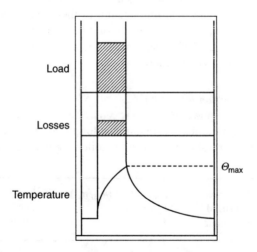

Figure C2.2   Load, losses and temperature for duty type S2

C2.4 Intermittent duty: S3

This form of duty rating refers to a sequence of identical duty cycles (Figure C2.3). Each cycle consists of an on-load and off-load period with the motor coming to rest during the latter. The on-load period during one cycle is too short for the motor to reach a steady thermal condition, and the off-load period is likewise too short for the motor to cool to the temperature of the cooling medium.

Starting and braking are not taken into account on the assumption that the times taken up by these events are too short in comparison with the on-load period, and therefore do not appreciably affect the heating of the motor.

The load torque during one cycle may be greater than the rated torque of the motor.

When stating the motor power for this form of duty, it is also necessary to state the cyclic duration factor:

$$\text{Cyclic duration factor} = (\text{on time}/\text{cycle time}) \times 100\%$$

The standards specify that the duration for one cycle must be shorter than 10 min. Cases where the duty cycle is longer than 10 min must be brought to the attention of the motor manufacturer.

With short cycle times starting and braking must be taken into account and the motor temperature rise must be checked (see Section C2.6, S5).

The standards specify certain preferred cyclic duration factors: 15, 25, 40 and 60 per cent.

The motor rating for intermittent duty S3 may be increased over that applied to S1 duty by a factor k_{S3}. This factor may be of the order of 1.4. Motor manufacturers can advise on this.

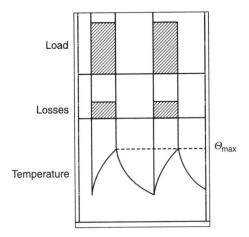

Figure C2.3   Load, losses and temperature for duty type S3

Example of motor selection for continuous duty S3

Torque $M_1 = 500$ N m for a period $t_1 = 10$ s
Torque $M_2 = 400$ N m for a period $t_2 = 30$ s
Torque $M_3 = 600$ N m for a period $t_3 = 5$ s

Cycle time $T = 120$ s
Motor speed $= 1\,200$ rpm

$$\text{Cyclic duration factor} = [(10 + 30 + 5)/120] \times 100\%$$
$$= 37.5\%$$

Select cyclic duration time of 40%

Rms motor torque $= \sqrt{\{[(500^2 \times 10) + (400^2 \times 30) + (600^2 \times 5)]/[10 + 30 + 5]\}}$
$= 450$ N m

Motor rating $P_N = (2\pi/60) \times 450 \times 1\,200$
$= 56.52$ kW @ 40% cyclic duration factor

If the motor had a k_{S3} of 1.4 then we could choose a motor with an equivalent continuous duty rating of $56.52/1.4 = 41$ kW @ $1\,200$ rpm.

C2.5 Intermittent duty with starting: S4

This is similar to S3 but takes a starting acceleration period into account (Figure C2.4).

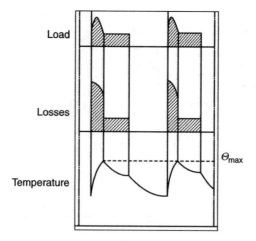

Figure C2.4 Load, losses and temperature for duty type S4

C2.6 Intermittent duty with starting and electric braking: S5

This is similar to S3 but takes starting acceleration and electric braking into account (Figure C2.5).

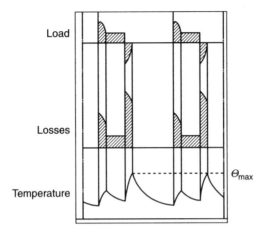

Figure C2.5 Load, losses and temperature for duty type S5

C2.7 Continuous operation periodic duty: S6

This is similar to S3, except the duty cycle is such that the motor has not returned to the temperature of the cooling medium by the end of the off period (Figure C2.6). The cycle should be repeated until the temperature at the same point on the duty cycle has a gradient of less than 2 °C per hour.

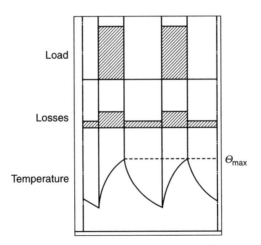

Figure C2.6 Load, losses and temperature for duty type S6

C2.8 Continuous operation periodic duty with electric braking: S7

This is the same as S6 but takes electric braking into account (Figure C2.7).

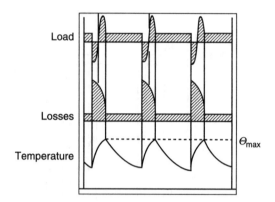

Figure C2.7 Load, losses and temperature for duty type S7

C2.9 Continuous operation periodic duty with related load speed changes: S8

This is the same as S7 but with defined and cyclic load speed changes (Figure C2.8).

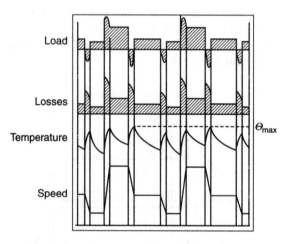

Figure C2.8 Load, losses and temperature for duty type S8

C2.10 Duty with non-periodic load and speed variations: S9

S9 duty cycles should be discussed with the motor manufacturer as the effects of this duty cycle are heavily dependent upon specific motor design philosophies (Figure C2.9).

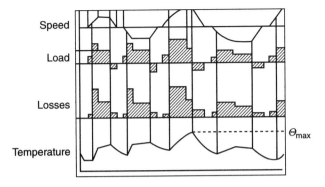

Figure C2.9   Load, losses and temperature for duty type S9

C2.11   Duty with discrete constant loads: S10

As with the S9 duty cycle, S10 duty cycles should be discussed with the motor manufacturer (Figure C2.10).

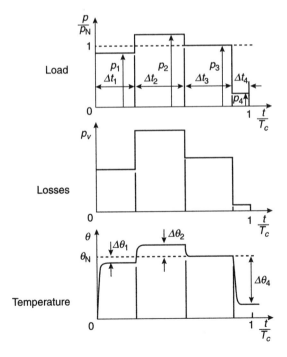

Figure C2.10   Load, losses and temperature for duty type S10

Chapter C3
Interfaces, communications and PC tools

C3.1 Introduction

The modern drive incorporates a large number of functions that were historically located in separate controllers. This trend is continuing apace with the result that there is an ever growing number of interfaces needed to the drive. It is also a sobering thought that the greatest cause of problems in many industrial plants is in the interconnecting wiring. There is clear need therefore to understand the interface types on a modern drive and how they should be used in order to ensure that the design of the system is correct. Furthermore, by understanding the different types of interface it is possible to simplify the physical interfaces by, for example, the use of a Fieldbus, and here the limitations and characteristics of such a digital communications system need to be understood if a system design is to be optimised.

Finally, the growing complexity of drive functionality, including the opportunity for users to write their own functions in a drive, has led to the development of advanced PC tools for the configuration and commissioning of individual drives and drive systems as well as data logging and fault-finding capabilities. An understanding of such tools and what benefits they can bring is important to getting the most out of a modern drive.

C3.2 Overview of interface types

For the purposes of simplicity we will consider three primary types of interface to a drive: analogue, digital and serial digital.

Analogue signals are used for transmitting reference information to the drive, and drive, motor or load data back to the process controller. When used as a reference, analogue signals may need to be of very high resolution and have excellence noise immunity. The analogue signals may have to be bipolar, with demanding characteristics in respect linearity, and dead-band near 0 V.

Simple digital interfaces can be in the form of simple interfaces directly to and from a transistor in the drive control circuit. Subject to appropriate isolation, these can be connected directly to a remote process or machine controller. Where isolation is needed, most drives have an on-board relay that can be programmed to switch on a given function threshold.

Serial digital interfaces are capable of transmitting considerable amounts of data to and from a drive. Serial communications can be used to fulfil virtually all interface functions to a drive, and this brings considerable scope for simplifying the system cabling. There are very many different forms of serial interface, from the relatively simple RS-232 or RS-485 systems, which were the early industrial standards, through to the Ethernet-based systems that predominate today. Serial communications is not easy to evaluate objectively, not only because of the multitude of physical and protocol alternatives, but because of the lack of unbiased comparative information in the public domain.

C3.3 Analogue signal circuits

C3.3.1 General

Before considering the performance of an analogue input or output circuit it is necessary to define the metrics by which they can be judged. To simply talk about accuracy is meaningless. Almost all drive controllers are digital. That is to say, the analogue input signals have to be converted into a digital form for use within the digital control algorithms. It is common for people to talk about resolution as the key measure of performance of a digital input. Resolution, which defines the incremental step of input that the controller will respond to, is shown in Table C3.1.

The above bit counts are common, as they relate to commercially available analogue to digital (A to D) converters. Some drives, however, use designs that utilise a voltage to frequency converter and derive the digital 'output' by counting the number of pulses. With a high-frequency output, this can give very high conversion resolution, but because the technique but not clearly definable in the form shown in Table C3.1, it may be described for conversion as giving 16-bit resolution.

For bipolar conversion the resolution will be defined in the form, for example, of a 16 bit plus sign, where the sign is an additional bit denoting the polarity of the signal.

Although resolution is important, it is not the only measure by which the performance of an analogue input should be measured. Figure C3.1 illustrates other important metrics: non-linearity, offset, dead-band, jump, gain asymmetry and monotonic characteristics. Each of these characteristics can be important for specific applications, and great care is necessary in defining these requirements.

Table C3.1 Resolution and impact on response step size

Bit count	Number of incremental steps	Step size for a 10 V signal
10 bit	1 024	1 mV
12 bit	4 096	244 μV
14 bit	16 384	61 μV
16 bit	65 536	15 μV

Interfaces, communications and PC tools 487

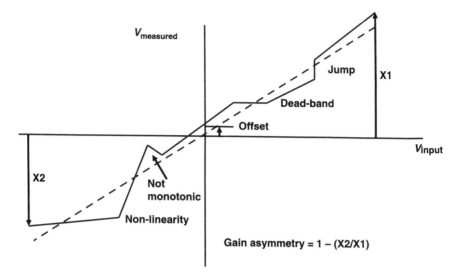

Figure C3.1 Typical inaccuracies and non-linearity of an analogue input

Figure C3.1 is shown with respect to a bipolar input, although on simple drives a unipolar analogue input is common. On drives with more than one analogue input it is common for some of the inputs to be unipolar.

The impedance of both input and output circuits is important to the user and should be carefully considered when considering the capability/compatibility with the equipment to which the drive is connected.

For simplicity we will consider voltage inputs through the bulk of this section and consider current inputs in Section C3.3.2.4.

Table C3.1 shows that high-resolution inputs are required in order to give, for example, a precise speed reference to the drive, and the signal will be sensitive to very small amounts of noise. Differential hardware circuits are therefore employed on such signal inputs and these, together with the necessary connection techniques, are described in the following sections.

C3.3.2 Hardware implementations and wiring advice

C3.3.2.1 General guidance on connecting analogue signal circuits

This introduction covers the correct use of screened cables, and connection for various types of analogue signal circuits. It is always possible that in certain specific situations behaviour might improve with an arrangement that contradicts the advice given here. Unfortunately, a combination of factors can result in unexpected effects. However, if these guidelines are followed then the chance of malfunctions is reduced, and if you understand the principles then you are more likely to be able to correct any malfunctions that do occur.

Although the advice given in this section contains some duplication of the information given in Chapter B6 on EMC, it is felt that given the number of practical problems that result from poor wiring and understanding of different types of control signals, it is appropriate to labour the point!

Basic principles
The correct application of screened cables and various signal circuit types is based upon a small number of essential principles:

- In any system where signals must be conveyed over distances of more than a few centimetres, noise voltages that exist between parts of the system can cause errors. Low-frequency noise that is within the frequency bandwidth of the signal circuits causes errors directly by changing the measured value, whereas high-frequency noise beyond the circuit bandwidth can cause errors indirectly because of non-linear effects, which can broadly be called 'noise rectification'.
- A signal circuit where both sending and receiving ends are single-ended and grounded cannot be made entirely immune to the noise voltages that exist between the circuits, i.e. common-mode voltages, unless one end is fully galvanically isolated, in which case it is effectively a differential circuit.
- A screened cable gives good rejection of high-frequency noise voltages that might exist between sending and receiving circuits. It can only do this if the screen is connected to the local reference point at both ends.
- A screened cable does not reject low-frequency noise voltages. The cable has a 'cut-off frequency', typically around 20 kHz, below which it does not reject common-mode voltage existing between its ends.
- Differential input and output circuits give good rejection of low-frequency noise voltages, provided they are within the common-mode rejection range.
- Differential circuits do not reject high-frequency noise voltages. Common-mode rejection typically becomes ineffective beyond about 100 kHz.

By correctly combining differential signalling with screened cables, noise rejection over the entire frequency range can be achieved.

In drive systems, because of the combination of high electrical power and fast-switching power semiconductors, noise can be expected to be present with a range of frequencies between the mains power frequency and many megahertz.

Terminology
In order to avoid confusion in what is a complex subject area, some key definitions are given below:

1. *Earthing (grounding)*. The main purpose of earthing is for electrical safety, to ensure that accessible metal parts cannot have a dangerous potential relative to

the ground beneath our feet. The connection to earth is usually by a relatively long conductor, which has enough inductance to make it not particularly effective for high frequencies. Unfortunately many engineers tend to use the term 'earth' or 'ground' for any connection that is intended to provide a common reference point for a number of interconnected circuits, as well as for the safety earth terminal on a product (PE).

Most mains powered equipment has an earth terminal, and accessible metal parts are connected to it. Signal circuits may or may not have one pole connected to the same terminal. Owing to common-mode noise effects, this detail can have an important effect on system noise immunity.

Correctly designed equipment does not have to be connected to earth in order to operate correctly. Battery-operated equipment and land-line telephones are two examples of classes of equipment that work well without earthing. However, the effect of earth potentials must be carefully considered for equipment where signals must be transferred between locations where earth potential differences may exist.

2. *Equipotential bonding.* This term is also used mainly in the realm of electrical safety. It is distinct from earthing to the extent that in a given area if all accessible surfaces have the same potential then there is no electrical hazard even if the potential is different from that of the earth outside.

3. *Return circuit (reference).* This term is used to refer to the conductor that completes an electrical circuit, and is referring to the fact that the return circuit may or may not be connected to earth, but vitally that it provides a common reference point relative to which voltages, usually signals, are generated and measured.

4. 0 V *terminal.* This terminology is used for the signal return circuit connection point on Control Techniques products. Within the product, this will be the common conductor to which the signals are referred. Often it is connected to a large plane on the printed circuit board, which is unfortunately referred to as a 'ground plane'. This is rather loose terminology for a local conductive surface that provides a reference point, but which may or may not be connected to ground.

The 0 V terminal has the following characteristics.

(a) It is usually isolated from earth within the product (i.e. from the PE terminal of the product).

(b) Most internal and external signals are measured relative to this point, with the exception of differential and isolated signals.

(c) It is shared by analogue and digital signals.

(d) It is permitted to be connected to earth or not, depending on the user's requirements.

(e) It carries some electrical noise because of stray capacitance between the internal circuits and the power switching parts of the drive. This means that connections to the 0 V terminal will experience noise voltage relative to earth if they are not earthed, or noise current if they are earthed.

C3.3.2.2 Single-ended circuits

The single-ended circuit shown in Figure C3.2 is the simplest possible signal circuit. It is adequate for many applications but it is vulnerable to errors and disturbance by noise in the low-frequency range below the cable cut-off frequency.

In this arrangement, common-mode noise voltages that exist between the sending and receiving ends are cancelled within the signal circuit by the screened cable, but only if they are above the cable cut-off frequency. This protects both sending and receiving circuits from the effects of high-frequency noise. The cable only provides this rejection where the screen fully surrounds the cores, so any earth or chassis connections must be made directly to the cable screen, and not to the pigtail connections to the 0 V terminals.

High-frequency noise can still enter through the pigtails, which must be kept as short as possible. Sometimes the 0 V connection is shared with other circuits that have high noise levels, and in that case the 0 V connection is best connected directly to earth so that the noise current does not flow in the pigtails.

This arrangement is vulnerable to disturbance by low-frequency currents flowing in the 0 V connection, i.e. the cable screen and any additional cores. Any voltage drop caused by this current appears in series with the wanted signal and causes an error. A large-area conductor with low resistance minimises the error, but it cannot be avoided. This is sometimes referred to as the problem of the 'earth loop'.

The kinds of error that may be experienced are as follows.

- Small static errors may be caused by d.c. currents. They may change if the d.c. current changes, for example if the 0 V conductor is shared with power circuits (e.g. contactors or lamps).
- There may be disturbance to the signal at 50/60 Hz or related frequencies. For inputs with bandwidth lower than 50/60 Hz this may be insignificant, but it may cause noise and vibration, for example in high-bandwidth servo applications.

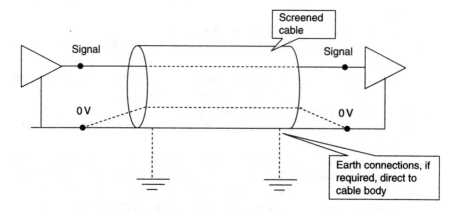

Figure C3.2   Single-ended signal circuit connection

If the 0 V connection at one end is not used as the return line for any other circuit, and is not earthed there, and is not subject to noise pickup from internal sources such as switch-mode power supplies, then its associated circuit works effectively as a differential circuit. In that case no low-frequency current can flow and there is no error. However, it is often not possible to know whether all these conditions will be met.

Low-frequency disturbances can be minimised by providing additional low-resistance earth return current paths. To prevent 'earth loops', which can actually encourage circulating mains-frequency current, additional return wires should follow the same physical route as the screened cable.

C3.3.2.3 Differential circuits

Differential circuits use dedicated, independent lines for the outgoing and return connections of every circuit, both of which are separate from the 0 V common or ground connection. The differential facility may be available at the receiver, or the sender, or both. It can be identified where there are two signal lines for each signal, which may be labelled '+' and '−' or 'non-inverting' and 'inverting', as well as a '0 V' or 'common' or 'ground' connection. However, there are several different forms of output circuit that must be carefully distinguished, as explained below.

Differential signalling has the capability to provide good immunity from all kinds of disturbance, but there are some important details that must be considered in order to gain the benefits. An important limiting factor for all differential circuits is the limited common-mode range, which is sometimes expressed as a maximum permitted voltage for any terminal relative to 0 V. If the common-mode voltage exceeds this range then the differential action, i.e. the common-mode rejection, ceases to work, and in extreme cases even physical damage may result. For this reason, it is always necessary to connect the 0 V terminals of the sender and receiver in order to limit and control the common-mode voltage. The only exception to this rule is in certain specialised circuits that use galvanic isolation to achieve differential operation, and which may have a very high common-mode range. A second factor is that often the differential action is only effective within the intended frequency range of the circuit, and common-mode noise outside this range may cause errors.

Differential receiver, single-ended sender
In this arrangement, shown in Figure C3.3, the differential input is used to measure the signal voltage from the sender, directly between its signal and 0 V terminals. The 'inv' (inverting) input has its own core connected to the sender 0 V, which has no other current in it that would cause an error voltage.

For low-frequency noise, any voltage drop in the cable screen or other 0 V core is seen as a common-mode voltage by the differential receiver and does not affect the received signal, provided it lies within the common-mode range of the receiver.

For high-frequency noise, where the differential function of the receiver is ineffective, the screening effect of the cable prevents common-mode noise voltage from appearing at the receiver input.

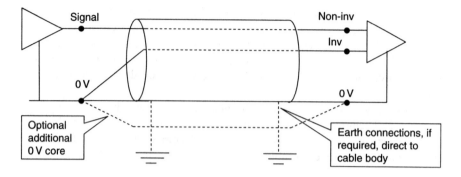

Figure C3.3   Correct connection of differential receiver with single-ended sender

Remote-reference sender, single-ended receiver

This arrangement, shown in Figure C3.4, is often encountered with motion controllers. The labelling of the output terminals suggests a differential output, but this is not the case and the application differs considerably from a true differential output. The manufacturer's instructions must be followed carefully.

In this arrangement, the sender is connected in such a way as to ensure that the intended signal voltage is applied directly to the receiver input, compensating for any common-mode voltage that appears between the sending and receiving ends.

For low-frequency noise, any voltage drop in the cable screen or other 0 V cores is seen as a common-mode voltage by the sender and does not affect the received signal, provided it lies within the common-mode range of the sender.

For high-frequency noise, where the sender differential action is ineffective, the screening effect of the cable prevents common-mode noise voltage from appearing at the receiver input or the sender 'negative' terminal.

(*Important note on analogue remote-reference outputs*. The remote-reference sender shown in Figure C3.4 is a feedback control amplifier, which sets the '+'

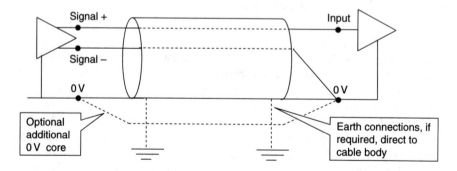

Figure C3.4   Correct connection of single-ended receiver with remote-reference sender

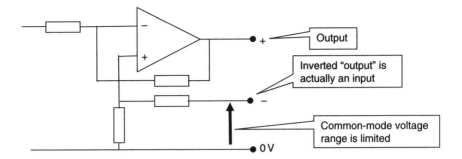

Figure C3.5 Internal arrangement of typical analogue remote-reference sender

output at a voltage equal to the sum of the desired signal and the voltage measured at its '−' output. Figure C3.5 illustrates the principle of the internal arrangement of a remote-reference output circuit. The '−' 'output' is in fact an input, which has a limited voltage range and frequency response, so the circuit will not work correctly if the common-mode voltage exceeds the common-mode range, or if there is excessive high-frequency noise that exceeds the limited high-frequency rejection capability of this input.)

One common error is to use the '−' output as if it were the 0 V terminal. In a drive application, this is very likely to result in excessive common-mode voltage, and errors in the sender. The 0 V terminals of the sender and receiver must be connected together to limit the common-mode voltage.

Remote-reference sender, differential receiver
The arrangement of Figure C3.5 works well when used in accordance with Figure C3.4 above. The fact that the '−' 'output' is in fact an input does, however, mean that a wrong connection can cause errors. Figure C3.6 shows a connection between a differential output and input that appears to be logical, but does not in fact work properly. The reason it does not work is that both the sender '−' terminal

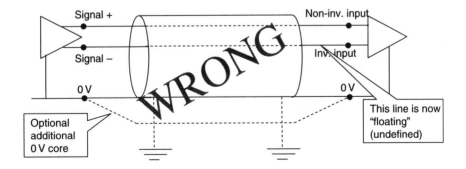

Figure C3.6 Wrong connection of analogue remote-reference sender to analogue differential receiver

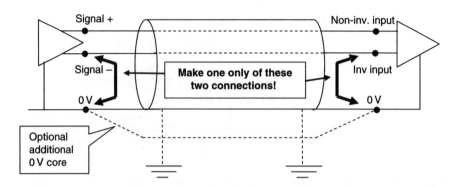

Figure C3.7   Correct connections of analogue remote-reference sender to analogue differential receiver

and the receiver 'inv' terminal are actually inputs with high impedance, so the voltage at that point is indeterminate.

The solution is to connect that circuit to 0 V at one point, as shown in Figure C3.7. It is actually not important to which end the connection is made – generally it should be at the end that has the least good immunity to high-frequency noise. Usually it will work either way round, but some trial and error may be necessary.

True differential sender and receiver
In these systems, the differential senders have two true complementary outputs, and the receivers have true differential inputs. They are connected together in the intuitively obvious way for differential or 'push–pull' signals, as shown in Figure C3.8. Note that the sender may still be sensitive to high-frequency noise, so the connection of the cable screen is still important. Clearly it is necessary to distinguish between a true differential sender and a remote-reference sender, because Figure C3.8 is wrong for the remote-reference sender.

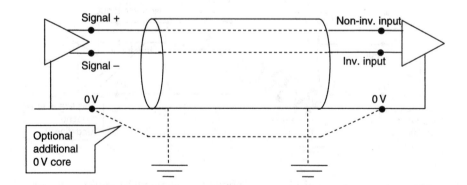

Figure C3.8   Connection of true differential analogue circuits

Isolated source
On occasion, the source may be fully isolated from earth or from local 0 V or reference terminals, as shown in Figure C3.9. The following are some such examples:

- non-powered sensors such as thermocouples or thermistors, isolated from their surroundings;
- an isolated analogue output;
- an analogue signal isolator unit.

This works in the same way as the differential source, but it may no longer be necessary to make the 0 V connection at the source because the isolation effectively gives a very high common-mode range.

Some care is required in such cases. The arrangement can have the same benefits as a differential source, but the source is exposed to electric fields from any common-mode noise voltage that may exist between the structure to which it is fixed and the receiving circuit. If the source is fully isolated and has little stray capacitance to the surrounding equipment or machinery – for example, a thermocouple that is physically small and requires no local power supply – then the immunity from common-mode noise can be extremely good. The same applies to an isolated analogue output (e.g. in a signal isolator unit) if it is well designed with good immunity to common-mode noise between its 0 V or earth terminal and its active output terminal. However, some such units do exhibit sensitivity to common-mode noise, and may require protection, typically by connecting the cable screen to a locally provided screen or reference terminal in the way shown dashed in Figure C3.9.

When connecting an isolated source to a differential receiver it is usually necessary to make the '0 V' connection shown in Figure C3.7 in order to prevent the possibility of uncontrolled common-mode voltage appearing at the input through stray capacitance.

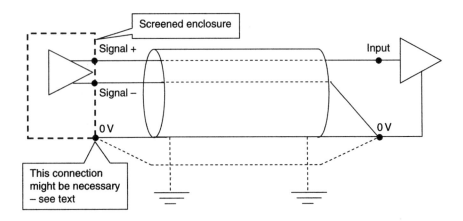

Figure C3.9   Isolated source with single-ended receiver

C3.3.2.4 The case for 4–20 mA and other current loop circuits

Current loop signalling is used particularly in process plant where long signal cable runs are common. This is because it is resistant to induced low-frequency noise potentials. The 4 mA offset is used to detect faults that result in a loss of the signal.

The ideal current loop transmitter behaves in the same way as the isolated source in Figure C3.9, except that it generates a controlled current (rather than a voltage) that is measured in the receiving circuit by passing it through a scaling resistor.

Exactly the same comments apply as above. Ideally the sender should be unaffected by noise voltage between its terminals and the local 0 V or earth. However, in some cases this sensitivity does exist, and the additional connection of the cable screen shown in Figure C3.10 is required to protect the sender from errors caused by high-frequency noise.

In some process applications it is common for an analogue current signal to be shared between two or more receivers by connecting them in series. With many drives this arrangement is not recommended, because it requires the 0 V terminal of at least one drive to be used as a signal terminal for another drive. As explained above, the 0 V terminal of a drive is a source of high-frequency noise caused by capacitive stray coupling within the drive. This is likely to cause disturbance to the other drive and/or the current sender. Splitter modules are available that provide multiple separate current loop outputs from a single input.

C3.3.2.5 The use of capacitors for connecting cable screens

It is best practice in most applications to connect the cable screen at both ends. There are occasions when this is not acceptable, or better results might be obtained without this connection. The following are examples where this may apply:

- the 0 V terminal at one end must not be connected to other 0 V terminals, e.g. with some series current loop inputs;

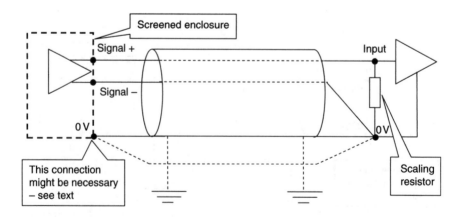

Figure C3.10   Current loop circuit

- sender and receiver are both single-ended, and low-frequency common-mode voltage causes a disturbance because of the voltage drop in the screen;
- the screen causes a current loop that injects unwanted low-frequency current elsewhere in the system.

In this case it is possible to obtain some of the benefit of the cable screen while maintaining low-frequency isolation at the point of connection, by using a capacitor to make the connection. The capacitor passes high-frequency current, allowing the screen to offer its beneficial magnetic coupling with the inner cores, while blocking low-frequency current. Note that one connection is still required between the 0 V terminals at each end. The capacitor is used to avoid multiple connections.

Typical capacitor values are of the order of 100 nF. There is, however, no precise rule for the value. Generally the value should be as high as possible, but not so that the capacitor becomes so physically large that its self-inductance reduces its benefit at high frequencies; i.e. the capacitor body should be physically no longer than a typical screen pigtail. Too small a value can sometimes be worse than no capacitor, because it resonates with the inductance of the earth loop to form a high-impedance parallel resonant circuit. If the resonant frequency happens to coincide with a characteristic frequency of the noise source then the screen connection becomes ineffective.

The principle is illustrated in Figure C3.11 using a simple single-ended circuit as an example.

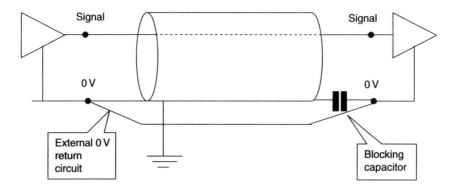

Figure C3.11   Use of a capacitor for connecting the cable screen while maintaining d.c. isolation

C3.3.3   Typical specifications for analogue inputs and outputs

As a general indication, the specifications in Tables C3.2, C3.3 and C3.4 are typical analogue input/output (I/O) specifications that can be found on commercial drives. This information has been included so the reader can appreciate the many elements that define the performance of these circuits.

Table C3.2 Typical precision analogue input specification

Type of input	Bipolar differential
Full scale voltage range	±9.8 V ± 1%
Absolute maximum voltage range	±36 V relative to 0 V
Working common-mode voltage range	±13 V relative to 0 V
Input resistance	100 kΩ ± 1%
Resolution	16 bit plus sign
Monotonic	Yes (including 0 V)
Dead band	None (including 0 V)
Jumps	None (including 0 V)
Maximum offset	700 μV
Maximum non-linearity	0.3% of input
Maximum gain asymmetry	0.5%
Input filter bandwidth	1 kHz
Sampling period	125 μs

Table C3.3 Typical general analogue input specification

Type of input	Bipolar single-ended voltage or unipolar current
Resolution	10 bit plus sign
Sampling period	250 μs
Operating in voltage mode	
Full-scale voltage range	±9.8 V ± 3%
Absolute maximum voltage range	±36 V relative to 0 V
Maximum offset	30 mV
Input resistance	>100 kΩ
Operating in current mode	
Current ranges (selectable)	0 to 20 mA ± 5%/20 to 0 mA ± 5%/4 to 20 mA ± 5%/20 to 4 mA ± 5%
Maximum offset	250 μA
Absolute maximum current	+70 mA
Equivalent input resistance	≤200 Ω at 20 mA
Absolute maximum voltage (reverse bias)	−36 V

Table C3.4 Typical general analogue output specification

Type of input	Bipolar single-ended voltage or unipolar single-ended current
Resolution	10 bit plus sign
Update period	250 μs
Operating in voltage mode	
Voltage range	± 9.6 V $\pm 5\%$
Maximum offset	100 mV
Maximum output current	± 10 mA
Load resistance	>1 kΩ
Protection	Short-circuit protection (35 mA)
Operating in current mode	
Current ranges (selectable)	0 to 20 mA $\pm$ 10%/4 to 20 mA $\pm$ 10%
Maximum offset	600 μA
Maximum open circuit voltage	+15 V
Maximum load resistance	500 Ω

C3.4 Digital signal circuits

C3.4.1 Positive and negative logic

Digital inputs and outputs are very straightforward as compared with analogue circuits and require very little discussion. Drives tend to follow the hardware specification for digital I/O in programmable logic controllers (PLC), which is defined in IEC 61131-2.

One significant area that can cause problems for users and needs to be appreciated is the difference between positive and negative control logic.

Positive control logic
The convention for positive logic is as follows:

$$\text{True} = \text{logic} \quad 1 = \text{input energized}$$
$$\text{False} = \text{logic} \quad 0 = \text{input NOT energized}$$

Positive logic is used by most controllers today as it does provide better protection/integrity in the case that a problem occurs with the machine wiring. It is usually the default setting in most commercial drives today.

Negative control logic
The convention for negative logic is as follows:

> True = logic   1 = input NOT energized
> False = logic   0 = input energized

C3.4.2 Digital input

Digital inputs are usually referenced to the drive control zero volt line (0 V). There is protection afforded by a series resistor, clamp diodes and a filter capacitor. A typical specification is given in Table C3.5.

Table C3.5   Typical digital input specification

Type	Positive or negative logic (selectable)
Voltage range	0 V to +24 V d.c.
Absolute maximum applied voltage range	±30 V d.c.
Input threshold	+10 V ± 0.8 V d.c.
Load	<2 mA at 15 V d.c.
Sample period	250 μs

When used as part of high-performance motion control systems a number of digital inputs are required to respond very quickly to events such as position registration signals. For such applications very fast inputs with sample periods as short as 67 μs can be required. It is common in such applications for 'freeze inputs' to be provided within the controller. This type of input captures the signal and stores it with a precise time stamp for processing. Some drive manufacturers are able to offer capability for

Table C3.6   Typical digital output specification

Type	Positive logic open collector or negative logic push–pull (selectable)
Voltage range	0 V to +24 V d.c.
Nominal maximum output current	200 mA
Maximum output current	240 mA
Load	<2 mA at 15 V d.c.
Update period	250 μs

registration control within their drive products. This type of application is specialized and should be discussed specifically with the drive or controller manufacturer.

C3.4.3 Digital output

Digital outputs are usually referenced to the drive control zero volt line (0 V). There is often integral over-current protection to ensure that damage is avoided if the output is inadvertently short-circuited. A typical specification is given in Table C3.6.

C3.4.4 Relay contacts

Many drives have a relay output that can be programmed to respond to specific functions such as drive healthy, or in more advanced drives the user can programme the function. A typical specification for such a relay is as given in Table C3.7.

Table C3.7 Typical control relay specification

Contact type	Normally open
Default contact condition	Closed
Contact voltage rating	240 V a.c. Installation over-voltage category II
Contact maximum current rating	2 A a.c. 240 V
	4 A d.c. (resistive load)
	0.5 A d.c. (30 V inductive load; $L/R = 40$ ms)
Contact minimum recommended rating	12 V, 100 mA
Update period	4 ms

C3.5 Digital serial communications

C3.5.1 Introduction

In a modern digital drive, adjustments such as PID gains and acceleration ramp times are software settings rather than the potentiometers and DIP switches found in their analogue predecessors. To permit adjustment of these software parameters, most drives provide a human–machine interface (HMI) consisting of an onboard display and buttons. Users manipulate the buttons to scroll to the parameter of interest and to alter its value. The problem is that many sophisticated drives may have several hundred parameters and the built-in HMI system can be cumbersome to use. Compounding this problem is the fact that modern automation applications usually involve multiple drives working in unison with each other. A field engineer may be faced with the daunting task of setting up the parameters for 30 drives, each one having over 400 parameters.

To address this challenge, most modern digital drives have a built-in communications port using the RS-232 or RS-485 serial standard. There are two principal reasons for the inclusion of a communications port on a digital drive: setup of the drive's parameters and real-time control of a number of drives in an automation application.

As factory automation has advanced there has been considerable effort made to integrate communications throughout a machine to improve all aspects from commissioning through process optimisation, data logging and fault finding. Serial communication or Fieldbus systems have played a pivotal role in realising this type of system. Although a significant number of hardware platforms and even more protocol variations have been developed for industrial application, it appears clear that Ethernet, dominant in the office, will be of growing importance in the industrial environment. What is also clear is that there will be many competing protocols operating on that platform for some time to come.

C3.5.2 Serial network basics

Communications networks have so much in common that it has become standard practice to relate their features and design elements to an internationally agreed model. The Open Systems Interconnection (OSI) Reference Model, developed by the International Standards Organisation (ISO), is used as a framework for organising the various data communications functions occurring between disparate devices that communicate (Figure C3.12). The complete OSI model defines seven component parts or layers; however, only three of the layers are usually developed for industrial communications.

Fieldbus systems often include special features related to device interoperability and the real-time nature of industrial automation applications. The Device Profile defines device interoperability features such as 'Electronic Data Sheets', which allow devices from different manufacturers to inter-operate without complex configuration or custom software. Cyclic Data is network data that bypasses parts of the software for efficiency.

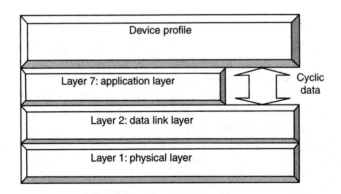

Figure C3.12   The OSI reference model structure

C3.5.2.1 Physical layer

Starting at the lowest layer in the OSI model, the physical layer is concerned with the actual transmission of raw bits. In a factory network, analogue quantities, switches, command codes and textual data are all converted into numeric information and transmitted as a stream of binary bits from the source node to the destination node. Typically, the bit stream is logically grouped into octets (or bytes).

The following sections describe the key facets of the physical layer.

Network cables and connectors

The physical layer requires a transmission medium for the data signals to flow. The principal transmission medium of factory networks today is either copper cable or fibre-optic cable. Twisted pair copper cable is preferred for cost and ease of installation. Fibre-optic cables are less susceptible to EM fields and offer higher bandwidth; however, they are more expensive and difficult to install. Also, an expensive hub is required for the nodes to be connected in a logical bus structure.

Interface circuits

Most PCs support the RS-232 data transmission system. The signal appears as a single-ended voltage with reference to a signal ground. The voltage swing of the RS-232 circuit shown in Figure C3.13 is ± 12 V (-12 V for logic one, $+12$ V for logic zero). When not transmitting the signal is held at logic one (-12 V).

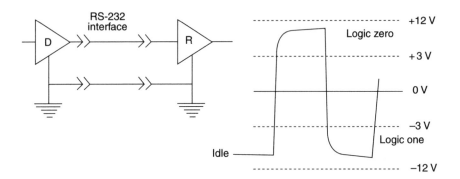

Figure C3.13   RS-232 interface

Observing that the logic zero/one detection thresholds are typically $+3$ and -3 V in most implementations, it is clear that a modest noise spike could trick the receiver into making a false bit determination, as shown in Figure C3.14. For this reason, RS-232 signalling circuits are almost never used in modern factory communications systems.

Adoption of differential signalling solves most of the problems inherent in single-ended RS-232 communications. In the RS-485 standard, two conductors are used to represent the bit: one carrying the original bit and the other carrying its logical inverse, as shown in Figure C3.15. The differential receiver at the receiving end

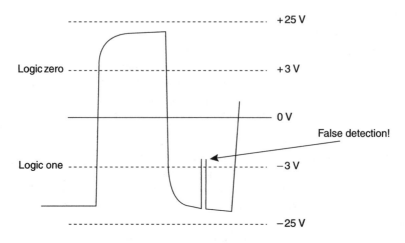

Figure C3.14   RS-232: effect of noise

subtracts the two signals to recover the original data. Any noise induced into one conductor is induced in the other conductor and the subtraction operation will thus cancel out the common-mode noise. Note: the signal levels are 0 and 5 V.

RS-485 also supports operation with multiple nodes on the same cable. When a transmitter is not actual transmitting it is disabled and presents a high impedance to the bus. Typically all receivers are enabled and see every message, but the protocol allows them to discard messages not intended for them.

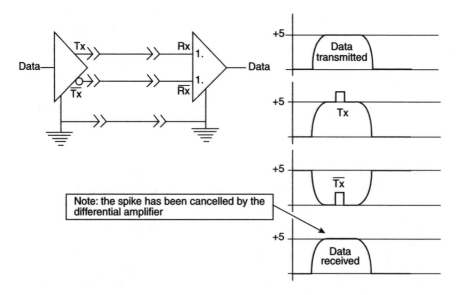

Figure C3.15   RS-485

Data encoding

The differential signalling described in the previous section allows us to reliably send a single bit down a network cable. Because the network is typically used for numeric data or characters, multiple bits are used to convey the information. These multiple bits must be encoded either asynchronously or synchronously.

In asynchronous encoding, it is up to the receiver to properly sample and detect the multiple bits. The standard PC COM port uses a method of encoding called NRZ (non-return to zero). In NRZ encoding, the voltage level determines the bit value (one or zero). In Figure C3.16a, the character hex 1C is transmitted. The receiving electronics must detect the change from line idle state to the start of a bit pattern (called the start bit). Once the start bit has been detected, each data bit must be sampled in the middle of its respective bit period. This assumes that the receiver has its own sample clock to do this and that the incoming bit rate (baud rate) is known.

The disadvantages of NRZ encoding is that the timing of the sampling is independent of the transmitted signal. Second, the start bit and stop bit carry no information and thus waste throughput.

Another encoding method called NRZI (non return to zero inverted) uses transitions to determine the bit values (zero is no transitions, one is a transition).

Finally, the Manchester encoding system allows the clock signal to be recovered from the transmitted data. In Manchester encoding, there is always a transition in the middle of a bit period. Logical zero is a downward transition and logic one is an upward transition. Because there is always a transition, a phase-locked loop circuit can be used to extract the clock signal. This makes it easy for the receiver circuit to sample and detect the bits.

Network topology

This is a description of how the nodes are connected together. The main topologies are the bus topology, where all nodes connect together onto a common medium, and a ring

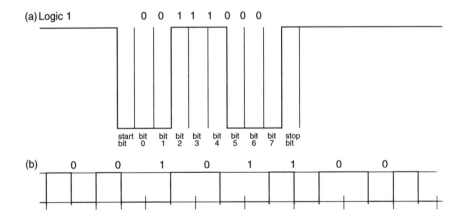

Figure C3.16 Data encoding, hex: (a) transmission of hex 1C; (b) receiving hex . C

topology, where nodes are interconnected in a unidirectional loop. Some networks are wired in a 'star' topology, which requires the use of a multi-output repeater called a hub. Ethernet is an example of this approach.

C3.5.2.2 Data link layer

The 'data link' layer is responsible for encapsulating the digital information into message frames and the reliable transfer of frames over the network.

Framing

An example of a data link frame is shown in Figure C3.17.

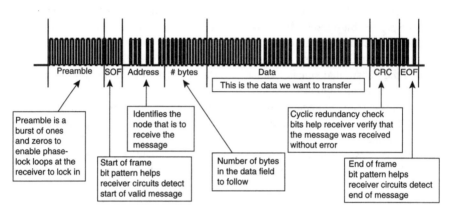

Figure C3.17   A typical data link frame

Data model

All industrial Fieldbuses work by connecting devices onto a shared medium and multiplexing data onto this medium in a serial fashion. The data model describes how messages or data are routed and identified on the network.

- *Source–destination.* Messages are identified by a single-node destination address. Most source/destination protocols also have the possibility for global addressing so that all nodes receive the message. However, it is not possible to select a group of nodes to receive the message (called 'multi-cast').
- *Producer–consumer.* A message 'produced' by a node is identified by its content (i.e. data identification) rather than a specific node destination. Any node may 'consume' this message if it detects that the data is required. Clearly, this is very powerful and makes best use of available bandwidth but does require a complex configuration phase when data identifications are allocated. CAN and ControlNet can successfully operate a producer–consumer model, although CAN only provides for a maximum of 2047 data objects.

Media access control

The media access control protocol defines how the access to the shared medium is arbitrated so that reliable data exchange can occur.

- *Carrier sense multiple access/collision detect (CSMA/CD)*. A node waits until the bus is idle and then transmits its message. While transmitting, the node senses its own transmission to determine whether a collision with another node occurs. If a collision is detected then several schemes can be employed to arbitrate. Ethernet specifies that the node jams the network and then 'backs-off' for a random time interval before trying again. As network loading increases the probability of collisions increases and the network ceases to operate efficiently. This stochastic behaviour is not suitable for the majority of automation applications.

    CAN is also CSMA but the collision is 'resolved' using bitwise arbitration according to the priority of the message. Significantly, this arbitration is accomplished on the physical layer itself using dominant bit signalling and does not result in wasted bandwidth. This signalling does have a bit rate and trunk length limitation of 40 m at 1 Mbit s^{-1} due to the finite propagation speed of the signal down the wire.

    CAN is deterministic for the highest priority message with a worst-case delay time of 130 μs (time for the maximum 8-byte message at 1 Mbit s^{-1}). Performance can be improved further by using the hardware 'transmission timestamp' to measure the delay and then transmit this value to the slaves so they can compensate for the jitter.

- *Token ring*. A node may only transmit a message if it is in possession of the token. Once the node has sent the message it must relinquish the token to its neighbour. This fair-share scheme provides a very flexible protocol for peer to peer communication without the need for a master or arbiter node. However, the worst-case time-window for a node to receive the token and transmit a message is large and occurs when, in one token pass, every node transmits a message of maximum length. ARCNET is a good example of an industrial network that uses token ring.

- *Master slave*. As the name suggests one node is designated the 'master'; usually, this is a unique node and also corresponds to the central controller of the system. The master controls all communication activity and the slaves only respond to a request from the master. This leads to deterministic behaviour but without any peer to peer communications. Much of the Fieldbus installed base is PLC systems and has a centralised architecture and a master slave protocol. The dominant network in Europe is Profibus-DP, which is firmly entrenched in the PLC-distributed slave architecture.

- *Time division multiplexing*. Each node is allocated a time slot on the network when it is permitted to initiate a transaction with a peer node. In the WorldFIP system the time slots are controlled by the 'bus arbiter' node, which stores a pre-defined list of data objects to be produced (Note: FIP uses a producer–consumer model). The bus arbiter cycles through the list and broadcasts the data identification onto the network, the node that is set up to produce this data object recognises the ID and then 'produces' the value on the network; any number of nodes then 'consume' this data. At the time of producing data the node may request a slot for a non-cyclic request. Once all the cyclic transmissions are complete the arbiter then cycles through all the cached non-cyclic requests. ControlNet uses similar media access control.

Error handling
In a factory environment, nearby lightning strikes, contact closures, power dips and other events may cause a transmitted message to be corrupted. A CRC polynomial inserted at the end of the message is used at the receiving end to determine if the message was corrupted. The data link layer detects this and schedules the packet for retransmission. If the retransmission is attempted too many times, then the message is discarded.

Conclusions
The producer–consumer model is very efficient if more than one node requires an item of data. However, in many automation applications the source–destination model with broadcast is adequate. Producer–consumer networks also need complex setup to allocate data Ids, although if tools are provided by the vendor this can be relatively painless. Master slave protocols only support centralised structures, although slower peer to peer connections for non-real-time data are possible on some networks. The producer–consumer capabilities of CAN without using a master arbiter make it an attractive solution for small networks. Token ring is the most flexible protocol for peer to peer communications but the deterministic behaviour is not as controllable.

C3.5.2.3 Application layer

The application layer defines and implements the 'services' that the network offers each device. The most common services are 'Read' and 'Write'. The flexibility of the application layer generally incurs large overheads and consequently slower execution. Indeed, if the dynamic performance is critical then many networks bypass this layer for real-time data. Network data handled in this way is often called cyclic data.

C3.5.2.4 Device profile

The communication system, layers 1 through 7, manage the transfer of data between nodes. The profile or companion standard is a detailed specification of how this data is interpreted or mapped onto device functions. A common misconception is to assume that with compatible communications, devices are interoperable: interoperability is only truly achieved if the profile layers are implemented.

C3.5.3 RS-232/RS-485 Modbus: A simple Fieldbus system

There are a number of protocols designed for RS-232 or RS-485 communications that are still widespread in automation applications. Modbus is perhaps the best known and most widely adopted, and is described here as a typical example.

The Modbus serial communications protocol is a de facto standard designed to integrate PLCs, computers, terminals, sensors and actuators. Modbus is a master/slave system meaning that one device, the master node, controls all serial activity by selectively polling the slave devices. Modbus supports one master device and up to 247 slave devices. Each device is assigned a unique node address.

There are two variants of Modbus: ASCII and RTU. ASCII mode uses a message format that is 'printable'. ASCII messages start with a colon and end with a carriage return.

RTU mode uses binary and is therefore not 'printable'. Eight-bit characters are sent as a continuous burst and the end of the message is denoted by 3.5 character times of silence. RTU mode messages use half the characters of an equivalent ASCII message.

Only the master initiates a transaction. The master is usually a host PC or HMI device because most Modicon PLCs are slaves and cannot initiate a Modbus transaction (the new Quantum PLCs can act as a Modbus master). Typically the host master will read or write registers to a slave. In each case, the slave will return a response message. For a read operation, the response will carry the requested data. For a write operation, the response is used to verify acceptance of the write command. A special case is the 'broadcast' operation where a write operation can be directed to all slaves. In this case, no response message is forthcoming.

The 8-bit address field is the first element of the message (1 byte for RTU, or 2 characters for ASCII). This field indicates the address of the destination slave device that should respond to the message; all slaves receive the message but only the addressed slave will actually act upon it.

The function code field tells the addressed slave what function to perform. Modbus function codes are specifically designed for interacting with a PLC on the Modbus industrial communications system.

Two error check bytes are added to the end of each message: ASCII mode uses a longitudinal redundancy check (LRC), and RTU mode uses a 16-bit CRC check.

In the example in Tables C3.8 and C3.9, the host PC is initiating a read request of three parameters starting with #1.08 from drive address 06. The starting holding register is 40108 but the '4' is dropped in the message string and the rest of the register address is entered as 'one less' (0108 becomes 0107, 0107 is entered as 006B in

Table C3.8   Query

Field name	RTU (hex)	ASCII characters
Header	None	: (colon)
Slave address	06	0 6
Function	03	0 3
Starting address Hi	00	0 0
Starting address Lo	6B	6 B
No. of registers Hi	00	0 0
No. of registers Lo	03	0 3
Error check	CRC (2 bytes)	LRC (2 chars)
Trailer	None	CRLF
Total bytes	8	17

CRLF = Carriage Return Line Feed.

Table C3.9   Response

Field name	TRU (hex)	ASCII characters
Header	None	: (colon)
Slave address	06	0 6
Function	03	0 3
Byte count	06	0 6
Data Hi	02	0 2
Data Lo	2B	2 B
Data Hi	00	0 0
Data Lo	00	0 0
Data Hi	00	0 0
Data Lo	63	6 3
Error check	CRC (2 bytes)	LRC (2 chars)
Trailer	None	None
Total bytes	11	23

hexadecimal). The response repeats the address and function code, but includes the values read from the drive.

C3.6   Fieldbus systems

C3.6.1   Introduction to Fieldbus

Fieldbus is the name given to a digital communication network that is used for automation and control; these networks have special properties that make them different from other networks and suitable for use in specific environments. Fieldbus networks will typically offer the following features:

- a high degree of electrical noise immunity,
- an ability to operate predictably – deterministic,
- good error handling and reporting,
- optimisation for small volumes of data, sent at a high frequency, and
- low lifetime costs.

In the past, PLC control systems were implemented, as shown in Figure C3.18, by hard wiring every switch, lamp, variable-speed drive, and solenoid back to a centralised controller that would read the inputs at the start of the PLC program and as a result set outputs at the end of the program in order to control the machine; this philosophy results in large looms of wiring that are expensive and time-consuming to wire, maintain and repair.

Interfaces, communications and PC tools 511

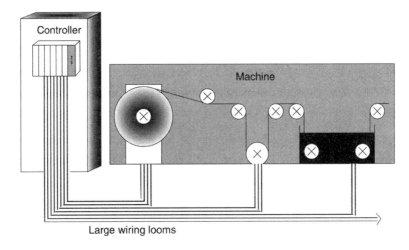

Figure C3.18   Traditional control system without Fieldbus

Fieldbus technology has revolutionised automation to the extent that all that now enters and exits a typical control cubicle are simple, low-cost communication cables. The I/O can be connected via Fieldbus and so located close to the source and destination of the signals, reducing the cabling and increasing the flexibility to add more I/O in the future (Figure C3.19).

Many actuators and sensors now connect directly to the Fieldbus network, allowing the status to be directly obtained by the controller, further reducing I/O and wiring.

In summary, Fieldbus reduces installation cost, reduces running costs, increases flexibility, reduces complexity, improves noise immunity, increases reliability,

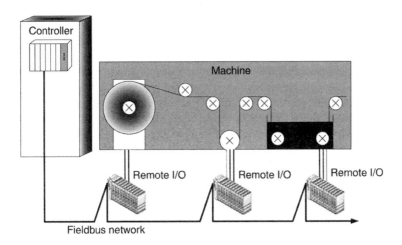

Figure C3.19   Modern control system using Fieldbus technology

512 *The Control Techniques Drives and Controls Handbook*

reduces downtime and improves diagnostics, resulting in widespread adoption and much more profitable machines. It is hard to think of any negative effects of implementing Fieldbus technology.

C3.6.2 Centralised versus distributed control networks

The terms centralised and distributed refer to the logical flow of data in the network and not how the network is physically wired. This may be referred to as the logical topology. The following sections describe some of the commonly used examples.

C3.6.2.1 Centralised network

Centralised networks require a master controller, typically a PLC, PC or motion controller. The master device is entirely responsible for controlling communications over the network, while the slave devices tend not to utilise any local intelligence, as indicated in Figure C3.20. Peer-to-peer communications is not generally well supported. For large networks, the master controller may need a substantial amount of processing power, even by today's standards, to handle the large amounts of data in a real-time application. In addition, the bandwidth required on the Fieldbus network for high-performance functions such as motion and positioning adds significant cost and performance penalties. The main advantage of centralised control is in its structural simplicity; the centralised controller is a one-stop location for configuration, maintenance and diagnostics.

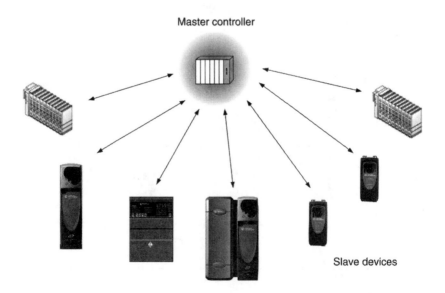

Figure C3.20   Centralised control

C3.6.2.2 Distributed network

Distributed networks, illustrated in Figure C3.21, require some local intelligence at each device, but no overall master device. Peer-to-peer communication is essential for these networks; it allows data transfer to be focused from one device to another, and has the effect of reducing the overall network traffic and increasing the available bandwidth for a given Fieldbus.

For high-performance applications such as motion, positioning and winding, distributed control offers distinct advantages. In distributed networks the physical and logical proximity of the controller to the actuator/drive enables the highest possible performance to be achieved. This is true because local control is able to completely remove the effect of Fieldbus network delays (all Fieldbus networks add some delay) and is able to immediately process the data from transducers and the drive parameters, unlike a centralised approach where the controller will usually schedule these tasks. Furthermore, decentralised control enables the position, motion or winder control to be precisely synchronised with the drives own control loops, ensuring that each and every control cycle is acted upon with the minimum of delay, again increasing performance.

A further advantage of distributed control is that it is easy to modify or expand the system. When adding additional devices to a network, additional processing power is also added, extending the capability, offering a future-proof solution.

One potential disadvantage with distributed control is in the added complexity of configuration, monitoring and maintenance across multiple devices; reference

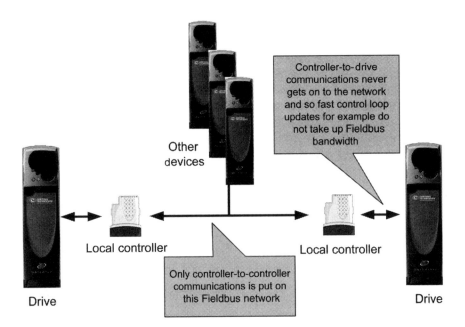

Figure C3.21   Distributed control

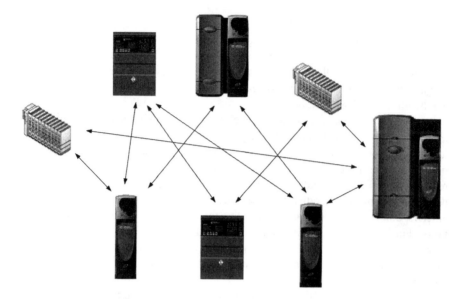

Figure C3.22   Distributed control: communications

to Figure C3.22 shows that with so much 'cross talk' things could easily get out of hand.

Structure can be brought to the communications by the use of high-level software tools that understand the network dynamics. A good example of this is the Control Techniques software tool SyPTPro, which has a full function network configuration suite, a network-wide monitoring tool and the ability to programme control across all the devices in a network. Figure C3.23 shows a typical screen shot of this tool showing a network comprising five drives, three remote I/O modules and a PC.

Figure C3.24 shows another screen shot of the same software tool in which the communications links between the system components are defined. This type of facility makes distributed systems very easy to configure and control as a unified system, while maintaining the advantages of a very flexible system.

C3.6.2.3 Hybrid networks

For some applications a very attractive network architecture is a hybrid approach, where a centralised controller is able to oversee and direct the activities of clusters of devices, such as a section of a machine (Figure C3.25). This allows the master controller to be modestly sized, with no requirement for fast control algorithms, and the Fieldbus that links the device clusters to the central controller is not required to pass fast updating data such as set points and feedback. Within the clusters, the distributed devices are able to provide the highest possible performance. Redundancy is possible and cost effective.

Hybrid systems offer the advantages of centralised and distributed control, and few disadvantages.

Interfaces, communications and PC tools 515

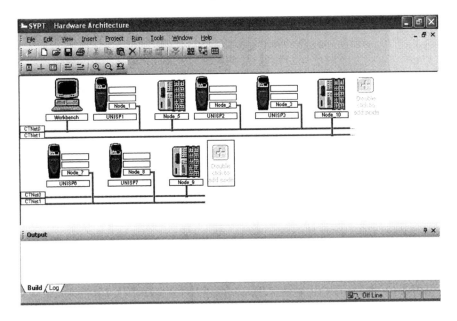

Figure C3.23   System configuration and monitoring tool for distributed control

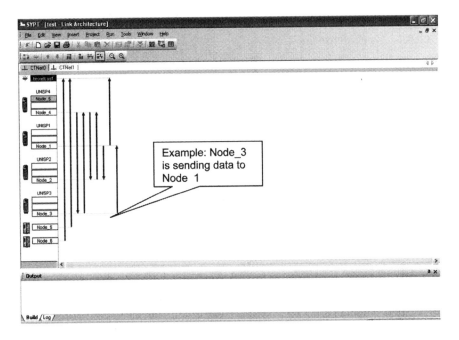

Figure C3.24   Communications configuration for a distributed control system

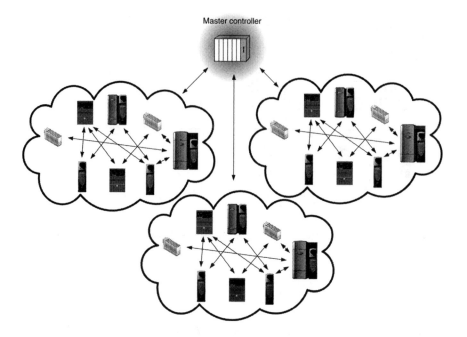

Figure C3.25   Hybrid control with master controller co-ordinating device clusters

C3.6.3   Open and proprietary Fieldbus systems

C3.6.3.1   Open networks

The term 'open' when used in the context of Fieldbus systems means that control of the specification and standards has been adopted or transitioned to an independent organisation that's membership is open to multiple vendors. Virtually all open networks allow vendors to add non-standard, vendor-specific functionality to allow them to differentiate themselves in the market, effectively offering an open network with some proprietary functionality. The advantage of open networks is that in theory devices can be purchased from multiple manufacturers and should be able to inter-operate. Most Fieldbus organisations provide a conformance certification programme to give end users the confidence that the devices are able to inter-operate.

It is good practice to always ensure that the devices that you purchase are certified by the relevant organisation.

C3.6.3.2   Proprietary networks

Proprietary networks are usually controlled and maintained by a single company or a small group of companies. The networks specifications are often tightly coupled to the features required for a particular company and their products. Although the closed nature of these buses limits the choice of components available to connect, these bus systems survive because they are able to offer a unique value proposition.

The specification is usually set by teams of individuals able to adapt dynamically to customer requirement and not having to satisfy committees with opposing agendas. The result is often a focused efficient solution to a particular problem or customer requirement; inter-operability is not an issue as all devices are purchased from the same vendor and so should have been tested. Communication standards such as OPC are able to provide a low-cost, low-investment mechanism to get access onto many proprietary networks, removing many barriers and increasing acceptance.

C3.6.4 OPC technology

OPC is an open, standard technology that allows PCs to connect to an industrial device and network for monitoring, configuration and data acquisition. The technology can be compared to your printer drivers in Windows^{TM}. The printer drivers allow the windows software to use any type of printer made by any manufacturer provided a valid driver is supplied for the product. Similarly, OPC allows computers to communicate with any industrial equipment provided the manufacturer can supply you with an OPC server for the particular device and network. In addition, multiple OPC servers can work together allowing your software to communicate with different devices from different manufacturers on different Fieldbus networks.

OPC was developed in conjunction with Microsoft; as a result, software tools such as Excel, Visual Basic, and Visual C++ etc. can be OPC clients. However, because of its obvious advantages, the major success has been within SCADA systems, and now all major suppliers supports OPC as a default.

OPC is managed by an organisation called the OPC Foundation. Control Techniques are long-standing members of the OPC foundation; our OPC server is available free of charge and supports communication with a single drive or a network of drives using either CTNet, Ethernet or the standard RS-485 communications available on the drive.

C3.6.5 Industrial Fieldbus systems (non Ethernet)

There are so many different Fieldbus systems that it would be impossible and of little value to list them all here. What is helpful is to consider the most popular systems.

C3.6.5.1 Profibus DP

Profibus started life in the 1980s, developed by academic institutions within Germany, and was quickly adopted as a German national standard; it is by far the most widespread Fieldbus in use, and is especially dominant throughout Europe. Profibus DP is optimised for speed and low cost. The DP version of Profibus is for communication between automation and control systems and distributed I/O at the device level. Its key statistics are shown in Table C3.10.

Table C3.10 Key statistics of Profibus

Speed	Fast: 12 Mbits s^{-1}
Physical topology	Typically bus
Logical topology	Centralised, master/slave and peer-to-peer communication is possible
Flexibility	Excellent
No. of devices	32 without repeaters, 124 with repeaters
Network length	200 m on RS-485 (can extend with repeaters/hubs)
Transmission method	Typically twisted pair RS-485 or fibre

Industry biases
Profibus is widely used throughout many industries, including the automotive, production machinery and metals processing industries.

Geographic biases
It is widely accepted throughout Europe and has a growing acceptance worldwide. For more information see www.profibus.org.

C3.6.5.2 DeviceNet

In the early 1990s, engineers at the US mechanical engineering company Cincinnati Milacron started a joint venture together with Allen-Bradley and Honeywell Microswitch regarding a control and communications project based on CAN. However, after a short while, important project members changed jobs and the joint venture fell apart. Allen-Bradley and Honeywell continued the work separately. This led to the two higher-layer protocols 'DeviceNet' and 'Smart Distributed System' (SDS), which are quite similar, at least in the lower communication layers. In early 1994, Allen-Bradley turned the DeviceNet specification over to the 'Open DeviceNet Vendor Association' (ODVA), which boosted its popularity. DeviceNet was developed especially for factory automation and therefore presents itself as a direct opponent to protocols like Profibus-DP.

DeviceNet uses a protocol called CIP (Common Industrial Protocol); other Fieldbus that use this protocol include ControlNet and Ethernet IP. At the moment these protocols are being enhanced to offer in-built safety features.

The key statistics of DeviceNet are shown in Table C3.11.

Industry biases
DeviceNet is widely used in a large range of industries, such as the general production machinery and automotive industries.

Geographical biases
DeviceNet is well accepted throughout the Americas. For more information about DeviceNet see www.odva.org.

Table C3.11   Key statistics for DeviceNet

Protocol	CIP
Speed	Slow: 500 kbits s^{-1}
Physical topology	Typically bus, but branching is also possible.
Logical topology	Centralised, master/slave, peer-to-peer and multi-master
No. of devices	64
Network length	500 m baud rate dependent (may be extended with repeaters)
Transmission method	CAN bus, twisted pair with power.

C3.6.5.3   CANopen

In 1993, a European consortium lead by Bosch had been developing a prototype of what should become CANopen. After the completion of the project, the CANopen specification was handed over to the CiA for further development and maintenance. In 1995, the completely revised CANopen communications profile was released and quickly became an important standardised network in Europe. CANopen does not only define the application layer and a communication profile, but also a framework for programmable systems as well as different device, interface and application profiles. This is an important reason why whole industry segments (e.g. printing machines, maritime applications, medical systems) decided to use CANopen during the late 1990s.

An important aspect of CANopen is its ability to close a position loop with reasonable update times despite a relatively modest bus speed. Its relatively loose specification may however hinder true inter-operability.

The key statistics for CANopen are shown in Table C3.12.

Industry biases
CANopen has gained widespread acceptance with OEMs who manufacture their own bespoke controllers. This is because of the low cost and simplicity of implementation.

Table C3.12   Key statistics for CANopen

Speed	Slow: 500 kbits s^{-1}
Physical topology	Bus
Logical topology	Centralised – Master/slave, peer-to-peer, multi-cast, multi-master
No. of devices	127
Network length	100 m (may be extended with repeaters)
Transmission method	CAN bus, twisted pair with power

Geographical biases
CANopen is popular in Europe. For more information see www.canopen.org.

C3.6.5.4 Interbus

The Interbus, or Interbus-S, protocol was developed in the mid-1980s by Phoenix Contact and several German technical institutions. The goal of the project was to simplify signal wiring in industrial applications. In 1987 Interbus was first release to the market but at this time it was a proprietary protocol. In 1990 Phoenix Contact decided to disclose the protocol specifications of Interbus and as such Interbus became the first manufacturer-independent Fieldbus system.

The Interbus protocol has continued to develop since that time and is now controlled by the Interbus Club, an organisation run by users of the protocol. Interbus has been adopted as a German standard (1994) and European standard (1998) according to DIN 19 258.

The key statistics for Interbus are shown in Table C3.13.

Industry biases
It has been widely adopted by automotive manufacturers.

Geographical biases
It is popular within Europe only. For more information see www.interbusclub.org.

C3.6.5.5 LonWorks

LonWorks is a flexible, distributed peer-to-peer networking architecture. The technology has been developed by a company called Echelon, who have handed control of the specification to an open body called LonMark. The technology, however, is heavily reliant upon an IC technology called a Neuron Chip, which is required in every device on the network; each of these are manufactured under licence from Echelon, who can also provide the configuration software for the network.

Table C3.13   Key statistics for Interbus

Speed	Good: 500 kbits s^{-1} full duplex[a]
Physical topology	Ring
Logical topology	Centralised
Flexibility	Good
No. of devices	512, with 4 096 I/O points
Network length	400 m between devices, up to 13 km network diameter
Transmission method	Twisted-pair and fibre

[a]Owing to protocol efficiency and full duplex data transmission the data throughput is excellent, despite a modest baud rate.

Table C3.14 Key statistics for LonWorks^a

Protocol	LonTalk
Speed	Slow: typically 78 kbits s^{-1}
Physical topology	Bus or free
Logical topology	Distributed
Flexibility	Excellent
No. of devices	64–128
Network length	2 200 m bus topology, 500 m free topology (may be extended with repeaters or hubs)
Transmission method	Various but typically twisted-pair. Power-line transmission is also possible

^{a}LonWorks makes the most from the limited bandwidth by reducing the amount of data on a network; this is achieved by only sending the data if it changes state.

Because virtually every LonWorks device uses the same Neuron Chips, and the LonMark certification process is fairly rigorous, LonWorks inter-operability is generally very good.

The key statistics for LonWorks are shown in Table C3.14.

Industry biases

LonWorks had originally aimed to provide a network solution to a wide spectrum of industries; however, it has found its home in the HVAC building automation industries, where fast updates are not of primary concern. In this industry ease of installation is essential and with LonWorks and free topology you would need to try hard to make it not work.

LonWorks distributed topology removes the requirement for a centralised controller; for this reason its main opposition has come from manufacturers of building management systems.

Geographical biases

LonWorks is used globally. For more information see www.lonmark.org.

C3.6.5.6 BACnet

BACnet stands for building automation control network. A data communication protocol developed by ASHRAE, its purpose is to standardize communications between building automation devices from different manufacturers, allowing data to be shared and equipment to work together easily.

In 1987, ASHRAE undertook the challenge to develop and put forward a standard set of rules (BACnet) governing communication between various devices used in building control systems. Now that BACnet is the accepted standard by ANSI and

Table C3.15 Key statistics for BACnet

Speed	Various
Physical topology	Typically bus
Logical topology	Centralised or peer-to-peer
Flexibility	Excellent
Transmission method	Commonly RS-485, Ethernet, LonTalk and Arcnet are also available.

ASHRAE, and is currently being considered for adoption internationally, the foundation has been laid for further industry cooperation.

The key statistics for BACnet are shown in Table C3.15.

Industry biases
BACnet has always been aimed at the building automation industry. With a master–slave configuration being possible it has found a home within many building management systems. The specification continues to develop to incorporate Ethernet-based communications.

Geographical biases
It is biased towards the Americas. For more information see www.ashrae.com.

C3.6.5.7 SERCOS II

SERCOS is an acronym for SErial Real-time COmmunications System. It defines an open, standardized digital interface for communication between digital controls, drives, I/O, sensors and actuators for numerically controlled machines and systems. It is designed for high-speed serial communication of standardized closed-loop data in real time over a noise-immune, fibre-optic cable.

Table C3.16 Key statistics for SERCOS

Speed	16 Mbits s^{-1}
Physical topology	Ring
Logical topology	Centralised
Flexibility	Good
No. of devices	Up to 255, but update rate is affected as device count increases
Network length	Depends on types of fibre used, but long lengths can be easily achieved
Transmission method	Fibre

The SERCOS interface takes full advantage of today's intelligent digital drive capabilities by not only replacing the de facto ± 10 V analogue interface standard, but also by allowing powerful two-way communications between control and drive.

The key statistics for SERCOS are shown in Table C3.16.

Industry biases
It is predominantly used in motion and CNC applications.

Geographical biases
It is used worldwide. For more information see www.sercos.de.

C3.6.6 Ethernet-based Fieldbuses

C3.6.6.1 General

For many years Ethernet has been considered for industrial applications, but non-deterministic performance and high costs have largely excluded it from widespread use on the factory floor, limiting its application to high-value controller-to-controller and controller-to-PC based communications. Now this has changed, with the advent of low-cost IP-ready microprocessors, cost-effective Ethernet switches and industrial protocols that optimise Ethernet for industrial applications. Ethernet offers a significant number of advantages to users, such as enterprise-wide connectivity, wireless and remote access, a mature technology and familiarity to many industrial and IT engineers.

Many industrial users expected Ethernet to be the answer to all of their problems in terms of standardisation and inter-operability between different vendors' products. The truth, however, is that standard Ethernet is designed for IT networking and is not suitable for accurate real-time control such as motion and position control. The Fieldbus organisations have met the challenge of making Ethernet suitable for industry, frequently by implementing a version of their standard protocol over Ethernet, sometimes with modification to the standard TCP/IP protocols and sometimes even the physical hardware.

The reality is that Ethernet will bring us no closer to achieving the open, inter-operability that is widely enjoyed in the world of IT, but does still provide important benefits to users.

Ethernet-enabled devices are able to provide easy-to-access web page interfaces (Figure C3.26), email generation, and make use of standard Ethernet network features such as obtaining time and date information, network addressing and so on.

The web page interface, as shown in Figure C3.3.26, allows the user to configure the Ethernet module, such as setting up email alerts and configuring the drive parameters.

C3.6.6.2 Modbus TCP/IP

An open TCP/IP specification was developed in 1999 encapsulating the standard Modbus packet within the TCP/IP messaging structure. Combining the versatile, scaleable, and ubiquitous physical network (Ethernet) with a universal networking

524 The Control Techniques Drives and Controls Handbook

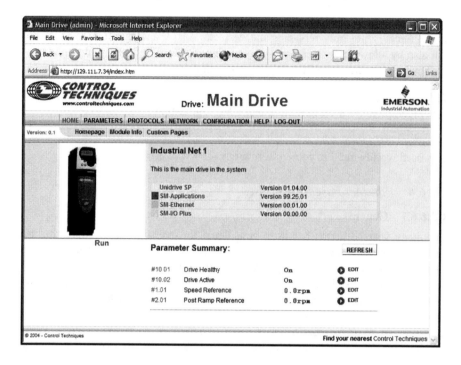

Figure C3.26   A web page embedded in a drive

standard (TCP/IP) and a vendor-neutral data representation (Modbus) gives a truly open, accessible network for exchange of process data. It is also extremely simple to implement for any device that supports TCP/IP sockets.

Modbus TCP/IP is currently the most popular industrial Ethernet protocol due to the simplicity of implementation using standard Ethernet, its compatibility with previous Modbus standards and its longevity in the market.

Industry biases
It is in general use.

Geographical biases
It is in use worldwide. For more information see www.Modbus.org.

C3.6.6.3   EtherNet IP

Ethernet/IP is an Ethernet-based protocol for industrial automation applications using the standard Ethernet TCP/IP protocol suite. The Ethernet/IP application layer protocol is based on the CIP protocol layer used in both DeviceNet and ControlNet Building on these protocols, Ethernet/IP provides a seamless integrated system for the enterprise network.

Ethernet IP is being developed to offer safety features and also real-time motion functionality; these are called CIPSafe and CIPSync.

Ethernet IP has a large and growing installed base due to the simplicity of implementation using standard Ethernet, its compatibility with previous CIP standards and its longevity in the market.

Industry biases
It is in use generally.

Geographical biases
It is in use in the Americas. For more information see www.odva.org.

C3.6.6.4 PROFINET

PROFINET is an industrial Ethernet standard for automation that includes plant-wide Fieldbus communication and plant-to-office communication. PROFINET can simultaneously handle standard Ethernet transmissions and real-time transmissions at 1 ms speeds. Engineers from more than 50 companies are working in 20 distinct working groups to advance PROFINET.

PROFINET communication is scalable at three levels. Component-based communication uses TCP/IP and enables cycle times on the order of 100 ms. It is preferred for communication between controllers. Real-time (RT) communication enables cycle times on the order of 10 ms and is well suited for use with distributed I/O. Isochronous real-time (IRT) communication enables cycle times less than 1 ms and is thus well suited for use in motion control applications. All these communication levels can co-exist on the same bus line together with IT communication.

Industry biases
PROFINET is an emerging technology and so has not yet established any industry bias. The German car industry has agreed to standardise on Profinet in future projects.

Geographical biases
PROFINET is likely to be more dominant in Europe than other countries. For more information see www.profibus.org.

C3.6.6.5 EtherCAT

EtherCAT is a real-time industrial Ethernet Fieldbus that has been developed by Beckhoff and now handed over to the open EtherCAT Technology Group to maintain and develop the standard. The bus system modifies the standard Ethernet hardware and uses a bus-type structure to pass the data from device to device. Protocol efficiencies are gained by embedding the data for many devices within one Ethernet frame, obtaining 90 per cent efficiency from the available bandwidth. EtherCAT is able to service 100-drive axis in 100 μs.

Industry biases
EtherCAT is used in specialised machinery utilising PC-based motion control.

Geographical biases
EtherCAT is used predominantly in Europe. For more information see www.ethercat.org.

C3.6.6.6 Powerlink

Developed and introduced by B&R in 2001, ETHERNET Powerlink is a now an open standard protocol enabling deterministic, isochronous, real-time data exchange using standard Fast Ethernet. It allows high-precision data communication with cycle times as low as 100 μs and network jitter well below 1 μs.

Industry biases
It is used predominantly in the textile and printing industry.

Geographical biases
Powerlink has a bias towards Europe, especially Germany and Switzerland. For more information see www.ethernet-powerlink.org.

C3.6.7 Company-specific Fieldbuses

A large number of companies have developed their own Fieldbus communications systems. Some use established physical layers, whereas others are entirely novel. Some appear to have no specific merit while others have characteristics well suited to either specific applications or types of equipment. It is interesting to consider two such networks.

C3.6.7.1 CTNet

Developed by Control Techniques, CTNet is a highly deterministic network designed for fully distributed control. Based on a 'ruggedised' version of ARCnet it was developed because no other off-the-shelf Fieldbus was able to provide the features required for fully distributed control. CTNet is available integrated into many Control Techniques drives, combining the network and the control processing capability within a drives own footprint. Together with SyPTPro (system programming tool), CTNet provides a compelling and cost-effective approach to high-performance drive-based control systems. CTNet provides connectivity to drives, I/O and operator panels.

This topology can be peer-to-peer, avoiding the need for a central or master controller. This allows machine builders to offer a standard off-the-shelf solution, and still provide a gateway between their machine and the customer's Fieldbus of choice. A CTNet OPC server is available.

The key statistics of CTNet are shown in Table C3.17.

Table C3.17   Key statistics for CTNet

Speed	Fast: 5 Mbits s^{-1}
Physical topology	Bus
Logical topology	Distributed
Flexibility	Excellent
No. of devices	255
Network length	Up to 250 m (may be extended through hubs and repeaters)
Transmission method	Twisted pair or fibre

Industry biases
CTNet has found a home within many industries, and particularly within machine builders, where the approach offers performance and cost advantages. For more information see www.controltechniques.com.

C3.6.7.2 CTSync

Developed by Control Techniques, CTSync is a unique protocol designed to address a specific problem of distributing master position information within a drives network. Many applications require multiple drives to synchronise to a master reference. In the past this has resulted in unwieldy wiring and low reliability. CTSync addresses this by utilising a low-cost RS-485 network running at high speed. The network hardware synchronisation provides update information on an accurate time base of 250 μs. This is also used to synchronise the speed, position and torque loop across multiple drives to provide the ultimate performance.

The key statistics are provided in Table C3.18.

Industry biases
It is use in printing, master–slave motion applications such as flying shear, rotary knife and packaging. For more information see www.controltechniques.com.

Table C3.18   Key statistics for CTSync[a]

Speed	Quite fast: 896 kbits s^{-1}
Physical topology	Bus
Logical topology	Master–slave, simplex communications (one direction only)
Flexibility	Poor: highly optimised for one task
No. of devices	16 devices (Can be extended with repeaters)
Transmission method	RS-485, twisted-pair

[a]CTSync uses a dedicated protocol that is highly optimised for the application; this results in cyclic updates of 250 μs synchronised with the drives' control loops.

C3.6.8 Gateways

In the sections above, a large number of Fieldbus variants have been discussed, and the requirement to communicate between these networks is an important practical issue. In general, each uses different hardware and/or data formats and so they may not be connected together. From an end-user perspective, this is a major disadvantage, as the Fieldbus selected limits the choice of devices available to be connected on the network. Sometimes this limitation must be overcome. As an example, consider a production machine that uses Profibus, but a packaging machine uses DeviceNet; for these two units to work together the buses must be linked. A device that links different types of Fieldbus is called a Fieldbus gateway. Typically a gateway is achieved by both Fieldbuses being resident within one device and sharing a common memory; however, traditionally these devices are expensive.

Some drives are available, such as the Control Techniques Unidrive SP (Figure C3.27) that have been designed to operate with more than one Fieldbus and as a result provide a very dynamic interface or gateway between the networks. As the Fieldbuses are implemented through option modules a wide range of gateways is possible. This allows machine builders to standardise on their favourite bus system, but still allows their customers to connect to the machine through their own Fieldbus of choice.

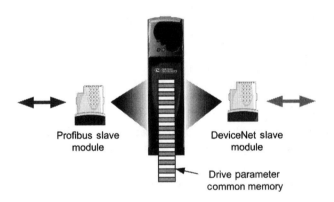

Figure C3.27   Drive acting as a Fieldbus gateway

C3.7 PC tools

In order to take full advantage of the functionality of a modern drive, manufacturers frequently supply a range of PC tools. These tools tend to be drive- or at least manufacturer-specific, and so a detailed description is not appropriate here. However, it is helpful to consider the types of tools that may be available either as an individual tool or as part of an integrated package.

C3.7.1 Engineering design tools

A wide variety of PC tools are available within this generic area. Some provide engineering calculations for specific applications, for example, calculating the motor rating or the torque requirement to follow a particular motion profile. This type of programme should be based around advanced models of specific motors and drives that will allow the user to be directed to the optimum solution/combination. Typically, mechanical system elements such as gearboxes, couplings and conveyors are included in such tools.

Other design tools are aimed at aspects of design such as calculating energy savings or the harmonic distortion on the supply due to an individual or multiple drives.

C3.7.2 Drive commissioning and setup tools

PC tools are available for configuring drives during commissioning and optimising. These tools frequently have facilities to monitor the drive during operation and so variables within the drive controls (or hardware such as temperatures) can be observed in real time. This type of drive may be available in different forms to suit the user type or application. It is common on simpler tools to see setup wizards that can facilitate the simple use of quite complex configurations.

Most tools allow the user to read, save and load drive configuration settings. Visualisation of the settings is often in the form of animated diagrams (Figure C3.28). An important aspect of this type of tool is to allow the user to manage the drive settings. They can be stored so that a known operational parameter set is available if it is suspected that the setup has been changed or in the event that a drive needs to be replaced.

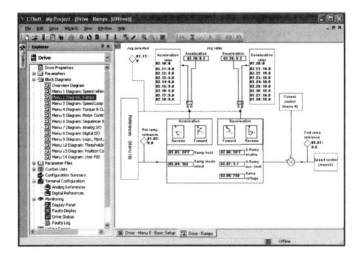

Figure C3.28 Typical commissioning and setup screen

Some manufacturers have such tools that can be operated over communications networks such as Ethernet so setup can be centrally managed.

C3.7.3 Application configuration and setup tools

The tool described in Section C3.7.2 allows the inbuilt functions within an individual drive to be configured. Some drives have additional programmability that allows the user to create their own functions.

Although there is no standard format for such tools, many closely follow IEC 61131-3, which is widely adopted for PLCs. The programming environment typically supports three industry standard programming languages: function block, ladder and structured text, and some tools offer sequential function block programming as well (Figure C3.29).

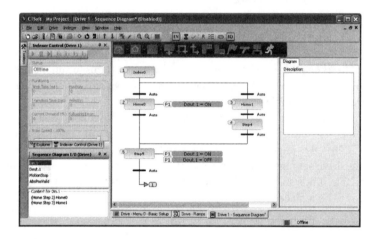

Figure C3.29   Typical sequential function block screen shot

Complex motion algorithms can be readily programmed using standard function blocks. These typically include various arithmetic, ramp, CAM and interpolation functions. A number of PLC and drive manufacturers support the PLCopen function blocks. Such adoption of standards does make life easier for the user in terms of common elements, but does not result in inter-operability of the finished code.

C3.7.4 System configuration and setup tools

The use of PC tools for system configuration has already been discussed in Section C3.6.2.2. The ability of such tools to provide a single programming and visualising environment for an entire system (Figs C3.23 and C3.24), even where those systems have highly distributed intelligence, is very significant and removes what used to be a major differentiation for PLCs and centralised control systems. SCADA screens can frequently be configured in such systems.

Interfaces, communications and PC tools 531

Such tools are typically integrated with those described in Section C3.7.2 and therefore use the same programming environment.

C3.7.5 Monitoring tools

Monitoring tools are provided in a variety of forms but a popular format is a software oscilloscope (Figure C3.30), which is useful for viewing and analysing changing values within the drive. The time base can be set to give high-speed capture for tuning or intermittent capture for longer-term trends. The interface is based on a traditional oscilloscope, making it very familiar to engineers.

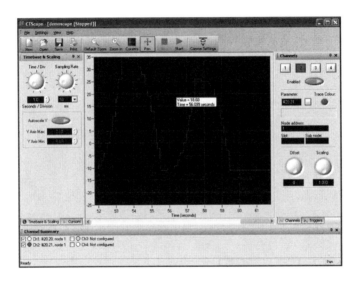

Figure C3.30   Typical monitoring/oscilloscope screen

Chapter C4
Typical drive functions

C4.1 Introduction

The functions available on a modern digital drive are so numerous that only the most studious of users will be aware of the capabilities of the drive. Although many of the functions will be common across all manufacturers, some will be unique and others subtly different. Intelligent drives offer the user the opportunity to write their own specific functions that will then run on the drive. Input and output terminals on many drives will also be configurable, that is to say that the function controlled by or driving the terminals can be set by the user.

It is important to note, however, that a well-designed drive will leave the factory with a default configuration that will cater for a large proportion of basic applications without the need for the user to make many or any changes to operate. Software tools are also available to guide the user through the configuration of their drive.

The objective of this section is to describe in general terms the sort of functions the user may expect to find on a typical industrial drive, and what they do. In demanding applications care needs to be taken about issues such as the sample/update times of given inputs and outputs (also relevant for data communicated through serial buses).

The user should consider the cost benefit of utilising such functions that exist within the drive. Less obvious is the potential improvement in performance obtained by using a function within the drive because communication time to the control loops is likely to be quicker than using an external controller. Sample and update rates must be taken into account for demanding applications.

For convenience the functions have been grouped in order to illustrate the types of functions that are available and to allow the reader to readily find specific functions. The list is not exhaustive.

C4.2 Speed or frequency reference/demand

The speed demand for a drive can come from a number of sources. It may be an analogue signal from a process controller (voltage or current) or traditional potentiometer. It may be a digital reference received through a serial interface/Fieldbus. It may be set through the drive's own keypad. Alternatively, it is possible on many drives to define a number of fixed or preset speeds, typically eight, and the drive can be programmed via digital signals to operate at any one of those speeds. Modern drives will usually allow the user to select the form of the speed demand. It would be usual for the default setting to be for the speed to be determined from the keypad.

Internally, advanced drives may also allow the reference to be modified with trims and offsets. The polarity of the signal can also be constrained if only one direction of rotation is to be permitted. Velocity feed-forward terms can also be found that can be used to improve the dynamic performance of the speed controller.

Clamps can also be applied to the maximum and minimum speed. On some drives a jog reference is possible for use on machines where it may be required to rotate the motor shaft at low speed or under manual control during setup or for investigating a system problem.

Skip frequencies/speeds are also available on many drives where a band of speed (typically up to three) can be defined where operation is not permitted. This function is most commonly used where operation at a particular frequency/speed could cause problems in the machine such as exciting a mechanical resonance.

Some drives contain a function typically referred to as a 'motorised potentiometer'. This gives functionality equivalent to the traditional controller of that name; digital inputs are used to raise or lower the reference signal. The rate of change can be defined and a reset to zero input is common. Unipolar or bipolar output range can be selected.

C4.3 Ramps

Reference or demand signals are not always applied directly to control loops and in many applications the rate of change of the demand signal is subject to a limit on the rate of change. Although very simple drives may have a single, though selectable/programmable, ramp rate advanced drives can allow different acceleration and deceleration ramp rates to be selected. Some drives have a range of preset acceleration and deceleration ramp rates that can be changed on the fly.

Although the ramps most commonly found are linear, some drives have advanced ramp profiles designed to reduce the jerk (or rate of change of rate of change of the reference). Such ramp profiles are frequently described as S-ramps. In detail, they can range from simple cosine-type functions through to exceptionally complex functions where the output function is controlled such that d^2x/dt is always controlled even when x is varying dynamically. Such control functionality is critical to applications such as elevators where the rate of change of acceleration is very noticeable to the passengers.

Introducing a ramp into the speed reference is in some ways analogous to adding inertia to the load. Conversely, because we know from first principles that in a linear system, torque is equal to inertia times acceleration, some drives take the derivative of the post ramp speed reference (which gives us desired acceleration) and use this as a scalable feed-forward term added into the torque controller; this gives us a very fast response without needing to have a fast speed controller, an important benefit in many applications.

C4.4 Frequency slaving

Frequency slaving is a function in which the output frequency of the drive is controlled to match precisely the frequency of a pulse train fed to the drive. The reference pulse train may be from a controller or an encoder mounted on a master motor that the drive is required to 'follow'. The frequency can usually be scaled and limited.

C4.5 Speed control

For open-loop drives with scalar V/f algorithms, speed controllers tend to be simple linear loops. Slip compensation would be available, which gives steady-state speed compensation for variations in load.

For closed-loop drives and high-performance open-loop drives, drives would typically contain a digital implementation of a PID speed loop. The gain values would be internally stored parameters, but it may be possible for these gains to be changed on the fly via external inputs. The requirement to change these gains dynamically is well illustrated by a robot arm, where the inertia relative to the motor shaft is dependent upon the arm position.

Closed-loop drives derive the actual speed signal from an encoder or resolver; d.c. drives would typically get their speed information from a tachogenerator (see Chapter A5).

Speed controllers usually incorporate clamps for maximum and minimum speed. Some users also require knowledge of when the motor is at the demanded speed; this is a normal feature of closed-loop drives, which typically allow the 'window' of monitoring to be set by the user.

The speed loop characteristics can be of critical importance in determining the performance of the user's application. In order to optimise the design and setup of the application the user may well wish to simulate the control loop performance. Competent suppliers will be able to provide the user with models of the drive controller, most helpfully in the s-domain. Key parameters are not simply the gain figures but critically the non-ideal delays due to sampling and update times in the digital controller.

See Chapter A4 for more information.

C4.6 Torque and current control

C4.6.1 Open loop with scalar V/f control

For open-loop drives with scalar V/f algorithms, current controllers tend to comprise a simple ramp to limit the di/dt and a PI controller. Current limits are applied and protection circuits included to monitor instantaneous current in the motor and converter. These limits may be different for motoring and braking operation. A thermal model of the motor and converter is usually incorporated, although the accuracy of these models varies greatly from one manufacturer to another.

The controller may react in different ways for different designs, but in most (but not all) applications it is desirable that the drive should avoid tripping and thereby losing control or shutting down the entire machine. Automatic selectable techniques exist to address such situations and good examples/good practice are suggested as follows.

- If an over-current limit is reached, reduce speed/frequency (for a motoring condition) in order to reduce the load.
- If a thermal limit is reached, on an a.c. drive reduce switching frequency.

C4.6.2 Closed-loop and high-performance open loop

Closed-loop and high-performance open-loop drives typically have a PI current controller acting upon the overall current fed to the motor. Details of the flux vector controller and alternative direct torque controller are given in Chapter A4, and will not be repeated here.

C4.7 Automatic tuning

An important aspect of good speed and torque performance is the accuracy of the motor models and the parameters used therein. Modern drives usually contain auto-tuning procedures, which, as the name suggests, will automatically measure motor and in some cases load parameters and optimise control loops.

These routines vary in content and accuracy from one manufacturer to another. Care must also be taken using some parts of the available routines as in order to obtain the most accurate parameters, the motor shaft is usually required to rotate, often up to say three-quarters of the rated load speed.

A stationary test can readily provide the following:

- the stator resistance by applying a d.c. voltage;
- the transient inductance by applying a voltage pulse to the motor winding.

A rotating test can provide the following:

- the stator inductance and the motor saturation break points;
- the load inertia;
- the encoder phase angle relative to the motor winding.

Other variations of routine involve limiting the motion of the shaft to a few degrees but still obtaining advanced parameter data. Advanced routines are also included in the auto-tuning of high-performance drives, but as these are very dependent upon the drive manufacturer they will not be considered further.

C4.8 Second parameter sets

In some applications such as a multiple pump system (Section C6.2.2) or some designs of crane (Section C6.5) it is a requirement to operate different motors with the same drive (not simultaneously). In such cases the parameter setup for the alternative motors and load are likely to be different and so different parameter sets are needed. Some drives allow more than one parameter set to be stored and changed quickly on command. This avoids the cost and time it would take to download another parameter set from an external controller.

C4.9 Sequencer and clock

The sequencer in a drive is simply the way that instructions and commands are handled. Some of these may be related to a real-time or run-time clock. Some of the commands will be linked through fixed or programmable I/O terminals on the drive or a serial connection such as a Fieldbus or keypad, whereas others can be internally generated signals, such as warning flags.

The user needs to consider carefully the functionality and behaviour he wants the drive to exhibit under normal and fault conditions. For example, the drive may be set to automatically try and restart after a trip. In the wrong application this could be a problem.

The clock function can be used to trigger particular functions at given times. For example, in a drive designed to operate in the building services industry, an ornamental fountain at a hotel may be required to switch on at 07.00 each morning and turn off at 23.30. This could be programmed using a real-time clock. Other functions such as calculating energy consumption or maintenance based on use of the drive could be programmed with a run-time clock.

C4.10 Analogue and digital inputs and outputs

Many modern drives allow the user to programme the functionality of some if not all analogue and digital input terminals. The hardware function may also be programmable between say a 0–10 V voltage input and a 4–20 mA current input.

C4.11 Programmable logic

Many drives have basic logic functions and threshold detectors that can be used to sequence basic application functionality without the need for creating user-defined functionality (see Section C4.13). In a similar way, configurable PID controllers may be available within the drive for the user to regulate any external variable. In some cases this control can be independent of the drive function, with inputs and outputs driven from the drive input and output terminals.

C4.12 Status and trips

Modern drives contain very comprehensive status monitoring, warning and trip functions. Typical status functions/indications include the following:

- drive healthy,
- drive active,
- zero speed,
- running at or below minimum speed,
- running below set speed,
- running at set speed,
- running above set speed,
- running at set load,
- drive current limiting the output current,
- motor is motoring or braking,
- brake resistor overload,
- direction of rotation (demand and actual),
- mains loss (supply),
- mains unbalance (supply),
- under-voltage (supply),
- drive over-temperature,
- external trip (trips marshalled through the drive),
- serial communications loss,
- encoder fault (various including line break),
- motor thermistor trip.

Furthermore, there are very comprehensive protection circuits within the modern drive including for example:

- soft start relay failed to close,
- thermistor fault,
- d.c. current transformer,
- auxiliary fan failure,
- power PCB over-temperature,
- control PCB over-temperature,
- heat sink over-temperature,
- power device over-temperature (based on thermal model),
- internal power supply fault,
- over-current on control inputs and outputs,
- d.c. bus over-voltage,
- d.c. bus under-voltage,
- memory errors,
- software errors.

C4.13 Intelligent drive programming: user-defined functionality

So far in this section the functionality has been ready-programmed into the drive by the drive designer and the user has configured the drive only through parameter settings.

Some drives, often referred to as 'intelligent drives', allow the user to design their own specific functions. Individual drive manufacturers tend to have their own design environment, but many tend to adopt the style of IEC 61131-3, an international standard for the programming of programmable logic controllers (PLC).

Within IEC 61131-3 there are a number of types of editors for the creation of user-specific functions and complete application programmes:

- *Ladder editor.* This editor emulates the operation of traditional relay systems with pseudo coils and contacts (Figure C4.1).
- *Text editor.* As the name suggests, this editor allows functions to be created using a textural programming language. This is very flexible but less intuitive than the other editors. Although error-checking routines are built into most programming routines, text editing is most prone to programmer error (Figure C4.2).
- *Function block editor.* Function block editors allow users to string together pre-programmed functions into a flow-type structure (Figure C4.3). This is a very powerful programming tool. It is intuitive and less prone to error than the text editor.

Some programming environments allow users to create their own function blocks either from a text editor or by combining standard function blocks. This

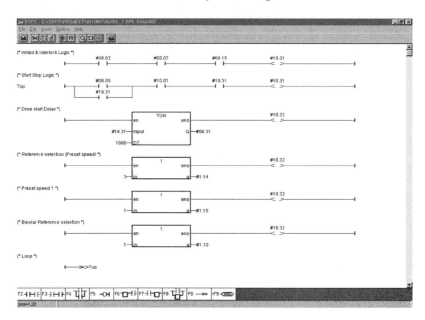

Figure C4.1   Typical Ladder Editor screen

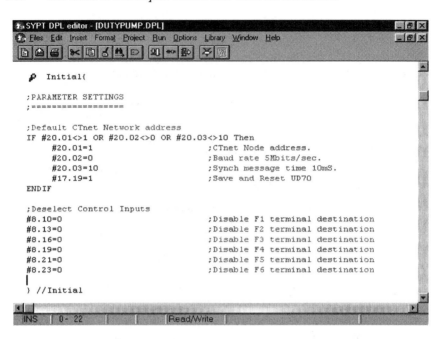

Figure C4.2   Typical Text Editor screen

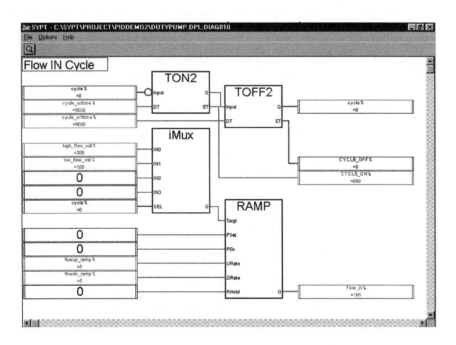

Figure C4.3   Typical Function Block Editor screen

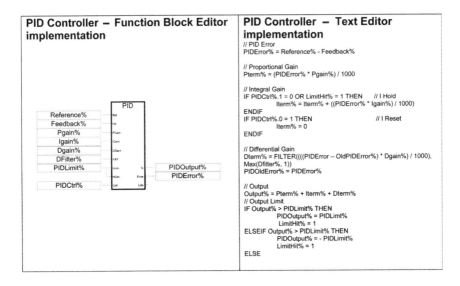

Figure C4.4   PID controller: Function Block and Text implementations

is very useful where users use a particular function many times. It also provides an additional level of security to protect IP in a situation where the user wants to give access to their programme to a customer, but wants to protect a specific function.

To give an indication of the relative complexity of the alternative programming editors, Figure C4.4 shows a PID controller function implemented as a function block and in text.

- *Sequential function block editor.* The sequential function block editor allows users to write down a sequence of events or moves in a logical timed sequence. This should be considered to be a higher-level programming environment than the other editors. Indeed the other editors would normally be used to programme the detailed events or moves described by the individual sequential block. Figure C4.5 shows the screen shot of a typical sequential function block editor, showing a simple indexing routine. The programme executes in a flow down the screen, starting with a homing routine then a number of prescribed movements before eventually looping to the start for another sequence.

Functionality of this type can be quite complex. For example, it may be possible to programme the entire functionality of an elevator including not only the anti-jerk ramps but also positioning to floor and all of the control functions associated with the call buttons and displays.

The examples above show how stand-alone drives can be programmed with user-defined functionality. Users may also programme 'intelligent drives' to operate together in a co-ordinated or even synchronised manner with interpolation. In such cases each drive would typically have its individual functionality programmed into it and the communications between drives determined through a Fieldbus.

542 The Control Techniques Drives and Controls Handbook

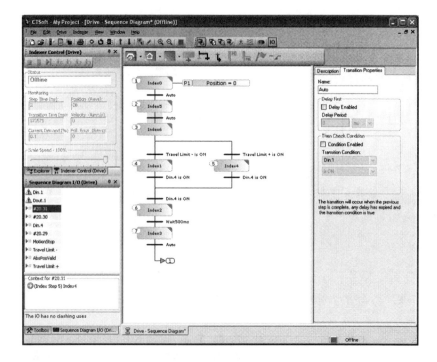

Figure C4.5 Typical Sequential Function Block Editor screen

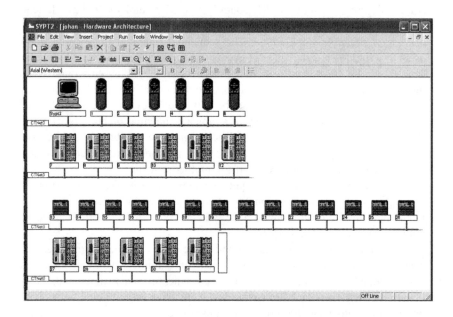

Figure C4.6 Multi-drive and distributed I/O system

Typical drive functions 543

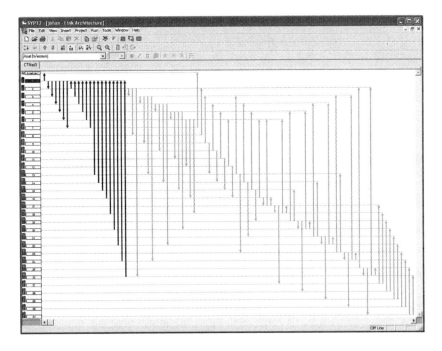

Figure C4.7   Defined system communication paths

Figure C4.6 shows a system with 6 a.c. drives, 14 d.c. drives and 11 banks of distributed I/O. Programming tools will typically allow users to define specific communications channels to be defined between the above nodes in the system in order to achieve high bandwidth of data transfer, as shown in Figure C4.7. Complex systems can be realised in this way.

By having the processing on the drive it is possible to readily synchronise the application programme to the drive control routines, and through the communications channel synchronise the control loops of all the drives in the system. This brings significant performance benefits over centralised systems, in regard to dynamic acquisition of, and response to, drive data and control jitter in high-performance systems.

C4.14 Functional safety

C4.14.1   Principles

'Functional safety' refers to functions of a drive, or other control equipment, that are designed in such a way that they can be used in applications where a malfunction could result in injury or death. Most commonly, such functions are used as part of the provisions for ensuring the safety of a machine or process, including allowing for the possibility of the control system itself failing. This means that they must have a very low probability of failing in a dangerous mode, and that it must be possible to demonstrate this by reference to a widely accepted technical standard.

Even with good-quality proven equipment, the inevitable background level of random hardware faults and the possibility of unrevealed design errors in complex systems with advanced hardware and software is unacceptable for a safety-related function. Almost invariably, special methods are required to ensure that the system is fault-tolerant to some degree, and that exceptional attention is paid to testing and assessing software.

The following is a brief account of the potential capabilities of drives in carrying out safety functions. When designing any system with functional safety impact, the possible failure modes of the drives must be considered as one potential source of hazard. Although a large proportion of faults in a drive result in a loss of output, which is usually a safe situation, there exists the possibility of a fault resulting in unexpectedly high torque or speed, or of wrong direction of motion.

This is a specialised subject that is covered by many publications, notably 'Control Systems Safety & Reliability' (William M Goble, ISA, ISBN 1-55617-636-8). Here, we consider only specific safety functions that can usefully be incorporated within drives.

The general approach to functional safety design covers the following basic aspects:

- the analysis of hardware failure modes and effects, choice of circuit structure and components to ensure that the likelihood of a dangerous failure is at an acceptably low level;
- the use of structured methods for generating well-defined software with a low risk of unexpected behaviour;
- design reviews and tests to minimise the risk of a design error (hardware or software) that could result in a dangerous failure, usually including the use of independent reviewers;
- production control to ensure that the product manufactured continues to meet the requirements, including after design changes;
- management of the processes of design, manufacture, installation and maintenance to ensure that the level of risk is maintained at the required level throughout the life of the equipment.

C4.14.2 Technical standards

Of the variety of available standards, the most widely used as the basis for general civilian applications outside of specialised areas such as traction are the IEC 61508 series. These use the concept of a safety integrity level (SIL) as an overall measure of the capability of a safety-related control system, although other data is also required when considering a sub-system such as a drive. We will focus here on standards that apply to machinery, where safety-related applications for drives are relatively common. The EC Machinery Directive has meant that the EU has a relatively well-structured set of machinery safety standards, but recent changes in the standards for control systems mean that there may be inconsistencies until the existing standards are revised.

Most new standards now begin life as international standards (IEC or ISO), which are then transposed into European Standards (EN), but where Europe has taken a lead the EN standard may have existed previously (e.g. EN 954-1) and been later converted into an international standard (ISO 13849-1).

- *EN 954-1 Safety of machinery – safety related parts of control systems – general principles for design.* This standard has been superseded by EN ISO 13849-1:2006, but there is much equipment in use that was designed according to it. It uses a system of categories ranging from 1 (lowest existing hazard) to 4 (highest hazard). These are associated with specified levels of overall risk, and have design requirements defined in qualitative terms, e.g. 'no single fault shall result in a loss of the safety function'. It does not lend itself to application in systems using complex hardware such as LSI microcircuits, nor software.
- *ISO 13849-1:2006 Safety of machinery – safety related parts of control systems – general principles for design.* This standard uses more quantitative methods for measuring hardware safety integrity, and includes provisions for the use of software, drawing on the methods of IEC 61508. It uses a measure of safety integrity referred to as 'Performance Level', with levels ranging from (a) to (e). A companion standard ISO 13849-2 'Safety of machinery. Safety-related parts of control systems. Validation', gives guidance on hardware failure analysis, including 'fault exclusions', which are component failure modes that do not need to be considered provided conditions are met.
- *IEC 62061 Safety of machinery – functional safety of safety-related electrical, electronic and programmable electronic control systems.* This standard implements the principles of IEC 61508 for machinery applications. It fully allows for the use of complex hardware and software in safety applications.
- *IEC 61800-5-2 Adjustable speed electrical power drive systems – safety requirements – functional.* This standard implements the principles of IEC 61508 for drives, with an emphasis on machinery applications. It is of particular interest here as it gives a list of specific safety functions that might be offered by drives. Because it builds on a common base with IEC 62061 it should be readily compatible with it. The measure of safety integrity offered by a specific drive safety function is the 'SIL capability', and this value must be accompanied by a probability of a dangerous random hardware failure (PFH – also referred to as PFH_D).

From the titles given above it is apparent that there may be a conflict between ISO 13849-1 and the others because of its use of a different measure of safety integrity. Strictly there is no simple equivalence between the SIL capability and the performance level, although the assessment processes applied are very similar so that both parameters can usually be derived.

There are many application-specific standards that might in some circumstances have an impact on drive requirements. The following are some particular examples:

- *IEC 60204-1* Safety of machinery – electrical equipment of machines,
- *EN 81-1* Safety rules for the construction of lifts, and
- *NFPA79* Electrical standard for industrial machinery.

All of these standards have recently been adapted to allow to some extent for the use of safety features in electronic drives, at least for the prevention of unintended operation (i.e. safe torque off), but they do not currently use the SIL system based on IEC 61508.

C4.14.3 Possible safety functions for drives

Because a drive may be responsible for providing motive power for many of the functions of a machine, and it is the motion that in many cases results in a potential hazard, there are some safety functions that appear to be naturally suited to implementation within the drive. Some possibilities are outlined in the following sections using the standardised terminology of IEC 61800-5-2.

C4.14.3.1 Safe torque off (STO)

An inverter drive is particularly suited to the implementation of a 'torque off' function, because inherently the d.c. link power is incapable of sustaining motion in an a.c. motor, and the inverter stage needs to be almost completely functional to produce sustained motion or torque. Therefore, if the inverter power devices can be securely decoupled from the pattern-generating control circuits then no credible fault or combination of faults can produce sustained motion, and the worst-case multiple device failure mode results in a brief torque transient in a permanent-magnet motor. In the terminology of safety systems, the inverter/motor interface uses the 'dynamic principle' to ensure a safe disable function, i.e. most of the power devices must be operative and actively switching in order to produce the rotating magnetic field required to produce sustained torque.

This contrasts favourably with the conventional approach to this function, in which the three-phase a.c. supply is separated from the motor by one or more contactors. This arrangement is not inherently self-testing. The contactor must be fitted with closely coupled auxiliary contacts in order to monitor its correct operation, and if a fault is detected then a second method for preventing torque is required (e.g. second contactor, brake, etc.). Implementing STO within the drive provides inherently high integrity while avoiding the use of contactors, which have relatively high wearout rates because of mechanical wear and electrical sparking.

In order to implement a complete STO function, means must be provided to convey the control command to the power device input terminals in a way that itself offers comparable safety integrity to the main inverter. Typically this is done by removing the source of power for the isolating gate drive devices. Some ingenuity is required to avoid introducing new unsafe failure modes at this point. The function can be implemented using simple discrete electronic devices with well-defined failure modes and well-substantiated failure rate data.

It is important for users to correctly understand the properties of the basic STO function.

- When operating, the function prevents the production of motor torque, with a very high integrity. In the event of a concurrent failure of two power semiconductors, in the worst case a permanent-magnet motor might produce a transient alignment torque.
- Because torque is prevented, no drive braking is possible unless additional functions are incorporated.
- The function does not provide electrical isolation.
- The integrity of the overall function cannot exceed that of the lowest-integrity part of the control chain (the weakest link).

C4.14.3.2 Advanced drive-specific functions

These are functions that are particularly and naturally suited to implementation within the drive, because they are closely associated with the measurements and variables available in the drive, e.g. position, speed and torque.

A useful list of functions is given in IEC 61800-5-2 and includes the following:

- *Safe torque off (STO)*. See Section C4.14.3.1.
- *Safe stop (SS)*. A number of specific stop functions are defined with different defined behaviour. All are defined and control deceleration, with or without subsequent prevention of movement.
- *Safe limited acceleration (SLA)*. The SLA function prevents the motor acceleration from exceeding the specified limit.
- *Safe limited speed (SLS)*. The SLS function prevents the motor from exceeding the specified speed limit. The motor is running within specified speed limits where all controlling functions (torque, speed, position, etc.) between the drive and the motor are active. The drive shall control the speed so as not to exceed the specified limit selected.
- *Safe limited torque (SLT)*. The SLT function prevents the motor torque from exceeding the specified limit.
- *Safe limited position (SLP)*. The motor is operating within specified position limit(s) (absolute or relative) where all controlling functions (torque, speed, position, etc.) between the drive and the motor are active. The drive shall control the position so as not to exceed the specified position limit(s). The SLP prevents the motor shaft from exceeding a specified position limit. Specification of the position limit value(s) should take into account the maximum allowable over-travel distance(s).
- *Safe direction (SD)*. The SD function ensures that the motor shaft can move only in the specified direction. Specification of the speed/position limit value(s) should take into account the maximum allowable travel in the wrong direction.

- *Safe motor temperature (SMT)*. The SMT function prevents the motor temperature from exceeding the specified limit.
- *Safe brake control (SBC)*. The SBC function provides a safe output signal to control an external safety brake.
- *Safe speed monitor (SSM)*. The SSM function provides indication when the motor speed is below a given limit.

For each function there must also be a defined fault reaction function, i.e. the action on detecting a fault and a fault reaction time.

In addition to providing the core function, provision must also be made for the following:

- the function should be enabled and disabled securely as required by the machine status, typically by redundant or self-checking input/output arrangements;
- the machine builder should configure the function and verify its configuration and operation to the required level of integrity.

In most cases these more advanced functions will be implemented in some form of two-channel architecture with cross-checking, and the action on detecting a failure will be to disable the drive. The presence of the STO function therefore facilitates the implementation of these functions.

For many of these functions there is a critical role for the motor shaft transducer. A conventional incremental shaft encoder can be used for the speed and relative position functions, and its natural two-channel operation gives some useful inherent redundancy. However, the probability of a mechanical failure affecting both channels cannot be discounted, so in many applications it is necessary to use two separate encoders.

C4.14.3.3 Other machinery safety functions

Because the safety-related drive forms part of a machine safety control system, in practice there will usually be other functions required to complete the safety-related control, as well as providing diagnostic information that is not directly safety-related. For example, there might be a machine guard with two independent detector switches, so that when the guard is open the machine speed must be limited to a value that is safe for the machine setter to make an adjustment. There would be a requirement to ensure that if a switch were to fail then the machine would remain in the safe state (safety function) and a diagnostic output would be generated (non-safety function).

The management of devices such as switches would conventionally be carried out in safety relays or in a safety PLC. It would be a natural extension of a drive with safety functions to provide these additional functions. However, they would not be usable unless there were also to be user-programmable features so that the correct logical connections could be made between the inputs and outputs. This leads to the logical next step of providing full safety PLC features in the drive. A family of proven user-configurable function blocks would be provided to link safety inputs (such as the switches discussed above) and safety variables to carry out complete safety functions.

C4.14.3.4 Safety bus interfaces

These are bus systems such as Profibus that have special protocols such as Profisafe for communicating safety-related data at a sufficient integrity level. They can be used in conjunction with any of the safety functions described above. They can eliminate the need for conventional wired I/O ports, and offer ready-made safety interfacing to safety controllers. However, ultimately most individual sensors require wired connections into a Profibus interface.

C4.14.3.5 Integration into a machine

When drive safety functions are used as part of the arrangement to ensure the safety of a machine or process, the responsibility for the overall safety must clearly lie with the machine designer. A well-defined process must be in place to ensure that the risks are correctly identified and controlled. An essential part of the machine design process is the clear definition and communication by the designer of the safety requirements for each sub-system, and by the equipment supplier of the safety functions offered by sub-systems such as safety-related controllers and drives. This depends strongly on the quality of the documentation provided.

C4.15 Summary

This section has attempted to give a flavour of the functionality, configurability and power of a modern drive. The functions described may be given alternative names and vary somewhat in their performance from one manufacturer to another, but the user should be on the look out to see how such functions can bring benefit to their application in cost and possibly performance as compared to using external control devices.

Chapter C5
Common techniques

C5.1 General

The core technology of motors and drives has been described in some depth. This section describes techniques that are commonly used in applications and systems in order to achieve specific system functionality such as load sharing between several motors and even tension in materials within process machinery. The functions described in this section are not specific to individual applications but find application in a broad range of applications, some examples of which can be found in Chapter C6.

In Part A of this book it has been shown that the shaft performance of an electric motor is defined by the speed at which it runs and the torque that it produces to drive the connected load. By combining the motor with a variable-speed drive controller we have shown how we gain control over both of these parameters, converting a relatively simple motor into a flexible means of delivering controllable power suitable for a wide range of applications.

Gaining full control of the motor speed and torque opens up the possibilities of control over an extended range of functions. Single- or multi-motor drive systems can be designed to maximise the flexibility and productivity of most industrial processes. These systems may involve multi-axis coordinated position control or precise speed control with complex referencing systems, possibly combined with motor load sharing and constant tension winding or simple systems for efficiency gains in pumping installations.

Precise control of shaft position allows the implementation of such systems as electronic gear boxes, position synchronisation, point-to-point indexing, CAM profiling and registration. Position-controlled drives find applications in many industrial processes, from the relatively simple X/Y table through to generating complex profiles on sophisticated machine tools.

Whether the drives are to be used in speed, position or torque control, the success of the application will depend just as heavily upon the engineering of the referencing and inter-drive communication systems as it does upon the speed controllers themselves. The design of these systems must be based upon a thorough understanding of the process requirements.

C5.2 Speed control with particular reference to linear motion

Many modern machines require the coordination of several motors, these multi-motor systems usually being required to follow a single speed reference ensuring that all sections of the driven machine accelerate and decelerate at the same rate and aim for similar target speeds. As these systems are usually applied to processes involving material being conveyed between successive processes, it is convenient to think of the various speed references in terms of linear values such as metres per minute. In these situations a common reference signal will be generated and passed to each drive within the process. The stability and resolution of reference signals is an important factor in the success of these systems; traditional analogue designs often used stabilised d.c. voltage sources with low-temperature coefficient devices. Modern digital systems are not subject to the temperature drift problems associated with analogue drives but must use numerical values with adequate resolution in order to avoid the possibility of reference granulation and the resultant step changes in final speed demands. A major advantage of using digital techniques to the system designer is that it allows representation of system variables such as speed and acceleration directly in meaningful application units, such as millimetres per minute for linear speed and millimetres per minute per second for linear acceleration. It needs to be understood that the resolution of this data is dependent upon the resolution provided within the drive controller. Any calculations can therefore be performed in these easily interpreted units rather being represented by analogue voltages, which must be referred to a maximum range in order to obtain actual values.

Good design dictates that the gear ratios selected when designing a multi-motor application will normally be arranged to ensure that in spite of differing roll diameters within the machine all motors will run at similar speeds, matching their speed range capabilities and power to speed characteristics to the process speed range. However, exact matching is clearly impractical and the motors will never run at absolutely identical rotational speeds. Using linear speed referencing between drives and converting from linear speed units to rotary motion within the individual drives overcomes this problem. It is essential that surface speeds track the speed reference irrespective of roll size and in-drive gearing, so by making all inter-drive speed reference transactions in linear units, any differences due to roll diameters and gear ratios are eliminated from the common reference system and will be accommodated within the individual drive controllers.

Most drive controller internal reference systems are revolution per minute based, meaning that their internal structure handles references in rotational speed units. This means that the internal 'ramp to speed' functions that are normally provided are also revolution based, making it difficult to use the built-in function to obtain predictable acceleration control when referred to surface speeds. Where independent matched acceleration is required, the ramp function should operate on values expressed in linear units. Normally where a group of drives is to accelerate in synchronism they will be arranged to follow a common ramped source.

To ensure smooth changes in speed and to minimise any speed lag errors between drives, the common speed reference value will be generated using a ramp generator.

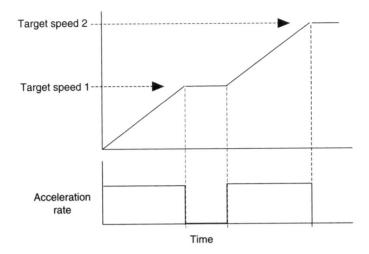

Figure C5.1   Acceleration under simple ramp control

Simple systems may use a linear ramp (Figure C5.1), which provides a constant rate of acceleration up to the target speed. As described in Section A6.2, a refinement of the linear ramp is the S-ramp (Figure C5.2), in which acceleration discontinuities are reduced by using jerk control. This type of ramp is standard in the elevator industry to ensure a smooth ride and is often applied in web-handling applications where smooth progression from one speed to the next is important.

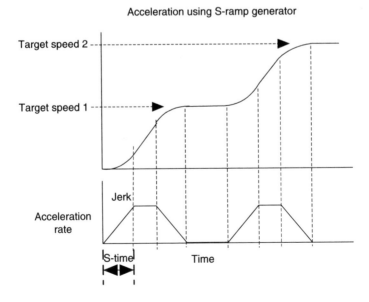

Figure C5.2   Acceleration under S-ramp control

When using a linear ramp, the unavoidable mismatches in torque to inertia ratios and varying frictional loads will make it impossible for all the drives to respond equally to the acceleration discontinuities, making it impossible to ensure that all drives follow exactly the same acceleration profile. Although jerk is not a problem in many simple applications, controlling jerk does have the advantage of removing rapid changes in torque demand on acceleration, which can have benefits in reducing mechanical wear and tear.

The S-ramp will have parameter settings not only for the normal acceleration rate but also for jerk or rate of change of acceleration. In a simple linear ramp the set value for acceleration is set and integrated to obtain the final speed reference. In the S-ramp a value for jerk is also set. This is then integrated to determine acceleration rate, which is again integrated to produce the final speed reference. In practical applications the jerk factor is often re-defined as the 'S-time', as users often relate more easily to time than a value expressed as acceleration/time.

Outputs from the S-ramp device will normally provide both a speed reference and acceleration rate signal. Having access to the acceleration rate can be useful in systems where acceleration torque calculations are to be performed.

Knowing the speed profile resulting from the ramp characteristic allows calculations to be made to predict stopping distance, this forms the basis of most position control systems where operational efficiency depends upon the system arriving at target position at the same instant as the speed reaches zero. A very simple example of this is stop on length for a process machine where the point of slow down is predicted such that the machine comes to standstill having processed exactly the required length of material.

When using a linear ramp the stop length is easily calculated using the equation

$$\text{Stop length (m)} = \frac{\text{current speed}^2 \text{ (m s}^{-1})}{2 \times \text{deceleration rate (m s}^{-2})}$$

Note that alternative but consistent units of speed and deceleration may also be used. More precise systems may use millimetres or even micrometres.

Introducing an S-ramp complicates this simple calculation as the rates of deceleration change during the S-ramp period. This is discussed in detail in Section A6.2 (equation A6.11) for the rotating reference.

Depending upon the values set for jerk, an S-ramp may increase the stopping distance quite considerably, leading to unacceptable errors in final stopping position if this simple approach to predicting slowdown length is used. These errors will be relatively small on low-speed machines, but as an example of this effect the stopping distance of a machine running at 2 000 m min^{-1} with a deceleration rate of 20 m min^{-1} s^{-1} will be 1 667 m, whereas adding a 10 s S-time will increase this to 1 833 m. Although such a level of error may be acceptable in some applications, they would not meet the requirements of most position control applications, and a more rigorous examination of the speed profile is necessary.

C5.2.1 Linear to rotary speed reference conversion

Speed referencing systems are frequently based upon linear speed units and must therefore be converted to rotary units (min^{-1}, which is revolutions per minute) before presentation to the drive for motor control. All applications rely on the same basic expression:

- For rolls

$$\text{Motor rpm} = \frac{\text{linear speed (m min}^{-1})}{\pi \times \text{roll diameter (m)}} \times \text{gear ratio}$$

- For the ball screw mechanism

$$\text{Motor rpm} = \frac{\text{linear speed (mm min}^{-1})}{\text{pitch (mm)}} \times \text{gear ratio}$$

- For a rack and pinion drive

$$\text{Motor rpm} = \frac{\text{linear speed (mm min}^{-1})}{\pi \times \text{pinion diameter (mm)}} \times \text{gear ratio}$$

To avoid rounding errors, which could result in cumulative errors, it is often convenient to express the gear ratio as two integers in the form

$$\text{Gear ratio} = \frac{\text{gear ratio numerator}}{\text{gear ratio denominator}}$$

C5.3 Torque feed-forward

The output from the speed error comparator/amplifier in a drive system is used as the demand for torque required by the motor to maintain the required speed, any reduction in speed resulting in a larger error and a greater demand for torque. A common feature in modern drives is to allow modification of this value from torque feed-forward demands, as discussed in Section A4.2.4.3. This allows the demand for torque to be increased in anticipation of a change in motor load rather than wait for a speed error to occur before the speed control loop reacts. Where changes in torque can be predicted, torque feed-forward may be used to advantage to improve the response of the system.

A typical example of this is the addition of inertia compensation, where the total torque demand comprises both the speed error torque demand and an additional accelerating torque calculated knowing the acceleration rate and load inertia; this approach can be used to improve the speed tracking of several high-inertia loads when required to follow a common speed reference.

A second example is the introduction of tension feed-forward in winding applications where the value of tension applied to an electrically driven unwind may be passed to the speed-controlled rewind as an indication of the level of required torque; a correctly scaled system will then sit in equilibrium with the torque generated by the rewind drive exactly balancing the tension torque from the unwind, and the speed error loop must then only exert a small effect for the system to run forward. A similar approach is used in the elevator industry to compensate for the weight of the car.

Introduction of offset torques such as tension feed-forward is relatively simple, only involving scaling of the signal from unwind to rewind. Inertia compensation is somewhat more complicated, requiring some online calculation to estimate the required torque levels.

Roll inertias may be calculated using the expression

$$\text{Inertia} = \text{mass} \times \text{radius of gyration}$$

For a solid roll this will be

$$\text{Inertia} = \frac{\text{radius}^2}{2} \times \text{mass}$$

For a thin hollow roll the following simplified expression will suffice in most applications:

$$\text{Inertia} = \text{radius}^2 \times \text{mass}$$

Where the wall thickness is large with respect to roll diameter the following more complex expression applies:

$$\text{Inertia} = \frac{(R_o^2 + R_i^2)}{2} \times \text{mass}$$

where R_o is the outer radius and R_i is the inner radius.

Acceleration torque is then calculated using

$$\text{Acceleration torque (N)} = \text{inertia (kg m}^2) \times \text{rotational acceleration (rad s}^{-2})$$

The effect of gear ratios must also be included when referring the results of these calculations back to the motor.

C5.4 Virtual master and sectional control

Where several drives are to follow a common reference an arrangement that has become known as 'virtual master' is used. The concept of the virtual master dates back to the old line shaft drive systems employed in the cotton mills and paper factories of the nineteenth century. The line shaft transmitted power to multiple take-off points, and any local variation in speed was obtained using cone pulleys with belt shifters. As technology progressed individual electric motors replaced the line shaft power take-off points, and schemes were devised to synchronise their speeds. These schemes mostly used manual adjustment, but some closed-loop systems based upon a form of mechanical differential were developed; these systems compared the required speed with the actual speed of the motor and adjusted a control device to change the speed accordingly. The required speed was usually provided from a low-power master drive turning a small reference line shaft, or in fully electric schemes driving a small master frequency generator.

Modern systems use a similar technique supplying a common speed reference to a series of motors that make up a multi-motor application. In simple systems one motor may be allocated as the master and its encoder used as the master frequency source for the remaining drives. This arrangement suffers a potential weakness; slaving one drive to follow the encoder of another results in any slight instability or jitter introduced by mechanical irregularity in the master's load (or poor drive tuning) being transferred to the slaves through disturbances in the encoder signal. Some early digital systems attempted to overcome this problem by reverting to the earlier solution of a master frequency generator using a small drive with no mechanical load, its encoder signal providing the master reference for the rest of the system.

Analogue systems used a variable d.c. voltage that was passed to each drive in the system as an analogue speed reference, speed ratio adjustments being made using potentiometers to slightly modify the signal to each drive. Unfortunately, analogue devices are prone to drift and noise and have now mostly been replaced by digital referencing schemes, where a numerical value is transmitted to each of the drive controllers.

Modern virtual master systems use totally numerical information and, in addition to the speed reference, may contain acceleration rate and task synchronisation data.

Invariably this speed reference must be modified as it passes from drive to drive to allow for changes in the material properties. For example, web-based processes suffer shrinkage or stretch, and rolling processes involve elongation such that the volume of the material entering and leaving a rolling nip remains constant. This results in differences in material length and hence speed as it passes through a process. These speed differences are usually expressed as ratios, which in the web-based industries are often termed 'draw' and in the metal rolling industries 'reduction factor'.

The actual relationship between motor speeds in rolling plant is somewhat different to that in web processing; draw occurs between two adjacent drives whereas the change in speed due to reduction factor occurs as the material passes through a reducing nip, the exit speed being somewhat greater than the entry speed in order to maintain constant volume of material before and after the reduction effect of the nip. This results in slightly differing schemes for the introduction of the speed ratios. Figures C5.3 and C5.4 illustrate the difference between draw and reduction factor.

The manner in which these speed ratios are applied must be arranged to mimic the process. In applications where speed ratios are expressed as draw, the speed of a section is determined by the shrinkage or stretch from the previous section, and draw is expressed as a simple multiplying factor and included in the roll speed reference. Draw ratios may be greater or less than unity, but are never more than a small percentage.

Rolling applications require a different arrangement. The mill speed is determined by the exit speed of the previous mill and is not affected by its own reduction factor.

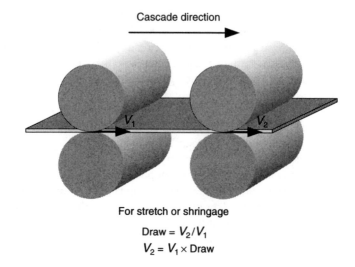

Figure C5.3 Effect of applying draw

Reduction factor is usually defined as exit speed divided by entry speed, resulting in a value greater than unity, and is applied as a divider to provide the reference for the preceding stand, which must always be slower. Some local non-cascaded adjustment of mill speed may also be required to match mill peripheral speed with the average surface speed of the material as it elongates in the nip. This adjustment has no

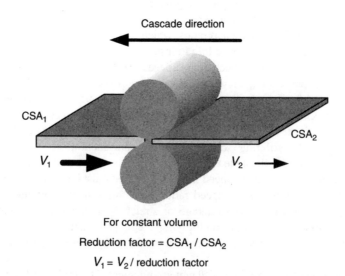

Figure C5.4 Effect of applying reduction factor (CSA, cross-sectional area)

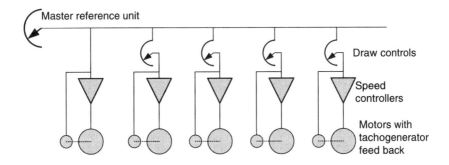

Figure C5.5 Typical analogue sectional drive configuration

relationship to the reduction factor and varies depending upon the profile of the material being rolled.

Because changes in speed applied to upstream sections affect the speeds of the subsequent parts of the machine, the speed references are often cascaded from one drive to the next to provide automatic compensation.

The original analogue systems used a common master reference source and individual draw adjusting devices (Figure C5.5). These were usually motor-operated potentiometers, and it was difficult to implement cascaded draw, although some attempts were made by sequencing draw adjustment signals to operate on multiple potentiometers.

Modern digital referencing schemes fall into two categories, using either distributed or central processing.

- *Distributed control.* Distributed speed ratio control systems introduce the speed ratios at the individual drives by multiplying or dividing the incoming reference by the required speed ratio, using the resulting speed as the reference for a drive and passing the modified value onto the next drive via an inter-drive communication (Figure C5.6).

  The distributed processing system can only be used where drives support some form of internal programmable logic and numerical processing such as the Control Techniques Unidrive. It has the advantage of combining all the speed calculations for the drive into a simple programme. Application of the speed ratio with its

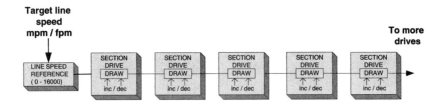

Figure C5.6 Distributed draw/speed processing

associated increase/decrease adjustment logic, together with the conversion from linear to rotational speed reference, involving mechanical data such as roll diameter and gear ratio, is all handled within the individual drive units. An identical programme can be used in each drive with a set of non-volatile storage registers allocated to contain the information such as roll diameter and gear ratios that is unique to each part of the machine.

Because the speed reference is transmitted from drive to drive, each processor forms a link in the referencing chain. This can be a disadvantage in some industries where it may be possible to bypass certain parts of the process and isolate the respective drives for safety or maintenance while the remainder of the system continues to operate. In systems where all the drives must be available at all times this restriction becomes less important and various advantages and disadvantages should be weighed against the saving achieved by negating the requirement for a powerful central processor.

Updates from drive to drive may not be synchronised, as each processor will handle the information at slightly different times. Their task scans will have been initiated at switch on and will not necessarily be synchronised. There will also be a propagation delay as the reference signal is processed and passed on down the chain, resulting in a ripple effect. However, except in the most demanding of applications this can be ignored, as it is unlikely to be greater than a few milliseconds and will be smoothed by the much slower responses of the drives due to the electrical/mechanical time constants of the various parts of the machine.

The mechanism by which these transient speed errors can occur is shown in Figure C5.7. Each drive will read its updated input at a different instant in time, due to delays in communication links and asynchronous operation of the various

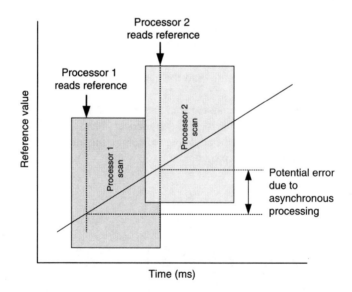

Figure C5.7   Inter-drive speed reference errors due to asynchronous processing

processes in the reference chain. However, as stated earlier it is unlikely that these errors will affect overall performance except in the most critical position control applications.
- *Centralised control.* Centrally processed systems perform exactly the same calculations as in a distributed system, but in a single processor generating a complete set of references, which are then passed to all the individual drives in the system using some form of high-speed communications system (Figure C5.8).

Centrally processed speed calculations overcome part of the problem of asynchronous processing in a distributed system, but at the cost of an additional

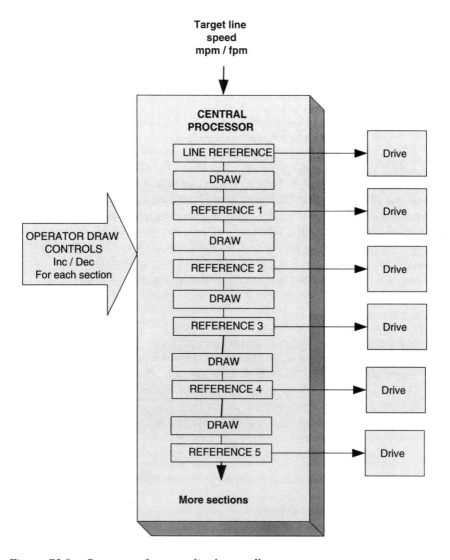

Figure C5.8 Structure of a centralised controller

processing system that can perform the speed-reference calculations for all the drives in a single scan. The resulting references can then be transmitted to the individual drive controllers at the end of the scan.

The most basic systems may use analogue means to achieve this, but most modern digital systems will use some form of cyclic block transfer over a high-speed Fieldbus. The central processing unit may be any device that supports simple logic processing combined with a capacity to perform numerical calculations together with a suitable I/O structure. This is often a PLC but any digital controller with access to a suitable interface I/O will perform this function.

What is not evident here is the non-deterministic nature of most industrial Fieldbus systems. Furthermore, to ensure that all drives in a system respond at precisely the same time to a new set of commands, all drive control systems need to be synchronised. Deterministic Fieldbus systems are on the market, for example EtherCAT and CTSync. Some drives are also able to synchronise their control loops through their Fieldbus interface to overcome these problems. Care needs to be taken when specifying the drives for this type of application to select the correct Fieldbus and to ensure that the drives are capable of synchronising their control loops through the Fieldbus.

The user must base his choice between distributed or centralised processing on the logistics of the total scheme. Factors to be considered are mainly commercial and about ease of installation and future support. There may be a preference for bringing all data and controls from the plant to a central location rather than to drive units distributed within the equipment. Alternatively, there may be a preference for all the individual setup and data storage associated with each part of the machine to be entered via a front panel keypad of that section's drive unit, with information relevant to that section such as roll diameter being stored in the drive non-volatile memory.

Where data is to be accessed by a third party for logging and display using a supervisory or control package, it may be more convenient if all the required information is available from a single centralised processor, although with modern Fieldbus systems access to remote data is rarely a practical problem. Large systems with many motors, each having a significant amount of associated stored data, such as diameter, gear ratio and motor characteristics, will require large amounts of non-volatile memory. This may be more economically provided using the processing and non-volatile storage provided in drive-based processors.

Care must be taken in specifying Fieldbus systems to ensure there is the bandwidth to provide the necessary data transmission. Drive control algorithms are normally in the 5 to 50 ms range, and to ensure deterministic updates should be matched by the commensurate data channels.

C5.5 Registration

Registration systems are used where variables exist that are outside the range of control of a basic position control system. Rather than rely on a known home position or a definite relationship with the master position encoder, registration systems rely on detecting some form of mark from which they can make the necessary moves to achieve alignment of workpiece and tool. Usually, detection is performed with both master and slave drives in motion.

Applications occur in many industries, including the following:

- rotary and linear printing,
- rotary knife or shear,
- flying shear,
- labelling machines,
- perforating machinery,
- sorting machines/conveyors,
- product-finishing machines e.g. edge cutting/trimming, and
- packaging machines.

A typical example of registration is seen in the printing industry where a printed mark on the surface of the paper is used by the control system to alter the path of the paper to ensure that the print impressions from several units all occur 'in register' or correctly aligned on top of each other. This may be to ensure page-to-page registration, where accuracy need only be to a millimetre or so, or for colour registration, where much greater accuracies are required, the final colour being produced by combining typically up to 12 different colours, each applied by a different unit. Accuracy here is critical, as the human eye can detect errors as small as 50 μm. The original technique used on line shaft driven machines involved lengthening or shortening the web run between units to ensure correct placement of the web as the impressions are applied (Figure C5.9).

Modern presses are now equipped with multi-motor sectional drives that are able to introduce register control by advancing or retarding the individual impression rolls using accurate position measurement and control (Figure C5.10).

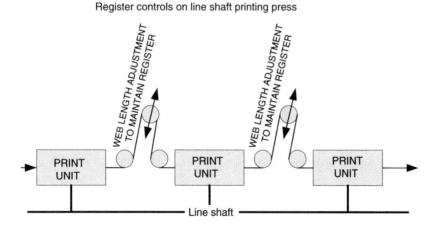

Figure C5.9 Registration mechanisms on a line-shaft-driven printing press

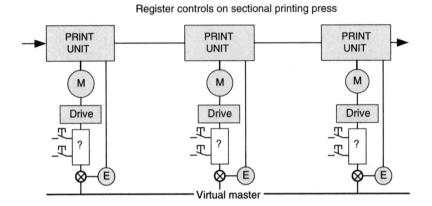

Figure C5.10   Registration system on a printing press sectional drive

The elements of a registration control system include position capture and detection of any error in length or distance travelled followed by calculation of the adjustments required to correct the error.

Position capture may be based upon edge detection, where the leading edge of the material is used as the reference point, or pattern recognition, where the system is looking for a mark on the material, or recognition of a pattern. Accurate position capture is a critical factor in achieving an operable registration control system. The sensor and associated software and circuitry must be as fast as possible and jitter-free to avoid errors at high speeds.

Consider, for example, a sensor with a response time of 40 μs, observing a single mark on a drum of diameter 500 000 μm. When the drum turns at 1 m min^{-1}, the position captured with respect to 1 revolution may provide a result of 200 000 μm. If the speed is increased to 600 m min^{-1} the response time in the sensor will cause an offset equal to (600 m min^{-1} × 1 000 000 × 40 μs)/(60 × 1 000 000) = 400 μm or 0.4 mm, giving a captured position of 200 400 μm. This will result in a change in registration position and could be easily detectable to the human eye if it applied to a printing system. Most modern registration control systems are capable of being programmed to compensate for this error.

To determine how much adjustment is required to correct any registration errors, a registration offset is generated. Calculation of this offset varies depending upon the application. However, there are two basic types. The first is as follows:

Registration offset = required position − actual captured position

This is a position-based calculation, and is used in applications such as printing or simple position profiled applications like edge-trimmers. The result will be used to estimate the amount by which the drive should be advanced or retarded to arrive at the correct position.

Alternatively, the following may be applied:

Registration offset = (current accepted event position − last accepted event position) − product length

This is a length-based calculation, and is used in more complex CAM-based applications like rotary knife/shear or flying shear to ensure that the correct cut length is maintained.

In addition to potential problems due to sensor response, further errors can be caused due to misreads of the position capturing device. The introduction of straightforward filtering introduces a delay that is not acceptable, as the trigger for capturing instantaneous position measurement must be as fast as possible. Several solutions to this problem are used, mainly techniques to validate the legitimacy of the incoming signal.

These techniques may involve setting minimum and maximum limits on the width of the trigger pulses that will be accepted. Where trigger pulses conform to the defined parameters they will be accepted. Those outside of the range will be rejected (Figure C5.11). An extension of this technique checks the distance between successive pulses again to ensure that they conform to the expected result (Figure C5.12). A third

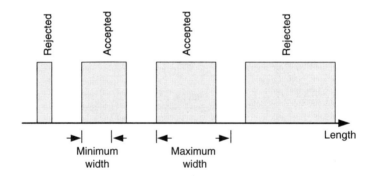

Figure C5.11   Signal validation based on pulse width

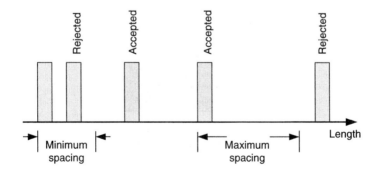

Figure C5.12   Signal validation based on pulse width and interval

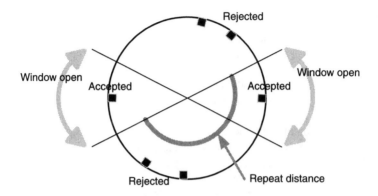

Figure C5.13   Signal validation based on window detection

option is to allow pulses to be accepted only if they are received within a specified window area (Figure C5.13).

The printing industry may often use pattern recognition as a means of registration control. A pre-printed pattern will include a sequence of marks that can be identified as the position marker together with additional marks for use by the width control units to ensure that the web is centrally located. Typical marks may appear as a sequence of black blocks at the side of the page (Figure C5.14). The registration detection system can be programmed to identify the particular sequence and width pattern.

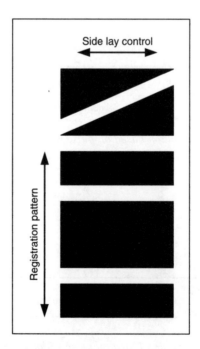

Figure C5.14   Typical printing registration pattern

Rotary knife systems frequently use registration techniques and are widely used in the converting and printing industries. To ensure the correct cut length, a rotary knife must speed up or slow down when it is not in contact with the material to ensure that the blade comes into contact with the material at the correct position. Rotary knife mechanisms consist of a cylinder with a knife blade (or blades) mounted along the longitudinal axis. As the cylinder rotates, it cuts the material passing under it. If the peripheral speed of the knife point matches the speed of the web material, the length of material cut will be equal to the circumference translated by the point of the knife. If the peripheral speed of the rotary knife is faster than the web speed, shorter cut lengths will be produced. Conversely, longer cut lengths will result if the peripheral speed is slower.

To obtain the highest quality cut possible and to avoid damaging the web material, the knife angular velocity must match or be very close to the linear speed of the web during the portion of the rotation that the knife is in contact with the web material (cut angle). Thus, in order to obtain cut lengths that are shorter than the circumference of the rotary knife, the knife must speed up when it is not contacting the web, and then slow down to match the web speed when the cut is performed. The converse is true when longer cut lengths are required, the knife rotation being decreased between cuts to allow more material to pass before the next cut occurs.

Rotary knife controls involve some complex calculations and are available as an application package from some drive suppliers. Typical features include the following:

- cut recipes, where several different cut lengths may be executed in sequence;
- on-the-fly cut length adjustment;
- park angle adjustment, where the acceleration angle may be altered;
- batch counting – to stop the cutter after the desired number of cuts
- master control to stop the master in an emergency or when the last cut has been done;
- user-defined units to allow any linear unit of measurement to be used;
- registration correction (continuous registration/phase correction);
- cut on registration;
- windowing or mark sensor blinding to mask out unused material registration marks.

C5.6 Load torque sharing

C5.6.1 General

When motors are connected in parallel or mechanically coupled, the drive system designer must ensure that the total load will be correctly shared between them.

Load sharing problems fall into two distinct categories: those where the motors are permanently mechanically linked together, examples being unit motors mounted in line on a line shaft such as are often used on news print presses, or tandem input gearboxes where two smaller motors are used instead of a single larger motor, often as a means of achieving pseudo 12-pulse by supplying them from independent phase-shifted converters. Applications falling into this category are usually based upon one

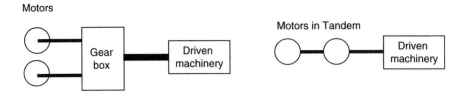

Figure C5.15   Typical mechanical arrangements requiring load torque sharing

drive acting as a speed-controlled master with the other being set to maintain its torque at a level equal to that of the master.

Typical mechanical arrangements requiring demanding continuous sharing are shown in Figure C5.15. The first involves two motors driving a double-input single-output gear box and the second two motors in tandem as typically used in rolling mill applications.

The other category in which load sharing is important is that in which motors may only be required to share load when their driven loads come into contact with another source of power (Figure C5.16). An example is where two motors driving each roll of a nipped press will only be required to share load when the press is closed. At all other times the two drives must operate individually to maintain the correct roll speeds.

Similarly, lead roll drives on a web processing machine will only need to share load when the web is in contact with the roll. When not sharing load these drives must maintain the driven roll at the correct speed in readiness for contact.

For reasons of economy, older systems based upon Ward Leonard power conversion were often based upon several motors being supplied from a single generator. This forced the designer to take care with his load sharing scheme. Modern systems tend to provide a controller for each motor, which allows the designer to take advantage of the torque control functionality available from the drive controller.

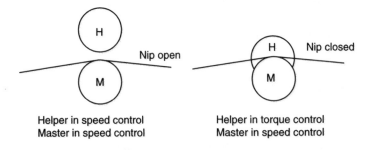

Figure C5.16   Load sharing when linked through the process

C5.6.2   Open-loop systems

Open-loop parallel motor load sharing systems rely on the sharing motors having matched torque–speed characteristics. For parallel connected motors to share load

Common techniques 569

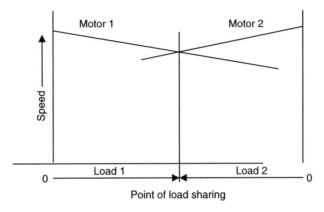

Figure C5.17   Load sharing based on speed droop

satisfactorily, their speeds must decrease as their share of the total load increases. Figure C5.17 shows how, due to natural speed regulation, two motors could assume their respective share of the total load.

Typically, early Ward Leonard d.c. 'helper drive' motors were specified to have a no-load to full-load droop of 8 to 10 per cent of maximum speed. This meant that as the load on the helper motor increased, its speed would reduce, eventually causing it to shed load. The actual load level was adjusted by trimming the motor field current to set the initial no-load speed in relation to its partners speed. Setting the helper speed higher by weakening the field would cause it to increase its share of the total load.

Parallel d.c. motor systems are not used very often these days, but load sharing can be achieved by field trimming the helper motors to set the correct armature current levels (Figure C5.18).

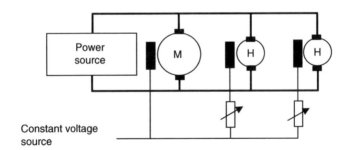

Figure C5.18   Load sharing of d.c. motors by field control

Unfortunately, load sharing by field trim is not satisfactory where the drive system must operate over a wide speed range. At low speeds the current taken by the individual motors is influenced more by the cable resistance and motor internal armature circuit resistance than by the effect of flux changes on the back-emf. This can result in a totally different load share profile at low speeds to that set up at medium to high speeds, where changes in the back-emf are more effective. The only solution is to use motors with suitably matched cable and armature circuit resistances or to add external resistors to correct any imbalance. This is not an acceptable solution by today's standards.

This technique is applicable to induction motors and is discussed in Section C5.6.4.

C5.6.3 Paired d.c. motors

If two d.c. motors are to be used for this type of application with only one converter, it is unlikely that they will share load satisfactorily if simply connected in parallel. In fact, it is quite possible for very large currents to circulate between the motors with no addition to the total shaft power produced. In extreme circumstances these currents could be considerably in excess of the motor maximum ratings.

Sharing between very-low-power motors can be achieved by adding series resistors to their armature circuits, but this is very inefficient. For larger systems, particularly where higher d.c. voltages may be used, the designer has the option of connecting the motors in series with the supply from a single converter or from two converters arranged to produce identical currents in a form of 'parallel-isolated' configuration.

When using the series configuration (Figure C5.19), the same armature current flows through both motors. Connecting the field windings in series ensures that any change in field current due to temperature rise will apply equally to both motors. A diverter resistance could be connected across one motor field winding to allow some trim to match their armature voltages. This arrangement could be suited to large systems where series-connected 12-pulse converters are to be used.

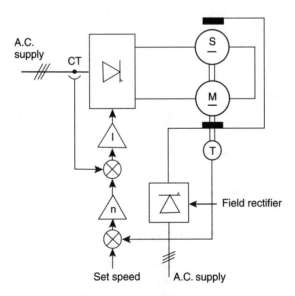

Figure C5.19 *Load sharing by mechanically coupled d.c. motors connected in series and fed from a common converter*

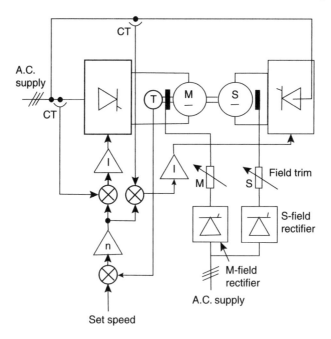

Figure C5.20   Load sharing by current slaving of converters

Alternatively, using two converters in the parallel-isolated format, one for each motor, allows one drive to be used as the master and the second as a torque or current slave, relying on the slave current loop to maintain the same current as that provided by the master drive (Figure C5.20). The result is equal armature currents in both motors. Again, field trim may be required to balance voltages if absolute power equality is essential.

The one converter per motor method of load sharing is the most flexible solution. Motors having differing characteristics and different ratings can be slaved in this way and arranged to take their proportion of the total load. A single master can provide the load demand for several current slave drives. Such an arrangement has been very popular in the printing industry, where each print unit was equipped with its own motor but all units were mechanically tied via a single line shaft.

There are the following advantages to the multi-converter system:

- ease of setup and commissioning;
- easy accommodation of multiple motor schemes;
- each motor may be individually protected by its associated converter system;
- motors of differing sizes can be handled in the same load share group;
- some degree of redundancy in multi motor systems;
- the two converters could be fed from phase-shifted transformer windings, resulting in pseudo 12-pulse operation, to reduce harmonic currents drawn from the supply.

The series arrangement may be more economical but will possibly involve working with higher voltage, which may affect the choice of the associated switchgear. Its potential advantages are as follows:

- simplicity;
- the possibility of using a series 12-pulse converter to reduce harmonic content;
- motor voltages do not necessarily have to be identical though their armature currents must be equal.

C5.6.4 Paired a.c. motors

C5.6.4.1 Parallel motors

Two a.c. induction motors that are solidly coupled together by a rigid mechanical link may be supplied from a single controller (Figure C5.21). In this case the system relies upon the motors being identical with similar slip values, and ideally their cable lengths should be matched in length and therefore impedance. The mechanical link between the motors must be absolutely solid. If vector control is to be used, then only one motor need be fitted with an encoder.

The advantages of the parallel motor configuration are as follows:

- simple to apply;
- suited to multi-motor applications.

There are the following problems associated with the parallel motor arrangement.

- The motor characteristics must be closely matched to ensure load sharing, and cable lengths and so on should be similar. (Note that motors of identical ratings from different manufacturers are unlikely to perform correctly.)
- Each motor must be individually protected against overload.

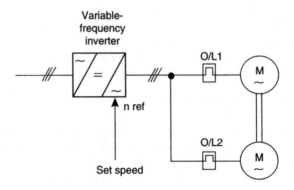

Figure C5.21   Induction motors fed from a common drive

C5.6.4.2 Frequency slaving

A development of the parallel motor configuration discussed previously is to frequency slave two inverters; this avoids the possibility of any parasitic currents circulating between the two parallel connected motors in the previous example (Figure C5.22). This arrangement provides some redundancy in the event of motor or inverter failure provided that the load remains within the capability of a single drive.

Advantages of this system are as follows:

- simple to apply;
- complete protection for each motor provided by its own inverter drive;
- suited to multi-motor systems.

Disadvantages include the following:

- additional cost of extra inverters;
- motor characteristics must be matched, and different manufacturers' products should not be mixed.

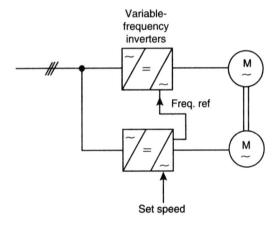

Figure C5.22   Frequency slaving of drives

C5.6.4.3 Current slaving

This system is similar to the d.c. torque master slave arrangement, where a signal proportional to load on the master drive is fed to the current control input of the slave inverter (Figure C5.23). This system forces equal currents in each motor and so may be used to overcome the difficulty of sharing load with motors from different manufacturers. The accuracy of this system is very dependent upon the drive used and the nature of the current signal used. If the current signal is the line current then load sharing typically within 10 per cent could be expected. If a flux vector drive is used

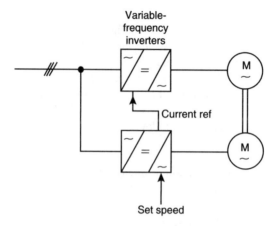

Figure C5.23   Current slaving of drives

and the torque 'current' is used, then the accuracy will be much higher. This is discussed in greater depth in Section C5.6.5.

Advantages of current shared inverters are as follows:

- easy to configure;
- good sharing possible with flux vector drives;
- motor characteristics less important in achieving correct performance;
- complete protection for each motor provided by its own inverter drive;
- could be applied to multi-motor systems.

Its disadvantages include the following.

- In basic V/f inverters, load referencing and control is based upon line currents, and torque sharing will not be precise, particularly if different manufacturers' products are used.

C5.6.5   Torque slaving systems

As mentioned above, because the modern approach is to supply each motor from its own controller, this removes any requirement to specify the motor design to meet load-sharing requirements. Any speed–torque characteristics can be obtained by changing the control characteristics of the drive controller. This means that from an electromagnetic perspective, standard stock motors can be used in virtually all applications where previously a special design would have been necessary.

Load sharing using individual controllers provides far more versatility and allows many more options for the designer to meet the application requirements. Some of the more frequently used options are outlined, together with suggestions as to where they may be applied. All these solutions apply equally to both a.c. and d.c. systems. Users should note, however, that load sharing of a.c. drives should be based upon 'real' or torque-producing currents, and the reactive part of the total current should not form part of the torque demand passed from master to helper.

Where there is a solid mechanical link between two or more drives, the simplest solution is to configure one drive as a speed-controlled master and the remainder as torque-controlled helpers. The speed-control amplifier of the master provides the current or torque demand for both the master and the torque helper.

This is very simply configured and may use either analogue signals or a serial link to pass the torque demand from master to helper drive. Because the individual current controllers ensure that the respective drive currents match the torque demand, there is no need for any feedback between helper and master drives. To ensure equal torque the controllers should have identical configuration/motor maps. This was the traditional arrangement used for multi-motor line-shaft-driven printing presses.

Using this system, sharing between unequal drives is possible if sharing is based simply on percentage torque demand, which is generated by the master drive and passed to each of the individual controllers. It is also quite acceptable to set up the system to share between a.c. and d.c. drives or to share between several drives, simply by repeating the number of helper current controllers fed from the master.

Note that as the helper drives will be torque controlled, if they become mechanically disengaged from the master their resultant speed will be undetermined. It is therefore suggested that helper drives be configured for torque control with some form of overriding speed limitation, easily achieved by operating in current limit with the limit being set by the torque demand.

The helper speed reference must be obtained from the same source as the master drive speed reference, but should include an offset to ensure that the helper attempts to run faster than the master, ensuring that torque will be maintained. The offset should be large enough to ensure that the helper speed error remains saturated at all times.

Where the master drive is a.c., the load share reference should be based upon real torque demand and not line current. A suitable signal will be sourced from the master drive speed error signal. A group of torque-sharing drives should operate as a single drive and must be interlocked such that they start and stop in synchronisation. Drive fault monitoring must be arranged in a manner that ensures that all drives are shut down in the event of a fault on any drive within the load-sharing group. This arrangement is commonly used on machinery having several in-drive points, such as a paper-making machine press section.

C5.6.6 *Speed-controlled helper with fixed torque*

Drives falling into this category are typically those in multi-motor web-processing systems where the linkage between helper and master is through the processed material. This means that before the web is established the helper drive must set its speed correctly to match the driven roll surface speed to that of the web. When the web contacts the roll the drive must smoothly accept any increase in load up to its rated limit.

A very simple arrangement can rely on the drive in-built current limit feature. The helper drive may be set to follow the line speed master reference but with a small speed

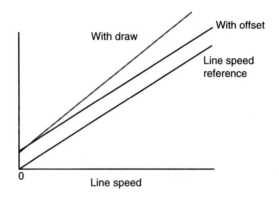

Figure C5.24   The effect of offset and draw

offset to increase its speed. Upon contact with the material the drive will slow down and go into current limit. Adjustment of the current limit setting may then be used to determine the current at which the drive will operate.

This arrangement does not provide true load sharing but allows the torque provided by the helper to be predetermined and maintained at a constant value. This approach, which is semi open-loop, is simple and requires no additional functionality. It may be well suited to applications such as bridle drives in the metal-processing industries or surface-driven winding applications. This solution was traditionally used for Pope Reelers at the end of paper-making machinery to provide a simple means of open-loop tension control before the advent of load cell tension control.

Note that speed offsets used in these types of applications should always provide a constant increase above referenced speed related to line speed (Figure C5.24). This ensures that the offset is just as effective at low speeds as it will be at maximum speed. Using a draw-type approach to achieve the increase in speed is not satisfactory, as the effective speed lead will reduce as the line speed is reduced.

C5.6.7   Speed-controlled helper with shared torque

Where the driven roll has intermittent contact with the web but must deliver its share of the total load, a simple speed-controlled helper with a built-in droop function may be used. This forms the modern equivalent of the Ward Leonard helper drive where the motor was specially designed to reduce its speed as the load increased.

In this case extra functionality can be built into the drive to reduce the drive speed reference in relation to increasing load. This will have exactly the same effect as using a motor with built-in droop.

The drive speed should be set slightly higher than web speed. When contact is made the drive will slow down, and final load levels can then be set by trimming the speed. A draw control arrangement is usually provided to obtain this trim. There will be a slight change in final speed for contact on and off, but load take-up will be smooth and this approach can be used where the web material will not sustain substantial increases in tension. This solution will typically be used in industries handling lighter-weight materials such as paper, plastics and film for the control of small path and lead roll drives.

In certain circumstances it may only be necessary to reduce the speed loop gains, in particular the integral term, of the helper drive in order to achieve the required droop. In more sophisticated applications some droop-generating software may be required. The droop-generating software should be arranged to reduce the helper line speed reference in direct proportion to any increase in helper load, probably with a maximum reduction of 10 per cent achieved at a 100 per cent load torque. The measurement of helper load can normally be obtained from the helper speed error signal, which may need some filtering to avoid any minor torque variations being reflected back to the motor speed. The designer should ensure that the introduction of droop at low speeds will not result in reversal of the helper drive rotation.

C5.6.8 Full closed-loop systems

The higher-performance solutions described above rely for their operation upon using individual closed current or torque loop controllers for each motor, but make no attempt to close the actual load share loop. Assuming that the shared load will always match its demanded value, there is therefore no automatic correction if the desired load is not achieved. It is possible to overcome this limitation by using an auxiliary controller to compare required and actual loads and adjust the helper accordingly. These solutions are usually based upon some form of auxiliary PID controller.

This solution is particularly suited to the category of applications where the mechanical tie between helper and master is indeterminate, either because for example a roll is not continuously in contact with the material or because one half

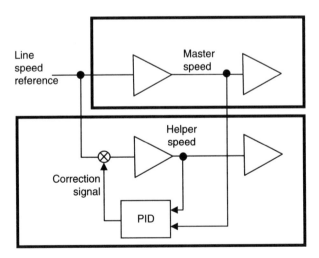

Figure C5.25   Outer PID loop to correct for sharing imbalance

of a nip press is made of compressible material, meaning that the effective diameter may change under pressure.

The configuration is essentially similar to that described in Section C5.6.7, but a PID controller is added to replace the manual draw trim with an automatic function. The PID is provided with a torque demand that is usually obtained from the master drive speed error, but could be subject to some operator adjustment to allow for load balancing. This is then compared with a helper torque feedback signal obtained from the helper speed error signal, and any mismatch between these two signals produces an output from the PID that is then used to adjust the helper line speed reference until the correct torque balance is achieved (Figure C5.25).

C5.7 Tension control

Many applications that involve web handling use some form of tension control, whether it is simply maintaining constant tension between adjacent parts of a machine or actually unwinding or rewinding the material.

Inter-section tension control, where a set tension value is to be maintained between adjacent parts of a machine, is usually achieved using some form of tension-measuring device to provide feedback. An auxiliary tension control loop then trims the speed of the downstream section to achieve the correct tension level. Measurement of the actual tension is performed using either load cells incorporated into the roll bearing support blocks or by a transducer attached to a moving or dancing roll.

The dancing roll must be pre-loaded to provide the correct amount of tension and the control system simply acts as a position controller to maintain the roll near to its mid position; adjustment of tension can only be achieved by altering the loading on the dancing roll (Figure C5.26).

When using load cells the actual measured tension may be compared against a reference value and variation of the set point becomes possible (Figure C5.27).

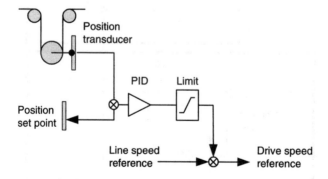

Figure C5.26 *Typical position control system: dancer proving correct tension*

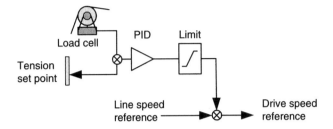

Figure C5.27   Closed-loop tension control loop

Most tension control loops are based upon PID controllers and as changes in tension are reflected by very small changes in speed the output of these PID loops are limited to provide a very small trim to the normal line speed reference supplied to the roll under control. It is not normally practical to rely solely on the tension control loop to provide the total speed reference signal.

Less sophisticated systems where the operating parameters and machine configuration are fixed (e.g. simple bridle roll applications in the metals industry) may achieve tension control simply by operating the drive under torque control with a limit applied to the speed should the bridle become unloaded. This arrangement may be considered to be a form of load share control.

As drive technology advances, the quality of the torque signal calculated in the drive is improving, but the linearity and accuracy in almost all drives is such that load cells are still used in the vast majority of applications.

C5.8   Sectional control

The various techniques described so far for virtual master (Section C5.4), load sharing (Section C5.6) and tension control (Section C5.7) are often combined in the design of sectional drive systems for modern industrial processes. Sectional drive systems probably first made an appearance in the paper industry, but the technology has now been widely applied. Possibly one of the most technically demanding is the application of sectional drives to high-quality printing presses. Not all sectional drive systems need the extreme accuracies required for print registration control. The simplest sectional drive may just be used to synchronise a set of conveyors where consistency of speed avoids peaks and troughs in production.

A modern sectional drive usually consists of a master speed reference generator, usually based upon some form of ramp generator that provides the speed reference for several drives. The drives may be required to accelerate and decelerate as a group or individually, but always to a target speed dictated by the master reference.

Developments in a.c. drive technology have now made it possible to obtain better control using a.c. induction motors than was achieved using d.c. separately excited motors, so the techniques originally developed for d.c. systems are now being actively applied to a.c. drive systems.

Sectional drive systems fall into two categories.

- The first category includes those in which the material is threaded through the machine at low speed, and once established from unwind to rewind the machine can be accelerated up to normal running speed. Systems in this group will always follow a ramped speed reference from a master or virtual reference system, and it is important that all sections track the main reference accurately to avoid inter-section speed errors. The majority of systems fall into this category.
- There are particular situations where material can only be produced at high speeds. Multi-stand bar and rod mills and paper-making machines fall into this category. Because material enters and passes through the machine from section to section at high speed it is essential that all drives are running at the correct speeds and that any speed changes due to load regulation are reduced to the absolute minimum.

Because individual sections may be stopped and started independently, each section controller must be provided with a ramp generator to control acceleration. However, once the section reaches the target speed set by the overall master reference system, the section ramp generator is no longer required and should be bypassed to avoid introducing velocity lags during overall machine speed changes under control of the master reference.

Modern digital section controllers are often multi-functional and can be configured as section masters contributing to a cascaded speed reference system and generating helper speed and torque references. Alternatively, they can be configured as helper drives to receive these references from the associated master and provide the required load-sharing functions. Typical features include local speed ratio adjustment, ramp generators, and speed detection for ramp bypass, together with PID controllers for load share or tension control. Some systems may provide 'bump-less' transfers from speed to tension control, where tension is monitored using a load cell.

Controllers will normally accept speed reference values in linear speed units such as millimetres per minute, and will include setup locations for essential data such as gear ratio and roll diameter to facilitate conversion from linear to rotational speed units within the package, ready for presentation to the drive controller.

C5.9 Winding

C5.9.1 General

Unwinding and rewinding schemes are usually quite complex, particularly if the winding machine is centre driven, as some form of diameter compensation must be built into the control scheme.

Winder systems may be based upon direct control of motor torque or fine speed trim under the influence of a tension-measuring feedback device as previously outlined (Section C5.7). Winding equipment is generally considered to be centre driven or surface driven. Coilers and uncoilers used in the metal processing industries and unwind brake systems used in the paper industry are perhaps the most well-known

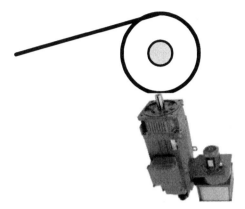

Figure C5.28   Centre-driven winder

applications employing centre drive techniques, although there are many instances where centre drives are used in much lower power applications (Figure C5.28).

Surface-driven machines are commonly referred to as drum winders, and are mostly used in web-handling processes. They rely on frictional contact between drum and material to create traction. Drum winders may use one or two driven drums. Single-drum winders (Figure C5.29) usually operate with the rewound roll alongside the drum, with contact maintained by pneumatic rams. Twin-drum winders (Figure C5.30) support the wound roll in the valley created between the adjacent drums and rely on roll weight to create contact between drums and material.

C5.9.2   Drum winders

Because the tractive effort is applied at the surface of the rewound material, the motor torque remains constant throughout the diameter range and is directly related to tension. The twin-drum winder is a particular case where the wound tension relies more on the torque differential between the two drums, the second drum being set

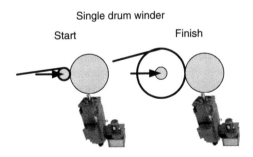

Figure C5.29   Single-drum (or surface-driven) winder

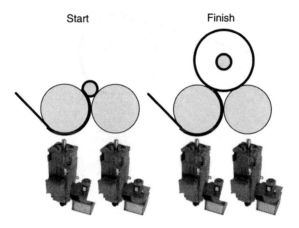

Figure C5.30   Twin-drum winder

to produce a greater torque than the first drum and effectively wrapping the material around the roll.

Some modern winding machinery may use a combination of centre and surface drive techniques to produce high-quality rewound rolls.

In all cases a tension balance must exist. The total tension generated by the rewind system can never be more or less than the tension created by the preceding equipment.

Twin-drum winders are often configured to profile the tension in the rewound roll to give a tight inner core with gradual reduction in tightness as the diameter increases. This is achieved by changing the ratio of torque distribution between the two drums as the diameter increases. This profile will often follow a non-linear relationship (Figure C5.31).

C5.9.3 Centre-driven winders

Centre-driven winding applications are more complex due to the effect of changes in the rewound roll diameter. Given that a winder system will normally operate at

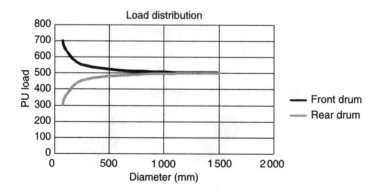

Figure C5.31   Load distribution between drums

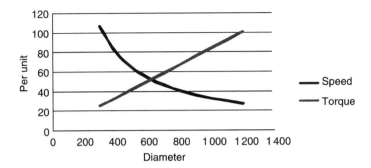

Figure C5.32   Relationship between torque speed and diameter for a centre-driven winder

constant line speed, the winder motor speed must be reduced as the diameter increases. If constant tension is to be maintained throughout the diameter range then the drive must also provide increasing motor shaft torque as the roll diameter increases (Figure C5.32).

Because power is related to the product of motor torque and speed, maintaining constant tension throughout the diameter range results in the centre-driven winder operating as a constant-power device over the chosen diameter range. The actual absorbed power is therefore directly related to the level of tension and the line speed of the system (Figure C5.33). Some early d.c.-driven winder systems took advantage of this relationship, using a field controller to keep the winder motor back-emf proportional to the line speed while keeping the armature current constant throughout the diameter range. This arrangement avoided any necessity to actually calculate or measure the diameter.

When sizing a motor for winder applications this constant power requirement implies that the maximum torque for a centre-driven system will occur at minimum rotational speed, an important factor when selecting a motor and drive. The power

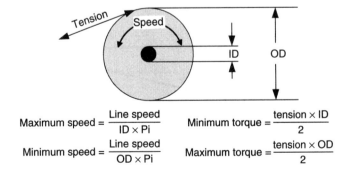

Figure C5.33   Speed and torque relationships for centre-driven winders

to produce tension at a known line speed is given by

$$\text{Tension power (kW)} = \frac{\text{line speed (m min}^{-1}) \times \text{tension (N)}}{60\,000}$$

or, using alternative units,

$$\text{Tension power (HP)} = \frac{\text{line speed (ft min}^{-1}) \times \text{tension (lbf)}}{33\,000}$$

When using this calculation to estimate motor size the requirement that the motor should produce the same power at minimum speed (maximum diameter) as it does at maximum speed (minimum diameter) results in a motor that is considerably larger than may have been anticipated, and the relationship between motor and drive size must be considered.

When choosing a motor two possibilities are presented. The motor may be specified to operate under constant power (Figure C5.34), in which case it is allowed to run above its base speed while its flux is controlled to match torque with diameter, the motor base speed having been selected to match the speed required at maximum diameter. In this situation,

$$\text{Centre wind motor (kW)} = \text{tension power (kW)}$$

Alternatively, the motor may be used only in its constant-torque mode (Figure C5.35), where base speed is chosen to match the speed at minimum diameter

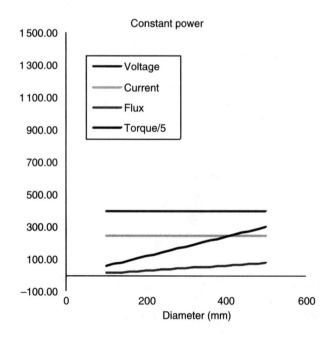

Figure C5.34   Constant-power winder characteristics

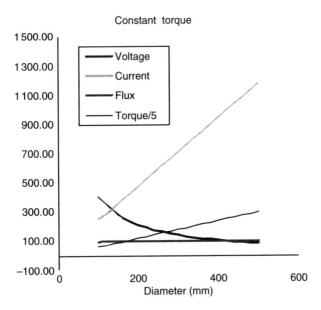

Figure C5.35   Constant torque winder characteristics

and the torque range is achieved by increasing the current. Maximum current will be selected to match the torque required at maximum diameter, resulting in an associated increase in drive current:

$$\text{Centre wind motor (kW)} = \text{tension power (kW)} \times \frac{\text{outer diameter}}{\text{inner diameter}}$$

Both options result in a larger frame size motor, but the constant-power solution avoids the necessity of increasing the drive size to match the increase in motor current.

Motor design restrictions impose a limit on the speed range over which constant-power systems can be operated. For d.c. applications the ultimate limit may be up to 5:1, but most motor manufacturers will limit their machines to less than 4:1. A.C. applications normally offer about 2.5:1 as standard. Higher ranges are possible, but this needs to be confirmed with the motor and drive manufacturer. In cases where the constant power range does not match the diameter range a combination of constant power and constant torque is often used, in which case some increase in drive size is unavoidable. It should be noted that any increase in drive size will be accompanied by an increase in supply current demand.

In addition to affecting the choice of motor, roll diameter plays an important part in most centre wind control systems, affecting the approach in both speed and torque control formats (Figs C5.36 and C5.37).

Apart from using direct measurement transducers or counting the number of revolutions at a known gauge, several methods of calculating the diameter within the drive system are possible. Perhaps the simplest is based upon the relationship

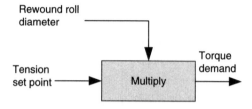

Figure C5.36 Winder torque calculation

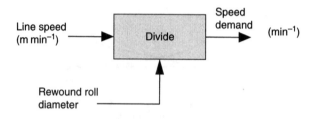

Figure C5.37 Winder speed calculation

between line speed and winder rotational speed.

$$\text{Diameter} = K \times \frac{\text{line speed}}{\text{winder speed}}$$

Because both these values are easily available it is a simple matter using modern digital techniques to generate a value for diameter. This approach may, however, produce inconsistent results at very low speeds and is normally combined with a track hold system, which locks the result when the signal levels fall below a set threshold.

A more sophisticated system that is independent of speed counts the length of material passing through the adjacent driven section during a full or known fraction of a revolution of the wound roll, the length measured during one rotation of the winder being equal to the circumference of the wound roll. This arrangement is not subject to errors at low speeds and produces an accurate measurement under all conditions.

Whereas speed-controlled winder systems calculate the speed at which the winder should run and then apply a small corrective trim obtained from the direct tension control loop to ensure that the correct torque is applied, torque-controlled winder systems rely upon the winder speed being set by the constraint of the wound material and simply predict what the torque should be, based upon tension set point and diameter.

Where wound rolls have high inertia values this results in problems during acceleration, as there will be insufficient torque to provide both tension and acceleration torque and so the predicted torque level must be adjusted accordingly. This technique is known as inertia compensation, where the accelerating torque is calculated and added to the tension-producing torque during changes in line speed. Calculation of driven inertia can be fairly complex as it is made up of motor, gear box and shafting inertia, which is constant in value, plus the rewound material inertia, which varies with wound diameter. It is therefore essential to know the correct diameter if the material inertia is to be calculated.

Most modern drive packages offer centre-driven winder control packages using on-board processing techniques where the following equations are easily processed:

$$\text{Acceleration torque (Nm s)} = \text{inertia (kg m}^2\text{)} \times \text{rotational acceleration rate (rad s}^{-2}\text{)}$$

where

$$\text{Inertia} = \text{fixed inertia} + \text{rewound roll inertia}$$

and

$$\text{Rewound inertia} = \frac{(OD^4 - ID^4) \times \text{width} \times \text{density} \times \pi}{32}$$

The acceleration rate may be measured by differentiating the line speed reference signal, or where available, extracting the jerk value from an S-ramp generator that sources the line speed reference.

High-speed winders may be subjected to acceleration torques that are comparative in value to the torque required to provide tension (Figs C5.38 and C5.39). In these cases the implementation of accurate inertia compensation becomes paramount. Incorrect values of acceleration torque will cause considerable errors in the tension produced during speed changes.

The total torque through the diameter range is a combination of tension and acceleration components whenever the line speed is changed (Figure C5.40). Inertia

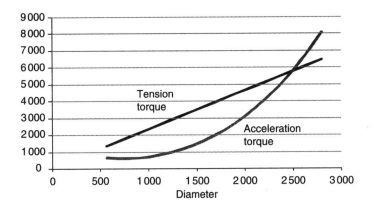

Figure C5.38 Torque versus roll diameter during winding

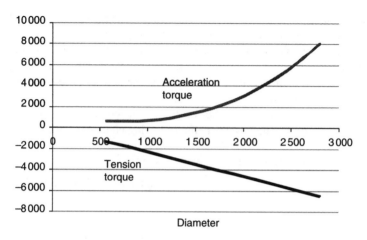

Figure C5.39   Torque versus roll diameter during unwinding

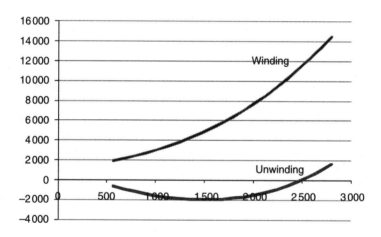

Figure C5.40   Combined tension and acceleration torque for a centre-driven winder

compensation has a considerable effect on the total torque of the winder during acceleration.

Winders with appreciable frictional losses in the mechanical transmission system may also require some form of loss compensation. In its simplest form this usually comprises separate static and dynamic components, but more sophisticated systems provide programmable function generators to attempt to accurately match loss profiles.

Both the inertia compensation and loss compensation torques are added to the tension-producing torque to provide a total predicted demand, which is used to set the drive torque. Load cell feedback may also be provided to provide a correction for any inaccuracies in torque prediction that affect the final tension.

An overview of a typical centre drive winder control scheme is shown in Figure C5.41.

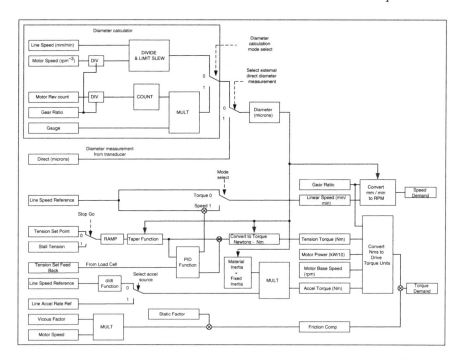

Figure C5.41   Typical centre-driven winder control scheme

C5.10   High-frequency inverters

C5.10.1   General

Advances in machine design have inevitably led to faster processing and machining of many materials. Additionally, adoption of high-speed machining techniques often gives a surface finish far superior to traditional 'rough-cut' and 'second-finish' techniques, to the point where further machining and polishing is rendered unnecessary.

Machines for processing man-made fibres, for example, are classed by their throughput speed, usually measured in metres per minute, with the faster machine producing the higher figure of merit. Indeed, many processes are essentially impractical unless high surface cutting speeds are available. Glass engraving and diamond polishing frequently use speeds in excess of $12\,000$ min^{-1}. Semiconductor wafer slitting saws often operate at $60\,000$ min^{-1}, and internal grinding (Barazon/Diamond) can use speeds of $90\,000$ min^{-1}.

Traditional approaches to attain high speed have been appropriate for a limited range of applications and were based on the use of small universal motors, air-powered motors and step-up belt and pulley arrangements. The a.c. induction motor, with its simple construction, offers itself as an ideal motor for this application if suitably designed for the rotational duty and if a cost-effective source of variable frequency and voltage can be provided.

C5.10.2 Frequency control of a.c. induction motors

It has been shown in Section A1.3 that the operational speed of the a.c. induction motor is determined primarily by the frequency of the connected supply voltage.

Ignoring slip, this is shown by the relationship

$$n = 60f/p$$

where n is the speed (min^{-1}), f the frequency of supply (Hz) and p the number of pole pairs.

For a two-pole (one-pole pair) machine connected to a 400 Hz supply, this would result in a synchronous running speed (i.e. ignoring slip) of

$$n = (400 \times 60)/1 = 24\,000\,\text{min}^{-1}$$

Varying the frequency of motor supply voltage is therefore the essential process to obtain control of motor speed. It should be noted that both increases and decreases of motor speed are achievable in this way. Where speed increases are required, the major constraints are mechanical and arise from motor design, construction and degree of balance.

In addition to variation of frequency, the motor supply voltage must also be controlled in order to achieve and maintain the correct magnitude of flux within the magnetic circuit of the motor. This control is described in Sections A1.3 and A4.4.1.5.

It is helpful to consider simple V/f control in order to gain an overview of the concepts, although the practical implementation is usually with an open-loop flux vector control strategy.

The control inverter must be configured to comply with the motor V/f ratio, because any deviation will affect the general system performance. Increasing the V/f ratio (over-fluxing) will increase motor torque availability but at the cost of much higher motor losses and consequentially higher operating temperatures. Increases will also occur in acoustic noise and possibly torque ripple, which can be superimposed on the motor shaft and eventually show as patterning on the workpiece.

Conversely, a lower than rated V/f ratio (under-fluxing) will result in reduced torque capability, which may result in inadequate torque for the duty. The motor will run both cooler and quieter in this condition.

A consequence of the theoretical V/f motor requirement is that two approaches to operation in the high-speed region are available. Figure C5.42 illustrates how a 50 Hz motor can be operated at higher speeds without electrical modification, and shows the resultant torque and power characteristics that may be expected. As the speed demand to the controlling inverter is increased from zero, both the voltage and frequency are increased on a linear basis until the output frequency reaches approximately 50 Hz. At this point a normally configured inverter will reach an output voltage approximately equal to that of the incoming line supply. Increasing the speed demand to the inverter further will continue to raise the output frequency, although the voltage supplied to the motor cannot increase further. The motor will increase in speed, responding to the

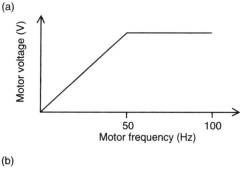
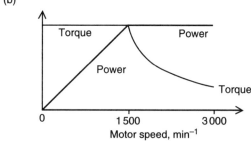

Figure C5.42   Characteristics of standard a.c. motors up to and above standard frequency: (a) V/Hz characteristic for high-speed operation; (b) motor torque and power characteristic

rising frequency, but the available shaft torque will fall away as the square of the effective voltage reduction. For example, a 10 per cent increase in frequency above 50 Hz would normally require a motor supply voltage increase of 10 per cent to maintain design torque; this 10 per cent deficiency in voltage will result in a 19 per cent reduction in torque capability. At higher frequencies the torque capability of the motor is further reduced, and all motors operating above base speed should be carefully checked with a competent motor supplier.

In this region the motor is essentially a constant-power device, as reducing torque while increasing speed results in a constant power characteristic (Figure C5.42). Clearly, mechanical limitations such as bearing performance must be considered when operating a standard induction motor above its rated speed.

Standard induction motors of 7.5 kW or less commonly have windings arranged in a six-wire configuration to allow connection for dual voltages such as 220 V in delta configuration and 400 V in star configuration. This makes it possible to extend the range of constant-power operation beyond the normal 50 Hz rating.

By connecting the motor in delta the motor is now rated at 220 V/50 Hz. Operating this inverter on a 400 V supply means that the motor can be operated up to $[(400/220) \times 50] = 87$ Hz, maintaining the same flux in the motor (Figure C5.43).

In the case shown in Figure C5.43, for a 1.5 kW, 400 V, 50 Hz, four-pole machine with a rated speed of 1 480 min^{-1}, by changing the stator windings from star to delta

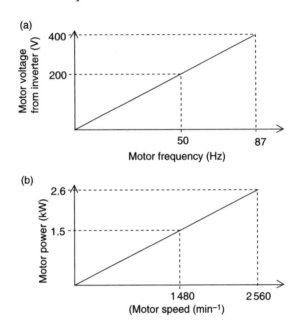

Figure C5.43 *Use of motor star and delta configuration to operate at increased frequency and power: (a) voltage supply to 200 V, four-pole, 1.5 kW motor from inverter; (b) additional kW operating 50 Hz motor at 87 Hz*

it would now be rated 2.6 kW, 400 V and 87 Hz. The increased losses are likely to be acceptable, as the effect of the shaft-mounted cooling fan improves at the increased rotational speed. In practice, frictional losses, magnetic losses and additional power absorbed by the cooling fan would detract from the rated power output, but a substantial benefit can be obtained by this method nevertheless.

It is important that the mechanical constraints of the motor and the driven system should not be overlooked, and the maximum speed allowable should always be checked with the motor supplier. The specification of rotor balancing should also be carefully specified.

C5.10.3 Purpose-designed high frequency motors

The vast majority of motors used in true high-frequency applications are specifically designed for the purpose. Motors rated up to $>180\,000$ min^{-1} are available, along with appropriate inverters with a $>3\,000$ Hz capability.

Motors designed for such speeds are normally of a slim construction in order to minimise the centrifugal forces and rotor inertia, offering a better dynamic response. Special bearings are invariably used and range from fairly standard deep-groove type up to 12 000 min^{-1}, oil-mist lubrication types up to 60 000 min^{-1}, and air-bearing or gas-type bearings for even higher rotational speeds.

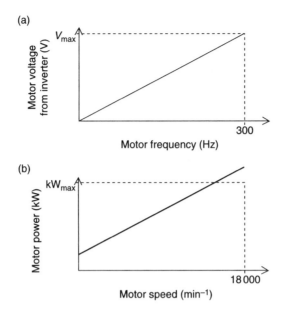

Figure C5.44 Characteristics of high-frequency purpose-designed motors: (a) high-frequency V/Hz characteristic; (b) high-frequency motor kW characteristic

Thermal considerations are also significant to the motor design, because the motor generally has a small physical size for its power rating and may run hot. Depending on the thermal reserve in the design, cooling may be surface only or extend to systems where air is drawn through the body of the motor. This, however, can degrade the protection integrity of the motor. In some cases motors may be water-cooled and fitted with elaborate water jackets and feed systems.

High-frequency motors are most often designed to offer a constant torque characteristic across the full operational speed range and to avoid the constant power/field-weakening region referred to earlier. Consequently, the motor winding is designed and wound for the highest operating voltage to coincide with the maximum design frequency. In other words, a constant flux is used across the speed range (Figure C5.44).

C5.10.4 High-frequency inverters

Traditionally, inverters for high-frequency applications were specifically designed. Today, inverters such as Control Techniques Unidrive are designed to operate up to 3 000 Hz in open-loop control.

For applications where full output voltage is to be achieved at 50 Hz, followed by constant output voltage as the frequency continues to rise, additional inverter design considerations are minimal, although control stability may be an issue with some loads. However, as mentioned previously, most true high-frequency motors operate with a constant torque characteristic right across the speed range, i.e. kilowatts increases with speed.

To assist motor braking, and to guard against inverter nuisance over-voltage tripping when the motor produces regenerative energy in its decelerating mode, it is normal practice to equip the inverter with a dynamic braking system.

C5.10.5 High-frequency applications

There are many applications for high-frequency inverters but a good example is woodworking machinery, a traditional area where high-speed cutting and finishing is essential to produce a satisfactory final product.

Traditionally, normal induction motors driving through a step-up belt drive system have been used. However, the problems associated with belt maintenance, and the desire for even higher throughput speeds and improved surface finish, have led to the adoption of high-frequency motors. These offer the additional benefit of compactness in the cutting head area where space is often at a premium. Normal speeds reached are in the range of $12\,000 - 18\,000$ min^{-1}.

Modern woodworking machines have to accommodate many shapes and profiles, and as a consequence employ numeric control (NC) systems to ensure the flexibility required. Inverters lend themselves readily to control by NC, and by using such techniques as load monitoring, the type or quality of wood can be evaluated automatically in order to set the optimum cutting speed. Tool changing is also a feature of these machines, and the NC can stipulate the speed and the torque available from the motor to suit the chosen tool.

Profiling and curvature machines are also quite common in the woodworking industry. Multiple tools are used, and these would include such functions as grooving cutters, facing drills and a circular saw for parting and slotting. These often operate at high speed, although they normally run at different speeds. For example, the grooving cutter may run at 300 Hz (18 000 min^{-1}) the facing drill at 120 Hz (7 200 min^{-1}) and the saw at 75 Hz (4 500 min^{-1}). Only one motor is used at any one time and this offers the possibility of using just one inverter but incorporating a suitably interlocked change-over system to connect each motor in turn to the inverter. At the same time the V/f ratio of the inverter is adjusted to pre-programmed values that suit each individual motor. Clearly, cost and space savings are attractive. During non-cutting periods the connected motor is ramped down under full control into a low-speed condition, drastically reducing machine noise but being ready to accelerate back to working speed within seconds.

C5.11 Special d.c. loads

The following are a few applications of d.c. thyristor drives in which the connected load is not the armature of a d.c. motor.

C5.11.1 Traction motor field control

Traction motors, such as those used in railway locomotives, are invariably of series-wound construction. This gives high starting torque, because the armature current passes through the field windings, resulting in maximum flux under heavy load conditions.

It is possible to duplicate this characteristic in a separately excited motor by controlling its field current by a thyristor drive configured as a current regulator, the reference being derived from the motor armature current through either a shunt or a d.c. current transformer.

The advantages of this technique include increased motor output (because the resistance of the field windings is not connected in series with the armature) and the facility to set maximum and minimum limits of field current, thus preventing saturation of the magnetic circuit and improving performance under light and overhauling load conditions, e.g. downhill running.

The speed amplifier needs to be used as a buffer amplifier by reducing its gain to unity. This allows the ramps, speed limits and lower current limits to be used to control rate of change, minimum and maximum current, respectively. A flywheel diode needs to be connected across the output terminals to provide a path for circulation of current (the load being inductive), although sometimes this is omitted in order to enable the drive to force field current down rapidly. The omission of the flywheel diode also makes field reversal possible without the use of contactors, should this requirement exist.

C5.11.2 Battery charging

The charging of secondary cells (e.g. lead-acid or nickel-iron accumulator batteries) is an application that calls for the control of current. The charging current is proportional to the area of the plates within a cell, multiplied by the number of cells connected in parallel. The voltage required for charging is proportional to the number of cells connected in series, the charging voltage per cell reaching a maximum when fully charged.

For this application a 'd.c. drive' needs to be configured as a simple current regulator, the reference input being configured to control the current reference. The current limit protects both drive and battery against over-current if a fully discharged battery is connected.

A more refined battery-charging system might use special application software to reduce the charging current when a predetermined voltage is reached, or after a period of time, giving boost and a trickle charge facility suited to such applications as uninterruptible power supplies and standby supplies for communications or medical equipment. The application programme could combine battery terminal voltage sensing with an adjustable delay before switching or ramping between two (pre-set) current levels, protecting the battery against overcharging, which could result in damage through loss of electrolyte or overheating.

Designers should select the converter and associated equipment with care, because, due to very low circuit impedance, the peak currents and resulting rms current may be higher than anticipated.

C5.11.3 Electrolytic processes

Examples of electrolytic processes include electroplating, refining of copper and other metals, and chlorination cells in which sodium hypochlorite is manufactured from brine. Deposition in such processes takes place at a rate that is proportional to current, and therefore the d.c. converter is configured as a current controller.

C.5.11.4 Electric heating and temperature control

A thyristor drive may be used for heating applications either in open-loop or closed-loop control configuration, and the application suits a d.c. drive and certain a.c. soft starts, which are phase-controlled.

Most heating elements consist of wire-wound or grid-type resistances supported on ceramic formers. When cold, the resistance of such elements is low in comparison with that at normal operating temperatures, and if connected directly to the mains supply a heavy current would flow, possibly causing localised over-heating or 'hot-spots', which may reduce the life of the element. Therefore, the current limit is used to set

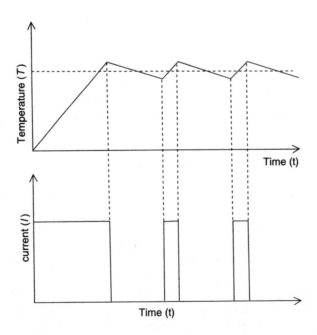

Figure C5.45   *Thermostat control, showing temperature and current with respect to time*

an upper limit to the output current of the converter to give controlled warm-up from the cold condition. Actual current, and therefore the rate at which heat is produced, is set by a set point potentiometer, and adjustments are made by the operator according to the final temperatures to be attained. Open-loop control would use an instrument connected to a thermocouple in contact with the process material to indicate actual temperature to the operator.

Such a system gives poor control of temperature, as it relies on the operator to monitor actual temperature and make the necessary adjustments. Better control can be obtained by controlling the 'heater on' contact by a thermostat. In this case, the current set-point reference potentiometer determines the rate of rise of temperature, but the actual temperature reached is controlled by the thermostat, which opens at the set temperature and switches off the drive, switching it on again as the temperature falls below the set point (Figure C5.45).

True closed-loop control of temperature requires a feedback signal proportional to temperature. This might be provided by a thermocouple amplifier or other temperature sensor, giving a linear $0-10$ V output over the operating range. Other signal ranges can be accommodated, using programmable offsets and scale factors, but linearity is important. In such a case, the outer control loop is configured as a temperature loop. By comparison with the thermostat control described above, which controls temperature by regulating the on/off time or duty cycle of the heaters at a constant current setting, the closed-loop system continuously regulates the current supplied to the heaters, sensing actual temperature and giving smoother and more precise control.

Chapter C6
Industrial application examples

C6.1 Introduction

Providing examples of all industrial applications of drives would be an impossible task, and provide the reader with little benefit. This section is therefore a glimpse into a small area of the drives world. Some of the most common applications have been selected, as well as some that show how drives have brought considerable benefit. Emphasis has been placed upon drive functionality and how that has an impact on the application.

It is also intended to show that required functionality and indeed the vocabulary of many applications is very specialised. A good knowledge of drives is not sufficient background to be able to design a good system. Experience helps, but the optimum design results when a drives engineer who knows what a drive can do works together with a customer who knows what functionality and performance is required and what problems he would like to see solved.

C6.2 Centrifugal pumps

C6.2.1 Single-pump systems

Control schemes for pumping systems usually involve a considerable degree of simple sequencing control: for instance, duty scheduling between several pumps to optimise the number of pumps in service and share running time. Various methods of controlling either flow or pressure may be used. Basic systems will operate from pressure- or level-detecting switches simply switching constant-speed pumps on and off depending upon the state of the detector switch. More sophisticated control can involve pressure transducers providing the feedback signal to a PID pressure control loop, in which the output determines the speed of the pump.

Various protective interlocking features may be incorporated into the pump control scheme. These will possibly include the following.

- There may be energy saving during times of low demand when the pump may be stopped and put into a 'sleep' mode until the pressure drops and the system 'wakes' to resume operation.
- There may be detection of no suction or a dry well to prevent pumps being operated when dry.
- There may be high or low pressure trips.

The energy savings obtainable using variable-speed pump control are quite significant, as demonstrated in the example below, particularly when compared to throttling to achieve required flow rates.

Figure C6.1 shows the characteristic curve of a typical centrifugal machine. It is based upon the performance characteristic of a real water pump with the following specification:

Rated flow = 10.6 m^3 min^{-1}
Head = 37.5 m
Rated speed = 1 480 min^{-1}

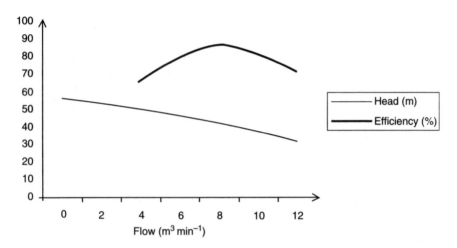

Figure C6.1   Characteristic curve of a typical centrifugal machine (pump, fan or compressor)

Figure C6.1 shows the head/flow relationship for constant speed. The curve of efficiency is also included and is seen to peak at a slightly lower flow than the rated flow of the pump. Efficiency at rated flow is 83 per cent, so the pump shaft power requirement is given by

$$\text{Shaft power (kW)} = \frac{\text{head (m)} \times \text{acceleration due to gravity (N kg}^{-1}) \times \text{flow (m s}^{-1})}{\text{efficiency}}$$

$$= \frac{37.5 \times 9.81 \times (10.6/60)}{0.83}$$

$$= 78.3 \,\text{kW}$$

If the pump speed is reduced from the rated speed of 1 480 min^{-1}, a similarly shaped curve exists relating the head/flow characteristic of the pump at any particular speed.

Figure C6.2 shows a family of curves generated for speeds over the range 1 480 to 960 min^{-1} (65 per cent).

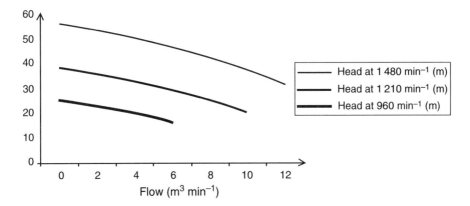

Figure C6.2   Characteristic curve of a typical centrifugal pump operating at different speeds

The head/flow curves for any speed can be readily estimated from the constant/rated speed characteristic. The following ideal relationships apply.

- If the rated speed is n_1 and the reduced speed is n_2, then

$$\text{Speed ratio is } (n_2/n_1)$$

$$\text{Flow is proportional to } (n_2/n_1)$$

$$\text{Head is proportional to } (n_2/n_1)^2$$

$$\text{Power is proportional to } (n_2/n_1)^3$$

- It follows that in respect of the shaft,

$$\text{Shaft torque is proportional to } (n_2/n_1)^2$$

Figure C6.3 shows these relationships graphically.

For pumping applications, the characteristics of the system into which pumping takes place must be considered and the back pressure of the pump consists of two components:

- static head
- friction losses

Figure C6.4 shows a system demand curve superimposed on the variable-speed curves of Figure C6.2.

The static head is seen to be about 22 m and represents a total lift from the pump inlet to elevation of the point of discharge. Friction losses are a function of pipe diameter, pipe length, inlet losses, pipe 'C' factor, specific gravity of the liquid, type and

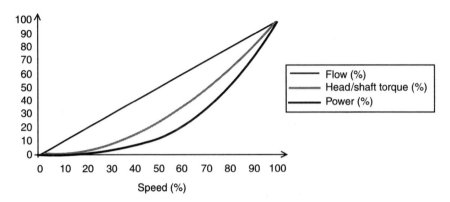

Figure C6.3 Characteristic relationship between flow, head/torque and power in a centrifugal pump operating at different speeds

number of bends, fittings, reducers, valves, and so on, and are approximately proportional to the square of the flow.

From Figure C6.4 it can be seen that at the maximum (rated) speed of 1 480 min^{-1} the system demand curve intersects the pump curve at 11 m^3 min^{-1}; this represents 100 per cent flow.

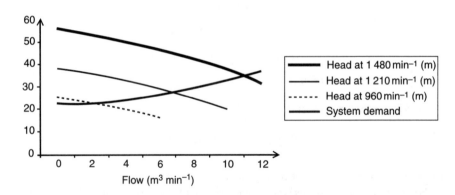

Figure C6.4 Characteristic curves of a centrifugal pump operating at different speeds with system demand superimposed

Consider the condition where flow is to be reduced to 6.8 m^3 min^{-1}, a reduction of 40 per cent. The pump speed required to operate at this point on the demand corresponds to the pump characteristic drawn for operation at 1 210 min^{-1}. This is shown in Figure C6.5.

The system back pressure at this point is 28 m, including friction losses of about 6 m:

$$n_2/n_1 = 1\,210/1\,480 = 0.82$$

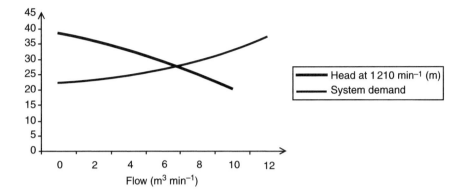

Figure C6.5 Characteristic curve of a centrifugal pump operating at reduced speed with the system demand superimposed

With this 18 per cent reduction in shaft speed the efficiency of the pump would typically fall by about 3 per cent. Figure C6.1 shows the pump efficiency to be 81 per cent at a flow rate of 6.8 m^3 min^{-1}. The efficiency at the reduced speed can therefore be assumed to be $0.81 \times 0.97 = 78.5$ per cent.

The pump shaft power requirement is given as

$$\text{Shaft power (kW)} = [\text{head (m)} \times \text{acceleration due to gravity (N kg}^{-1})$$
$$\times \text{ flow (m s}^{-1})]/\text{efficiency}$$
$$= [28 \times 9.81 \times (6.8/60)]/0.785$$
$$= 39.7\,\text{kW}$$

If, instead of using variable-speed control, a mechanical throttle were used, the system curve would be modified as shown in Figure C6.6.

The operating condition for the pump is now as follows:

Head = 45 m
Flow = 6.8 m^3 min^{-1}
Efficiency = 81%

The pump shaft power requirement is given as

$$\text{Shaft power (kW)} = [\text{head (m)} \times \text{acceleration due to gravity (N kg}^{-1})$$
$$\times \text{ flow (m s}^{-1})]/\text{efficiency}$$
$$= [45 \times 9.81 \times (6.8/60)]/0.81$$
$$= 61.8\,\text{kW}$$

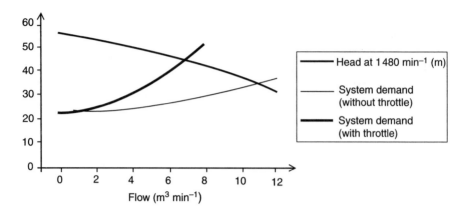

Figure C6.6 Characteristic curve of a centrifugal pump operating at fixed speed with the system demand with throttling superimposed

The 'excess' head dropped across the valve therefore represents the energy loss due to throttling.

The energy saving by using a variable-speed drive in this example, at that duty point, is $(61.8 - 39.7)$ kW $= 22.1$ kW $= 35.7\%$.

This example does assume that the pump selection and valve sizing have been optimised for the application. It should be noted that even at the design flow rate, some 'excess' head must exist across the valve for it to be able to control flow. Typically this would be 5–10 per cent of the rated pump head.

So far only the pump has been considered. In order that overall losses are included in the calculation (i.e. from the electrical supply to the pump output), it is necessary to take into account the efficiency of the drive system. For the fixed-speed motor rated in 75 kW, the rated efficiency could be expected to be 94 per cent. For a modern a.c. variable-speed drive, the converter efficiency can be conservatively assumed to be 97 per cent. The motor under variable-frequency control could be conservatively assumed to lose two points of efficiency.

The power requirements in the above example, at the reduced flow, can be restated as

Mains power for variable speed solution $= 39.7/(0.97 \times 0.92) = 44.5$ kW

Mains power for the throttling solution $= 61.8/0.94 = 65.7$ kW

The power saving is 21.2 kW/32.2 per cent.

If we consider such a pump running for say 8 h a day for 250 days a year, then the energy saving would be 42.4 MW h, which is equivalent to over 18 tonnes of CO_2. Many pumps are left operating continuously, so the application of variable-speed drives can lead to substantial cost savings and a considerably improved carbon footprint.

C6.2.2 Multiple pump systems (duty-assist control)

Duty-assist control is an effective method of controlling multiple pumps or fans in parallel, to maintain the required process demand, while using only a single drive. Pumps and fans are often used in parallel banks to

- avoid motor overload,
- guaranteed security of supply (system redundancy),
- reduce running cost due to system load cycles, and
- provide a wide range of control and flexibility.

The system consists of a duty drive and assist starters. The assist starters can be of any type (e.g. DOL, star-delta, auto-transformer, soft starter or inverter); the choice is dependent on the system limits.

The duty drive can control one dedicated motor (fixed duty), or with additional external switchgear could be selected to control other motors within the parallel configuration (flexible duty).

The duty drive is controlled by a PID loop, which will maintain the process level required, if the demand exceeds the capacity that can be derived from the duty, then an assist starter will be enabled to assist with the delivery. If the demand continues to increase, additional starters will be called. When the demand decreases, then each assist starter will be deselected to reduce the delivery to the required demand.

Typically, in a parallel bank of four, it is usual to find three running and one standby. The duty assist can typically control up to four assist starters with a 'fixed duty', and three assist starters with a 'flexible duty' selection.

Some intelligent drives can have the entire control functionality for a duty-assist system embedded within the drive without the need for an external controller. The standard and optional I/O on the drive provides the I/O interfacing to the other starters and the process.

Typical processes that can be controlled with parallel configurations using the duty-assist functionality are

- pressure,
- flow,
- temperature,
- tank/reservoir level, and
- sewage aeration.

A typical configuration for controlling pressure with a multiple-pump system is shown in Figure C6.7.

C6.2.2.1 Note on parallel operation of pumps

Contrary to commonly held beliefs, the flow does not double on addition of a second similar pump in parallel. In fact, each successive pump or fan adds a smaller amount to the total system pressure and flow, although the total flow is split equally between each pump/fan.

Typical system curves for parallel pump/fan operation are shown in Figure C6.8.

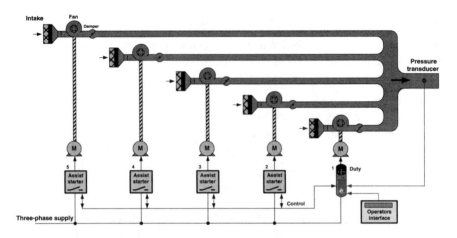

Figure C6.7 Typical duty-assist configuration for multi-pump pressure control

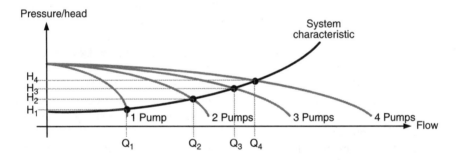

Figure C6.8 System characteristic for parallel pump systems

C6.3 Centrifugal fans and compressors

Centrifugal fans obey the same basic characteristics as the centrifugal pumps described in Section C6.2, and the shape of their pressure/volume curves is therefore similar to those of pumps. The calculations and the rational for variable-speed operation are the same.

The pressure/volume characteristic for centrifugal compressors, although generally similar to the pump, differs in that a pressure peak occurs between low flow and rated flow. The range of flow reduction cannot therefore extend beyond the limit where 'surging' takes place. Good flow control can be achieved with variable-speed operation over typically a 2:1 speed range, although care must be taken to avoid surge problems.

C6.4 Heating, ventilation, air conditioning and refrigeration (HVAC/R)

C6.4.1 Introduction

HVAC/R controls are common in residential, industrial and commercial applications to provide comfortable environments for occupants or precision environments for manufacturing and process control. One of the largest market sectors for drives, specifically open-loop a.c. drives, is the commercial building sector. Examples of types of commercial buildings that incorporate drives into their HVAC environmental controls are

- hotels,
- shopping malls,
- office buildings,
- car parks,
- prisons,
- casinos,
- theatres,
- museums, and
- educational establishments.

Refrigeration applications are most commonly found in retail facilities such as supermarkets and restaurants. Figure C6.9 summarises the applications found in the HVAC/R market.

APPLICATIONS	HVAC	Cooling Towers	Supply Return Fans	Chill Water Pumps	Hot Water Pumps	Condenser Fans	Stairwell Pressurization	Exhaust Fans	Fume Hoods	REFRIGERATION	Compressors
Supermarkets			✓	✓	✓	✓		✓			✓
Schools & Colleges		✓	✓	✓	✓	✓		✓	✓		✓
Hospitals & Healthcare		✓	✓	✓	✓	✓	✓	✓			✓
Stadiums & Arenas		✓	✓	✓	✓	✓		✓			✓
Shopping Malls		✓	✓	✓	✓	✓		✓			✓
Hotels		✓	✓	✓	✓	✓	✓	✓			✓
Office Buildings		✓	✓	✓	✓	✓	✓	✓			✓
Apartments			✓	✓	✓	✓	✓	✓			✓
Prisons & Correctional Institutions		✓	✓	✓	✓	✓		✓			✓
Manufacturing Facilities		✓	✓	✓	✓	✓		✓	✓		✓

Figure C6.9   HVAC/R solutions matrix

The specifications for drives vary globally by region and sometimes by city, depending on the local experiences and preferences. In Europe a significant proportion of the drives installed are rated to degree of protection IP54 in order to protect the control electronics from dust and dripping water when the drives are installed and during the construction phase of the building. By way of contrast, nearly all drives installed in the United States are UL type 1 (IP20) with provision for metal conduit connections that are not typically used elsewhere.

Drives are often specified to be plenum rated, which allows the drive to be safely mounted in an air stream; this can save building space and help cool the drive. Another frequent but specifically North American requirement is for some form of bypass, or back-up circuit, in the case of a drive failure. This requirement is somewhat historical in nature as drives were widely applied to HVAC systems in the United States many years before modern a.c. drive technology matured and attained the high levels of reliability seen today. However, it should also be noted that the extreme heat experienced in the southern states of the United States and also seen in countries such as Australia dictate that redundant air-conditioning systems such as bypass are installed for human safety in these areas.

Some of the product requirements for drives in the commercial HVAC/R market are different from industrial applications. For example, a safety feature of dedicated HVAC drives is 'fire mode'. When activated this feature is used to control the speed of a fan in order to purge the air from a fire and therefore aid in the evacuation of a burning building.

Most HVAC applications provide some level of either air or water pressure or flow control. Many systems are centrally controlled by a building management system (BMS), some are controlled by a drive, and some have hybrid control schemes. Dedicated HVAC drives incorporate one or more process PID control loops. For more demanding applications it is sometimes necessary to use multiple feedback devices in order to provide redundancy, averaging or zone control.

It is helpful to consider the application of drives in the HVAC/R market in the following segments:

- new commercial buildings,
- retail facilities, and
- original equipment manufacturers.

C6.4.2 Commercial buildings

The majority of the drives used in the HVAC/R market are purchased by contractors for installation in new buildings. Nearly all new large buildings and large facility retrofits have some form of total building automation system that interfaces with the drive via a serial communications network. Depending on the local requirements most drives installed in new buildings have advanced control algorithms for local process control and the latest safety features such as fire mode.

C6.4.2.1 Building automation systems

A building automation system (BAS) controls and monitors the building's mechanical and electrical equipment such as air handling and cooling plant systems, lighting,

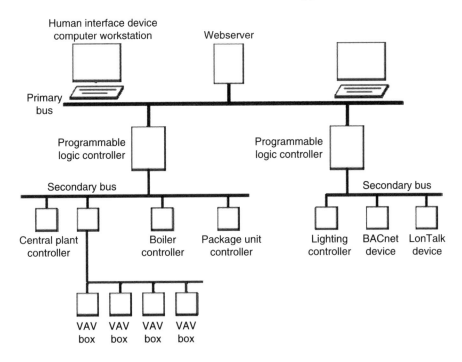

Figure C6.10   Typical building automation system (BAS) architecture

power systems, fire systems and security systems. A BAS consists of software and hardware and typically comprises the elements and connections shown in Figure C6.10. BAS core functionality keeps the building climate within a specified range, provides lighting based on an occupancy schedule, and monitors system performance and device failures and provides email and/or text notifications to building engineering staff. The BAS aims to optimize building energy and maintenance costs when compared to a non-controlled building. A building controlled by a BAS is often referred to as an intelligent building.

Such a BAS would be connected to drives using either discrete analogue and digital wiring connections or some form of serial communications bus. The communication networks that are commonly found in commercial HVAC systems are typically different to their industrial counterparts with the exception of Modbus RTU.

The most common networks used are

- BACnet (see Section C4.6.5.6);
- LONworks (see Section C4.6.5.5);
- Metasys N2 (developed by Johnson Controls Inc., Metasys N2 is similar to BACnet although the standards are different).

C6.4.2.2   HVAC applications

Most commercial HVAC drive applications involve control of centrifugal fans and pumps. One of the key market drivers for the industry is the fact that variable-speed control of fans and pumps has tremendous energy savings potential. In nearly all applications speed control is the most energy-efficient method of process control. Alternative devices that provide less efficient process modulation are mechanical devices such as dampers and inlet vane guides. Other practical benefits of drives in HVAC applications include reduced maintenance costs due to soft starting and audible motor noise reduction at reduced speeds.

Examples of common commercial building HVAC applications are

- variable air volume (VAV) fan control,
- constant air volume (CAV) fan control,
- cooling towers,
- chilled/hot water pumps,
- water pressure booster pumps,
- stairwell pressurization, and
- parking garage ventilation.

Examples of VAV and CAV applications will be considered in some detail below, utilizing advanced control features available in dedicated purpose HVAC drives. In many cases these control functions are provided by an external BAS controller and the drive simply follows a speed command and is started and stopped remotely. Cooling tower drives will also be discussed, as they often require specific environmental packaging and options.

Variable air volume example
Variable air volume (VAV) systems are the most common form of air conditioning and ventilating spaces in commercial buildings. Variable-frequency drive control of the VAV system supply and return fans provide the most efficient and mechanically straightforward method of control of air flow and pressure.

There are the following drive application requirements:

- *Supply fan*. This is used for the control of supply duct pressure (supply fan) with varying loads caused by VAV box modulation.
- *Return fan*. This is used for the control of supply and return fan flow differential to maintain optimum building pressurization.

Figure C6.11 shows a typical VAV system. The supply fan drive provides basic PID closed-loop pressure regulation using typically a single 0–100 psi pressure transducer for feedback. The return fan drive provides PID closed-loop differential flow regulation using two flow transducers.

Constant air volume example
Constant air volume (CAV) systems are ventilation systems that are used to supply fresh conditioned air to large building zones and were widely installed before VAV systems were available. Like the VAV system they incorporate supply and return

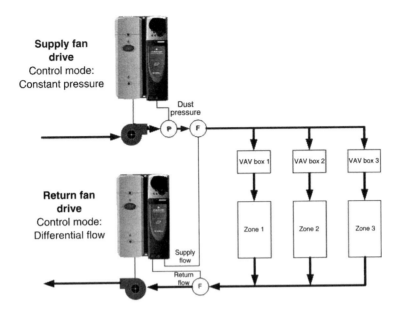

Figure C6.11   Typical VAV drive control system

fans. Older systems can benefit in performance with the addition of drives to control indoor air quality or temperature.

There are the following drive application requirements:

- *Supply fan.* This is used for the control of fresh air supply with varying loads caused by occupancy changes.
- *Return fan.* This is used for the control of return fan static pressure to maintain optimum building pressurization.

Figure C6.12 shows a typical CAV system. The supply fan drive provides basic PID closed-loop air quality control usually using a carbon dioxide monitor for feedback. The return fan drive provides basic PID closed-loop pressure regulation.

Cooling tower example

Cooling towers (Figure C6.13) are commonly used in HVAC systems, particularly in dry climates, as an energy-efficient means of cooling condenser water as part of a chilled water system. An integral part of the cooling tower is a fan that blows air across the condenser water at a rate determined by the condenser water temperature. This application also typically demands the avoidance of resonant running frequencies and motor sleep functionality for additional energy savings during extremely low load demand.

Cooling tower drives are often packaged in outdoor NEMA3R enclosures, such as that shown in Figure C6.14, with accessories such as panel and basin heaters to prevent freeze problems during extreme cold weather conditions.

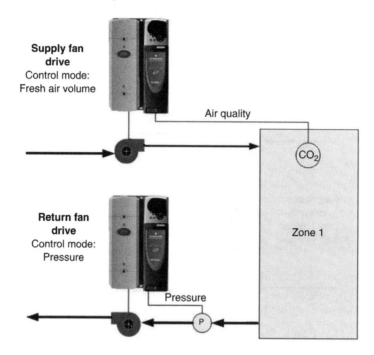

Figure C6.12   Typical CAV drive control system

Figure C6.13   Outdoor cooling tower

Industrial application examples 613

Figure C6.14 Cooling tower drive package for outdoor installation

Commercial HVAC drive requirement summary
In general, drives for large commercial buildings have the following specific attributes:

- BAS serial connectivity,
- fire mode operation, and
- multiple process PID control loops.

These specific features are provided by definite purpose HVAC drives that also incorporate general-purpose fan and pump functions such as

- variable torque mode,
- energy optimization,
- mains dip ride through (reduces trips during supply brown outs),
- real-time clock,
- power metering,

- auto reset,
- resonant frequency avoidance,
- S-ramp (reduces mechanical shock on start up and shut down), and
- auto-tune.

These features provide energy savings and maintenance cost reduction benefits.

C6.4.3 Retail facilities

HVAC/R drives are frequently applied to both HVAC and refrigeration applications in a modern retail facility such as a supermarket or convenience store (Figure C6.15). Many facilities incorporate dedicated building automation systems that are application-specific and that control food freezers and display cases as well as HVAC and lighting, typically on a smaller scale than the systems provided in a large commercial building. HVAC applications include air handling unit fans and ventilation fans. The refrigeration control applications that incorporate drives are the refrigeration compressors and the condenser fans.

The concept of the intelligent store has emerged: a comprehensive, integrated site-management system, providing intelligent monitoring and control of HVAC, refrigeration, lighting, plus many other devices and systems (Figure C6.15).

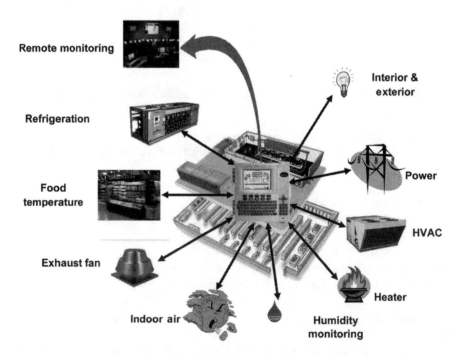

Figure C6.15 Intelligent store architecture [photograph courtesy of Emerson Climate Technologies]

C6.4.3.1 Refrigeration applications

Condenser fan control

Most commercial refrigeration drive applications involve control of centrifugal fans and refrigeration compressors. The same rules for energy savings apply to variable-speed control of fans as in HVAC applications. The fans are used to vary air flow over condenser coils, which are usually installed on the roof of a retail facility such as a supermarket. Typical installations consist of multiple numbers of fans (Figure C6.16) that were historically switched on and off using conventional contactor control. When drives are applied to condenser fan systems, one drive controls multiple fans simultaneously, resulting in significant energy savings over the on–off fan cycling mode. Practical installations have returned up to 40 per cent savings. Additional benefits include noise reduction, which is very important for retail facilities located in densely populated urban areas.

Figure C6.16   Multiple fan condenser unit

Refrigeration compressor example

Drives are used to vary the speed of refrigeration compressors in order to achieve optimum system performance. In multi-compressor rack systems common in the Americas one drive is applied to the lead compressor and the remaining units operate at fixed speed. This arrangement is a very cost-effective method of accurate temperature and refrigerant pressure control. A secondary but very important side benefit is the reduced cycling of the fixed-speed compressors, which reduces system stress and wear and tear on contactors.

The potential speed range for compressors is less than that of fans and pumps and it is extremely important not to under-speed compressors as they are typically self-lubricating and require minimum speed settings in the 75 per cent range.

Figure C6.17   Multiple pump installation

Conversely, it is possible to operate compressors above the base speed of the motor and gain 10–15 per cent output.

C6.4.4   Original equipment manufacturers

Original equipment manufacturers (OEMs) supply equipment packages such as complete air handling units, cooker hoods and condenser units that may well incorporate drives. In many cases OEMs provide stand-alone controllers that include unit controls and also communicate directly to a BAS. Drives specified by OEMs tend to be basic motor control devices without the dedicated HVAC function set required for commercial buildings. For air handling equipment one common requirement is that the drives are 'plenum rated', which means that they can be physically mounted in an air duct.

Pump manufacturers who supply complete pump systems on a skid or container, commonly use multiple drives that operate in multiplexing modes such as lead/lag and pump alternation (Figure C6.17). Dedicated pump drives incorporate these functions as standard and operate in constant-pressure mode with direct inputs from pressure transducers and flow switches. See Section C6.2.2 for more information on duty-assist systems.

C6.5   Cranes and hoists

C6.5.1   General

There are many different types of crane and a complete review is not possible here. However, it is interesting to consider some of the crane types and the way they are controlled.

C6.5.2 Overhead cranes

Overhead cranes come in a wide variety of forms, starting with very simple systems that may comprise a basic hoist and two travel functions. The long travel motion moves the crane itself along the tracks built alongside the work area, while the cross-travel motion moves the hoist position across the crane beam.

Overhead cranes are often equipped with long travel drives at both ends of the boom; this can create a potential problem due to slightly unequal travel speeds at either end resulting in crabbing. The specification for the drive system must take this into account. All motions may be controlled from a simple hand-held push-button box, either hard wired or connected via a radio link to control for the individual motions. In the most simple applications the control of all axes may be with simple open-loop a.c. drives and induction motors. On more demanding applications closed-loop induction motors prevail.

The operation of such cranes can require a very skilled operator, as jerky motion can cause the load to sway. In more sophisticated overhead cranes, anti-sway control algorithms are incorporated into the crane controller. Typical anti-sway algorithms use information about the travel speed profile, the weight of the load (and hook) and the length of the cable from the hoist drum to the load centre of gravity.

C6.5.3 Port cranes

The transportation of loads in ports, terminals, industry, steel works, power stations and shipyards requires high-performance cranes that operate with great availability, reliability, precision and safety. A.C. squirrel-cage motors with a.c. drives are the optimum solution for high-performance cranes. A.C. drives with open-loop vector control typically provide 1 per cent speed regulation and are acceptable for some low-performance applications. Closed-loop vector control provides 0.1 per cent speed regulation and is used in most applications.

On-board intelligence with high-speed digital bus communications on modern a.c. drives can provide application solutions, enhancing crane performance.

C6.5.3.1 Ship-to-shore container cranes: grab ship unloaders

Figure C6.18 shows a typical ship-to-shore container crane. The hoists, grab, trolley, boom and gantry movements are usually carried out by closed-loop a.c. drives and a.c. induction motors. A single drive is often switched as required between the boom and trolley. In order to achieve this, the drive must support alternative parameter sets to ensure stable operation.

Active supply converters are also frequently used to provide good harmonic supply conditions and to regenerate energy.

The hoist and grab functions comprise two motors in closed-loop control. In order to allow for rapid operation in light load conditions and between loads, the drives and motors must have a good constant power control range. The boom comprises a single motor in closed-loop control, again with a good constant power control range. The trolley comprises either two or four motors in closed-loop speed control. The gantry comprises 16 or as many as 20 motors, usually fed from an open-loop

Figure C6.18 Ship-to-shore container crane

drive. The high-voltage cable reel comprises a single motor controlled by a closed-loop a.c. drive with a simple centre winder torque control algorithm. Different torque offsets are applied when working as a winder or an unwinder.

The rated powers of the motors and drives in this application are typically in the range 75 to 1 000 kW.

C6.5.3.2 Rubber-tyred gantry cranes

Figure C6.19 shows a typical rubber-tyred gantry (RTG) crane. It is important to note that an RTG is powered by a diesel generator set. This generator supplies all the equipment on the crane, including all the drives and motors, the crane controller, lights and other auxiliary devices. The hoists, trolley and gantry movements are controlled using a.c. induction motors fed from closed-loop drives. The drives tend to be configured with a common d.c. bus arrangement with a single/bulk supply converter.

The hoist comprises a single motor operated in closed loop capable of operating with a good constant power control range. The trolley comprises two motors operated in closed loop. The gantry comprises either four or eight motors typically operated in open-loop mode fed from only two drives.

The rated powers of the motors and drives in this application are typically in the range 37 to 300 kW.

C6.5.3.3 Rail-mounted gantry cranes

Figure C6.20 shows a typical rail-mounted gantry crane. The hoists, trolley and gantry movements are controlled using a.c. induction motors. The drives are linked with a PLC controller via a Fieldbus with a remote input/output rack in the drive cabinet and the driver's cabin.

Figure C6.19 Rubber-tyred gantry crane

The hoist comprises two a.c. induction motors with separate drives, each operated in closed-loop and capable of operating with a good constant power control range. The trolley comprises two or four a.c. induction motors operated in closed-loop speed control. The gantry comprises 16 or 20 a.c. induction motors operated in open-loop control from two drives. The high-voltage cable reel comprises a single motor controlled by a closed-loop a.c. drive with a simple centre winder torque control algorithm. Different torque offsets are applied when working as a winder or an unwinder.

The rated powers of the motors and drives in this application are typically in the range 45 to 300 kW.

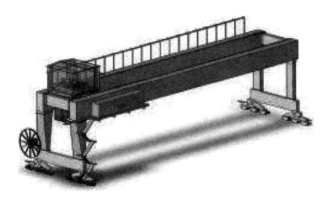

Figure C6.20 Rail-mounted gantry crane

C6.5.4 Automated warehousing

An entirely different type of crane is that used in automated warehousing, where motions involve travel, hoist and fork extension. These cranes are often automated following instructions to visit, pick and deliver from locations arranged in an $X-Y$ matrix on either side of the aisle in which the crane runs. Automatic warehouse cranes are normally equipped with an on-board position control system that receives target location instructions in $X-Y$/left–right format from a ground-based warehouse automation system. The fork carriage is then moved using both travel and hoist motions to align it with the required pallet location to allow pallet retrieval. Because of the difficulties in assembly, these large rack systems operate to strict dimensional specifications. Owing to the possibility of distortion under loading, final positioning is usually performed using proximity devices rather relying on the overall position counters that are driven from the travel and hoist drive systems.

C6.5.5 Notes on crane control characteristics

C6.5.5.1 Hoisting control

All cranes have a hoisting function and depending upon the required performance, this is usually controlled with a closed-loop voltage source inverter feeding an induction motor. Open-loop drives are used but low-speed operation tends to be poor and unsuitable for demanding applications.

Many crane hoist drives have a light hook mode of operation where the motor is run into its constant power range to allow rapid hoisting or lowering when unloaded and also usually include some form of pre-torqueing associated with the hoist brake control.

C6.5.5.2 Slewing control

Historically, the slewing movement on cranes was undertaken with a.c. slip-ring induction motors in combination with rotor resistors. This provided good motor torque control for acceleration and deceleration and it is possible to coast when the controller is moved to zero. However, this method of control was very poor at low speeds, with sudden steps in torque between resistor steps. It was also very inefficient with respect to power usage, and the system required very regular and intensive maintenance.

Today a.c. cage induction motors are used, mostly in closed-loop control. Figure C6.21 shows the behaviour of a conventional speed-controlled drive system showing the problem of controlling swaying of the load. To counter this effect it is possible to design a program to run within a programmable drive or external PLC that gives the crane driver optimal control over the swaying load. Such a control system would typically provide the driver with control over both the speed and the motor torque. Speed control is important for accurate positioning at low speed. It also provides compensation for the wind forces on the load. Torque control is crucial for controlling the sway. In this way, the driver is always able to anticipate the movement of the load and compensate for it, as shown in Figure C6.22. By

Industrial application examples 621

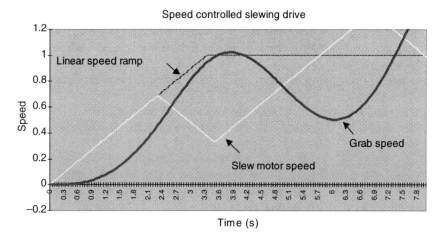

Figure C6.21 Load/hook speed compared with motor speed for an uncompensated system

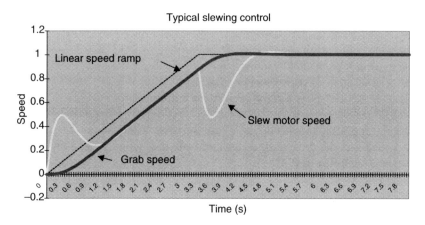

Figure C6.22 Load/hook speed compared with motor speed for a compensated system

bringing the controller back to zero, the movement is effectively coasting, which gives a major damping effect on the sway of the load.

C6.5.6 Retrofit applications

In most retrofit applications with modern drive control systems, there are several issues that need careful consideration.

- There is a need to install the equipment in an existing environment that may be far from ideal for electronic equipment.

- There is a need to provide control characteristics appropriate for the user of the crane to have appropriate 'feel' when operating the crane. These characteristics can be difficult to quantify and require considerable experience to understand.
- There is a possibility of improvements in performance available using modern drive equipment.
- The maximum capacity of the crane should be considered.
- The latest safety rules should be applied; these should be constantly being revised and tightened.

With regard to the maximum capacity of the crane, this is only valid if all of the controls for the crane movements are tuned in such a way that the driver is able to control the load in an easy and safe manner.

When designing a drives system for a retrofit, or for a new crane, it is essential that the driver's needs are fully considered. The ability to build in particular control characteristics is very important. Some drive manufacturers have available programmable functionality for both a.c. and d.c. drives, to customise the performance and features of the drive to exactly meet the requirements of the user.

Requirements typically include

- slewing control on a cargo crane,
- grab control on the hold-and-close drive,
- synchronisation on two or more motors, and
- load-dependent speed control on the hoist movement.

By utilising a programmable drive, it is possible to provide all the movement-related control in the drive itself thereby, in most instances, avoiding the need for a stand-alone controller or PLC.

C6.6 Elevators and lifts

C6.6.1 Lift system description

Historically, lift systems were operated using motor generator sets and Ward–Leonard multi-motor control. With the development of drive technology, motor generator sets were replaced with d.c. motor and drive converters incorporating 6-pulse and 12-pulse systems (i.e. two drives in parallel or series configuration).

The advantages of the parallel 12-pulse system over the 6-pulse system include the following:

- reduction of electrical interference on the electrical supply,
- reduction in audible noise from the motor and thus reflection into the lift shaft,
- reduced harmonics in the line currents, and
- less ripple current thus less torque variation (producing better lift ride performance).

Today, the d.c. lift market is declining and has largely been replaced by a.c. drives feeding either induction motors or where very compact motor designs are required, permanent-magnet synchronous motors.

The lift market can be split into three basic types of lift control systems:

- *Hydraulic.* These are fitted in low-rise buildings and are based on the extension and retraction of a hydraulic ram fitted to the bottom of the lift car. The lift speeds vary, typically in the range 0.5 to 1.25 m s^{-1}.
- *Geared.* These are fitted on low- to intermediate-sized buildings and are based on a motor gear arrangement (the gearing used being of helical, planetary or worm type) with lift speeds in the range 1.0 to 2.5 m s^{-1}. The speed of these systems is limited by the losses/audible noise created in the gearbox at higher speeds.
- *Gearless.* Lift systems of this type are generally found in taller buildings with lift speeds in the range 2.5 to 10 m s^{-1} for passenger elevators and even faster for goods elevators. However, with the latest development in technology in motor design and a.c. drives, and as energy efficiency becomes more and more important, gearless technology is taking an ever greater share of the market.

The main components of a lift system as shown in Figure C6.23 are as follows:

1. *Variable-speed drives.*
2. *Lift controller.* The lift controller's main functions can be split into the following sections:
    (a) Handling car/landing calls. A separate group controller is often used to control the car/landing calls when more than one lift is used in a building.
    (b) Learn and store floor positions. Floor sensors are set up at each landing level. The lift is moved at a controlled speed from the bottom floor to the top floor. Positional information is read from an incremental encoder as the lift passes by the floor plates, which indicate floor level. The position is then stored in memory. This positional storing is continued until the top floor of the building is reached.
    (c) Generating lift speed patterns (16 bit and 32 bit). (These speed patterns reference the drive speed loop.)
    (d) Calculating lift position.
    (e) Providing general control/safety functions.
    (f) Motor brake control.
3. *Safety gear.* This is a mechanical safety device that is positioned underneath the car and works in conjunction with the governor. It is activated if the speed of the lift exceeds the contract lift speed by a fixed percentage set at the commissioning stages of the lift. The function of the safety gear is to bring the lift to a halt in over-speed conditions.
4. *Counterweight.* This is a counterbalance weight that is approximately the weight of the lift car plus 40 to 50 per cent of the lift carrying capacity. Long-term studies of people traffic in lift cars have shown that the car on

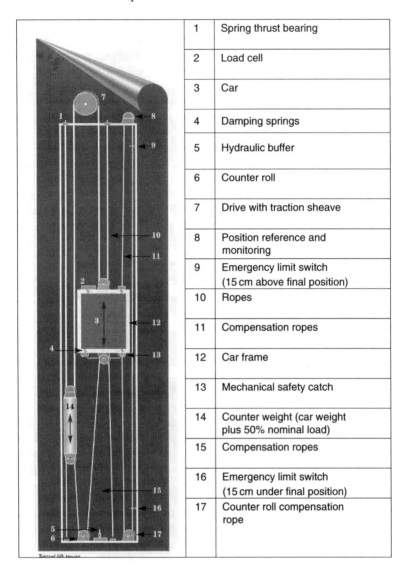

1	Spring thrust bearing
2	Load cell
3	Car
4	Damping springs
5	Hydraulic buffer
6	Counter roll
7	Drive with traction sheave
8	Position reference and monitoring
9	Emergency limit switch (15 cm above final position)
10	Ropes
11	Compensation ropes
12	Car frame
13	Mechanical safety catch
14	Counter weight (car weight plus 50% nominal load)
15	Compensation ropes
16	Emergency limit switch (15 cm under final position)
17	Counter roll compensation rope

Figure C6.23   Key components of an elevator system

average is mostly 40 to 50 per cent full on most of the lift journeys. With this factor in mind, from the above it can be seen that having a counterweight reduces the amount of power required to run the lift.

5. *Governor.* This is a pulley that links via a rope to the lift car and runs at the same speed as the lift. The pulley has a configuration of masses on board that move out in proportion to the lift speed due to centrifugal force. If the lift should overspeed by a fixed percentage, the masses come out enough to trip the safety gear.

6. *Buffers.* These are hydraulically filled rams that act as dampers to the lift and counterweight should the lift reach the extreme limits of the shaft. There is a buffer for the counterweight and one for the lift car.
7. *Positional encoder.* This sends positional information back to the lift controller.
8. *Floor sensors.* These are sensors in the shaft that show the true floor level.
9. *Motor/brake.*
10. *Shaft peripherals.* Shaft limits are positioned at the extremes of the lift shaft as electrical safety checks to the controller.
11. *Selector.* This shows the lift floor position. The selector increments and decrements as the lift passes the floor sensors in the lift shaft.

The lift controller's main function is to generate optimum speed profiles for every lift journey with minimum floor-to-floor times, floor level accuracy typically better than 3 mm and, most importantly, passenger comfort.

C6.6.2 Speed profile generation

As mentioned previously, the speed profile generation is normally done in the lift controller but can be incorporated within the drive if an applications module is available. Different profiles are generated dependent on different floor distances travelled, as shown in Figure C6.24.

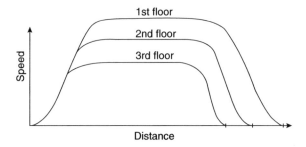

Figure C6.24   Lift speed profiles for different journeys

Jerk is the rate of change of acceleration and can be optimised in the profile on setup to give smooth take-off of the lift from floor level. This is a very important factor in lift control. The different jerk acceleration rates can be modified on the run by the controller, or the lift variables can be entered as fixed values on less complex systems.

Traditionally, the profile was slightly modified from the above with the final sections of the slowdown dropping to a slower speed called the creep speed before aiming into the floor position (Figure C6.25).

Sensors in the lift shaft indicate to the controller when to start slowing down and also when to start different sections of the profile. The trend is towards direct to floor positioning and this is now more widely used in new elevator systems. The system works on real-time pattern generation using third-order positioning algorithms.

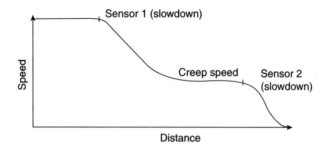

Figure C6.25   Typical velocity profile

This allows optimum profiles to be generated depending on the shaft distance to be covered, giving optimum flight times and passenger comfort.

C6.6.3   Load weighing devices

In some lift applications load cells are used in the control system. This helps in removing drop-back when the lift brake and then lift patterns are released. It does this by adding a torque feed-forward term into the torque loop of the drive that is proportional to the load in the car. The motor is thus pre-excited with the torque before the brake is released, thus improving zero speed holding. However, with the improved speed/current loops in today's drives this is becoming a thing of the past.

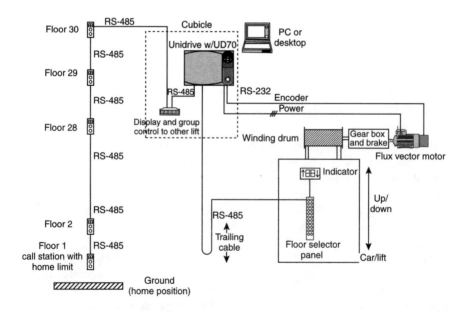

Figure C6.26   Simplified block diagram of a typical lift electrical system

C6.6.4 Block diagram of lift electrical system

A typical lift system is shown in Figure C6.26. In this example the elevator control is undertaken within an intelligent drive, and communication between the elements made with a Fieldbus.

C6.7 Metals and metal forming

C6.7.1 Introduction

The metals industry has been a traditional user of variable-speed drives from the very earliest days, starting with Ward–Leonard systems in their various guises, initially open loop then gradually progressing through the various developments in closed-loop control until today where high-performance solid state drives using both a.c. and d.c. motor and drive technology are commonplace.

The industry incorporates initial metal production through to conversion and fabrication uses drives in all their formats, high-power reversing drives for hot and cold rolling, coiling and uncoiling, slitting, levelling, cut to length, and servo drives to control the X–Y axis of various types of cutting equipment.

The processing of metal is a classical sectional process, with closely coordinated speeds, tension control and so on. Winders and un-winders are common.

C6.7.2 Steel

The rapid expansion of the world economy has seen an increasing use of reinforced-concrete building techniques, with a resultant increase in the demand for rolled-steel reinforcing bar. This is often produced from recycled steel, which is cast into ingots and then run through a multi-stand hot rolling mill, eventually producing long lengths of small-diameter bar.

A typical installation will comprise perhaps 20 or so stands, each driven by its own motor, which will form part of a co-ordinated line, the speed of each stand being increased as the diameter of the bar is reduced. Features such as tensionless rolling and looper control are a standard requirement of drive systems applied to this type of application.

Inter-stand tension in a rolling mill can occur due to a mismatch between the surface speeds of adjacent stands, temperature differentials along the bar, and possibly incorrect screw down settings, resulting in an accumulation or deficit of material between the stands. Fluctuations in inter-stand tension can result in variations in bar cross-sectional area, resulting in dimensional variations in the final product. Additionally, variations in inter-stand tension will increase the probability of inter-stand stock buckling and cobbling caused by compression (negative tension) between adjacent stands.

Drive designers are expected to provide solutions to these potential problems. Where the gauge of the material is such that the material is inflexible, one solution is to use current comparison as the bar enters each stand. Stand speeds are trimmed to revert to the measured parameters that existed before the tail entered the nip, and hold those values through the length of the bar.

As the material passes through the mill and becomes flexible a similar effect can be achieved by adjusting inter-stand speed ratios to control the looping device to a constant height.

A typical hot mill application is described in Section C6.7.2.1.

C6.7.2.1 Main mill drives

The main mill motors (Figure C6.27) utilise d.c. drives rated at 1 850 A d.c., 600 V supplied from 12-pulse converters. Localised control boxes enable operators to run/jog reverse/stop the stands and to lock off the power if work needs to be done on the stands themselves. For safety reasons, the mill cannot be started from the main pulpit, but the 'cascaded speed control' is carried out from this position, providing the operators with adjustment over inter-stand speed ratios based upon the required reduction factors. The speed reference cascading system will allow individual stands to be bypassed if not in service.

The menus for the range of products intended to be rolled are stored in a PC located in the main pulpit. When a particular product is selected for rolling, the information concerning the cross-sectional area of the billet and the mean roll diameter of each stand are downloaded to a PLC.

The control system is based on the principle of constant-volume rolling, where the speed of each stand is determined by the speed and reduction factor of the following stand, the overall speed of the line is dictated by the desired finishing speed set into the master finishing stand.

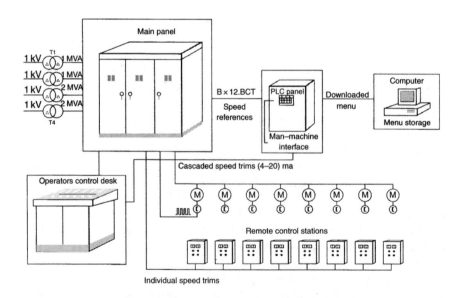

Figure C6.27 Layout of the main mill d.c. drives on a combination merchant bar and rod mill

Calculations concerning the motor speed reference for each drive are carried out continuously in the PLC and transmitted to the drives, where the linear speed references are converted to rotational speeds calculated from knowledge of the effective roll diameter and the gear ratios. All stand speeds are simultaneously updated, thus avoiding tension or compression occurring between the stands during speed changes.

If the speed of any section is adjusted by an operator using the desk-mounted 'cascaded speed trim' controls, the PLC will calculate the new speed of all upstream drives and simultaneously update them. Individual local speed trims for the drives are also provided, but their effect will not form part of the total cascaded reference. The speed of the master finishing stand will be set from the main control pulpit as it relates to the complete system.

C6.7.2.2 Auxiliary drives

The principal auxiliary drives are shown in Figure C6.28. The roughing mills are both 750 kW a.c. induction motors running at constant speed. The conveyor systems for the roughing mills are equipped with d.c. motors and are controlled by d.c. drives. These conveyors are manually controlled via joysticks, and are required to accelerate rapidly up to speed to guarantee entry of the bar into the roughers. The pinch roll drive at the entry to stand 2 and the exit from the disc shear are both d.c. drives running at bar speed.

The disc shear, after the final stand cuts the bar into preset lengths that are suitable for handling on the cooling bed and also to avoid wastage, will be multiples of the final lengths to be cut at the cold shear. The speed of the bar leaving the finishing stand, and its instantaneous position, is determined by measuring the elapsed time as the nose of the bar passes two hot metal detectors located at the finishing stand and set a known

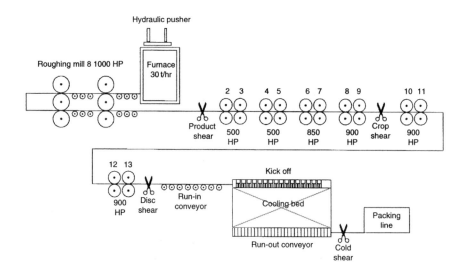

Figure C6.28   Overall mill layout from reheat furnace to packing line

distance apart at a known distance apart, situated after the finishing stand. Using this information, the sequencing of the disc shear can be controlled to cut exactly the correct length required for the cooling bed.

The disc shear itself is driven by three d.c. drives, one for each disc blade and one to orientate the shear mechanism itself. The run-in conveyor to the cooling bed is driven by 32 d.c. motors connected in parallel across a single d.c. drive.

The kick-off shaft is driven by five 28 kW d.c. motors distributed along the shaft and controlled by five four-quadrant drives arranged to share the load. This provides enough redundancy to allow the kick off to continue to operate with only three drives in the event of failure. Slow down and stopping is by regenerative braking with a holding brake to ensure that the system remains stationary when stopped. The slow down and stop position are determined by shaft-mounted CAM switches.

The 80-m-long cooling bed operates in a similar manner to the kick-off. It is driven by five 37 kW d.c. motors controlled from five four-quadrant d.c. drives that accurately control the slow down and stopping.

Shuffle bars, driven by five 12 kW d.c. motors controlled by single-quadrant d.c. drives, carry the bars from the cooling bed to the run-out conveyor. This in turn is driven by three 24 kW drives operated manually from a joystick controller at the cold shear desk. Three more d.c. conveyor drives carry the cut bars down to the packing line, which is driven by four open-loop inverter drives and controlled by a PLC.

The cold shear is a brake/clutch unit with a flywheel driven by a 37 kW a.c. induction motor. Three more d.c. conveyor drives carry the cut bars down to the packing line, which is driven by four open-loop a.c. drives and controlled by a PLC mounted in the control desk. In total, there are 45 d.c. drives and 6 a.c. drives controlling more than 100 motors and the mill.

C6.7.2.3 Strip rolling mills

The object of the initial rolling operation is to convert ingots into usable formats, including beams, sections, plate or strip. The production of beams and sections requires profiled rolls to shape the metal into the required shape.

Flat material will initially be in the form of plate, but further rolling can convert this to strip. So initial rolling will reduce the gauge until the material can be wound into a coil, at which point it becomes transportable to other equipment for additional rolling.

In the case of strip handling mills a separate de-coiler will usually be provided on the ingoing side of the mill to feed this wound coil for the first pass. Material passes from the de-coiler or un-coiler through the mill and is reduced in gauge. It is then passed to the coiler for rewinding (Figure C6.29). At the end of this first pass the expiring tail is transferred to the un-coiler and the de-coiler can be made ready to receive the next coil. The process continues, passing the material back and forth through the mill using the coiler and un-coiler until the required gauge is achieved.

The coiler drive systems require all the normal functionality of a centre driven winder, with diameter measurement, tension torque control, loss compensation and

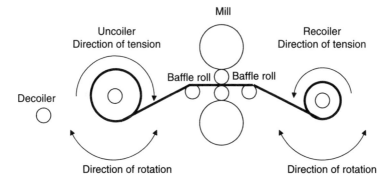

Figure C6.29   General arrangement of reversing rolling mill

inertia compensation. Systems handling larger-gauge materials may also require compensation for material bending moment.

The operators will be provided with controls for

- thread,
- run,
- normal stop,
- fast stop,
- hold on speed changes,
- emergency stop.

These controls are illustrated in Figure C6.30.

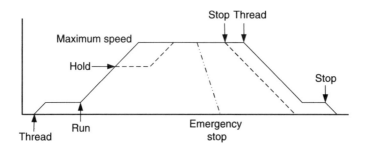

Figure C6.30   Mill speed time diagram to demonstrate control functions

A typical rolling mill drive control system is shown in Figure C6.31. The mill drives often extend the speed range of the motors by running into the constant power region as the gauge reduces and hence the rolling torque reduces. Large systems may use tandem motors for both the mill and coiler drives. Two smaller motors are often more economical than single large machines and lend themselves to quasi 12-pulse or 24-pulse operation if a d.c.-motor-based system is used. Where wide tension

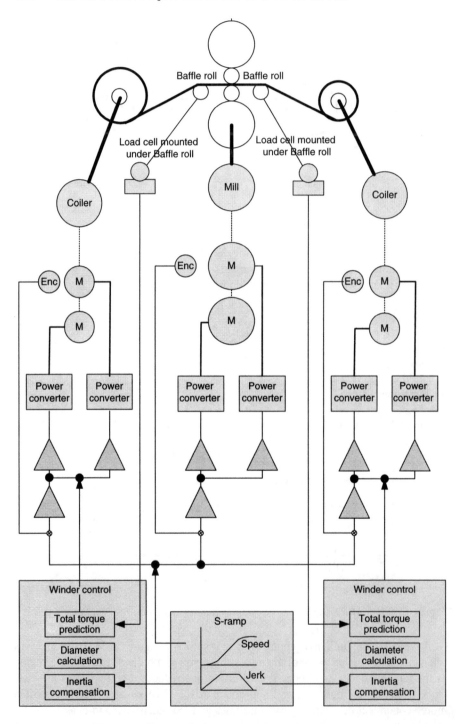

Figure C6.31 Basic rolling mill drive control scheme

ranges are to be handled by the coilers, the use of tandem motors can be used to advantage to reduce the torque regulating range by using a single motor when running at lower tensions.

Most mill systems are equipped with automatic slow down and stop on diameter. This feature continuously checks the change in diameter of the un-winding coil against the time it would take to stop and applies the stop signal when the reducing diameter reaches that value. If the tail end of the material was to run completely out at high speed there is a considerable risk of damage to the equipment and potential loss of production. Inclusion of this feature removes the burden from the operator.

Another common feature is automatic gauge control (AGC), where adjustments are made to ensure that the rolled material exits the mill at the correct thickness or elongation. AGC is a complex process involving many variables. The mill roll conditions are monitored by both position and pressure transducers. Other variables are coiler tension and mill speed. An AGC system normally consists of an outer loop based upon thickness measurement or elongation, which provides the set point for the mill roll position or pressure control systems. The AGC systems may influence the action of the drive system but do not form part of it and this is therefore beyond the scope of this book.

The following are among the many parameters required when setting up the control system.

1. For the mill drive which also serves as the line master:
   - S-ramp acceleration rate including normal, fast and emergency stopping,
   - corresponding S-ramp jerk limits,
   - thread speed,
   - maximum speed.
2. For the coiler drives:
   - stall and run tension set points,
   - minimum and maximum diameter,
   - starting diameter,
   - strip width,
   - strip density,
   - strip gauge,
   - static and dynamic friction coefficients,
   - fixed inertia components, for example values for the motor and mandrel;
   - pay-out and take-up speeds.

C6.7.2.4 Continuous casting

The continuous casting process is used in the steel and non-ferrous metal industries to produce metal sections with a strictly controlled cross-section. A typical continuous caster is shown in Figure C6.32. Molten metal is usually fed from a double-ladle system into a small reservoir, the reservoir providing a buffer store to keep production flowing during ladle changes. The metal passes from the reservoir into a mould. The outer surface cools and solidifies but the inner core remains molten, forming a 'strand'.

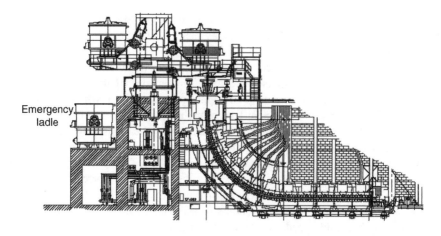

Figure C6.32   Continuous caster (side elevation)

This strand is supported by closely spaced, water-cooled rollers, providing support for the outer walls of the strand against the ferrostatic pressure of the still molten liquid within the strand. As the section passes between these supporting rollers it is water cooled to assist with solidification.

The strand is subsequently cut into suitable lengths and passed to the next process, which is usually hot rolling. The process can be used to produce flat plate and strip, sections for beams or even the stock for wire drawing. From a drive point of view it is imperative that the supporting rollers share the load equally and maintain the

Figure C6.33   View of the roller drive control panel

correct set speed. Any errors in this area can lead to break-out of the strand and considerable costs in machine down time. The rollers are driven by motors in the range 7.5–10 kW, each separately controlled from their own inverter. A typical drive panel for a continuous caster is shown in Figure C6.33.

Owing to the continuity of the process the drives and drive control system must be designed for high reliability with schemes in place to allow continued production in the event that one drive should fail, including a standby main controller (Figure C6.34).

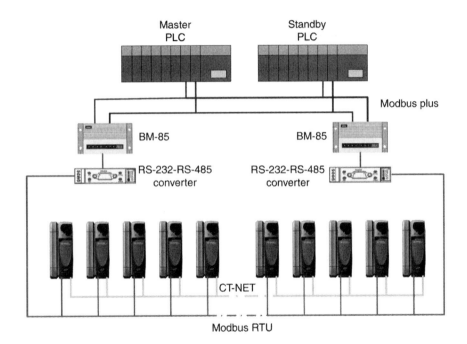

Figure C6.34   Typical control scheme with redundancy

All drives are continuously in contact with the master and standby controllers. One drive must act as the master to determine speed and total load, but facilities must be in place to ensure that should this drive fail any of the others will immediately take its place, allowing the system to remain operational.

C6.7.3   Wire and cable manufacture

C6.7.3.1   Wire drawing machine

The wire-drawing process typically reduces the wire diameter from 12 to 1 mm using up to 14 tungsten carbide lined dies. As the wire diameter reduces, the speed of successive blocks' drives have to increase to handle the increased material length and to maintain a constant tension between each stage.

Between each block, there is a dancer arm, which takes up the slack wire. From each of these a non-contact proximity detector sends an analogue signal to the intelligent drive. Using a control program with a PID-type function, the drive speeds up or slows, with the objective of keeping the dancer arm in its central position.

In practice, the control is such that, after the start-up, during which the software works out the sizes of the dies in use, the dancer arms have minimal movement from the optimum central positions. The speed ratio between drives is daisy-chained from the master drive (the final block drive of the sequence) and each of the slave drives. This gives the optimum speed, which, if followed, prevents wire breakage, even during a fast stop.

With the dancer arms running in central positions, more trim is available before reaching limit stops, so there is less disruption to production and far fewer incidents of wire breakages. At the first limit in either direction, the drive accelerates or stops ramping up and at either second limit, the machine comes to a controlled stop. Drives can be set to allow ramping up from standstill at a pre-set ratio, with the speed of the fastest drive being limited to allow slower drives on earlier blocks to ramp up in the correct ratio, without exceeding their current limits. Any drive approaching maximum speed can hold the speed of the other drives, thus preventing wire breakage due to incorrect drafts (die sizes). During threading of a new wire, each new block becomes the speed master, the master 'pilot-block' moving down the machine automatically.

Additional drives are installed for the spoolers, typically 30 kW, and rotors for the formers, typically 56 kW.

A PC with custom-written software, placed in the operator's desk, allows improved operator feedback and system monitoring. This shows him the normal operation mode, speed, current (torque) and dancer arm position of each block. The operator is able to trim each block between viable limits with real feedback.

Figure C6.35 Wire drawing machine [photograph courtesy of Bridon International]

Other key aspects include system status icons and a time-to-run calculator. Maintenance is incorporated into the system, with user explicit alarm captions and a 12-monthly logging system. Further maintenance features allow an automatic 'gateway' for engineers to use software tools to monitor the drives and the system's performance while the line is in production. These tools enable the operator to quickly identify faults using on-board diagnostics such as 'guards up', 'dancers back', and so on.

A wire-drawing machine is shown in Figure C6.35.

C6.7.3.2 Twin carriage armourer

Many cable-making machines (Figure C6.36) combine materials from several sources, which often involves close positional synchronisation of several rotating parts at adjustable predetermined ratios, either twisting single strands to form a cable or applying the outer case, which often involves wrapping. These outer covers must be applied with a controlled degree of overlap. Typical of this type of application is the armouring process, where tape is drawn from bobbins mounted upon a rotating carriage. The carriage rotates around the cable to create the wrap and the speed ratio between the carriage and the forward motion of the cable determines the lay or degree of wrap with which the tape is applied. A typical machine layout is shown in Figure C6.37.

Figure C6.36   Twin carriage armouring machine [photograph courtesy of JDR Cables]

Traditionally these machines were driven by a single motor via a line shaft, and changes in pitch or lap were achieved by changing gear ratios. This restricted the possible lay settings to the number of gear settings available. Ranges could be extended using adjustable PIV gear boxes, but a major disadvantage of PIVs is that they can only be changed on the run, precluding the possibility of presetting a machine for a new run while it was still shut down. PIVs are costly devices that are easily damaged and can involve fairly high maintenance.

A sectional electric drive using electronic gear box technology offers the ideal solution (Figure C6.38). It can be set up to an almost infinite range of ratios with a very high resolution, typically to eight decimal places. Setup is quick, as menus for different products can be stored and downloaded from a screen-based master processor at the touch of a button, with the added benefit of removing the possibility of operator error.

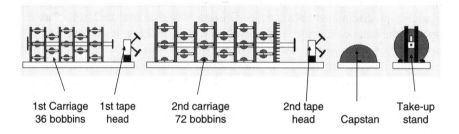

Figure C6.37   Typical armouring machine layout

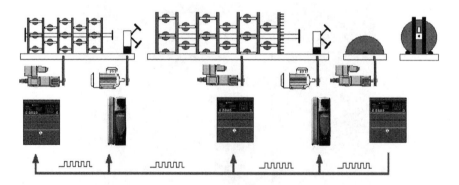

Figure C6.38   Sectional electronic implementation

C6.8   Paper making

C6.8.1   General

Since it was first invented by the Chinese, paper has always been made in essentially the same way. Paper is manufactured by pulping wood or similar cellulose material with water to produce a stock. In the simplest process, this stock is laid on a gauze base to allow the water to drain. More water is then extracted by squeezing, and finally the paper is dried by evaporation.

Paper was originally made sheet by sheet, but a modern paper-making machine combines all these processes to produce paper at continuous speeds of up to 1 000 or 2 000 m min^{-1} with web widths up to 5 m or more. These machines often run under computer supervision with many interactive control loops to achieve the required production schedules and to maintain consistent quality. A very important part of such a machine is its electrical drive system, which may consist of up to a dozen main motors and many smaller helper motors, with a total installed capacity of several hundred kilowatts.

C6.8.2 Sectional drives

The speed of each section of the machine must be accurately controlled to maintain the correct inter-section speed differentials or 'draws', to cater for the changes in characteristic of the product as it makes its way from the wet end of the process, through the press section and drying sections, to be reeled at the end of the machine ready for the next process.

The shunt wound d.c. motor has been the traditional choice for driving paper machines. It has good speed-holding characteristics, a high torque availability at very low speeds, essential when starting high-inertia sections of the machine. When combined with a modern thyristor power converter it provides an efficient and reliable means of obtaining the degree of control and response that is essential to the paper-making process.

Today the a.c. drive has almost completely replaced the d.c. system. They provide the benefits of lower-motor costs and low maintenance and are now the preferred choice for new machines, although d.c. controllers still have a good market for upgrades of existing machines already utilising d.c. motors. Both provide an efficient and reliable means of obtaining the degree of control and response that is essential to the paper-making process.

Early machines were fitted with a line shaft driven by a large d.c. shunt motor. Cone pulleys were provided to adjust intersection speeds, and clutches allowed sections to be stopped and started as required. Modern paper machine drive systems are sectional, a motor and controller being provided for each section of the machine, as shown schematically in Figure C6.39. The following are typical motor sizes for the medium-sized machine making paper at 300 m min^{-1} and a width of 3 m:

Wire section	100 kW
First press	50 kW
Second press	50 kW
First dryer	50 kW
Second dryer	50 kW
Third dryer	50 kW
Calender	100 kW
Reel	50 kW

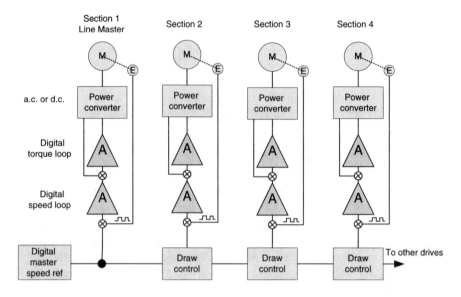

Figure C6.39   Sectional paper machine drive layout

This makes a total installed capacity of 500 kW.

Selecting the motor sizes for a sectional paper machine drive involves checking the normal running load (NRL) and recommended drive capacity (RDC) calculated for each section and usually rationalising around two or three frame sizes of motor to reduce the number of spare parts to be carried. Each motor drives its respective section via a speed-reducing gearbox, which is carefully chosen with the arduous 24 h 7-day operating requirements in mind. For smoothness of transmission, helical gears are usually specified. Often, a restricted range of different gearbox sizes is used with interchangeable wheels and pinions, again to reduce the cost of spare parts stock.

Each section motor is fitted with a high-accuracy, temperature-compensated tacho-generator or high-grade digital encoder to provide accurate speed feedback to the section power controller.

The motors must operate under very adverse conditions: very wet at the wire section, and hot and humid at the dryers. D.C. motors, where used, are therefore often ventilated by a clean air supply ducted from outside the machine room, air flow switches normally being provided to monitor the flow of air through each motor and to shut the equipment down if the air supply fails. In contrast, because of their inherently high degree of protection, a.c. motors can be used without any special provisions.

C6.8.3 Loads and load sharing

The maximum power that may be transmitted into a section (i.e. to drive the felt or the wire) by any particular roll is limited by (a) the arc of wrap of the medium around the roll and (b) the coefficient of friction between the roll and the driven medium. Where this power is less than the total power required to drive the section, additional power is provided by helper motors driving other rolls in the section.

Typical examples of such arrangements are the suction couch roll driven by a helper motor to assist the wire turning roll motor as shown in Figure C6.40, or the multiple drive points of a machine felt section. These helper motors are usually tied mechanically to the main section speed-controlled motor by the wire or felt that passes around the section.

Because a paper machine operates over a very restricted speed range, many older systems relied upon parallel operation of helper motors supplied from a single power source. However, modern practice is to provide each motor with its own drive. The helper drive is arranged to operate in some form of torque control,

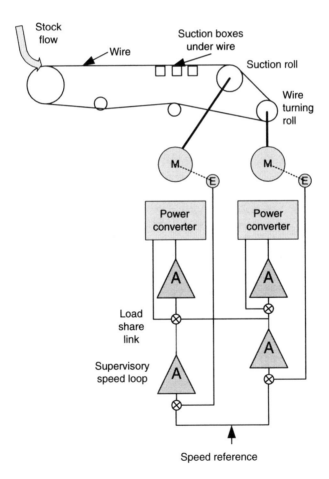

Figure C6.40 Section motor drives wire-turning roll has encoder fitted for speed control; section couch roll driven by helper motor under current control

usually with a speed override function provided by its own supervisory speed loop as shown in Figure C6.40.

Although a paper machine operates over a very narrow speed range, probably never more than 3:1, low speeds are required for fitting and running in the wire and felts and when cleaning. Unlike other types of process machine, the paper is threaded through the machine at normal paper-making speed, each section being brought in as the web finds its way through the machine. The stock is fed onto the wire and recycled through large 'broke pits' situated below the wire section until its consistency is correct. A thin tail of paper is then picked off the wire and threaded through the machine section by section. When the paper tail reaches the reel section and is running between the reel drum and the empty reel shell its width is gradually increased at the wire until a full width is passing right through the machine. At this stage the paper is wrapped around the reel shell and reeling commences. While the threading process is going on, the section motor speeds must be held constant in spite of rapid load changes as the paper enters the various sections of the machine and as its width increases. Any slight variation in speed at this time will cause the fragile web of paper to break.

The paper reel is a surface-driven system (Figure C6.41). The reel shell with the paper winding on it is pressed against the reel drum. When the reel is complete a new shell is lowered on top of the reel drum and the paper is transferred from the filled shell to the new shell, allowing paper to be reeled continuously. The reel section drive normally operates in speed control while threading and during reel changes, but tension control possibly using a load cell is normally used while reeling.

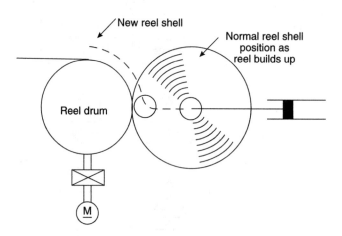

Figure C6.41 Arrangement of reel section

C6.8.4 Control and instrumentation

The basic elements of a paper machine sectional drive control system normally comprise a set of a.c. or d.c. drives, each housed in its own cabinet and supplied through individual main isolators, motor contactors, excitation and control circuits to provide inch, crawl and run functions. Each section has its own stop, crawl and run operator interface. Electrical power is normally distributed to the cabinets by an enclosed three-phase a.c. bus bar system for d.c. drives or a d.c. system fed from a common rectifier for a.c. drives.

The traditional instrumentation and controls for the operator are now commonly integrated on touch-screens, which allow customisation for touch inputs, plant displays, 'push-button' machine adjustment etc., all incorporated in one small box. Serial communications to drives cut down on both wiring and installation costs and simplify maintenance.

As the paper travels along the machine, passing through various pressing and drying sections, it undergoes changes in length. The section speeds must be adjusted to allow for this.

The inter-section speed differentials thus created (draw) may vary between 0 and ± 5 per cent of maximum speed. All the drive sections receive a speed reference signal from a master speed reference unit. This signal is generated as a digital value and is transmitted to the drives over a high-speed local area network.

Individual section speeds are trimmed with respect to the master reference to obtain the required draws. Normally, one section is tied solidly to the master reference with no draw adjustment. Draws may be individually and separately adjusted, or cascade adjustment may be provided in which an adjustment to one section is automatically passed on to all succeeding sections, so removing the necessity for the operator to reset all draws following an adjustment at one section.

A slack take-up system operated from a push-button on each section is normally provided, for all except the master section, to allow the operators to pull out any slack in the tail as it is fed from section to section while threading up.

Arranging for all the drives to follow a single master reference signal ensures that no velocity lags occur between sections, as would be the case if each drive took its reference from the preceding section; it also ensures that slight instability on one section is not transferred to succeeding sections. The master section is normally the wire section, as this is the first section, and all draws may be expressed with reference to it. This arrangement is particularly useful where cascaded draw systems are required, as all draws can then be considered to be positive.

The rate of change of the master reference is usually chosen to be very slow, thus limiting the rate at which changes in overall machine speed may be made. This ensures good speed tracking between sections that may have very different torque-to-inertia ratios. It also ensures that the process control side of the machine can keep up with machine speed changes and is able to maintain the correct flow of stock to the wire section and the correct amount of steam feed to the dryer section.

Individual section drives are completely independent as regards starting and stopping. The rates of acceleration applied to the various sections, when running up to the pre-set master reference speed, should be limited according to the maximum power that can be mechanically transmitted into the section. The couch roll and forward drive rolls, which are the main drive rolls on the wire section, must not be allowed to accelerate so rapidly that they lose traction and slip inside the wire.

Two possible solutions are used here. The first is to accelerate under reduced torque limit and switch to a higher value once running speed is attained. Alternatively, it is possible to use a ramp generator to control the rate of acceleration of the section up to running speed. This is then disabled to avoid the introduction of any additional velocity lag during overall changes in machine speed.

Some sections have large inertia-to-torque ratios, the dryers in particular, which consist of a dozen or more 1.6 m (5 ft) cylinders to each section. Accelerating these up to speed from start requires substantial overload capacity: at least 2 to 2.5 times full-load torque should be catered for. Older machines may also have plain bearings, requiring break-away torque of up to four or five times normal running torque. These factors should always be given careful consideration when planning to update an old system, particularly if the original drive was a Ward–Leonard system with essentially no limit on the current available.

With the advent of microprocessor-controlled drive systems, paper machine drives have become more sophisticated. Modern drives allow many control possibilities to be explored, such as speed and draw menu storage and immediate setup by product code. When used in conjunction with a modern paper-machine process control computer, such programmes allow very quick and efficient changes in product grade, and improved start-up times with less time required for the machine to settle in to the new grade. Management reporting on drive performance can now be provided easily via serial communications. Assistance to the electrical maintenance departments in fault location and correction can be provided by recording and storing, in the drive memory, the historical trends of various important drive parameters prior to a shut-down condition.

C6.8.5 Winder drives

Once the paper has been wound into reels on the paper machine, it must be reduced into manageable sizes of rolls. This function is carried out on a slitter rewinder, which converts the large machine roll into several smaller diameter and reduced width rolls. This normally consists of an unwind stand (into which the reel is loaded as it comes from the paper machine), a set of driven slitter knives, and a rewind system.

The rewind usually consists of two drums, mounted side by side, upon which the slit and rewound roll of paper sits. The distance between the two drum centres is only slightly greater than their individual diameters, so ensuring that the rewound roll is supported by both drums. A rider roll, which sits on top of the rewound roll is arranged to rise in a vertical slide as the diameter of the rewound roll, increases. The pay-off stand is controlled by a braking generator and is centre driven, whereas the rewind drums and rider roll provide a surface drive to the rewound roll.

Each motor is provided with its own a.c. or d.c. power controller. The pay-off normally runs in tension control and the two drums run in a master and slave configuration in speed control, the second drum load-sharing with the first. The rider roll also runs under torque control, its control strategy depending upon the particular winder type. A two-drum winder of this type will wind paper at speeds up to

2 000 m min^{-1} and handle reels from the paper machine up to 2.5 m diameter and 5 or 6 m wide. The rewound rolls will be from 1 to 1.5 m in diameter, necessitating several stops in the course of rewinding one parent reel. The winder is therefore designed to run at two or three times the speed of the paper machine in order for it to keep up with the production rate. An installation of this type is a very large capital investment and it is not acceptable for the paper machine to be held back by under-capacity of the winder.

Control of the rewind drum motors is fairly straightforward. The front drum (the one nearest the unwind stand) runs under speed control and sets the speed of the rewinding process. The rear drum runs under torque control, its load being determined as an adjustable percentage of the front drum load; a load-trimming adjustment is provided for the operator to make adjustments. The torque differential so set between the two drums is used to control the tension, which is wound into the roll and therefore determines its hardness. Most modern winders employ profiling of this hardness by automatic adjustment of the load sharing in relation to the rewound roll diameter. This is easily measured by a transducer attached to the rider roll. Additional hardness control may also be provided by adjusting the rider roll torque. Both these current-controlled motors normally have some form of speed override to prevent excessive over-speed when they are not in contact with the rewound roll.

The slitter knives consist of rotating discs with ground faces. They can be moved laterally across the machine to provide different widths and are usually individually motorised and provided with plugs and multi-position sockets to allow selection to suit the job on the winder at the time. The slitter knives must run at a speed slightly in excess of the paper speed to provide a cut. But if this speed difference is too great, excessive knife wear will occur; if too small, the knives will not cut. The slitter motors are normally supplied from a common a.c. drive, which is referenced from the drum drive but given a fixed positive bias to ensure the correct positive differential in speed over the paper.

The unwind brake generator at the pay-off operates as a constant-power braking system to maintain constant tension in the paper fed from the unwind. Constant tension in this part of the machine is essential to ensure correct operation of the slitter knives. Any variation will cause interleaving of the slit drums at the rewind.

The unwind brake generator will normally be required to control tension over a diameter range of about 4:1 or 5:1 and also deal with a range of sheet tensions, which depends on the variety of grades to be handled. A typical tension range may be 5:1, resulting in a torque regulating range of 25:1.

C6.8.6 Brake generator power and energy

The size of the brake generator is relatively simple to decide: it must provide only the tension power required by the paper. This may be calculated from the winding speed and maximum permissible sheet tension:

$$\text{Tension power (kW)} = \frac{\text{total tension (N)} \times \text{linear speed (m min}^{-1})}{60\,000}$$

Calculations should also be made to check on the overload capacity required to accelerate and decelerate the unwind with a large reel of paper present and, if necessary, the brake generator should be increased in power to cater for this. This is particularly important where low tensions are to be used with materials such as newsprint, where the acceleration power may be substantial in relation to the tension power requirement.

Acceleration and deceleration torques depend upon the inertia of the rotating system. Examination of the dynamics of the unwind show that the inertia consists of two components, a fixed inertia related to the motor, shafting and reel shell upon which the paper is wound, and a variable component related to the diameter of the unwinding roll:

$$\text{Paper inertia} = \frac{(OD^4 - ID^4) \times \text{width} \times \text{density} \times \pi}{32}$$

where OD and ID are the outer and inner diameters, respectively. Obviously fixed inertia values can be obtained from the equipment suppliers, or estimated using standard formulae.

It is common practice for the motors on a winder of this type to be arranged for direct drive. This is of particular importance on the unwind, as it reduces frictional losses in the system to a minimum and makes control of sheet tension more accurate.

Typical parameters for a medium-sized winder are as follows:

Speed	2 000 m min^{-1}
Parent reel diameter	1.7 m
Reel shell diameter	0.33 m
Paper sheet width	4.25 m
Sheet tension	9 g mm^{-1} width

This gives a parent reel weight of about 6 000 kg and a total sheet tension of 382 kg, resulting in a tension power of 93 kW to be absorbed by the brake generator. The variation in stored energy of the expiring parent reel and the reel shell as the diameter reduces may be seen from Table C6.1.

This clearly shows the effect of the fixed-inertia component provided by the reel shell and brake generator armature, and the varying inertia component of the parent reel as it is unwound and reduces in weight and diameter. The total stored energy reduces as the reel diameter reduces, but increases at lower diameters as the rotational speed increases.

While accelerating and decelerating, the torque in the braking generator must be adjusted to compensate for the stored energy values outlined. This is a technique known as inertia compensation. The actual change in torque demand must be predicted, bearing in mind the total stored energy of the unwind and field flux level of the motor at the operating diameter and the rate of change in speed.

Table C6.1 Comparison of stored energy

Diameter (m)	Reel (kW s)	Shell (kW s)	Total (kW s)
1.52	366	20	386
1.27	241	29	270
1.00	68	46	114
0.76	58	84	141
0.60	0	180	180

Figures C6.42 and C6.43 show the relationship between inertia compensation torque-producing current and diameter for the values listed in Table C6.1, taking the effect of motor excitation into account.

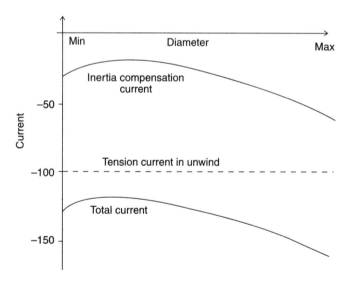

Figure C6.42 Effect of inertia compensation on unwind brake generator: decelerating

C6.8.7 Unwind brake generator control

The drive control system for an unwind brake generator is quite complex. The system must be capable of operating under speed control during threading of the winder and under tension control once the machine has been threaded with paper and put into the normal tension control operating condition.

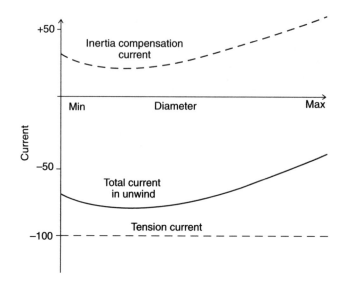

Figure C6.43 Effect of inertia compensation on unwind brake generator: accelerating

C6.8.8 Coating machines

As the name implies, these machines coat the paper. They are used to impart a special finish that cannot be done on the paper machine itself. Coated paper products include high-gloss magazine papers, NCR paper and other such products. A coating machine consists of an unwind stand, often arranged to handle two reels, supporting them on turrets and rotating them into position to perform a splice as the old reel expires, thereby maintaining continuous processing. Nip rolls pull the paper from the unwind. There are usually one or two coating heads, depending on whether both sides of the paper are to be coated, together with their respective drying sections, which may be ovens or heated cylinders or a combination of both. Hold-back rolls control the paper before it enters the rewinder, which, again, like the unwind, may be arranged for continuous operation with rotating turrets and reel changing on the run.

As the paper passes through the machine it undergoes considerable changes in moisture content and hence the tension levels between the various sections vary considerably. Dancer rolls or load cells are usually fitted between important sections to allow the state of the paper to be controlled more readily.

Each section is provided with its own motor, which must be accurately controlled to maintain correct speed or tension. The powers involved in a coating machine are considerably less than those on a paper-making machine or winder, and the control problems are somewhat different.

The pay-off must control sheet tension accurately through a 4:1 or 5:1 diameter range, but it must also accelerate a new reel up to match the speed of the expiring reel to enable a flying splice to be successfully completed. Once the splice has been made the pay-off must smoothly and quickly transfer to tension control, which is maintained until the reel is finished, at which point it must bring the empty reel shell quickly to rest after the next splice transfers control back to the other payoff.

The speed-controlled sections in the machine must track speed accurately throughout the full-speed range of the machine, which may be 50 or 100 to one. Trims from dancer rolls or load cells must also be incorporated into some of the section speed controllers. These tension control loops may have gain-adjusting circuits to increase their effect as the machine speed increases.

The twin rewinds are normally centre driven, each with its own motor and, like the pay-offs, must accelerate a new reel core up to speed in readiness for a changeover at speed when the previous reel becomes full, changing to tension control once the splice is complete.

Some turret unwinds and rewinds have complex mechanical power transmission systems involving concentric shafts. Considerable attention must be paid to providing adequate compensation for frictional losses within the tension control systems. Both static and dynamic friction compensation circuits should be incorporated into the controller together with inertia compensation. Also, the paper surface, once coated, becomes very smooth. This results in telescoping of the rewound reels unless the tension is carefully monitored and controlled. Usually, coated papers require some degree of taper tension to prevent telescoping and the rewind tension control system must invariably include this feature. The tension taper control reduces the web tension as the diameter increases by some pre-set ratio under control of the operator. As with a paper machine, each section is provided with its own motor and drive, which all follow a master speed reference signal and are controlled from the master line run and stop controls.

The machine is threaded with paper at crawl speed and the unwind and rewind set to tension control. The machine is then accelerated up to operational speed and the drive system must maintain control over the paper throughout this process.

C6.9 Plastics extrusion

C6.9.1 General

The concept of a single-screw extruder as a pump of tubular design has been in existence for several thousand years. However, ram extruders to manufacture lead pipe were used in the early nineteenth century. Towards the end of the nineteenth century the demands of the rubber processing industry rationalised design to a machine of a type that formed the basis of modern extruders. The extrusion of PVC in 1925, and later polythene, marked the commencement of modern extrusion technology and by the late 1940s a machine of today's basic layout had arrived.

The thermoplastic extrusion industry experienced rapid growth in the 1950s, and since that time the development of new polymers, the development of new extrusion technology, and almost total penetration of many product areas has ensured an upward trend in the industry.

Extrusion processes involving the forcing of molten thermoplastic through a die to form a continuous product are used to make a wide range of products such as those shown in Table C6.2. Also, with the use of additional machinery, other products can be manufactured such as those shown in Table C6.3. Secondary products are manufactured by extrusion, such as packaging bags from film and thermoformed articles from sheet.

Table C6.2   Products made by extrusion

Film	Tubular or flat
Sheet	Rolls and cut sheet, laminated and co-extruded, embossed
Profiles	Window frames, gaskets
Pipe	Rigid pressure pipe, gas pipe

Table C6.3   Products made using additional machinery

Coated substrates	Paper, aluminium foil
Insulated materials	Wire and cable
Filaments and strapping	Uniaxially orientated for ropes, textiles, sacks

One of the main problems in extrusion is surging. This usually appears as a fluctuating extruder output, giving a variable product geometry. It is caused by inadequate mixing, melt instability, feed temperature effects, and solid bed break-up variations. The result is dimensional variations and take-off problems, such as drift.

Possible remedies are to run the extruder more slowly, increase the pressure, cool the screw, and lower the metering zone temperature to achieve a 'bite' limit. Bite limit is where the melt reaches the correct consistency, and is not too soft (when the screw would fail to drive the melt).

The process of extrusion has certain basic requirements:

1. The extruder should be capable of processing polymer at high and consistent output rates. To aid this, a drive with a high degree of torque/speed stability is required.
2. The polymer produced should be within an acceptable melt temperature range, and the temperature should not vary. A good-quality temperature controller with PID control will enhance temperature control.
3. The pressure developed in the extruder should be consistent. This requires a combination of good control of process variables and speed. The use of a process

controller and a carefully considered choice of variable-speed drive will aid consistent production quality.
4. The polymer should be sufficiently well mixed and not contain any low-temperature volatiles that would spoil product appearance. Mixing is assisted by the screw design, both with single screws and twin screws.

The optimising of screw designs has made lasting changes in extrusion technology. These developments have included many features from those listed in Tables C6.4 and C6.5.

Table C6.4   Single screw design features

Type	Characteristic feature
Rubber screw	Reducing root diameter, decreasing pitch
Plastic screw	Reducing root diameter, constant pitch
Metering screw	Zoned screw, with feed, compression and metering zones
Two-stage or double metering screw	Two metering screws connected by a decompression vent
Mixing and barrier sections	e.g. Pins and Madelock heads
Solid-channel/melt-channel screws	Various designs of barrier flight screws Grooved feed-sections

Table C6.5   Twin screw design features

Type	Characteristic feature
Non-intermeshed twin screws	
Intermeshed contra-rotating twin screws	
Contra-rotating twin screws, self-wiping type	
Modular twin screws	Interchangeable screw components for different formulations
Conical twin screws	Better feed of dry blend; better bearing support

Other machine and process technology developments have given large gains in output and improvements in cost/performance ratio. Such developments include improvements in the very critical area of compound formulation to hold polymers to specific tolerances.

There is a trend towards open modular design, enabling components to be changed without stripping down the machine, leading to greater operational efficiency and

Table C6.6 Typical extruder performance ranges

Nominal torque (kP m)	Screw diameter (mm)	Maximum screw speed (min^{-1})	Power drive (kW)	Dynamic load of thrust bearing (MPa)
80	45	105–322	7.5–18.5	47.5
160	45, 50, 60	84–280	11–30	47.5/18.5
220	60	110–278	23–45	81.5
350	60, 70, 80, 90	72–275	23–60	81.5/118
700	70, 80, 90, 100, 120	64–230	37–110	118/153
1 300	90, 100, 110, 120, 150	58–207	68–189	153/224
3 000	120, 140, 150, 160, 200	45–146	110–280	224/375
7 000	180, 200, 220, 250	32–100	230–600	585/810

maintenance economy. In addition to this, some machine manufacturers offer, for any given application, the option of selecting the torque, screw diameter, screw speed range, axial pressure and drive power most appropriate for their series of machines. Table C6.6 shows a typical range of figures.

C6.9.2 Basic extruder components

As with most industries, extrusion has its own technical terms, although different terms are not always consistently used in the industry. At the end of the screw is the adapter region and filter pack and the head and die region. Beyond this are the various take-off units, which may be devices such as accumulators, conveyors, cutters, winders or others not relevant to this discussion.

In some of the downstream equipment, variable-speed drives are used to match up to the extruder speed or, for example, to introduce a specific ratio required for the finished product from its extruded size to the finished size. Stretching and heating produces orientation of the polymers and makes the end-product much stronger.

Regardless of the physical design of the machine used, the extrusion process is broken down into four basic steps:

- solid feeding,
- melting,
- mixing,
- melt conveying.

These operations are identified by the zones of the screw (Figure C6.44). The feed zone is a zone of constant feed depth whose function is to preheat the bulk material before it is conveyed by screw rotation to the next zone. The temperature rise is produced by heaters around the barrel, and also by frictional heat generated by the

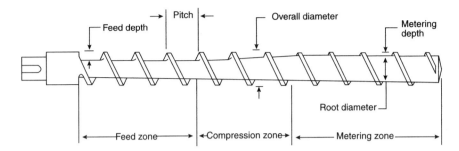

Figure C6.44   Principal zones of typical feed-screw

granules travelling along the barrel. The compression zone is the part of the screw where the root diameter is gradually increased while pitch and overall diameter remain constant, thus reducing the volume. This accelerates heat generation and therefore melting, and a change in the bulk density of the polymer also occurs; as more granules melt, any air trapped in the barrel is squeezed out through the hopper.

The melted polymer is transferred to the pumping or metering zone, which is shallow and of constant depth. This zone delivers the polymer melt at a uniform flow rate, composition and temperature.

The length of the metering section must be such that it allows it to achieve the melt while the compression zone develops sufficient pressure to pump the melt through the die at a constant output unaffected by small fluctuations in pressure. It is clear that the design of the screw fixes certain parameters at the design stage, but the operator can improve mixing and melting in several ways.

1. *Reduction of the screw speed.* This will reduce the channel length required for melting to be completed; the obvious drawback is reduced production rate.
2. *Increase of pressure at the screw exit.* This can be done by reducing die temperature, or by using a finer screen pack. Temperature within the polymer would increase.
3. *Use of screw cooling.* This delays the break-up of the polymer solid bed. Output is reduced, but this can usually be overcome by increasing the screw speed. This has some benefits in the melt in terms of mixing, but power consumption is increased.

C6.9.3   Overall extruder performance

For this to be assessed, some knowledge of the die characteristics is required. A die may be characterised by a pressure drop and a flow rate and therefore the output/pressure characteristics should be used with those of the screw to achieve an overall picture of the performance. The limits to output are summarised in Figure C6.45.

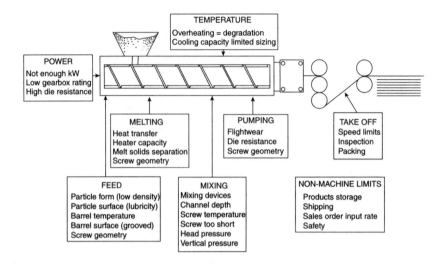

Figure C6.45   Factors imposing output limitations

C6.9.4 Energy considerations

Most operations of polymer processing involve the heating and cooling of solid or liquid polymers. Usually the net energy change from solid granules to the extrudate is very small compared with the energy that flows into and out of the polymer, as shown in Figure C6.46.

On the whole there is not much that can be done to affect the basic form of this diagram once extrusion is chosen as the means of process. Nevertheless, a number of points are noteworthy.

1. The cost of different forms of energy, in particular directed energy (electrical and mechanical), which is obtained from thermal energy, is relatively expensive. The mode of heating to be used needs to be chosen with care.
2. The capital cost of providing the needs of A (Figure C6.46), have to be matched by the cost of enabling the energy to be removed, B; i.e. cooling air or water and associated installations. Therefore a saving in A will lead to equipment and services saving in B also.
3. In extrusion, the rejected energy offers little potential for useful recovery, being largely to air or water.
4. Figure C6.46 shows that for the net input A to the polymer, heat losses to the machine and then to the atmosphere offer some potential for savings.

As the screw diameter transmits the power to the material its diameter governs the power of the drive, so

$$\text{Power (at max speed)} = 2D^{2.5}$$

where D is the screw diameter.

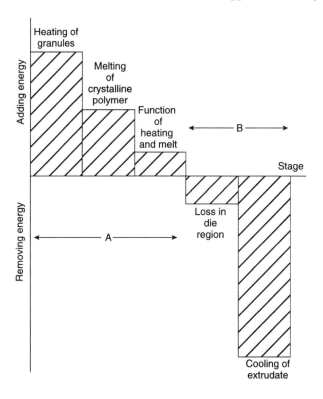

Figure C6.46   Energy considerations

Minimum power required (for 80 per cent drive efficiency)

$$P_{min} = Y \times Q_W \times C_P(T_1 - T_2)$$

where P_{min} is the minimum power, Q_W the output by weight, C_P the specific heat (average, between T_1 and T_2), $Y = 22.2 \times 10^{-4}$ (SI units) or 5.3×10^{-4} (Imperial units), T_1 the melt temperature and T_2 the feed temperature.

The following are some typical C_P values of common polymers are:

Nylon   0.65
PVC   0.85
HDPE   0.85

Approximate losses of energy in a typical small extruder are shown in Figure C6.47. As can be seen, some saving could be made by effective lagging and by precise application and effective control of electrical heating.

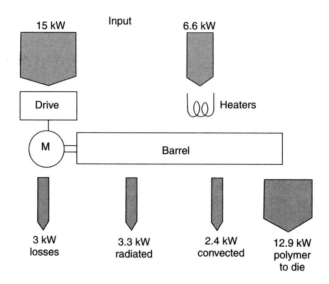

Figure C6.47   Typical extruder energy balance

C6.9.5 Motors and controls

Motors have historically been either slip-ring motors, where speed variation is obtained by varying the effective rotor resistance, or Schrage motors with rotating multiple rings of brushes or the NS motor combined with an induction regulator to increase and decrease speed. The rotation of the brush rings or adjustment of the induction regulator was usually done by a pony motor. This type of motive power is expensive to maintain, quite apart from its high first-cost.

D.C. drives were dominant in this application, but today, a.c. drives are generally used. The use of power electronic controllers offers important benefits including the following:

1. Power consumption becomes a function of applied load, thus creating a power saving as compared with the old a.c. drives.
2. Extremely high resolution of speed holding is possible by using an encoder/ tachometer feedback from the motor. This helps when a constant throughput is required from the extruder.
3. As the screen becomes progressively obstructed, the pressure increases in the barrel causing an increase in the torque required to drive the extruder. A converter compensates automatically, to aid in achieving a constant throughput.
4. For extruders that need to run at different speeds for different products, the speed is easily varied by the use of a potentiometer or increase/decrease push-buttons.
5. Modern drives have serial communication ports to connect with computers and other drives. Computers enable automatic control of integrated systems, and as

more extruder manufacturers move towards continuous production lines, fully integrated controls become essential. There is the further advantage that drive parameters can be changed 'on line'. The communication ports can also be used for data collection. This is a valuable benefit in view of the increasing application of ISO9000, which requires that a record of production characteristics should be kept on a data logger or other such device.

C6.10 Stage scenery: film and theatre

Electrical variable-speed drives are used extensively throughout the world of theatre and film. They facilitate rapid and precisely controlled movement of stage scenery as well as enabling stunts and effects to be performed. The applications are very diverse, but perhaps an unusual application, where the technology was understated, is of interest as it illustrates the flexibility and accuracy of drives.

C6.10.1 The Control Techniques orchestra

Trade exhibitions are in many respects a stage, and an eye-catching exhibit can be memorable and technically challenging. With the launch of Unidrive, Control Techniques was looking for something with great visual impact, but with the high performance of the product as the underlying theme. What resulted was an exhibit that dominated press and television coverage of the Hanover Messe where it was exhibited (Figure C6.48). It played on national and international television and for the Queen by special invitation.

The idea of drives creating music was not new. At an earlier exhibition, Control Techniques Engineers had played 'God save the Queen' on a d.c. motor, the note or

Figure C6.48   The Control Techniques orchestra

tone being based on the speed of the motor. The faster the motor rotated, the higher the pitch of the note.

The brief for the orchestra was to build nine motors and drives in three sections of three. The sections were defined by the mode of operation of the Unidrive:

1. Closed-loop control of a PM servo motor
2. Closed-loop control of an induction motor
3. Open-loop control of an induction motor

There was a 5 m diameter circular area allocated for the exhibit.

The first problem was how to create good-quality music. The University of York was involved and it was agreed that it was practical for the pitch of the note to be determined by the rotational speed of the motor. However, the idea of using its inherent noise was seen as low quality and frankly bad publicity, because many applications, including stage scenery, require the motors to be as quiet as possible. The idea of mounting something such as a siren to the motor shaft was discussed, but discounted as its scope was very limited.

In the end, for reasons of versatility and simplicity, a rotating disc with an electronic pickup to produce the basic electrical waveform at the required frequency was used (Figure C6.49). This signal was then 'processed' and fed to a loudspeaker, a principal reminiscent of the Hammond 'tone wheel' organ.

The requirement for the music was wide – ranging; it could be anything from classical to pop, from rock and roll to film themes. Before starting on the music arrangements it was necessary to establish the musical characteristics of the 'instruments' and see what limitations existed. The three main issues were accuracy

Figure C6.49   *Toothed wheel and electronic pick-up*

(getting the notes in tune), stability (will there be any unwanted vibrato), and dynamic response (getting quickly from one note to the next). The results of the early trials exceeded all expectations. Accuracy was not an issue. Even the open-loop drives were accurate to 1 or 2 min^{-1}. Stability was good and dynamic response was amazing. It was agreed to settle on a working range of two-and-a-half octaves for each instrument. The open-loop drives would play bass with a top note of middle C. The closed-loop induction motor drives would be tenor with a range an octave above the basses. Finally, the closed-loop PM servo drives would be treble, an octave above that.

So the core technology and hardware were established, but control still had to be considered. In the end a PC based MIDI system was used. The PC contained a Music Quest MQX-32M MIDI interface card with built-in timecode facility. The musical arrangements were commissioned in MIDI format and transferred directly. Nine microprocessor units were built, one for each musician, to first filter out and act upon the locally relevant MIDI commands. Second, they had to process the continuous square-wave output of the optical pick-up produced by the spinning toothed disk on the end of the motor shaft. The unit sent a speed demand signal to the drive via a high-speed serial link.

The 'processing' comprised chopping up the signal into individual notes and adding side bands. This avoided a 'Stylophone' type of sound.

The choice of the music catered for all tastes: Beethoven's 'Ode to Joy', Mozart's 'Eine Kleine Nachtmusik', Handel's 'Water Music'. Then there were a few Beatles numbers, as well as big band swing numbers such as the 'Pink Panther' and 'New York, New York' as well as some military brass band arrangements. As the original exhibition was in Germany a number of Bavarian 'Umpah' tunes were included for good measure.

Having established the technology of the orchestra, what was it going to look like? The orchestra had to look like an orchestra, and from the early days of creating the key constituents were in place. The Unidrive itself was to be the head, below that the torso would be a box containing the loudspeaker and any other electronics. In front of the torso was a motor representing the instrument being played. Arms and legs were made of Anglepoise light fittings and there would be a variety of hairstyles fashioned from cables and a selection of bow ties and necklaces (Figure C6.48).

To reinforce the connection between motor speed and pitch of the note, each musician would be equipped with a rope light thermometer-type indicator, calibrated in semitones, as well as displaying the speed numerically on the 'face' of the drive. To make the orchestra visually captivating there had to be some animation. If there was too much it would detract from the drives. In the end there was a liberal sprinkling of moving hands and tapping feet, swaying heads and revolving bow ties. The musicians were arranged on a 4.5 m tiered podium, with a small stage area at the front for a human performer or demonstrator.

To complete the stand a 'conductor' was hired. His speciality was to be able to lean his body forwards about 45°, anchored to bolts in the floor. Who says trade exhibitions cannot be fun! The orchestra still performs today, on special occasions such as carol concerts at local schools.

Part D
Appendices

D1 Symbols and formulae
D2 Conversion tables
D3 World industrial electricity supplies (<1 kV)

Appendix D1

Symbols and formulae

D1.1 SI units and symbols

The following formulae are based on the International System of Units, known as SI (Systeme Internationale d'Unites), which is used throughout this book. SI was adopted in February 1969 by a resolution of the CGPM (Conference Generale de Poids et Mesures) as ISO Recommendation R1000.

A base unit exists for each of the dimensionally independent physical quantities: length, mass, time, electric current, thermodynamic temperature and luminous intensity. The SI unit of any other quantity may be derived by appropriate simple multiplication or division of the base units without the introduction of numerical factors.

The system is independent of the effects of gravity, making a clear distinction between the mass of a body (unit of mass = kilogram) and its weight, i.e. the force due to gravity (unit of force = newton).

Example
A force of 1 N acting on a mass of 1 kg results in an acceleration of 1 m s^{-2}.

Conversion factors from non-SI units to SI units are to be found in Section D2.

D1.1.1 SI base units

SI base units are described in Table D.1, and decimal multiples and sub-multiples in Table D.2.

Table D.1   SI base units

Quantity	Unit symbol	Unit name
Length	m	Metre
Mass	kg	Kilogram
Time	s	Second
Electric current	A	Ampere
Temperature	K	Kelvin
Luminous intensity	cd	Candela

Table D.2 Decimal multiples and sub-multiples

Factor	Prefix	Symbol
10^{12}	Tera	T
10^{9}	Giga	G
10^{6}	Mega	M
10^{3}	Kilo	k
10^{2}	Hecto	h
10	Deca	da
10^{-1}	Deci	d
10^{-2}	Centi	c
10^{-3}	Milli	m
10^{-6}	Micro	μ
10^{-9}	Nano	n
10^{-12}	Pico	p
10^{-15}	Femto	f
10^{-18}	Atto	a

D1.1.2 Derived units

Geometrical, time-related and mechanical units are described in Tables D.3, D.4 and D.5, respectively.

Table D.3 Geometrical units

Symbol	Quantity	Unit symbol	Unit name
l, s	Length, distance	m	Metre
A	Area	m^2	Square metre
V	Volume	m^3	Cubic metre
α, β, γ, etc.	Plane angle	rad	Radian
		°	Degree
α, β, γ, etc.	Solid angle		Steradian

Table D.4 Time-related units

Symbol	Quantity	Unit symbol	Unit name
t	Time	s	Second
τ	Time constant	s	Second
u, v	Velocity	$m\,s^{-1}$	Metre per second
a	Acceleration	$m\,s^{-2}$	Metre per second per second
ω	Angular velocity	$rad\,s^{-1}$	Radian per second
α	Angular acceleration	$rad\,s^{-2}$	Radian per second per second
f	Frequency	Hz	Hertz
n	Rotational frequency	s^{-1}	(Revolution) per second

Table D.5   Mechanical units

Symbol	Quantity	Unit symbol	Unit name
m	Mass	kg	Kilogram
F	Force	N	Newton
$G\,(W)$	Weight	N	Newton
J	Moment of inertia	kg m^2	Kilogram metre squared
$M\,(T)$	Torque	kg m	Kilogram metre
$W\,(E)$	Work (energy)	J	Joule
P	Power	W	Watt
p	Pressure	Pa	Pascal
E	Modulus of elasticity	Pa	Pascal
σ	Stress	Pa	Pascal
ρ	Density	kg m^{-3}	Kilogram per cubic metre
δx	Rate of flow	m^3 s^{-1}	Cubic per metre second
k, k$_1$, etc.	Any constant factor		

D1.2   Electrical formulae

D1.2.1   Electrical quantities

Electrical quantities are described in Table D.6.

Table D.6   Electrical quantities

Quantity	Symbol	Unit	
		Name	Symbol
Electromotive force	E, e^*	Volt	V
Potential difference	V, v^*	Volt	V
Current	I, i^*	Ampere	A
Magnetic flux	Φ	Weber	Weber
Frequency	f	Hertz	Hz
Flux linkage	λ	Weber-turns	–
Resistance	R	Ohm	Ω
Inductance	L	Henry	H
Capacitance	C	Farad	F
Impedance	Z	Ohm	Ω
Reactance	X	Ohm	Ω
Power, d.c., or active	P	Watt	W
Power, reactive	Q	Volt–ampere reactive	VAr, var
Power, total or apparent	S	Volt–ampere	VA

(*Continued*)

Table D.6 Continued

Quantity	Symbol	Unit	
		Name	Symbol
Power factor angle	φ	–	°, deg
Angular velocity	ω	Radians per second	rad s^{-1}
Rotational velocity	n	Revolutions per second	s^{-1}, rev s^{-1}
		Revolutions per minute	min^{-1}, rpm
Efficiency	η	–	
Number of pairs of poles	p	–	

*Capital and small letters designate rms and instantaneous value, respectively.

D1.2.2 A.C. three-phase (assuming balanced symmetrical waveform)

All quantities are r.m.s. values:

$$V_1 = \text{line-to-line voltage}$$
$$V_p = \text{phase voltage (line-to-neutral)}$$
$$I_1 = \text{line current (star)}$$
$$I_p = \text{phase current (delta)}$$

In a star-connected circuit $V_p = V_1/\sqrt{3}$, $V_1 = \sqrt{3}V_p$, $I_1 = I_p$.
In a delta connected circuit $I_p = I_1/\sqrt{3}$, $I_1 = \sqrt{3}I_p$, $V_1 = V_p$.
Total apparent power in VA $= \sqrt{3}V_1I_1$
Active power in watts, W $= \sqrt{3}V_1I_1 \cos\varphi$
Reactive power in VAr $= \sqrt{3}V_1I_1 \sin\varphi$
Power factor (pf) $= \cos\varphi$
$\phantom{\text{Power factor (pf)}} = $ active power/apparent power
$\phantom{\text{Power factor (pf)}} = $ W/VAr

D1.2.3 A.C. single-phase

All quantities r.m.s. values

$$V = IZ$$

Total or apparent power in VA $= VI = I^2Z = V^2/Z$
Active power in watts, W $= VI \cos\varphi$
Reactive power in VAr $= VI \sin\varphi$

D1.2.4 Three-phase induction motors

All quantities are rms values:

$$kW_{mech} = horsepower \times 0.746$$

$$kW_{elec} = \sqrt{3} V_1 I_1 \cos \varphi \text{ at rated speed and load}$$

where V_1 is the supply voltage, I_1 the rated full-load current and $\cos \varphi$ the rated full-load power factor.

Efficiency, $\eta = (kW_{mech}/kW_{elec}) \times 100\%$

Phase current, $I_p = I_1$ for star connection and $I_p = I_1/\sqrt{3}$ for delta connection.

D1.2.5 Loads (phase values)

Resistance R, measured in ohms (no energy storage).

Inductive reactance $X_L = \omega L = 2\pi f L$ ohms (stores energy), where f is the frequency (Hz) and L the inductance (H).

Capacitative reactance $X_C = 1/(\omega C) = 1/(2\pi fC)$, where f is the frequency (Hz) and C the capacitance (F).

D1.2.6 Impedance

Impedance is the algebraic sum of the separate load values, so

$$Z = \sqrt{(R^2 + X_L^2)} \text{ or } \sqrt{(R^2 + X_C^2)}$$

If R, X_L and X_C are present in series in the same circuit then X_L and X_C may be summated, treating X_C as negative, so

$$Z = \sqrt{[R^2 + (X_L - X_C)^2]}$$

D1.2.7 A.C. vector and impedance diagrams

If a voltage V is applied to an impedance Z (Figure D.1a), the current I will be phase-displaced by an angle ϕ (Figure D.1b). The current vector may be resolved into two component vectors at right angles (Figure D.2). The component in-phase with the voltage represents the value of the current due to resistance $= I \cos \phi$, and the quadrature component represents current due to reactance (which may be inductive or capacitive) $= I \sin \phi$.

If each vector component is multiplied by the voltage V, the resulting triangle is similar, ϕ is unchanged, and the vector diagram Figure D.3 shows the total power S, active power P and reactive power Q.

Because $V = IZ$, then $S = VI = I^2 Z$.
Similarly, $P = VI \cos \phi = I^2 Z \cos \phi$ and $Q = VI \sin \phi = I^2 Z \sin \phi$.

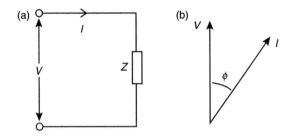

Figure D.1 (a) Impedance circuit; (b) voltage and current vectors

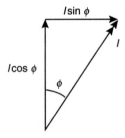

Figure D.2 Current vector diagram

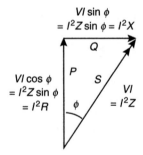

Figure D.3 Power vector diagram

If these vectors are divided by I^2, the resulting triangle is similar, x is unchanged, and Figure D.4a shows the impedance Z and the resistance as the other two sides:

$$Z \cos \phi = R$$
$$Z \sin \phi = X$$

Furthermore

$$Z = \sqrt{(R^2 + X^2)}$$

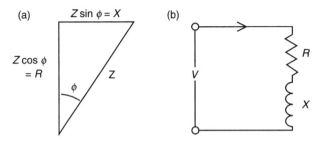

Figure D.4 (a) Impedance triangle; (b) equivalent circuit

and

$$\cos\phi = R/Z$$
$$\sin\phi = X/Z$$
$$\tan\phi = X/R$$

Thus the impedance circuit of Figure D.1a, may be represented by a circuit containing a resistance and a reactance in series (Figure D.4b). All equations hold good whether X is inductive or capacitive. If capacitive, the current (Figure D.1b) would lead the voltage and the quadrature vectors of Figures D.2 and D.3 should be drawn in the opposite sense.

D1.2.8 Emf energy transfer

Emf is the internal driving force, symbol E, of an energy source. It is equal to the sum of all the voltage drops in the whole circuit, including any voltage drop attributable to the energy source itself (Figure D.5):

$$E = IZ + Iz$$

where E is the source emf, I the current caused to flow by E, Z the sum of all impedances external to the source and z the impedance of the source itself.

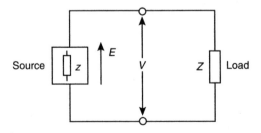

Figure D.5 Elements of a power circuit

In the part of the circuit external to the source, the voltage drop across the load is equal to the terminal voltage,

$$V = IZ$$

Substituting,

$$E = V + Iz$$

and

$$V = E - Iz$$

In a.c. circuits, quantities must be evaluated vectorially. If the source is d.c., Z, z become R, r and evaluation is arithmetical.

Maximum power transfer from a supply source to a load occurs when the resistance of the load is equal to the internal resistance of the source. Varying values of load resistance are plotted against power developed by the source (of fixed emf and internal resistance) in Figure D.6, showing that power dissipated reaches a peak value.

Note that practical power sources of low internal resistance are usually unable to transfer maximum power due to current overload.

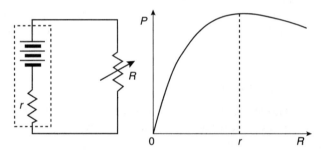

Figure D.6   Maximum power transfer

D1.2.9   Mean and rms values, waveform

D1.2.9.1   Principles

By definition a symmetrical alternating quantity oscillates about a zero axis and its mean value, and is therefore zero.

Each half cycle, however, has a definite mean value. For practical purposes, the mean of a half cycle represents the d.c. value of a rectified alternating quantity.

The power delivered by a symmetrical alternating voltage or current is proportional to the square of the quantity at any instant. Negative values, when squared, are positive, so the sum of the squares of the instantaneous values has a definite mean value. The square root of this mean represents that value of a symmetrical quantity that produces the same power or heating effect as if it were a direct quantity. It is called the root mean square (rms) value, being the square root of the mean of the sum of the squares of the instantaneous values of the voltage or current wave.

D1.2.9.2 Mean d.c. value

For a sine wave (Figure D.7a), the instantaneous value is given by

$$i = A \sin \alpha$$

The area is the sum of the instantaneous values in one half cycle

$$\sum_0^\pi (i) = \int_0^\pi A \sin \alpha \, d\alpha$$

Period (base length) is π.

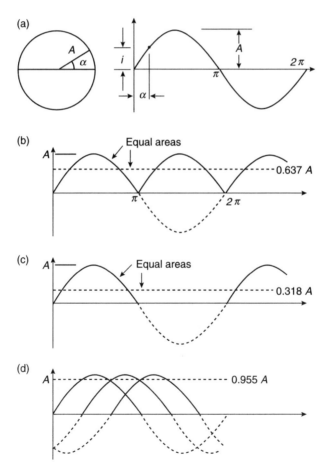

Figure D.7 Sine wave: (a) instantaneous value; (b) rectified single-phase full-wave; (c) rectified single-phase half-wave; (d) rectified three-phase full-wave

The mean height of one half cycle,

$$\text{Area/base} = \left(\int_0^\pi A \sin\alpha \, d\alpha\right) / \pi$$

$$= (A/\pi) \int_0^\pi \sin\alpha \, d\alpha$$

$$= 2A/\pi = 0.637A \text{ (3 sf)}$$

If the half wave is not repeated in the next half cycle, from π to 2π (i.e. half-wave rectification), the mean value is halved because the area enclosed is the same but the period is double (Figure D.7c):

$$\text{Half-wave mean} = A/\pi = 0.318A \text{ (3 sf)}$$

If the supply is three-phase, the effective value of the three phases added vectorially is 1.5 times the value of one phase (Figure D.7d):

$$\text{Three-phase mean} = 3A/\pi = 0.955A \text{ (3 sf)}$$

Other waveforms (Figure D.8)

$$\text{Square wave mean} = AB/B = A$$
$$\text{Semicircular wave mean} = \pi A^2/(2 \times 2A) = \pi A/4$$
$$\text{Triangular wave} = AB/(2 \times B) = A/2$$

D1.2.9.3 rms value

For a sine wave (Figure D.9a), where the instantaneous value

$$i = A \sin\alpha$$

the square of the instantaneous value (Figure D.9b) is given by

$$i^2 = A^2 \sin^2\alpha$$

The area is the sum of the squares of the instantaneous values in one half cycle:

$$\Sigma_0^\pi (i)^2 = \int_0^\pi A^2 \sin^2\alpha \, d\alpha$$

The period (base length) is π.

Symbols and formulae 673

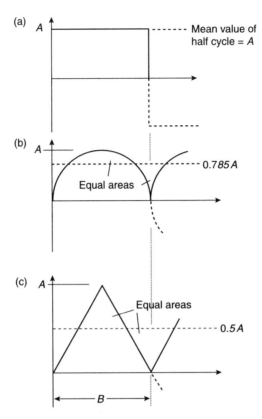

Figure D.8 Other waveforms: (a) square wave; (b) semicircular wave; (c) triangular wave

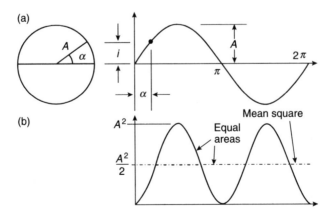

Figure D.9 Sine wave: (a) instantaneous value; (b) squares of instantaneous values

The mean height of one half cycle is given by

$$\text{Area/base} = \left(\int_0^\pi A^2 \sin^2 \alpha \, d\alpha\right) / \pi$$

$$= A^2/\pi$$

The representative value is the square root of this mean value:

$$A_{rms} = \sqrt{(A^2/2)} = A/(\sqrt{2}) = 0.707A \text{ (3 sf)}$$

Other waveforms
Square wave:

$$A_{rms} = A$$

Semicircular wave:

$$A_{rms} = \sqrt{2}A/\sqrt{3} = 0.816A \text{ (3 sf)}$$

Triangular wave:

$$A_{rms} = A/\sqrt{3} = 0.577A \text{ (3 sf)}$$

D1.2.9.4 Form factor

The ratio of the rms to the mean value is called the form factor. A square wave has a form factor of unity. All others are greater than unity and the sharper or more peaked the wave shape, the higher the form factor:

Sine-wave form factor = $\pi/(2\sqrt{2}) = 1.11$ (3 sf)
Square-wave form factor = 1
Semicircular-wave form factor = $(4\sqrt{2})/(\sqrt{3}\pi) = 1.039$ (3 sf)
Triangular-wave form factor = $2/\sqrt{3} = 1.554$ (3 sf)

D1.3 Mechanical formulae

Mechanical formulae are shown in Table D.7.

D1.3.1 Laws of motion

Sir Isaac Newton (1642–1727) is well known for his work and discoveries in optics, gravitation and many other fields of mathematics and the physical sciences. In 1687, Newton completed his monumental work on mechanics, in which the concepts of force, velocity and acceleration were for the first time accurately inter-related.

Table D.7 Mechanical formulae

Term	Description	Unit
d	Diameter	m
F	Force	N
g	Acceleration due to gravity	m s^{-2}
J	Total inertia	kg m^2
J_L	Load inertia	kg m^2
J_M	Motor inertia	kg m^2
m	Mass	kg
M	Motor torque	N m
M_a	Accelerating torque	N m
M_L	Load torque	N m
n	Rotational frequency	rpm*
n_1	– input	rpm*
n_2	– output	rpm*
Δn	Change of rotational frequency	rpm*
p	Pitch	m
P	Motor power	kW
P_a	Accelerating power	kW
P_L	Load power absorbed	kW
r	Radius	m
s	Distance	m
t	Acceleration time	s
Δt	Acceleration period	s
v	Linear velocity	m min^{-1}*
Δv	Change of linear velocity	m min^{-1}*
V	Traction capacity	m^3 s^{-1}
W	Energy	J (joule)
η	Efficiency	–
μ	Coefficient of friction	–

Note: For practical convenience, some of the units in the formulae are not SI units; for example, rotational frequency is commonly measured in revolutions per minute, although the SI unit is revolutions per second. In these Servo Formulae, the terms used are as tabulated above. Those that are in non-SI units are marked *.

The Laws of Motion, for which Newton is best known, are among the fundamental principles of mechanical engineering, and are expressed as follows:

1. A body continues in its state of rest or uniform motion in a straight line, unless impressed forces act upon it.
2. The impressed force is proportional to the rate of change of momentum which it produces, and the change of momentum takes place in the direction of the straight line along which the force acts.
3. Action and reaction are equal and opposite.

The First Law states that a force must be applied to a body in order to make it move from rest, or to change its speed or direction if it is moving.

The Second Law relates the applied force to the rate of change of momentum that it produces. Momentum is the product of mass and velocity, therefore, for a constant mass, force is proportional to the rate of change of velocity, i.e. to acceleration.

The Third Law can be appreciated by considering, for example, the way in which a rowing boat is propelled forward by the rearward force exerted by the oars upon the water.

D1.3.1.1 Linear motion

Consider a body mass m acted upon by a single force F (Figure D.10). The body accelerates in the direction in which the force is acting, at a rate given by

$$A = F/m$$

After a time t has elapsed, the body has achieved a velocity v, where

$$v = u + at$$

and where u is the initial velocity before the force F was applied. If the body was initially at rest, u is zero.

The distance s travelled by the body during time t is given by

$$s = ut + at^2/2$$

Distance and velocity are related by the following equation, derived from the two previous ones:

$$v^2 = u^2 = 2as$$

The work done by the force in accelerating the body is the product of force and distance:

$$W = Fs$$

The kinetic energy of the body, i.e. the energy it possesses by virtue of its motion, is the product of its mass and the square of its velocity:

$$E_k = mv^2/2$$

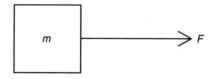

Figure D.10   *The action of a single force on a body*

Furthermore, because energy is conserved, the work done by the force is equal to the change in the body's kinetic energy (neglecting losses):

$$W = m(v^2 - u^2)/2$$

Power is the rate at which work is done, so it is the product of force and velocity:

$$P = Fv$$

D1.3.1.2 Rotational or angular motion

A force acting perpendicular to a pivoted lever (Figure D.11), causes a turning effect or torque at the fulcrum. The torque is the product of the force and the radius at which it is applied.

$$M = Fr$$

If a torque is applied to a body that is free to rotate, as in Figure D.12, an acceleration results in a way analogous to the example of linear motion above. Indeed a similarity will be noticed between the equations of motion.

Any body that is capable of rotating possesses a property known as moment of inertia, which tends to resist acceleration in the same way as does the mass of a body in linear motion. The moment of inertia is related not only to the mass of the body, but also to the distribution of that mass with respect to radius.

The moment of inertia of a solid cylinder of radius r is given by

$$J = mr^2/2$$

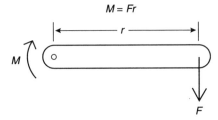

Figure D.11   *The concept of torque*

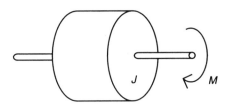

Figure D.12   *The action of torque on a body*

By comparison, the moment of inertia of a hollow cylinder is

$$J = m(r_o^2 + r_i^2)/2$$

where r_o and r_i are outer and inner radii, respectively. It can be seen that, for a given outer radius, the moment of inertia of a hollow cylinder is greater than that of a solid cylinder of the same mass. In Figure D.12, a body having a moment of inertia J is acted upon by a torque M. Its angular acceleration is given by

$$\alpha = M/J$$

After a time t has elapsed, the angular velocity ω (rate of change of angle) is given by

$$\omega = \omega_o + \alpha t$$

where ω_o is the initial angular velocity before the torque M was applied. If the body was initially at rest, ω_o is zero.

The angle γ through which the body rotates in time t is given by

$$\gamma = \omega_o t + \alpha t^2/2$$

Angle and angular velocity are related by the following equation:

$$\omega^2 - \omega_o^2 = 2\alpha\gamma$$

The work done in accelerating the body is the product of torque and angle of rotation:

$$W = M\gamma$$

The kinetic energy of the body is the product of its moment of inertia and the square of its angular velocity:

$$E_k = J\omega^2/2$$

Because energy is conserved, the work done is equal to the change in kinetic energy (neglecting losses):

$$W = J(\omega^2 - \omega_o^2)/2$$

Power is the product of torque and angular velocity, i.e. the rate at which work is being done:

$$P = M\omega$$

D1.3.1.3 Relationship between linear and angular motion

Consider a body of mass m moving in a circle of radius r with an angular velocity ω (Figure D.13). When the body has rotated through an angle γ, it has covered a distance s along the circumference of the circle, where

$$s = \gamma r$$

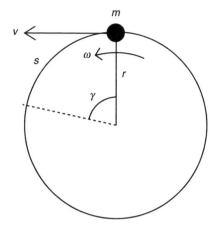

Figure D.13 Relationship between linear and angular motion

Similarly, the tangential velocity or peripheral speed v, being the quotient of distance and time, is given by

$$v = s/t = \gamma r/t$$

Angular velocity ω is the quotient of angle and time;

$$\omega = \gamma/t$$

Therefore

$$v = \omega r$$

Similarly, for acceleration,

$$a = v/t = \omega r/t$$
$$\alpha = \omega$$

Therefore

$$a = \alpha r$$

The moment of inertia is given by

$$J = mr^2$$

D1.3.1.4 The effect of gearing

When calculating the torque required to accelerate or decelerate the moving parts of a machine, it is necessary to take into account any gearing that introduces a ratio between the speeds of different parts. It is unusual to calculate the moment of inertia referred to the motor shaft, because this figure may be added arithmetically

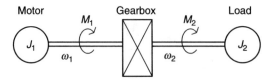

Figure D.14 The effect of gearing between motor and load

to the motor inertia to arrive at a figure for the total inertia of the system. Figure D.14 illustrates a motor having a moment of inertia J_1 driving a load with inertia J_2 via a gearbox.

If the gearbox has a ratio k, then the relationship between input and output angular velocities is as follows:

$$\omega_1 = k\omega_2$$

Neglecting losses, the input and output torques are related by

$$M_1 = M_2/k$$

The load inertia reflected back through the gearbox to the motor shaft is reduced by a factor equal to the square of the gear ratio. Therefore the total inertia that the motor has to overcome is given by

$$J = J_1 + J_2/k^2$$

D1.3.1.5 Linear to rotary speed reference conversion

Speed referencing systems are frequently based upon linear speed units and must therefore be converted to rotary units (min^{-1}, revolutions per minute) before presentation to the drive for motor control. All applications rely on the same basic expression:

- For rolls

$$\text{Motor rpm} = \frac{\text{linear speed (m min}^{-1})}{\pi \times \text{roll diameter (m)}} \times \text{gear ratio}$$

- For the ball screw mechanism

$$\text{Motor rpm} = \frac{\text{linear speed (m min}^{-1})}{\text{pitch (mm)}} \times \text{gear ratio}$$

- For a rack and pinion drive

$$\text{Motor rpm} = \frac{\text{linear speed (m min}^{-1})}{\pi \times \text{pinion diameter (m)}} \times \text{gear ratio}$$

To avoid rounding errors that could result in cumulative errors, it is often convenient to express the gear ratio as two integers in the form

$$\text{Gear ratio} = \frac{\text{gear ratio numerator}}{\text{gear ratio denominator}}$$

D1.3.1.6 Friction and losses

Friction is the name given to the force acting tangentially to the surfaces of two bodies in contact, which opposes their relative motion or tendency to relative motion. Consider a body of mass m at rest on a horizontal surface (Figure D.15). If a small force F is applied as shown, parallel to the surface on which the body is resting, an equal and opposite frictional force F_1 is set up, preventing motion. If the applied force is gradually increased the opposing force increases with it, up to a point beyond which no further increase occurs, and the body begins to move. The maximum value of the opposing force is called the limiting frictional force. It can be shown to be independent of contact area, dependent on the nature of the surfaces in contact, and proportional to the normal reaction (the force perpendicular to the surfaces in contact – in this case the weight of the body, mg, where g is the gravitational constant).

In general, where F_L is the limiting frictional force and F_N is the normal reaction, the ratio of the two forces is constant for a particular combination of surfaces in contact:

$$F_L/F_N = \mu$$

The constant μ is known as the coefficient of static friction between the two surfaces. Friction is present whenever there is relative motion between two surfaces in contact, although it may be considerably reduced by means of lubricants which, by forming an interposing layer, keep the surfaces apart.

In all machines, part of the power supplied is used to overcome friction, so the power available to do useful work is diminished. Other losses occur due to the viscous friction of lubricants (the oil in a gearbox, while reducing friction between gears, introduces other losses), air resistance, and so on.

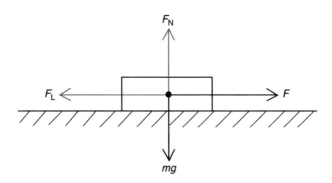

Figure D.15   Friction between a body and a surface

The usable output power of a system is equal to the input power minus the system losses. The ratio of output power to input power is called the system efficiency η:

$$\eta = P_{out}/P_{in} = (P_{in} - \text{losses})/P_{in}$$

The mechanical power lost in overcoming friction etc is converted into heat, and the disposal of this heat is an important consideration in large systems.

Consider, for example, a machine driven by a 100 kW motor via a gearbox having an efficiency of 0.9 (usually expressed as 90 per cent):

$$\text{Losses} = P_{in} - \eta P_{in}$$
$$= 100 - 90 = 10\,\text{kW}$$

Therefore 10 kW of heat is generated when the machine works at full power. An oil cooler or heat exchanger may be required to prevent overheating of the gearbox, and the gearbox manufacturer will allow for this in the design.

D1.3.1.7 Fluid flow

The term 'fluid' is used to describe any medium that is capable of flowing. Gases and liquids are typical fluids. Others include foams, slurries and some granular solids (e.g. bulk grain). For flow to take place, a pressure gradient must exist. Therefore fluid will flow through a pipe connecting two vessels only if a difference in pressure exists between them. Flow takes place from the higher-pressure vessel to the lower. In the case of liquids, pressure is directly proportional to depth, so in Figure D.16 liquid flows from vessel A to vessel B until the levels are equal, when the flow ceases. The rate at which the fluid flows through a pipe or duct depends on four factors:

- the pressure gradient,
- the viscosity of the fluid,
- the cross-sectional shape and area of the pipe or duct,
- whether the flow is laminar (Figure D.17a) or turbulent (Figure D.17b).

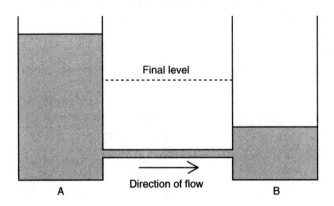

Figure D.16   Flow of liquid between two vessels

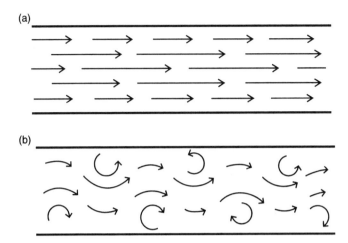

Figure D.17   Fluid flow: (a) laminar, (b) turbulent

If turbulence occurs, a greater pressure is required to achieve a given rate of flow than when the flow is laminar. Pipework and ductwork should therefore be designed to avoid abrupt changes in direction or cross-section, and without obstructions that might give rise to turbulence. Consider the arrangements shown in Figure D.18, where a fan forces air through a heat-exchanger matrix. The fan inlet is open to the atmosphere, as is the outlet from the matrix, so these two points are at atmospheric pressure. The pressure between the fan and the matrix exceeds atmospheric pressure, so air flows through the matrix as shown.

If p is the pressure difference (due to the fan) that gives rise to a rate of flow δ, the power by the fan is given by

$$P = p\delta$$

Similarly, Figure D.19 shows a pump being used to raise liquid from a tank to one at a higher level. The pressure difference between inlet and outlet of the pump is

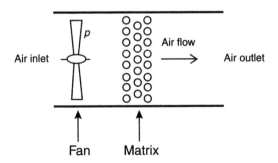

Figure D.18   Example of air flow

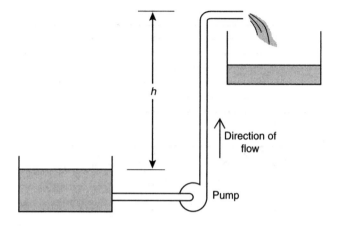

Figure D.19   Example of liquid flow

proportional to the height h to which the liquid is being pumped with respect to the level in the lower tank. The actual pressure depends upon the density ρ of the liquid:

$$p = \rho g h$$

For a rate of flow δ, the power delivered by the pump is given by

$$P = p\delta = \rho g h \delta$$

In practice the pressure will be increased by viscous friction and turbulence within the pipework, but under normal circumstances these effects are small in comparison with the pressure due to the height (head) of the liquid column.

D1.4   Worked examples of typical mechanical loads

Note: Data is typical, not necessarily SI units. Calculations are correct to three significant figures.

D1.4.1   Conveyor

A horizontal conveyor (Figure D.20) carries a loose material (not sensitive to shock). The conveyor is required to start when fully loaded. Find the rating of a standard

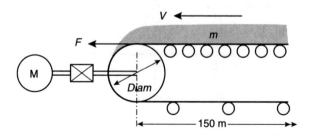

Figure D.20   Conveyor

Table D.8   Data for conveyor example

Required delivery rate	900 tonnes per hour	(250 kg s^{-1})
Density of load material	2 kg dm^{-3}	(2 000 kg m^{-3})
Width of conveyor belt	800 mm	(0.8 m)
Length of conveyor	150 m	(150 m)
Maximum safe depth of load on belt to prevent spillage	100 mm	(0.1 m)
Weight of conveyor belt	15 kg m^{-1}	
Diameter of belt pulley	400 mm	(0.4 m)
Coefficient of friction of conveyor, static	0.09	
Coefficient of friction of conveyor, moving	0.07	
Manufacturer's information:		
Breakaway torque, fully loaded	6 850 N m	
Operational requirement:		
Desired acceleration time	4 s	

three-phase 400 V a.c. induction motor to drive the conveyor, assuming typically four-pole, 50 Hz, 1 475 rpm at full load. From the same data, find the motor power if the conveyor raises the load through a height of 20 m. The data are shown in Table D.8.

Method
1. Ensure all data is in SI units.
2. Determine the linear and rotational speeds and gearing ratio.
3. Determine the loading and acceleration, and from them the forces and torques to accelerate and to run.
4. Calculate the power ratings of the motor and the drive.

To calculate speeds and gearing ratio:
Volume of the required delivery rate:

$$\text{Required delivery rate per hour} = \text{volume} \times \text{density}$$

From data:

$$\text{Volume delivered} = \text{mass delivered}/\text{density} \, (\text{m}^3 \, \text{s}^{-1})$$

$$\text{Volume per second} = 250/2\,000 = 0.125 \, (\text{m}^3 \, \text{s}^{-1})$$

Linear velocity v of the conveyor belt:

$$v = \text{volume per second}/\text{cross-sectional area}$$
$$= 0.125 \times 1/(0.8 \times 0.1)$$
$$= 1.56 \, \text{m s}^{-1}$$

Rotational speed ω of the belt driving-pulley:

$$\omega = \text{linear speed/radius}$$
$$= 1.56/0.2$$
$$= 7.8 \text{ rad s}^{-1}$$

Converting this to revolutions per minute gives

$$\text{rpm} = \omega \times 60/(2\pi) = 74.5 \text{ min}^{-1}$$

Therefore

$$\boxed{\text{Gearbox ratio will be } 1475{:}74.5 \approx 20{:}1}$$

To determine load, force and torque
1. Data typically states 'weight' of the conveyor belt, but gives the value in kg m^{-1}, which is mass per unit length and correct in the SI system.
2. A factor of 2 is applied to account for the return run of the belt.

Load = mass of charge + mass of belt

$\quad$ = (volume of charge × density) + 2(conveyor length × mass of belt per metre)

$\quad$ = [(150 × 0.8 × 0.1)(2 000)] + 2(150 × 15)

$\quad$ = 28.5 × 10^3

Breakaway torque to start the conveyor from rest in the fully loaded state is given in the data as 6 850 N m.

Horizontal force required to accelerate the loaded conveyor against rolling friction after breakaway is given by

F_a = (load mass × acceleration) + (load mass × g × coefficient of rolling friction)

$\quad = ma + mg\mu$

$\quad = m(a + g\mu)$

Furthermore, torque to accelerate the conveyor against rolling friction is 'force × radius of the pulley':

$$M_a = F_a r$$
$$= [m(a + g\mu)]r \text{ (N m)}$$

Now, determine the linear acceleration from the calculated linear velocity (1.56 m s^{-1}), and the required acceleration time (4 s):

$$\text{Linear acceleration} = 1.56/4 = 0.39 \text{ m s}^{-2}$$

Substituting this into the torque equation

$$M_a = 28.5 \times 10^3 [0.39 + (9.81 \times 0.07)] \times 0.2$$
$$= 6\,140 \text{ N m}$$

The horizontal force required to run the fully loaded conveyor at full speed against rolling friction (assuming wind resistance to be negligible) is

$$F_s = \text{(load mass} \times g \times \text{coefficient of rolling friction)}$$
$$= mg\mu$$

Torque to run the conveyor against rolling friction is 'force × radius of the pulley':

$$M_s = F_s r$$
$$= mg\mu r$$
$$= 28.5 \times 10^3 \times 9.81 \times 0.07 \times 0.2$$
$$= 3.91 \times 10^3 \text{ N m}$$

It is often helpful to plot the torque/speed demand as in Figure D.21.

To calculate the power ratings for the motor and drive

Because power is equal on both sides of a gearing mechanism, there is no necessity to refer torque and acceleration to the motor. The power demand at the belt pulley is the power demand at the motor.

$$P = M\omega$$

As power is proportional to speed, this seems to present a problem because at the instant of starting, speed is zero and so power is zero also. Furthermore, during acceleration the speed increases from zero to the normal running speed, and therefore the power demand appears to have no particular value. But in fact it does, because the acceleration torque continues up to the point where the motor reaches full speed, and so the accelerating torque at full speed can be used to calculate the accelerating power demand.

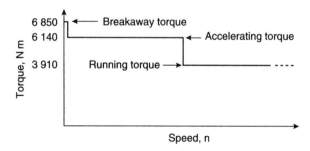

Figure D.21   Conveyor torque/speed demand

According to the data given, the starting and acceleration period is to be 4 s. If a motor is selected that is typically rated for 150 per cent overload for 30 s, use can be made of this capability because, at 4 s, the demand period is comfortably within the overload period of 30 s.

Motor power to accelerate the conveyor is the accelerating torque at the conveyor pulley multiplied by the rotational speed of the pulley at full speed, taking the efficiency η of the gearing into account. For this example, η is assumed to be 98 per cent, but in practice it must be verified for the gearbox installed:

$$P_a = M_a \omega / \eta$$
$$= 6\,140 \times 7.8/0.98$$
$$= 48.9\,\text{kW}$$

This would be the rating of a motor able to deliver accelerating torque continuously. If full use is to be made of the 150 per cent overload rating, the actual full load rating of the motor is $48.9/1.5 = 32.6$ kW

> A standard motor rated 37 kW will be satisfactory

This rating must now be verified for the drive module required to operate the motor. The reason is that the drive must be able to deliver the current demanded by the breakaway torque.

As current is approximately directly proportional to torque, the current demand for breakaway is in the ratio $6\,850/6\,140$ of the current demand for acceleration.

Note, however, that the current demand during acceleration must be based on the 48.9 kW overload rating for acceleration, not the selected motor rating (37 kW).

Calculate the currents involved and, having found the staring current demand, find the rating of the drive. The starting current is based on the motor rating of 48.9 kW and is calculated from

$$P = \sqrt{3} \times V_L \times I_L \times \text{power factor} \,(\text{kW})$$

where the working power factor is assumed to be the full-load, full-speed, full-voltage rated power factor of the motor, 0.85 (from data):

$$\text{Total } I_L = P/(\sqrt{3} \times V_L \times \text{power factor})$$
$$= 48.9 \times 10^3/(\sqrt{3} \times 400 \times 0.85)$$
$$= 83\,\text{A}$$
$$\text{Starting current} = 83 \times 6\,850/6\,140$$
$$= 92.6\,\text{A}$$

The drive itself will have a short-time over-current rating in the same way as the motor. For the Control Techniques Unidrive, this is 150 per cent for 60 s. The drive efficiency must be taken into account (98 per cent). The required full-load rated current of the drive is $92.6/(1.5 \times 0.98) = 63$ A.

In practice, the full load current, the short-time rating and the efficiency must be verified. It is inadvisable to attempt to calculate, using the formula as for a motor, a theoretical value for the rating of the drive module. Users of drives are recommended to consult the data supplied by the manufacturers of drives.

D1.4.2 Inclined conveyor

An inclined plane adds a component of force to the forces required to start, accelerate and run a conveyor. This component is the force required to raise the load (Figure D.22). The total force is the hypotenuse of the vector diagram (above) of which the horizontal component is the force required either to start, accelerate or drive the load at constant speed as already calculated, and the additional vertical component is mg multiplied by the sine of the angle of the slope, γ (in this case, $20 \div 150$).

The total force, F is calculated from

$$F = \sqrt{[F_h^2 + (mg \cdot \sin \gamma)^2]}$$

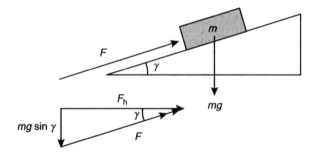

Figure D.22   Forces on an inclined plane

D1.4.3   Hoist

As an example, a hoist is to lift sensitive loads of 2 tonnes maximum. The block-and-tackle arrangement has two pulley sheaves at the top and two at the hook. The fixed end of the rope is secured at the top sheave block. The hoist is to be operated by a standard three-phase four-pole 400 V 50 Hz induction motor, full-load speed 1 475 rpm, controlled by a VSD. A suitable rating needs to be selected for the motor.

The data are shown in Table D.9.

Table D.9 Data for hoist example

Diameter of winding drum	200 mm (0.2 m)
Gearbox ratio, motor to drum	11:1
Coefficient of hoisting friction	0.095

Method

Velocity ratio (VR)

This is a simple numerical ratio determined by the total number of falls (ropes passing between pulleys) or alternatively by counting the total number of pulley wheels in both sheaves (Figure D.23).

This example, with four falls, has a ratio of 4:1.

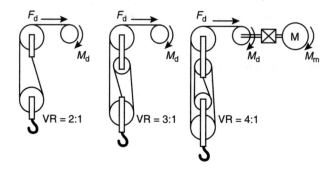

Figure D.23 Pulley ratios

Speed and acceleration of the hook

The speed of the lifting hook is a function of the motor speed, the gearbox ratio, the winding drum diameter and the VR:

Drum speed = motor speed/(gear ratio × 60) (min^{-1})

Hook speed = drum speed × drum circumference lifting ratio

$= [1\,475/(11 \times 60)] \times (0.2\pi/4)$

$= 0.351\,\mathrm{m\,s^{-1}}$

As the load to be raised is sensitive, acceleration from rest must be slow. An acceleration time of 8 s should be adequate. Derive the linear acceleration of the load from

$v = u + at$ where $u = 0$

$a = v/t = 0.351/8 = 4.39 \times 10^{-2}\,\mathrm{(m\,s^{-2})}$

The lifting force and torque to accelerate from rest to full speed

The linear force F_a required at the hook to raise the load from rest up to full lifting speed is the algebraic sum of the forces required to

- suspend the load at rest (mg),
- accelerate the load from rest to lifting speed (ma), and
- overcome rolling friction ($mg\mu$):

$$F_a = mg + ma + mg\mu$$
$$= m(g + a + g\mu)$$
$$= m[a + g(1 + \mu)] \text{ (N)}$$

The equivalent tangential force F_{da} at the winding drum is F_a multiplied by the inverse of the lifting ratio (the inverse, because the calculation is proceeding form the output to the input):

$$F_{da} = F_a/4 = m[a + g(1 + \mu)]/4 \text{ (N)}$$

The accelerating torque at the drum is given by

$$M_{da} = F_{da} \times \text{drum radius}$$
$$= \{m[a + g(1 + \mu)]/4\} \times d/2$$
$$= \{2 \times 10^3 [(4.39 \times 10^{-2}) + 9.81(1 + 0.095)]/4\} \times 0.2/2$$
$$= 539 \text{ N m}$$

The lifting force and torque to maintain full speed

The linear force F_S required at the hook to maintain the load at full lifting speed is the algebraic sum of the forces required to

- suspend the load at rest (mg) and
- overcome rolling friction ($mg\mu$):

$$F_S = mg + mg\mu$$
$$= mg(1 + \mu)$$

The equivalent tangential force F_{ds} at the winding drum is F_S multiplied by the inverse of the lifting ratio:

$$F_{ds} = F_S/4 = mg(1 + \mu)/4 \text{ (N)}$$

The full-speed torque at the drum is given by

$$M_{ds} = F_{ds} \times \text{drum radius}$$
$$= [mg(1+\mu)/4] \times d/2$$
$$= [2 \times 10^3 \times 9.81(1+0.095)/4] \times 0.2/2$$
$$= 537 \text{ N m}$$

The required motor power rating

Power is the multiple of torque and speed. For the purpose of finally calculating the required motor power it is immaterial whether power is calculated at the winding drum or at the motor, provided that the efficiency η, of the gearbox is taken into account. For this example, η is assumed to be 98 per cent, but in practice it must be verified for the gearbox installed.

During acceleration the speed increases from zero to the normal running speed, and therefore the power demand appears to have no particular value. However, it does, because the accelerating torque continues up to the point where the motor reaches full speed, and so the accelerating torque at full speed can be used to calculate the accelerating power demand:

$$P = M\omega/\eta$$

Motor power related to the accelerating torque is given by

$$P_a = M_{da} \times \omega_d/\eta$$

where M_{da} is the accelerating torque at the drum (539 N m), ω_d the rotational speed of the drum in radians per second and η is the efficiency of the gearbox.

Note that the inverse of the efficiency is applied; greater torque and power are required from the motor than the calculated torque at the winding drum:

$$\text{Drum speed} = [\text{motor speed}/(\text{gear ratio} \times 60)] \times \pi$$
$$= [1\,475/(11 \times 60)] \times \pi$$
$$= 7.02 \text{ rad s}^{-1}$$

Accelerating power is given by

$$P_a = 539 \times 7.02/0.98$$
$$= 3.86 \text{ kW}$$

This would be the rating of a motor able to deliver accelerating lifting torque continuously. Standard motors are usually designed with a short-time overload rating of 150 per cent of full-load rating for 30 s. According to the data, accelerating time is 8 s, comfortably within the maximum for which a standard motor is designed. If use is to be made of the overload rating, the actual rating of the motor is $3.86/1.5 = 2.58$ kW.

From inspection of the figures calculated earlier, it is obvious without calculation that the motor rating for acceleration will be adequate for continuous lifting at full speed.

> A standard motor rated 3 kW will be satisfactory

Drive module power rating

In hoisting applications, drive module ratings correspond typically to standard motor ratings, and the short-time overload rating of a drive is typically 150 per cent for 60 s. A drive rating of 3 kW will therefore be appropriate. Some applications require a significantly higher overload for the acceleration and deceleration periods, or conditions with a low duty cycle. In such cases the drive manufacturer should be consulted in order to obtain an optimised solution.

Notes

1. In this example, where the acceleration time is relatively long, the motor rating has been calculated taking advantage of the short-time overload ratings. If the duty cycle is one that demands rapid repetition of acceleration from rest, the overload factor is not an option for either the motor or the drive due to the integration of I^2 with time.
2. A well-designed drive for a.c. motors will possess a torque feedback control feature as well as a programmable acceleration ramp. If the torque feedback is used to control the ramp, the motor can be made to run at a very low speed as soon as the initial torque demand appears. This enables the motor to collect the inevitable slack and backlash in the winding system quickly and then pick up the load slowly, before accelerating under acceleration ramp control.

D1.4.4 Screw-feed loads

Loads such as that shown in Figure D.24 are, in principle, the same as the conveyor example. A mass is moved horizontally. The force F is related to acceleration and mass as in a conveyor. In many instances the mass concerned may be quite low, and the force required to overcome friction may be an insignificant part of the total. However, in machine-tool applications, acceleration times may be especially short, and there may be repetitive duty cycles.

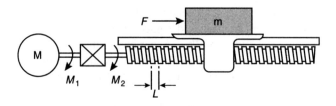

Figure D.24 Screw feed

Heavy machine tools can exploit drive-control of conventional a.c. squirrel-cage motors. Calculation of motor power is essentially the same as for conveyors. Remember the following:

1. Ensure all variables are converted to SI units.
2. Determine the rotational and linear speeds from the motor full speed and the gearing ratio, taking note that the displacement of the load is the 'lead' L of the feed screw per single revolution of the feed screw.
3. Determine the loading and acceleration, and from them the forces and torques to accelerate and to run.

Appendix D2
Conversion tables

D2.1 Mechanical conversion tables

Mechanical conversion tables are provided for length (Table D.10), area (Table D.11), volume (Table D.12), mass (Table D.13), energy (Table D.14), inertia (Table D.15), torque (Table D.16), force (Table D.17) and power (Table D.18).

Table D.10 Mechanical conversion table for length

	mm	cm	m	inch	foot	yard	km	mile
mm	1	10^{-1}	10^{-3}	3.937×10^{-2}	3.280×10^{-3}	1.093×10^{-3}	10^{-6}	6.213×10^{-7}
cm	10	1	10^{-2}	3.937×10^{-1}	3.280×10^{-2}	1.093×10^{-2}	10^{-5}	6.213×10^{-6}
m	1 000	100	1	39.3701	3.28084	1.09361	10^{-3}	6.213×10^{-4}
inch	25.4	2.54	2.54×10^{-2}	1	8.333×10^{-2}	2.777×10^{-2}	2.54×10^{-5}	1.578×10^{-5}
foot	304.8	30.48	3.048×10^{-1}	12	1	3.333×10^{-1}	3.048×10^{-4}	1.893×10^{-4}
yard	914.4	91.44	9.144×10^{-1}	36	3	1	9.144×10^{-4}	5.681×10^{-4}
km	10^6	10^5	1 000	39 370.1	3 280.84	1 093.61	1	6.213×10^{-1}
mile	1.609×10^6	16 0934	1 609.34	63 360	5 280	1 760	1.609	1

Conversion tables 697

Table D.11 Mechanical conversion table for area

	cm^2	m^2	are	hectare	km^2	inch^2	foot^2	yard^2	mile^2	acre
cm^2	1	10^{-4}	10^{-6}	10^{-8}	10^{-10}	1.55 × 10^{-1}	1.076 × 10^{-3}	1.196 × 10^{-4}	3.861 × 10^{-11}	2.471 × 10^{-8}
m^2	10 000	1	10^{-2}	10^{-4}	10^{-6}	1 550	10.7639	1.19599	3.861 × 10^{-7}	2.471 × 10^{-4}
are	10^6	100	1	10^{-2}	10^{-4}	155 000	1 076.39	119 599	3.861 × 10^{-5}	2.471 × 10^{-2}
hectare	10^8	10 000	100	1	10^{-2}	1.55 × 10^7	107 639	11 959.9	3.861 × 10^{-3}	2.47105
km^2	10^{10}	10^6	10 000	100	1	1.55 × 10^9	1.076 × 10^7	1.196 × 10^6	3.861 × 10^{-1}	247.105
inch^2	6.4516	6.4516 × 10^{-4}	6.4516 × 10^{-6}	6.4516 × 10^{-8}	6.4516 × 10^{-10}	1	6.944 × 10^{-3}	7.716 × 10^{-4}	2.491 × 10^{-10}	1.594 × 10^{-7}
foot^2	929.03	9.2903 × 10^{-2}	9.2903 × 10^{-4}	9.2903 × 10^{-6}	9.2903 × 10^{-8}	144	1	1.1111 × 10^{-1}	3.587 × 10^{-8}	2.295 × 10^{-5}
yard^2	8.36127	8.36127 × 10^{-1}	8.36127 × 10^{-3}	8.36127 × 10^{-5}	8.36127 × 10^{-7}	1 296	9	1	3.228 × 10^{-7}	2.066 × 10^{-4}
mile^2	2.589 × 10^{10}	2.589 × 10^6	25 899.9	258.999	2.58999	4.014 × 10^9	2.787 × 10^7	2.0976 × 10^6	1	640
acre	4.046 × 10^7	4 046.86	40.4686	4.04686 × 10^{-1}	4.04686 × 10^{-3}	6.272 × 10^6	43 560	4 840	1.5625 × 10^{-3}	1

Table D.12 Mechanical conversion table for volume

	cm^3	dm^3 (= litre)	inch^3	foot^3	yard^3	US fl oz	Imp fl oz	US gal	Imp gal	Imp pint
cm^3	1	10^{-3}	6.102 × 10^{-2}	3.531 × 10^{-5}	1.308 × 10^{-6}	3.3814 × 10^{-2}	3.519 × 10^{-2}	2.641 × 10^{-4}	2.199 × 10^{-4}	1.759 × 10^{-3}
dm^3	1 000	1	61.0237	3.531 × 10^{-2}	1.308 × 10^{-3}	33.814	35.1951	2.641 × 10^{-1}	2.199 × 10^{-1}	1.75975
inch^3	16.3871	1.638 × 10^{-2}	1	5.787 × 10^{-4}	2.143 × 10^{-5}	5.541 × 10^{-1}	5.767 × 10^{-1}	4.329 × 10^{-3}	3.604 × 10^{-3}	2.883 × 10^{-2}
foot^3	28 316.8	28.3168	1 728	1	3.7037 × 10^2	957.506	996.614	7.48052	6.22884	49.8307
yard^3	764 555	764.555	46 656	27	1	25 852.7	26.9086	201.974	168.179	1 345.43
US fl oz	29.5735	2.957 × 10^{-2}	1.80469	1.044 × 10^{-3}	3.868 × 10^{-5}	1	1.04084	7.8125 × 10^{-3}	6.505 × 10^{-3}	5.204 × 10^{-2}
Imp fl oz	28.4131	2.841 × 10^{-2}	1.73387	1.003 × 10^{-3}	3.716 × 10^{-5}	9.6076 × 10^{-1}	1	7.506 × 10^{-3}	6.25 × 10^{-3}	5 × 10^{-2}
US gal	3 785.41	3.78541	231	1.336 × 10^{-1}	4.951 × 10^{-3}	128	133.228	1	8.326 × 10^{-1}	6.66139
Imp gal	4 546.09	4.54609	277.149	1.605 × 10^{-1}	5.946 × 10^{-3}	153.772	160	1.20095	1	8
Imp pint	568.261	5.682 × 10^{-1}	34.6774	2.0068 × 10^{-2}	7.432 × 10^{-4}	19.2152	20	1.501 × 10^{-1}	1.25 × 10^{-1}	1

Table D.13 Mechanical conversion table for mass

	g	kg	oz	lb	US ton
g	1	10^{-3}	3.5274×10^{-2}	2.204×10^{-3}	1.102×10^{-6}
kg	1 000	1	35.274	2.20462	1.102×10^{-3}
oz	28.2495	2.835×10^{-2}	1	6.25×10^{-2}	3.125×10^{-5}
lb	453.592	4.536×10^{-1}	16	1	5×10^{-4}
US ton	907 185	907.185	32	2 000	1

Table D.14 Mechanical conversion table for energy

	J	Wh	kp m	k cal	BTU
J	1	$2\,778 \times 10^{-4}$	1.019×10^{-1}	2.388×10^{-4}	9.478×10^{-4}
Wh	3 600	1	367.098	8.598×10^{-1}	3.41214
kp m	9.80665	2.724×10^{-3}	1	2.342×10^{-3}	9.295×10^{-3}
k cal	4 186.8	1.163	426.935	1	3.96832
BTU	1 055.06	2.931×10^{-1}	107.586	2.519×10^{-1}	1

Conversion tables 699

Table D.15 Mechanical conversion table for inertia

	kg cm²	kp cm s²	kg m²	kp m s²	oz in²	oz in s²	lb in²	lb in s²	lb ft²	lb ft s²
kg cm²	1	1.019×10^{-3}	10^{-4}	1.019×10^{-5}	5.46748	1.416×10^{-2}	3.417×10^{-1}	8.850×10^{-4}	2.373×10^{-3}	7.375×10^{-5}
kp cm s²	980.665	1	9.806×10^{-12}	10^{-2}	5361.76	13.8874	335.11	8.679×10^{-1}	2.32715	7.233×10^{-2}
kg m²	10^4	10.1927	1	1.019×10^{-1}	54674.8	141.612	3417.17	8.85075	23.7304	7.375×10^{-1}
kp m s²	98066.5	100	9.86065	1	536176	1388.74	33.511	86.7962	232.715	7.23301
oz in²	1.829×10^{-1}	1.865×10^{-4}	1.829×10^{-5}	1.865×10^{-6}	1	2.590×10^{-3}	6.25×10^{-2}	1.6188×10^{-4}	2.340×10^{-4}	7.349×10^{-5}
oz in s²	70.6155	7.201×10^{-2}	7.061×10^{-3}	7.200×10^{-4}	386.089	1	24.1305	6.25×10^{-2}	1.675×10^{-1}	5.208×10^{-3}
lb in²	2.9264	2.984×10^{-3}	2.9264×10^{-4}	2.984×10^{-5}	16	4.144×10^{-2}	1	2.590×10^{-3}	6.944×10^{-3}	2.1548×10^{-4}
lb in s²	1129.85	1.15212	1.29×10^{-1}	1.152×10^{-2}	6177.42	16	386.089	1	2.68117	8.333×10^{-2}
lb ft²	421.401	4.297×10^{-1}	4.214×10^{-2}	4.2971×10^{-3}	2304	5.96754	144	3.729×10^{-1}	1	3.108×10^{-2}
lb ft s²	13558.2	13.8255	1.355	1.382×10^{-1}	74129	192	4633.06	12	32.174	1

Table D.16 Mechanical conversion table for torque

	N cm	N m	kp cm	kp m	p cm	oz in	in lb	ft lb
N cm	1	10^{-2}	1.019×10^{-1}	1.019×10^{-3}	101.972	1.41612	8.850×10^{-2}	7.375×10^{-3}
N m	100	1	10.1972	1.019×10^{-1}	10197.2	141.612	8.85075	7.375×10^{-1}
kp cm	9.80665	9.806×10^{-2}	1	10^{-2}	1000	13.8874	8.679×10^{-1}	7.233×10^{-2}
kp m	980.665	9.80665	100	1	10^5	1388.74	86.7962	7.23301
p cm	9.806×10^{-3}	9.806×10^{-5}	10^{-3}	10^{-5}	1	1.388×10^{-2}	8.679×10^{-4}	7.233×10^{-5}
oz in	7.061×10^{-1}	7.061×10^{-3}	7.200×10^{-2}	7.200×10^{-4}	72.0078	1	6.25×10^{-2}	5.208×10^{-3}
in lb	11.2985	1.129×10^{-1}	1.15212	1.152×10^{-2}	1152.12	16	1	8.333×10^{-2}
ft lb	135.582	1.35582	13.8225	1.382×10^{-1}	13825.5	192	12	1

Table D.17   Mechanical conversion table for force

	N	kp	p	oz	lb f
N	1	1.019×10^{-1}	101.972	3.59694	2.248×10^{-1}
kp	9.80665	1	1 000	35.274	2.20462
p	9.806×10^{-3}	10^{-3}	1	3.5274×10^{-2}	2.204×10^{-3}
oz	2.780×10^{-1}	2.835×10^{-2}	28.3495	1	6.25×10^{-2}
lb f	4.44822	4.536×10^{-1}	453.592	16	1

Table D.18   Mechanical conversion table for power

	kW	PS	hp	kp m s^{-1}	kcal s^{-1}
kW	1	1.35962	1.34102	101 972	2.388×10^{-1}
PS	7.355×10^{-1}	1	9.8632×10^{-1}	75	1.756×10^{-1}
hp	7.457×10^{-1}	1.01387	1	76.0402	1.781×10^{-1}
kp m s^{-1}	9.806×10^{-3}	1.333×10^{-2}	1.3515×10^{-2}	1	2.342×10^{-3}
kcal s^{-1}	4.1868	5.69246	5.61459	426.935	1

D2.2   General conversion tables

General conversion tables are provided for length (Table D.19), area (Table D.20), volume (Table D.21), mass (Table D.22), force and weight (Table D.23), pressure and stress (Table D.24), linear velocity (Table D.25), angular velocity (Table D.26), torque (Table D.27), energy (Table D.28), power (Table D.29), moment of inertia (Table D.30), temperature (Table D.31), flow (Table D.32), and linear acceleration (Table D.33).

Table D.19   General conversion table for length

SI unit: metre (m)		
To convert from	**To**	**Multiply by**
Mile	m	1 609.344
Nautical mile	m	1 853
km	m	10^3
cm	m	10^{-2}
mm	m	10^{-3}
yd	m	0.9144
Ft	m	0.3048
In	m	2.54×10^{-2}
mil	m	2.54×10^{-5}

Conversion tables 701

Table D.20   General conversion table for area

SI unit: square metre (m²)

To convert from	To	Multiply by
Square miles	m^2	2.59×10^6
Acre	m^2	4 047
Hectare ha	m^2	10^4
km^2 (sq. km)	m^2	10^6
cm^2	m^2	10^{-4}
mm^2	m^2	10^{-6}
yd^2	m^2	0.8361
ft^2	m^2	9.29×10^{-2}
in^2	m^2	6.45×10^{-4}
mil^2	m^2	6.45×10^{-10}

Table D.21   General conversion table for volume

SI unit: cubic metre (m³)

To convert from	To	Multiply by
yd^3	m^3	0.765
ft^3	m^3	2.83×10^{-2}
in^3	m^3	1.64×10^{-4}
dm^3	m^3	10^{-3}
Litre	m^3	10^{-3}
Gallon (Imperial)	m^3	4.55×10^{-3}
Gallon (US)	m^3	3.79×10^{-3}
Pint (Imperial)	m^3	5.68×10^{-4}
Pint (US)	m^3	4.73×10^{-4}

Table D.22   General conversion table for mass

SI unit: kilogram (kg)

To convert from	To	Multiply by
Ton (Imperial)	kg	1 016
Ton (US)	kg	907.2
Tonne (metric)	kg	10^3
Slug	kg	14.59
lb	kg	0.4536
oz	kg	2.84×10^{-2}
g	kg	10^{-3}

Table D.23   General conversion table for force and weight

SI unit: Newton (N)

To convert from	To	Multiply by
Tonf (ton wt)	N	9 964
lbf (lb wt)	N	4.448
Poundal	N	0.1383
ozf (oz wt)	N	0.2780
kp	N	9.807
p	N	9.81×10^{-2}
kgf (kg wt)	N	9.807
gf (g wt)	N	9.81×10^{-2}
dyn	N	10^{-5}

Table D.24   General conversion table for pressure and stress

SI unit: Pascal (Pa)

To convert from	To	Multiply by
at (technical atmosphere)	Pa	9.81×10^3
in WG	Pa	248.9
mm WG	Pa	10.34
in HG	Pa	3 385
mm HG (torr)	Pa	131.0
$kp\,cm^{-2}$	Pa	9.81×10^3
$N\,m^{-2}$	Pa	1.0
bar	Pa	10^5
$lbf\,ft^{-2}$	Pa	47.88
$lbf\,in^{-2}$	Pa	6 895
$kgf\,m^{-2}$	Pa	9.807
$kgf\,cm^{-2}$	Pa	9.81×10^4

Table D.25   General conversion table for velocity (linear)

SI unit: metre per second ($m\,s^{-1}$)

To convert from	To	Multiply by
mph (mile per hour)	$m\,s^{-1}$	0.4470
$ft\,min^{-1}$	$m\,s^{-1}$	5.08×10^{-3}
$ft\,s^{-1}$	$m\,s^{-1}$	0.3048
$km\,h^{-1}$	$m\,s^{-1}$	0.2778
$m\,min^{-1}$	$m\,s^{-1}$	1.67×10^{-2}
Knot	$m\,s^{-1}$	0.5145

Table D.26   General conversion table for velocity (angular)

SI unit: radians per second (rad s^{-1})

To convert from	To	Multiply by
rpm (revolutions per min)	rad s^{-1}	0.1037 ($2\pi/60$)
r s^{-1} (revolutions per sec)	rad s^{-1}	6.283 (2π)
$^\circ$ s^{-1} (degrees per sec)	rad s^{-1}	1.75×10^{-2} ($2\pi/360$)

Table D.27   General conversion table for torque

SI unit: Newton metre (N m)

To convert from	To	Multiply by
lbf ft (lb ft)	N m	1.356
lbf in (lb in)	N m	0.1129
ozf in (oz in)	N m	7.062×10^{-3}
kgf m	N m	9.8067
kp m	N m	9.8067
N m	lb ft	0.7375
N m	lb ft	8.857
N m	oz in	141.6

Table D.28   General conversion table for energy

SI unit: Joule (J)

To convert from	To	Multiply by
Btu	J	1.055×10^3
therm (10^5 Btu)	J	1.055×10^8
cal	J	4.187
ft lbf (ft lb wt)	J	1.356
ft poundal	J	0.0421

Table D.29 General conversion table for power

SI unit: kilowatt (kW)

To convert from	To	Multiply by
HP	kW	0.7457
ps	kW	0.7355
ch, CV	kW	0.7355
Btu s^{-1}	kW	1.055
kcal s^{-1}	kW	4.1868
ft lbf s^{-1}	kW	1.36×10^{-3}

Table D.30 General conversion table for moment of inertia

SI unit: kilogram metre^2 (kg m^2)

To convert from	To	Multiply by
GD^2 (in kp m^2)	kg m^2	0.25
kp m s^2	kg m^2	9.807
lb ft s^2	kg m^2	1.356
lb in^2	kg m^2	2.926×10^{-4}
oz in^2	kg m^2	1.829×10^{-5}
lb in s^2	kg m^2	0.113
oz in s^2	kg m^2	7.06155×10^{-3}
kg m^2	lb in s^2	8.85075
kg m^2	oz in s^2	141.612
kg cm^2	kg m^2	10^{-4}

Table D.31 General conversion table for temperature

SI unit: Kelvin (K)

To convert from	To	Factor
°C	K	$\times 1$
t°C	K	t + 273.15
°F	K	$\times 0.5555$
t°F	K	(t − 32) × 0.5555

Table D.32   General conversion table for flow

SI unit: cubic metre per second ($m^3\ s^{-1}$)		
To convert from	**To**	**Multiply by**
Gallon per hour (Imp)	$m^3\ s^{-1}$	1.26×10^{-6}
Gallon per hour (US)	$m^3\ s^{-1}$	1.05×10^{-6}
Litre per hour	$m^3\ s^{-1}$	1.67×10^{-5}
Litre per second	$m^3\ s^{-1}$	10^{-3}
cfm	$m^3\ s^{-1}$	4.72×10^{-4}
$m^3\ h^{-1}$	$m^3\ s^{-1}$	2.78×10^{-4}
$m^3\ min^{-1}$	$m^3\ s^{-1}$	1.67×10^{-2}

Table D.33   General conversion table for linear acceleration

SI unit: metre per second^2 ($m\ s^{-2}$)		
To convert from	**To**	**Multiply by**
$in\ s^{-2}$	$m\ s^{-2}$	2.54×10^{-2}
$ft\ s^{-2}$	$m\ s^{-2}$	0.3048
$m\ s^{-2}$	$in\ s^{-2}$	39.37
$m\ s^{-2}$	$ft\ s^{-2}$	3.2808

D2.3 Power/torque/speed nomogram

A power/torque/speed nomogram is shown in Figure D.25.

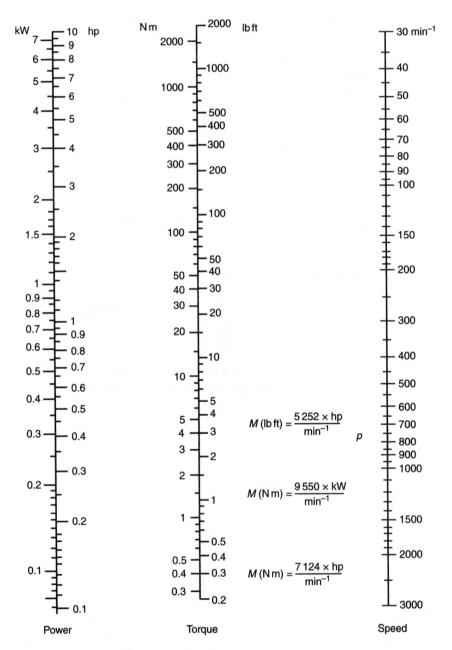

Figure D.25 Power/torque nomogram

Appendix D3
World industrial electricity supplies (<1 kV)

Data for world industrial electricity supplies (<1 kV) are shown in Table D.34.

Table D.34 World industrial electricity supplies (<1 kV)

Country	Industrial three-phase supply voltages below 1 000 V (V)	Supply frequency (Hz)
Afghanistan	380/220	50
Albania	380/220	50
Algeria	380/220	50
Andorra	400/230	50
Angola	380/220	50
Anguilla	415/240	60
Antigua and Barbuda	400/230	60
Argentina	380/220	50
Armenia	380/220	50
Aruba	220/127	60
Australia	400/230	50
Austria	400/230	50
Azerbaijan	380/220	50
Bahamas	480 240/120	60
Bahrain	400/230	50
Bangladesh	400/230	50
Barbados	400/230	50
Belarus	380/220	50
Belgium	400/230	50
Belize	440–220	60
Benin	220–380	50
Bermuda	208/120 240/120	60

(*Continued*)

Table D.34   Continued

Country	Industrial three-phase supply voltages below 1 000 V (V)	Supply frequency (Hz)
Bhutan	400/230	50
Bolivia	400/230	50
Bosnia and Herzegovinia	380/220	50
Botswana	400/230	50
Brazil	380/220	60
Brunei Darussalam	415/240	50
Bulgaria	380/220	50
Burkina Faso	400/230	50
Burundi	400/230	50
Cambodia	380/220	50
Cameroon	380/220	50
Canada	600/347 416/240	60
Canary Islands	380/220	50
Cape Verde Islands	380/220	50
Cayman Islands	480/277 480/240	60
Central African Republic	380/220	50
Chad	380/220	50
Chile	440 380/220	50
China	380/220	50
Colombia	240/120	60
Comoros	380/220	50
Congo	380/220	50
Cook Islands	415/240	50
Costa Rica	480/277 240/120	60
Croatia	400/230	50
Cuba	400/230	60
Cyprus	400/230	50
Czech Republic	400/230	50
Denmark	400/230	50
Djibouti	400/230	50
Dominica	400/230	50
Dominican Republic	480 220/110	60

(*Continued*)

Table D.34 Continued

Country	Industrial three-phase supply voltages below 1 000 V (V)	Supply frequency (Hz)
Ecuador	440/220	60
Egypt	380/220	50
El Salvador	440/220 240/120	60
Eritrea	380/220	50
Estonia	380/220	50
Ethiopia	380/230	50
Falkland Islands	415/240	50
Faroe Islands	415/240	50
Fiji	415/240	50
Finland	690/400 400/230	50
France	690/400 400/230	50
French Guiana	380/220	50
French Polynesia	220/127	60
Gambia	380/220	50
Georgia	380/220	50
Germany	690/400 400/230	50
Ghana	415/240	50
Gibraltar	415/240	50
Greece	400/230	50
Greenland	400/230	50
Grenada	400/230	50
Guadeloupe	380/220	50, 60
Guam	480/227 240/120	60
Guatemala	220/110	60
Guyana	415/240	50, 60
Haiti	220/110	60
Honduras	480/277 240/120	60
Hong Kong	380/220	50
Hungary	400/230	50
Iceland	400/230	50
India	400/230	50
Indonesia	380/220	50

(*Continued*)

Table D.34 Continued

Country	Industrial three-phase supply voltages below 1 000 V (V)	Supply frequency (Hz)
Iran	400/230	50
	380/220	
Iraq	380/220	50
Ireland, Northern	400/230	50
Ireland, Republic of	400/230	50
Israel	400/230	50
Italy	400/230	50
Ivory Coast	400/230	50
Jamaica	220/110	50
Japan	380/220	50, 60
	200/100	
Jordan	415/240	50
	400/230	
Kazakhstan	400/230	50
	380/220	
Kenya	415/240	50
Korea, North	380/220	60
Korea, South	380/220	60
Kuwait	400/230	50
Kyrgyzstan	380/220	50
Laos	380/220	50
Latvia	380/220	50
Lebanon	380/220	50
Lesotho	380/220	50
Liberia	400/230	50
	380/220	
Libya	380/220	50
Liechtenstein	400/230	50
Lithuania	400/230	50
Luxembourg	400/230	50
Macau	400/230	50
Macedonia	380/220	50
Madagascar	380/220	50
Malawi	400/230	50
Malaysia	415/240	50
Maldives	400/230	50
Mali	380/220	50
Malta	400/230	50

(*Continued*)

Table D.34 Continued

Country	Industrial three-phase supply voltages below 1 000 V (V)	Supply frequency (Hz)
Martinique	400/230	50
Mauritania	380/220	50
Mauritius	430/230	50
Mexico	480/227	60
	220/127	
Moldova	380/220	50
Monaco	400/230	50
Mongolia	380/220	50
Montserrat	400/230	60
Morocco	380/220	50
Mozambique	380/220	50
Mustique	400/230	50
Myanmar	400/230	50
Namibia	380/220	50
Nepal	400/220	50
Netherlands	400/230	50
Netherlands Antilles	380/220	50, 60
	230/115	
	220/127	
	208/120	
New Zealand	400/230	50
Nicaragua	480/240	60
	240/120	
Niger	380/220	50
Nigeria	400/230	50
	380/220	
Norway	690	50
	400/230	
Oman	415/240	50
Pakistan	400/230	50
Panama	480/277	60
	208/120	
Papua New Guinea	415/240	50
Paraguay	380/220	50
Peru	380/220	60
Philippines	440/240	60
Poland	690/400	50
	400/230	
Portugal	400/230	50

(Continued)

Table D.34 Continued

Country	Industrial three-phase supply voltages below 1 000 V (V)	Supply frequency (Hz)
Puerto Rico	240/120	60
Qatar	415/240	50
Romania	400/230 660/380 440/220	50
Russian Federation	660/380/220	50
Rwanda	380/220	50
St Kitts and Nevis	400/230	60
St Lucia	415/240	50
St Vincent and the Grenadines	400/230	50
Samoa	400/230	50
San Marino	380/220	50
Saudi Arabia	380/220	60
Senegal	380/220 220/127	50
Seychelles	400/230	50
Sierra Leone	400/230	50
Singapore	400/230	50
Slovak Republic	400/230	50
Slovenia, Republic of	660 500 400/230	50
Solomon Islands	415/240	50
Somalia	440/220 220/110	50
South Africa	525 400/230	50
Spain	400/230	50
Sri Lanka	400/230	50
Sudan	415/240	50
Suriname	440/220/127	60
Swaziland	400/230	50
Sweden	400/230	50
Switzerland	690/400 400/230	50
Syrian Arab Republic	380/220	50
Taiwan	380/300	60

(*Continued*)

Table D.34   Continued

Country	Industrial three-phase supply voltages below 1 000 V (V)	Supply frequency (Hz)
Tajakistan	380/220	50
Tanzania	400/230	50
Thailand	380/220	50
Togo	400/230	50
Tonga	415/240	50
Trinidad and Tobago	400/230	60
Tunisia	400/230 380/220	50
Turkey	380/220	50
Turkmenistan	380/220	50
Uganda	415/240	50
Ukraine	380/220	50
United Arab Emirates	415/240 380/220	50
United Kingdom	400/230	50
United States of America	460/265 208/120 575	60
Uruguay	380/220	50
Uzbekistan	380/220	50
Vatuatu	380/220	50
Venezuela	480/277 208/120	60
Vietnam	380/220	50
Yemen, Republic of	440/250	50
Yugoslavia	380/220	50
Zambia	400/230	50
Zimbabwe	390/225	50

Bibliography

The documents listed here have been selected to provide the reader with useful sources of information and further reading relating to electrical variable-speed drives and their application. Some of the books are out of print but can be obtained through internet searches.

General electrical engineering

G.R. Jones, M.A. Laughton and M.G. Say, *Electrical Engineer's Reference Book*, Butterworth-Heinemann, 1993, ISBN 7506 1202 9. ('Everything' in one book.)

Electric motors

B.J. Chalmers, *Electric Motor Handbook*, Butterworths, 1988, ISBN 0 408 00707 9. (A practical reference book covering many aspects of characteristics, specification, design, selection, commissioning and maintenance.)

C.V. Jones, *The Unified Theory of Electrical Machines*, Butterworths, 1967. (Classic machine theory.)

P. Vas, *Electrical Machines and Drives*, Oxford University Press, 1992, ISBN 0 19 859378 3. (A definitive book on space vector theory of electrical machines.)

Drives: General

G. Moltgen, *Converter Engineering*, John Wiley, 1984, ISBN 0 471 90561 5. (A reference for fundamental power converter operations and relationships.)

K.K. Schwarz, *Design of Industrial Electric Motor Drives*, Butterworth-Heinemann Ltd, 1991, ISBN 0 7506 1141 3. (Useful, motor drive and system design information.)

B. Wu, *High Power Converters and AC Drives*, John Wiley & Sons Inc., 2006, ISBN 13 978 0 471 73171 9 / ISBN 10 0 471 73171 4. (Good practical book on high power a.c. drive topologies.)

Drives: Control

P. Barrass and M. Cade, 'PWM Rectifier Using Indirect Voltage Sensing', *IEE Proceedings Part B*, September 1999, pp. 539–44. (Detailed description of PWM rectifier control.)

M. Cade, 'Improvement of Induction Machine Stability by Modulation Techniques', *IEE Proceedings Part B*, November 1994, pp. 347–52. (Detailed description of space vector modulation.)

M. Kaufhold et al., 'Failure mechanism of the inter-turn insulation of low voltage electric machines fed by pulse-controlled inverters', *IEEE Electrical Insulation Magazine*, Vol. 12, No. 5, 1996.

P. Vas, *Sensorless Vector and Direct Torque Control*, Oxford University Press, 1998, ISBN 0198564651. (General background to the theory of vector control of motors.)

P. Vas, *Vector Control of AC Machines*, Oxford University Press, 1990, ISBN 0 19 859370 8. (Forerunner of the previous book – good advanced-level introduction.)

G.W. Younkin, *Industrial Servo Control Systems*, Marcel Dekker Inc., 2003, ISBN 0 8247 0836 9. (Good insight into servo control loop analysis and application.)

System design

F.T. Brown, *Engineering System Dynamics*, Marcel Dekker Inc., 2001, ISBN 0 8247 0616 1. (Good book on the application of control to practical systems.)

M. Neale, P. Needham and R. Horrell, *Couplings and Shaft Alignment*, Professional Engineering Publishing Limited, 1991, ISBN 1 86058 170 6. (A practical guide to coupling selection and problems of shaft alignment.)

A. Wright and P.G. Newbury, *Electrical Fuses*, 2nd edition, IEE Power Series 20, ISBN 0 85296 825 6. (A clear guide to fuse design, performance, application and good practice.)

Index

a.c. commutator motor: *see* a.c. motor
a.c. converter: *see* a.c. drive
a.c. drive
   current source inverter 99–102
      converter-fed synchronous machine (LCI) 100
      converter-fed induction motor drive 101
      forced commutated induction motor drive 101
      static Kramer drive 102
   direct a.c. to a.c. converter 103–8
      cycloconverter 104–5
      matrix converter 106
      static Scherbius drive 107
      soft starter/voltage regulator 103
   performance summary 6–9
   voltage source inverter with intermediate d.c. link 91–102
      four quadrant operation 253–60, 400–2
      multi level 97–9
      pulse width modulation (PWM) 6–7, 96–7, 302, 318, 336–8, 342–9, 376–8, 400–2, 415–24, 435
a.c. induction motor: *see* a.c. motor
a.c. inverter: *see* a.c. drive
a.c. motor 36–70
   a.c. commutator motor 63–4
   a.c. induction motor
      construction 50–2
      control: *see* drive control
      DOL starting 43–5
      equivalent circuit 39–41
      pole amplitude modulated (PAM) speed change 50
      parameter determination 41–2
      performance 210
      slip 39
      slip ring induction motor 48–9
      speed changing motor 50
      steady state performance 43–5
      torque production 38–42
   theory 36–50
a.c. synchronous motor 52–61
   back-emf constant 54
   construction 58–61
   control: *see* drive control
   equivalent circuit 53–6
   starting 61–2
   theory 52–7
   torque constant 55–6
   see also a.c. motor/PM motor
balancing 352–3
bearing current 349–52
cooling 383–8
direction of rotation 363–70
enclosure degree of protection 355–6
insulation 337–48
installation and maintenance: *see* installation and maintenance/motors
mounting 360, 363
noise 373–7
overspeed 352, 624
permanent magnet (PM) motor
   construction 58–61
   control: *see* drive control
   equivalent circuit 55
   limits of operation 57, 386–8
   performance 209
   theory 53–5
   see also a.c. motor/a.c. synchronous motor
reluctance motor 62
rotating magnetic field 37–8
terminal markings 363–5, 371
thermal protection 438–40, 538
see also switched reluctance motor drive

Index

a.c. motor control: *see* drive control
a.c. supply: *see* power supply system
a.c. synchronous motor: *see* a.c. motor
a.c. to a.c. converter: *see* a.c. drive
a.c. to d.c. converter 72–85
   single-phase supply 73–7
   three-phase supply 78–81
   see also d.c. drive
absolute encoder: *see* encoder
acceleration control: *see* control/motion control
accuracy
   encoder: *see* encoder
   analogue signal 486
acoustic noise 373–7
active input stage: *see* a.c. drive/voltage source inverter/four quadrant operation
air conditioning: *see* HVAC
air filter: *see* thermal management *and* installation and maintenance
air flow: *see* thermal management
altitude: *see* ambient conditions
ambient conditions
   altitude 380
   corrosive gasses 380–1
   humidity and condensation 373, 455
   temperature 371–2
   vibration 378–80
ambient temperature: *see* ambient conditions
analogue signal 486–98, 537
angular motion: *see* control/motion control
anti-condensation heater: *see* thermal management
armature: *see* a.c. motor/construction *and* d.c. motor/construction
armature reaction 22–4
auto-transformer starting: *see* a.c. motor/synchronous motor
autotune 536

back-emf constant: *see* a.c. motor/a.c. synchronous motor *and* a.c. motor/PM motor
backlash 172
battery charging 595
bearings 32, 222, 349–51, 373–5, 444, 449–52, 592, 644,
bearing current: *see* a.c. motor

bipolar junction transistor (BJT): *see* power semiconductors
Bode plot 174–5, 182–4
brake motor 64
brake resistor 256–7
braking 253–60
brown out: *see* power supply system/voltage dips and short interruptions
brush gear: *see* d.c. motor
brushless PM motor: *see* a.c. motor/PM motor
brushless servo motor: *see* a.c. motor/PM motor

cabling
   cable screening 300, 414–25, 426–31, 487–97
   long motor cable 337–49
CAM function: *see* control/motion control
CAN: *see* communication networks
CANopen: *see* communication networks
centre winder: *see* control/winding
chemical industry 466
chopper: *see* d.c. to d.c. converter
closed-loop control: *see* drive control *and* control
cogging 33, 95–6, 210,
commissioning 251, 449–56
common d.c. bus 393–409
communication networks 501–28
   BACnet 521
   CANopen 519
   CTNet 526
   CTSync 527
   DeviceNet 518
   EtherCAT 525
   Ethernet 523–6
   Ethernet IP 524
   Fieldbus systems 516–28
   gateways 528
   Interbus 520
   LonWorks 520
   Modbus TCP/IP 523
   OPC 517
   OSI model 502–8
   Powerlink 526
   Profibus DP 517
   PROFINET 525
   RS-232/RS-485 508
   SERCOS II 522

commutator: *see* d.c. motor
compliance angle 180, 183–5
compressor 606
  see also HVAC
concrete pipe manufacture 467
condensation: *see* ambient conditions
continuous casting 633–5
control 171–210
  centralised control 512–13
  distributed control 513–14
  load sharing 567–77
  hybrid control 514–15
  motion control 233–52
    CAM profile 243–7
    electronic gearbox 248
    indexer 250
    position, speed, acceleration and jerk 234–8
    registration 562–6
    time-based profile 239–44
  position control 186–91, 249
  sectional control 556–61, 579
  speed control: *see* drive control
  tension control 578
  torque feed forward 555
  torque slaving 574
  virtual master 556–61
  winding 580–8
  see also drive control
conversion tables 695–705
converter-fed induction motor drive:
  see a.c. drive
converter-fed synchronous machine:
  see a.c. drive
converter 71–108
  see also a.c. drive
  see also a.c. to d.c. converter
  see also d.c. drive
  see also d.c. to d.c. converter
conveyor 684–9
cooling: *see* thermal management
  see also a.c. motor/cooling *and* d.c. motor/cooling
corrosive gasses: *see* ambient conditions
cranes 467, 616–21
critical speed 444
CTNet: *see* communication networks
CTSync: *see* communication networks
current source inverter: *see* a.c. drive
cycloconverter: *see* a.c. drive

d.c. drive
  single-converter 82–3
  chopper: *see* d.c. to d.c. converter
  dual converter 84
  performance summary 4–5
d.c. motor
  back-emf 26
  brush gear 34–5
  commutator 21–34, 385, 445, 450–4
  compound d.c. motor 30–1
  construction 32–5
  control: *see* d.c. motor control
  cooling 383–6
  enclosure degree of protection 34, 355–6
  direction of rotation 363–70
  installation and maintenance:
    see installation and maintenance/motors
  mounting 34, 360
  noise 373–7
  permanent magnet d.c. motor 31–2, 35
  separately excited d.c. motor 25–8
  series d.c. motor 29–30
  shunt d.c. motor 30
  terminal markings 365–70
  theory 20–36
  torque production 26
d.c. motor control: *see* drive control
d.c. to d.c. converter 86–90
  see also d.c. drive
degree of protection (IP rating)
  drive 356–9
  motor: *see* a.c. motor *and* d.c. motor
DeviceNet: *see* communication networks
digital signal 499–528, 537
digital slaving: *see* frequency slaving *and* load sharing
diode: *see* power semiconductors
diode rectifier: *see* a.c. to d.c. converter
direct on line (DOL) starting: *see* a.c. motor/a.c. induction motor
direct torque control (DTC): *see* drive control
  see also a.c. motor control
direction of rotation: *see* a.c. motors *and* d.c. motors
distributed control: *see* control
drive characteristics 473–6

720　*Index*

drive control
  a.c. drive
    a.c. induction motor
      open loop 205
      performance comparison 208–9
      rotor flux control (RFC) 204–6
      torque and flux control 203–4
    PM motor 197–202
    direct torque control 206–8
  d.c. drive
    flux control 194
    torque control 192–3
  drive controllers (generic)
    flux control 179
    general principles 171–85
    position control 186–91
    speed control 179–85
    torque control 176–78
  see also control
drive converter effects: *see* drive-motor interaction
drive converter topologies 71–108
  see also a.c. drive
  see also a.c. to d.c. converter
  see also d.c. drive
drive-motor interaction 335–54
  a.c. motor 336–52
  d.c. motor 335–6
  motors in hazardous conditions 353
drive mounting 360–2
  see also enclosures
  see also installation and maintenance/electronic equipment
  see also thermal management
dual-converter: *see* d.c. drive
duty cycle 477–84
dynamic braking resistor: *see* brake resistor

earth leakage current 420, 426
earth fault 435
earthing 350–1, 416–19, 428–31, 488–9
efficiency 7, 20, 36, 43–4, 61, 95–6, 109, 282, 336–7, 353, 371, 372, 393–4, 438, 600–4, 623, 651, 667, 682, 688–93
electric heating 596
electrolytic process 596
electromagnetic compatibility: *see* EMC
electromagnetic principles 11–19
electromechanical energy conversion 16–19
electromotive force (emf) 26, 46–8

electronic gearbox: *see* control/motion control
elevator 467, 622–7
EMC 411–32
  behaviour of variable speed drives 414–15
  emission 414–15
  filter 389–92, 409, 420–26
  immunity 414
  installation rules/cabling and wiring recommendation 416–32
  regulations and standards 411–13
  risk assessment 416–17
  theory 422–6
emission: *see* EMC
EN 60034-5 355–60
EN 60034-6 385
EN 60034-7 360–3
EN 60034-8 363–6
EN 61000-3-2 301–2
EN 61800-2 446
enclosures
  thermal design 389–92
  degree of protection: *see* degree of protection
encoder 211–32
  absolute 226–30
  BiSS 230
  commutation signals 223–4
  EnDat 228
  environment 215
  Hiperface 229
  incremental 221–4
  interpolation 224–6
  maximum speed 215
  noise immunity 215
  SINCOS 224–30
  specification 211–16
  SSI 229
  wireless 231–2
Ethernet: *see* communication networks
EtherCAT: *see* communication networks
explosion protection: *see* hazardous locations
extruder 464, 466, 471, 472, 649–57

fans 467–8, 606
feedback devices 211–31
field control 85
field weakening 23, 28, 57, 83, 85–6, 210, 254, 469, 473–6, 593
fieldbus: *see* communication networks

Fleming's left hand rule 18
Fleming's right hand rule 19
flux controller: *see* drive control
flying shear 463, 527, 563, 565
  see also rotary knife
following error 187
food, biscuit and confection 472
forced-commutated induction motor drive:
  see a.c. drive
four-quadrant operation/braking 84–5,
  89–90, 99, 105, 107, 192–3,
  253–60, 399–402, 463, 464,
  466, 467, 468, 469, 470, 471,
  472, 476, 630
frequency reference 533
frequency slaving 535
functional safety 543–9
  advanced drive-specific functions 547
  integration into a machine 549
  machinery safety functions 548
  principles 543
  safe torque off (STO) 546
  safety bus interface 549
  standards 544–5
fuses 433–4, 455

gate turn-off thyristor (GTO): *see* power
  semiconductors
geared motor 64
Gray code 227

harmonic filter 106, 318, 322,
harmonics: *see* power supply system
hazardous locations 65–70
high frequency inverter/motor 589–94
hoist: *see* cranes
humidity: *see* ambient conditions
HVAC 607–16
hybrid control: *see* control
hydraulics 471

IEC 60034-5 355–60
IEC 60034-6 385
IEC 60034-7 360–3
IEC 60034-8 363–6
IEC 61000-3-2 301–2
IEC 61800-2 446
IEEE c62.45 436–8
IEEE Std 519 307–8, 321
immunity: *see* EMC
incremental encoder: *see* encoder

indexing: *see* control/motion control
induction motor: *see* a.c. motor/a.c.
  induction motor
inertia compensation 555–6, 587–9,
  631–2, 646–9
installation and maintenance
  electronic equipment 454–5
  motor 449–53
  see also EMC
insulated gate bipolar transistor (IGBT):
  see power semiconductors
integrated gate commutated thyristor
  (IGCT): *see* power semiconductors
integrated motor drive 363
intelligent drive programming 539–42
interaction between drive and motor: *see*
  drive-motor interaction
Interbus: *see* communication networks
interpolation: *see* encoder
IP rating: *see* degree of protection *and* a.c.
  motor *and* d.c. motor
IT supply: *see* power supply system

jerk: *see* control system/motion control
junction field effect transistor (JFET):
  see power semiconductors

Kramer drive: *see* a.c. drive

laws of motion 674–84
lift: *see* elevator
linear motion: *see* control/motion control
load characteristics 461–71
load sharing: *see* control

machine tools 464
magnetic circuit 11–16
magnetic materials 14
magnetic saturation 13–15
maintenance: *see* installation and
  maintenance
marine applications 435
master-slave control 507–27, 573–4
material handling 466
matrix converter: *see* a.c. drive
maximum encoder speed: *see* encoder
maximum motor speed 4–9, 25, 234,
  352–3
metals industry 462–3, 627–37
Modbus: *see* communication networks
Moiré principle 221

722 Index

MOS controlled thyristor (MCT): *see* power semiconductors
MOS turn-off thyristor: *see* power semiconductors
MOSFET: *see* power semiconductors
motion control: *see* control
motor: *see* a.c. motor *and* d.c. motor
motor cooling: *see* a.c. motor *and* d.c. motor
motor mountings: *see* a.c. motor *and* d.c. motor
motor protection: *see* a.c. motor *and* d.c. motor
motors for hazardous locations: *see* a.c. motor
multilevel converter: *see* a.c. drive
multiple parameter sets 537

network topology: *see* communication networks
noise 373–7
notching: *see* power supply system

OPC: *see* communication networks
open-loop a.c. drive: *see* a.c. drive
open-loop control: *see* drive control
open-loop induction motor: *see* a.c. drive
open-loop inverter: *see* a.c. drive
OSI model: *see* communication networks
operating temperature: *see* ambient conditions
over-current protection 114, 159, 318, 434, 440, 501, 536, 538, 689
overspeed 352
over-temperature protection 357, 455, 538
overvoltage 254, 327, 333–4, 414, 437, 501, 538, 594

packaging industry 469
paper manufacturing 468, 638–48
PC tools 528–32
  application configuration and set-up tools 530
  design tools 529
  drive programming tools 529
  monitoring tools 531
permanent magnet (PM) motor: *see* a.c. motor
phase angle control: *see* a.c. to d.c. converter *and* d.c. drive
PI controller 173–5, 177, 179–83, 220, 254, 259, 535–6

plastics industry 464–5
plastic extrusion 649–56
point of common coupling: *see* power supply system/harmonics
pole amplitude modulated (PAM) speed change: *see* a.c. motor/a.c. induction motor
position control: *see* control
position lock 244
position sensor: *see* encoder *and* resolver
power factor 54, 63, 72, 76–7, 81, 87, 92, 96, 100–2, 254, 257–8, 300, 323, 324–6, 329, 400
power semiconductors 109–70
  bipolar junction transistor (BJT) 159
  diode 114–21
  gate turn off thyristor (GTO) 130–5
  insulated gate bipolar transisor (IGBT) 147–59
  integrated gate commutated thyristor (IGCT) 135–7
  junction field effect transistor (JFET) 162
  materials 162
  MOS controlled thyristor (MCT) 160
  MOS turn-off thyristor 161
  MOSFET 137–47
  packaging 163–70
  performance/application summary 110–12
  thyristor (SCR) 122–9
  triac 130
power supply system
  d.c. supply: *see* common d.c. bus
  fault level 311
  flicker 323
  frequency variation 326
  generator supply 310
  harmonics 299–323
  imperfections 326–34
  interharmonics 321
  IT supplies 435
  notching: *see* power supply system/voltage notching
  over-voltages 327
  power factor 324–6
  voltage dips and flicker 323
  voltage dips and short interruptions 329–30
  voltage notching 331–2, 322
  voltage transients 327, 436
  voltage variation 326

Index 723

voltage unbalance 327–9
worldwide voltages 707–14
power/torque/speed nonogram 706
printing 469
Profibus DP: *see* communication networks
PROFINET: *see* communication networks
programmable logic 537
programming tools: *see* PC tools
protection
    drive 538
    motor: *see* a.c. motor
pump 468, 599–605
pulse width modulation (PWM):
    see a.c. drive

ramps 62, 104, 186–8, 205, 234–43, 247–9,
    293, 469, 472, 501, 534, 535, 541,
    552–4, 579–80, 587–9, 594, 614,
    620–1, 632–6, 693
rectifier: *see* power semiconductors *and* a.c.
    to d.c. converter *and* d.c. drive
reference pulse 535
refrigeration: *see* HVAC
registration: *see* control system/motion
    control
regenerative braking: *see* braking
reluctance motor: *see* a.c. motor
reluctance torque 16–18
resistive braking: *see* braking
resolver 218–20
resolver to digital conversion 220
resonance 444
rolling mill 568, 627–33
root mean square (rms) value 672
rotary knife 249, 463, 527, 562–7
    see also flying shear
rotating magnetic field: *see* a.c. motor
rotor copper loss 38–41, 43, 104, 210, 329
RS-232: *see* communication networks
RS-485: *see* communication networks
rubber 465

safety functions: *see* functional safety
Scherbius drive: *see* a.c. drive
Schrage motor 63
screened cable: *see* EMC and installation
    and maintenance
sectional drive system: *see* control
servo motor: *see* a.c. motor/PM motor
sequencer 537
shaft vibrations 443–4

SI units 663–4
silicon carbide 122, 162
silicon-controlled rectifier (SCR):
    see power semiconductors
SINCOS encoder: *see* encoder
single phase converter: *see* a.c. to d.c.
    converter
sinusoidal filter 340–2, 377
six pulse converter: *see* a.c. to d.c. converter
six step converter/inverter 94–5
slip: *see* a.c. motor/a.c. induction motor
slip-ring induction motor: *see* a.c. motor/a.c.
    induction motor
snubber network 99, 111, 118, 126–7, 131–6,
    156, 332–3
soft start 103–4
software comissioning tool: *see* PC tools
space vector modulator: *see* drive
    control/a.c. drive
speed accuracy 214–5, 217
    see also drive control
speed controller: *see* drive control
speed following 446
speed range 4–9, 30, 34, 46–9, 50, 96, 103,
    107, 210, 261, 282, 286, 335, 352,
    386, 462–73, 473–6
speed reference 533
speed response 181, 189
speed sensor: *see* tachogenerator *and* encoder
    and resolver
square-wave voltage source inverter:
    see a.c. drive
squirrel-cage induction motor: *see* a.c.
    motor/a.c. induction motor
stage scenery 657–9
star-delta starting 439, 605
    see also a.c. motor/induction motor/DOL
    starting
static Kramer drive: *see* a.c. drive
static Scherbius drive: *see* a.c. drive
stator copper loss 38–41, 43
steel industry 627–34
step response 173–5, 180–7
stepper motor drive 284–95
storage 449
supply frequency: *see* power supply system
supply system: *see* power supply system
supply voltage: *see* power supply system
supply harmonics: *see* power supply system
switched reluctance (SR) motor drive
    261–83

synchronous motor: *see* a.c. motor/a.c. synchronous motor

tacho: *see* tachogenerator
tachogenerator 216–17
temperature control 596
tension control: *see* control
terminal markings: *see* a.c. motor *and* d.c. motor
test rig 470
textiles 472
theatre 657–9
thermal management 383–92
  drive cooling 389–92, 455
    enclosure design 389–92
  motor cooling 383–8, 450
thermal overload
three phase converter: *see* a.c. to d.c. converter *and* d.c. drive
thyristor: *see* power semiconductors
token ring: *see* communication networks
torque constant: *see* a.c. motor/a.c. synchronous motor *and* a.c. motor/PM motor
torque control: *see* drive control
torque feed forward: *see* control/speed control
torque motor 64–5
torque ripple 7, 31, 33, 59, 183, 210, 224, 272, 282–3, 444–6
  see also torsional dynamics
torsional dynamics 441–7
torsional stiffness 180–90, 441–3
total harmonic distortion (THD) 76, 93, 301, 305, 319–21, 325

transient voltage suppressor 425–6
trench gate: *see* power semiconductors/IGBT
triac: *see* power semiconductors
trips 538
twelve pulse converter 377, 397–9, 567–8, 570–2, 622, 628–32

user-defined functionality 539–42
uncontrolled bridge: *see* a.c. to d.c. converter
uncontrolled converter: *see* a.c. to d.c. converter
uncontrolled rectifier: *see* a.c. to d.c. converter
unipolar switching 288–90

variable frequency inverter: *see* a.c. drive
variable speed drive: *see* a.c. drive *and* d.c. drive
variable reluctance motor 261
vector control: *see* a.c. motor control
ventilation: *see* thermal management and HVAC
vibration 378–80, 441–6
virtual master: *see* control
voltage boost 48
voltage dips and flicker: *see* power supply system
voltage notching: *see* power supply system
voltage source inverter: *see* a.c. drive
voltage transients: *see* power supply system
voltage unbalance: *see* power supply system
volts per Hertz (V/f) 45–7

winder: *see* control
wire and cable 635–7, 470–1

Control Techniques - a market leader in drives and drive systems

A part of Emerson Industrial Automation, Control Techniques has a single minded focus on the design and manufacture of variable speed drives. Our drives are used to control motors in a wide range of applications, from precision machines to high performance elevators and from cranes to fans. Whatever the application, Control Techniques drives deliver an effective solution to increase productivity and reduce energy consumption.

Business without borders
Control Techniques is a global player, with manufacturing and E&D (Engineering and Design) facilities in Europe, USA and Asia. Our 91 subsidiary Drive Centres and resellers in 67 countries offer customers local technical sales, service and design expertise; many also offer a comprehensive system design and build service.

Our expertise is in your industry
Our experience and expertise in a broad range of applications, allows us to work with you to maximize the performance of your machinery. We integrate the best available drive technology to enhance your existing application and redefine the possibilities for your new investments.

> "Market research suggests our customers choose Control Techniques because they have confidence in our ability to provide solutions where product performance and quality support are most highly valued."

What's in a drive?

Always a step ahead

Many of the drive technologies that are common today were first developed into high volume commercial products by Control Techniques, such as the first digital DC drive and the first AC closed loop vector drive. We continue to be an innovator delivering a drives range that technologically excels.

Unidrive SP

High performance universal drive for AC motors from 0.37kW to 1.9MW

Unidrive SP is a universal AC drive that allows you to control virtually any type of AC motor with maximum performance and total flexibility.

Proven in thousands upon thousands of applications Unidrive SP adds value wherever it is installed. With powerful onboard automation & motion control functionality, total communications flexibility and the widest power range to cover every application.

Commander SK

General purpose drive for AC motors from 0.25kW to 132kW

Commander SK is simple to use, compact and offers great value. It has excellent open loop performance, optional onboard PLC functionality and click-in modules for I/O, Ethernet and Fieldbus communications. Commander SK allows you to do much more than you would ever expect with a general purpose drive.

Mentor MP
High performance drive for DC motors
25A to 7400A
Two or four quadrant operation

Mentor MP is Control Techniques' new DC drive and integrates the control platform from the world's leading intelligent AC drive, Unidrive SP. This makes Mentor MP the most advanced DC drive available, giving optimum performance and flexible system interfacing capability.

Affinity
Dedicated drive for building automation and refrigeration
1.1kW to 132kW (1.5hp to 200hp)
IP20 (NEMA 1) and IP54 (NEMA 12)

Affinity is Control Techniques' dedicated HVAC/R drive, designed specifically to meet the needs of consultants, contractors and owners of commercial buildings. A comprehensive product line incorporates special drive functionality and accessories to simply and efficiently solve your HVAC/R applications.

Digitax ST
Intelligent, compact and dynamic servo drive range
0.72Nm to 19.3Nm (57.7Nm peak)

Meeting the demands of modern lean manufacturing requires smaller and more flexible machinery. Digitax ST is the first ever drive designed to help machine designers and system integrators meet these challenges; optimized for servo applications requiring high peak torque, dynamic response, ease of use and flexible integration features.

Solutions Modules

Control Techniques Solutions Modules are common across our drive families, providing the ultimate flexibility to customize the drive features to meet the needs of your application.

Intelligence: High performance programmable automation and motion Solutions Modules.

Ethernet and Fieldbus communications: Solutions Modules for flexible system integration.

Feedback: Encoder and Resolver Solutions Modules for a wide variety of incremental and absolute feedback devices.

Inputs and Outputs: Additional analog and digital input and output Solutions Modules for additional connectivity.

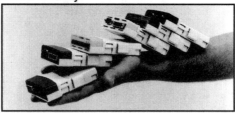

Technology and innovation
System engineering

Many Control Techniques Drive Centres offer a comprehensive drive system design, build and commissioning service. This allows you to take maximum advantage of our expertise and experience in drive based automation.